Edible Alliums

Botany, Production and Uses

Edible Alliums

Botany, Production and Uses

Edited by

Haim D. Rabinowitch

The Hebrew University of Jerusalem, Israel

Brian Thomas

University of Warwick, UK

CABI

CABI is a trading name of CAB International

CABI
Nosworthy Way
Wallingford
Oxfordshire OX10 8DE
UK
Tel: +44 (0)1491 832111
E-mail: info@cabi.org
Website: www.cabi.org

CABI
200 Portland Street
Boston
MA 02114
USA
Tel: +1 (617)682-9015
E-mail: cabi-nao@cabi.org

A catalogue record for this book is available from the British Library, London, UK.

ISBN-13: 9781789249972 (hardback)
9781789249989 (ePDF)
9781789249996 (ePub)

DOI: 10.1079/9781789249996.0000

Commissioning Editor: Rebecca Stubbs
Editorial Assistant: Emma McCann
Production Editor: Marta Patiño
Typeset by SPi, Pondicherry, India
Printed and bound in the UK by Severn, Gloucester

Contents

List of Contributors

Mostafa Abdelrahman
Botany Department, Faculty of Science, Aswan University, Egypt; Faculty of Science, Galala University, Egypt;
Avinash Ade
Savitribai Phule Pune University, Pune, India
Sarah Almulla
Department of Chemistry, Idaho State University, Pocatello, Idaho, USA
Edivânio R. Araújo
Epagri – Estação Experimental de Ituporanga, Estrada Geral Lageado Águas Negras, Ituporanga, Santa Catarina, Brazil
edivanioaraujo@epagri.sc.gov.br
Ashwini Benke,
ICAR-Directorate of Onion and Garlic Research, Rajgurunagar, India
Andrés Bermúdez
Dep. Microbiología, Campus Rabanales C6-Planta Baja-Oeste, Universidad de Córdoba, Córdoba, Spain
Abdul Karim Darkal
Division of Bioorganic Chemistry, School of Pharmacy, Saarland University, Saarbruecken, Germany
Gabriel Dorado
Dep. Bioquímica y Biología Molecular, Campus Rabanales, Campus de Excelencia Internacional Agroalimentario (ceiA3), Universidad de Córdoba, Spain
bb1dopeg@uco.es
Nikolai Friesen
Botanical Garden of the University of Osnabrück, Germany
nfriesen@uni-osnabrueck.de
Suresh Gawande
ICAR-Directorate of Onion and Garlic Research, Rajgurunagar, India
Pranjali Gedam
ICAR-Directorate of Onion and Garlic Research, Rajgurunagar, India
Irwin L. Goldman
Department of Horticulture, University of Wisconsin-Madison, Madison, Wisconsin, USA
ilgoldma@wisc.edu
Pilar Hernández
Instituto de Agricultura Sostenible (IAS-CSIC), Consejo Superior de Investigaciones Científicas (CSIC), Alameda del Obispo, Córdoba, Spain

Yoshihiro Imahori
Osaka Prefecture University, Osaka, Japan
imahori@omu.ac.jp
Claus Jacob
Division of Bioorganic Chemistry, School of Pharmacy, Saarland University, Saarbruecken, Germany
c.jacob@mx.uni-saarland.de
Rina Kamenetsky-Goldstein
Agricultural Research Organization, The Volcani Institute, Rishon-LeZion, Israel
vhrkamen@volcani.agri.gov.il
Anil Khar
Division of Vegetable Science, ICAR-Indian Agricultural Research Institute, New Delhi, India
anil.khar@gmail.com
Kiran Khandagale
Savitribai Phule Pune University, Pune, India
kirankhandagale253@gmail.com
Mirtes F. Lima
Embrapa Hortaliças, Fazenda Tamanduá, Brasília, Brazil
Neelima Mahato
School of Chemical Engineering, Yeungnam University, Gyeongsan, South Korea
Jesús Martín
Dep. Microbiología, Campus Rabanales C6-Planta Baja-Oeste, Universidad de Córdoba, Spain
Sameena Mateen
School of Chemical Engineering, Yeungnam University, Gyeongsan, South Korea; Biomedical and Pharmaceutical Sciences, College of Pharmacy Kasiska Division of Health Sciences, Idaho State University, Pocatello, Idaho, USA
Muhammad Jawad Nasim
Division of Bioorganic Chemistry, School of Pharmacy, Saarland University, Saarbruecken, Germany
Yannick Ney
Division of Bioorganic Chemistry, School of Pharmacy, Saarland University, Saarbruecken, Germany
Srinath Pashikanti
Department of Chemistry, Idaho State University, Pocatello, Idaho, USA; Biomedical and Pharmaceutical Sciences, College of Pharmacy Kasiska Division of Health Sciences, Idaho State University, Pocatello, Idaho, USA
Ross Peters
Agricultural Research Organization, The Volcani Institute, Rishon LeZion, Israel
rossthyra@gmail.com
Samjhana Pradhan
School of Chemical Engineering, Yeungnam University, Gyeongsan, South Korea
Haim D. Rabinowitch
The RH Smith Institute of Plant Science and Genetics in Agriculture, The RH Smith Faculty of Agriculture, Food, and Environment, The Hebrew University of Jerusalem, Rehovot, Israel
haim.rabinowitch@mail.huji.ac.il
Pragati Randive
ICAR-Directorate of Onion and Garlic Research, Rajgurunagar, India
Renata S. Resende
Epagri – Estação Experimental de Ituporanga, Estrada Geral Lageado Águas Negras, Ituporanga, Santa Catarina, Brazil
Rahman Shah Zaib Saleem
Department of Chemistry, SBA School of Science and Engineering (SBASSE), Lahore University of Management Sciences (LUMS), Lahore, Pakistan
Muhammad Sarfraz
Division of Bioorganic Chemistry, School of Pharmacy, Saarland University, Saarbruecken, Germany

Kavita Sharma
Department of Chemistry, Idaho State University, Pocatello, Idaho, USA; Biomedical and Pharmaceutical Sciences, College of Pharmacy Kasiska Division of Health Sciences, Idaho State University, Pocatello, Idaho, USA
kavitasharma@isu.edu

Einat Shemesh-Mayer
Agricultural Research Organization, The Volcani Institute, Rishon-LeZion, Israel
shemeshe@volcani.agri.gov.il

Masayoshi Shigyo
Laboratory of Vegetable Crop Science, College of Agriculture, Graduate School of Sciences and Technology for Innovation, Yamaguchi University, Yamaguchi City, Japan
shigyo@yamaguchi-u.ac.jp

Dhananjay Shirsat
ICAR-Directorate of Onion and Garlic Research, Rajgurunagar, India

Major Singh
ICAR-Directorate of Onion and Garlic Research, Rajgurunagar, India

Brian Thomas
School of Life Sciences, University of Warwick, Coventry, UK
Brian.thomas@warwick.ac.uk

Hikaru Tsukazaki
Tohoku National Agriculture Research Center/NARO, Morioka, Japan

Turgay Unver
Ficus Biotechnology, Ostim Teknopark, Yenimahalle, Ankara, Turkey

Atsushi Yamasaki
Kyushu Okinawa National Agriculture Research Center/NARO, Kurume, Japan
yamasaki@affrc.go.jp

Dedications

First and foremost, this book is dedicated to my family:

To my wife, Shoshie, for being such a great source of inspiration; whose love, support and endless patience made this endeavour possible

To my children and their spouses, Ithai and Talchen, Shlomit and Dan, Noa and Asaf

To my grandchildren, Noam, Beeri, Eitam, Tomer, Maya, Neta, Yuval, Yael, Gilad, and Dekel; for being such a source of encouragement and joy

I dedicate this book to my mentor, the late Nachum Kedar, to my students and to my technical staff

Finally, I dedicate this book to the memory of the late James L. Brewster, a brilliant alliumologist, a wonderful colleague and a very close friend, who has gone too soon but won't be soon forgotten.

Haim D. Rabinowitch

....

This book is also dedicated to Carolyn, who has been a constant source of support, encouragement and love during its preparation.

Brian Thomas

1 Introduction to Edible Alliums: Evolution, Classification and Domestication

Nikolai Friesen*
Botanical Garden of the University of Osnabrück, Germany

1. The Genus *Allium* L.

1.1. General characteristics

The Angiosperm Phylogeny Group (APG) classified *Allium* L. as the only member of the monotypic tribe Allieae within the subfamily Allioideae of the Amaryllidaceae (Chase *et al.*, 2016; Fig. 1.1).

Allium Linnaeus (1753, p. 294) is one of the largest monocotyledonous genera with ~1000 accepted species (Goevarts *et al.*, 2020), and the numbers keep growing with annual additions of over 10 new species.

Allium is a genus of perennial, mostly bulbous, plants characterized by:

- underground storage organs: bulbs, rhizomes or swollen roots;
- bulbs: often on rhizomes; true bulbs (1–2 extremely thickened prophylls (bladeless 'true scales') or false bulbs (thickened basal sheaths and prophylls); several membranous, fibrous or coriaceous tunics; annual or perennial roots;
- rhizomes: rarely runner-like condensed or elongated with very diverse branching patterns;
- leaves: basally arranged, consisting of a basal sheath and terminal blade, often covering the scape and thus appearing cauline;
- bracts: two to several, often fused into an involucre ('spathe');
- inflorescence: fasciculate to often umbel- or head-like (rarely spicate), (one) few-to many-flowered, loose to dense;
- flowers: pediceled, actinomorphic, hypogynous, trimerous of very diverse shape;
- tepals: in two slightly differentiated whorls, free or basally united;
- stamens: two whorls, sometimes basally (or up to two-thirds) connected, the inner often widened and/or toothed;
- ovary: trilocular, three septal nectaries of various shapes, two or more curved (campylotropous) ovules/locule, sometimes diverse apical appendages (crests and horns) developing into a loculicidal capsule dehiscing along the carpels' midrib;
- style: single, with slender, capitate or, rarely, trilobed stigma;
- seeds: angular to globular, black (phytomelan epidermal layer), extremely variable shape, and coat ornamentation;
- karyology: predominant basic chromosome numbers x = 8 and x = 7 (rarely x = 9 or x = 10) with polyploidy in both predominant series; chromosome morphology differs with taxonomic groups.

*Address for correspondence: nfriesen@uni-osnabrueck.de

(eds H.D. Rabinowitch and B. Thomas)
DOI: 10.1079/9781789249996.0001

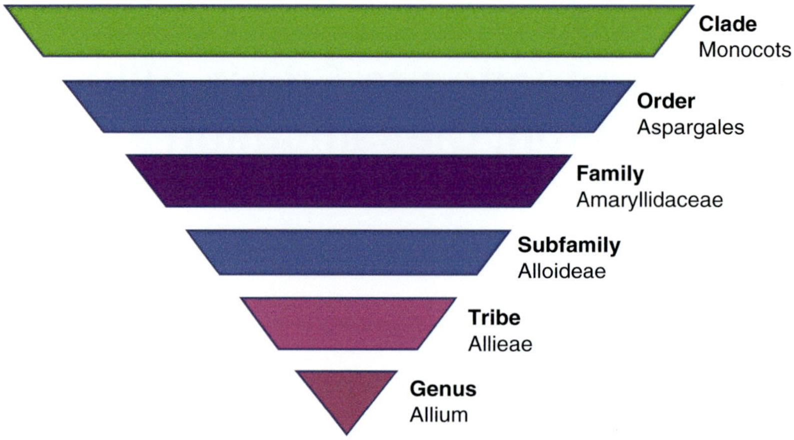

Fig. 1.1. APG IV taxonomic classification of *Allium* hierarchy. (Angiosperm Phylogeny Group; Chase *et al.*, 2016)

Shape, size, colour and texture of rhizomes, bulbs, roots, leaf blades (e.g. flat, channelled, terete or fistulose, sheath/lamina ratio), scapes, spathes, inflorescences, tepals (mostly white, rose to deep purple/violet, rarely blue or yellow), stamens, ovaries and seeds vary considerably with species. The same is true for the anatomy, cross-sections and internal structure of all plant parts.

Basal bulblets and topsets are important vegetative propagules.

Most *Allium* species are allogamous and spontaneous interspecific hybridization is not rare, but strong crossing barriers exist in some groups, even between morphologically similar species.

1.2. Distribution and ecology

Natural distribution of the genus *Allium* occurs in the northern hemisphere over the holarctic region from the dry subtropics to the boreal zone. Its major centre of diversity stretches from the Mediterranean basin to Central Asia and west China and a second but smaller one (~100 species) in North America (from Alaska to Mexico) (Fig. 1.2).

One or two species inhabit the subarctic belt (e.g. *Allium schoenoprasum* L.), and a few are scattered in subtropical and tropical mountains or highland (e.g. Vietnam and Myanmar: *Allium wallichii* Kunth (Quang *et al.*, 2020); Sri Lanka: *A. hookeri* Thwaites; east Sudan to northern Somalia: *A. spathaceum* Steud. ex A. Rich. (De Wilde-Duyfjes, 1976)). Indications of some Chilean (*A. juncifolium* – invalid name) and Brazilian (*A. sellowianum* – synonym of *Nothscordum bivalve* (L.) Britton) *Alliums* (Block, 2010) are not supported by others and are erroneous.

A single *Allium* (*A. synnotii* G.Don (syn. *A. dregeanum* Kunth)) of unknown origin has been described in South Africa, either a modification of *A. ampeloprasum* L. introduction by early European and North African settlers (De Wilde-Duyfjes, 1976) or South African indigenous plants (Mathew, 1996; Germishuizen and Meyer, 2003), which is rather unlikely (Friesen, 2007) as it exhibits the highest number of chromosomes for the genus (2n=64 and 80; De Sarker *et al.*, 1997) and because it has an almost identical intergenic spacer of the ribosomal DNA (ITS) to the European species *Allium scorodoprasum* L. and *A. rotundum* L., section *Allium* (Friesen *et al.*, 2006; Hirschegger *et al.*, 2010; NCBI GenBank Accession AJ411962). *Allium synnotii* is possibly a descendent of *Allium scorodoprasum* and/or related species introductions by early European settlers followed by hybridization, polyploidization and/or other manipulation. This question remains open.

1.3. Phylogeny and classification

Advances in molecular phylogenetics have revolutionized our understanding of *Allium* taxonomy and evolution. The overwhelming morphological diversity is mirrored by a complicated taxonomic structure consisting of 15 subgenera

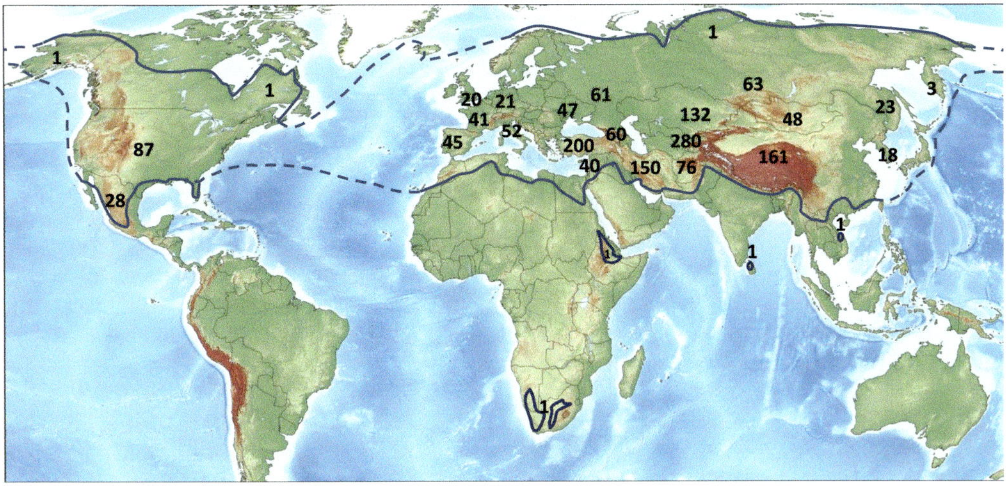

Fig. 1.2. World distribution of wild *Allium* species. The figures indicate the number of local species/regions.

and 72 sections (Friesen *et al.*, 2006) of three evolutionary lineages (Fritsch and Friesen, 2002; Friesen *et al.*, 2006).

The oldest lineage, subg. *Nectaroscordum*, subg. *Microscordum* and subg. *Amerallium*, has one row of vascular bundles and subepidermal laticifers in the leaf blades (Fritsch, 1988). Both first mentioned subgenera show bulbose species and there is a great morphological resemblance in many features (also karyological with x = 9 or x = 8). The subgenus *Amerallium* includes also rhizomatous groups of American and European species and its basic chromosome number is x = 7; secondary there is also x = 8 and x = 11.

The second lineage, subg. *Anguinum*, subg. *Vvedenskya*, subg. *Porphyroprason*, subg. *Caloscordum* and subg. *Melanocrommyum*, has a basic chromosome number of x = 8. Subgenus *Anguinum* is the only rhizomatous representative here. Blades are mostly flat (cylindrical blade occurs in the monotypic subg. *Vvedenskya* and subg. *Caloscordum*), and the laticifers are generally on the inner border of the palisade parenchyma (Friesen *et al.*, 1986). In the leaf blades of subg. *Anguinum* and *Porphyroprason*, one or two rows of vascular bundles occur, while two oppositely oriented rows of vascular bundles are present in subg. *Melanocrommyum*, at least on the ventral side (Fritsch, 1988).

The youngest evolutionary lineage with seven subgenera (x = 8) have laticifers on the inner border of the palisade parenchyma. Leaf blades are often flat, and have two oppositely oriented rows of randomly distributed vascular bundles at the top, except for subgenus *Cyathophora* (with one row of vascular bundles). Less common are cylindrical leaf blades as in the common onion, with a ring of vascular bundles and thread-like leaf blades. The species-poor rhizomatous subgenera *Butomissa*, *Cyathophora* and *Rhizirideum* in the narrowest sense are less specialized.

More specialized is the bulbous subgenus *Allium* with three subgroups, corresponding to the classic sections *Allium*, *Codonoprasum* and the inhomogeneous *Scorodon s.l.* The latter is divided into several small, partly oligotypic sections: *Avulsea*, *Brevidentia*, *Coerulea*, *Costulatae*, *Crystallina*, *Eremoprasum*, *Kopetdagia*, *Longivaginata*, *Mediasia*, *Minuta*, *Multicaulia* and *Pallasia* (Khassanov, 1996, 2000, 2018; Fritsch *et al.*, 1998; Friesen *et al.*, 2006). Most derived molecular features are shown by the rhizomatous subgenera *Reticulatobulbosa*, *Polyprason* and *Cepa*.

This classification employed almost exclusively known names, even if several taxonomic groups were united, others received higher rank and were applied in the narrow sense.

Lately, more *Allium* sections were combined and described: *Longibidentata* (R.M. Fritsch in Khassanov and Fritsch, 1994, p. 974); R.M. Fritsch (2009, p. 465), *Decipientia* (Omelczuk, 1962, p. 71), *Asteroprason* and *Procerallium* (Fritsch *et al.*, 2010, pp. 168, 184, 199), *Unicaulia* and *Haneltia* (Khassanov *et al.*, 2011, p. 174), *Rechingeria* (Khassanov *et al.*, 2013,

p. 214), *Kingdonia* and *Trifurcatum* (Huang *et al.*, 2014, pp. 283, 284, respectively), *Tulipifolia* (Friesen *et al.*, 2021). Some sections have been assembled according to the latest phylogenetic data (Friesen *et al.*, 2020).

All subsequent phylogenetic studies (Li *et al.*, 2010; Wheeler *et al.*, 2013; Li *et al.*, 2016; Hauenschild *et al.*, 2017; Costa *et al.*, 2020; Jimenez *et al.*, 2020; Xie *et al.*, 2020) confirmed the genus division into three major evolutionary lineages and the monophyletic origin of all subgenera included in the two older evolutionary lineages. Only Namgung *et al.* (2021) showed a paraphyletic situation in subgenus *Melanocrommyum*, based on the very doubtful position of *Allium nigrum*.

The phylogenetic relationships in the youngest lineage are less clear (Li *et al.*, 2016; Hauenschild *et al.*, 2017; Friesen *et al.*, 2020; Xie *et al.*, 2020), as analyses of most species from subgenera *Allium*, *Rhizirideum*, *Cepa* and *Polyprason* show paraphyletic characters, as demonstrated by the SplitsTree network with over 300 nrITS sequences, with species representatives from every accepted section of the genus *Allium* (Fig. 1.3).

Little is known of the taxonomy and genetic diversity within some well-established subgenera and sections (Friesen, 2007). However, reliable taxonomic and bio-geographic analyses are available for the sections *Schoenoprasum* (Friesen and Blattner, 2000) and *Cepa* (Gurushidze *et al.*, 2007), subgenus *Melanocrommyum* (Gurushidze *et al.*, 2010; Fritsch 2012, 2016), section *Sacculiferum* (Choi and Oh, 2011), the American species of subgenus *Amerallium* (Nguyen *et al.*, 2008; Wheeler *et al.*, 2013; Mashayekhi and Columbus, 2014), section *Oreiprason* (Seregin *et al.*, 2015), section *Coerulea* (Khassanov *et al.*, 2013), subgenus *Anguinum* (Herden *et al.*, 2016), subgenus *Cyathophora* (Li *et al.*, 2016), section *Rhizirideum* (Sinitsyna *et al.*, 2016), section *Daghestanica* (Xie *et al.*, 2019, 2020), *Caloscordum* (Yang *et al.*, 2017), section *Rhizomatosa* (Friesen *et al.*, 2020), sections *Decipientia* and *Tulipifolia* (Friesen *et al.*, 2021),

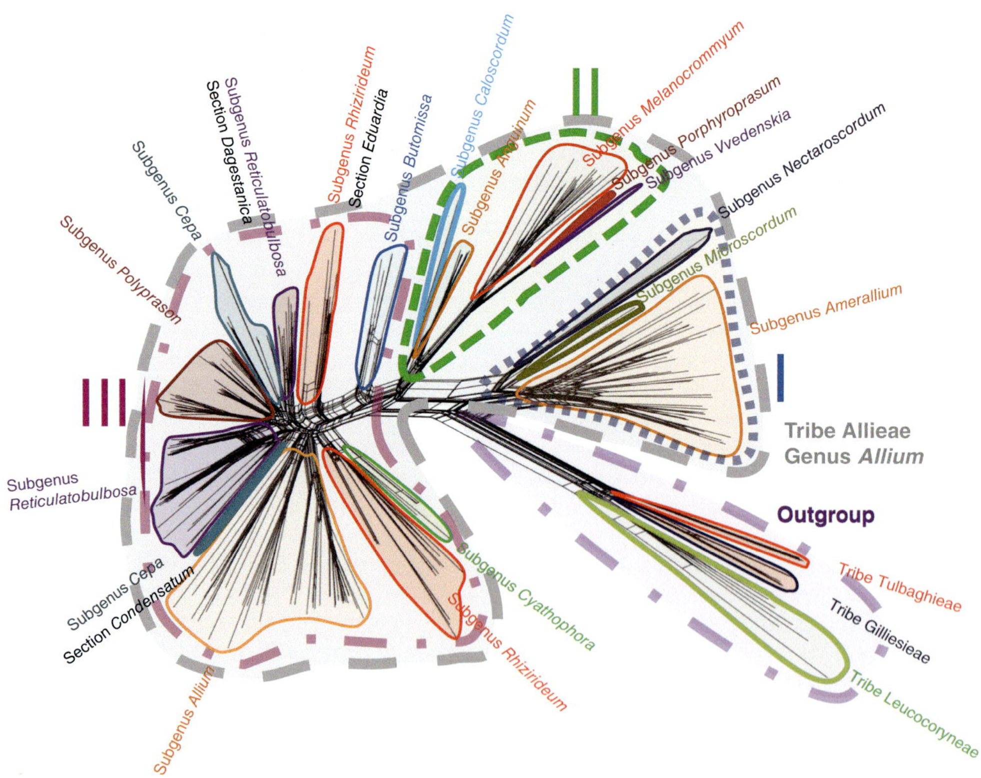

Fig. 1.3. The SplitsTree network of the tribe Allieae, based on nrITS sequences of >300 *Allium* species.

the Eurasian species of subgenus *Amerallium* (Friesen *et al.*, in preparation).

It is important to include all species in phylogenetic analysis to solve the paraphyly issue in some third evolution lineage taxa, and results from both nuclear and plastid genomes are very useful in tracking down the evolutionary hybridogenic events within the genus.

The enormous size of *Allium* nuclear genomes precludes full sequencing soon, while the entire plastid genome has already been published for >60 *Allium* species (Filushin *et al.*, 2016, 2018; Lee *et al.*, 2017; Xie *et al.*, 2019, 2020; Yang *et al.*, 2019; Yusupov *et al.*, 2020; Namgung *et al.*, 2021), as well as the transcriptome of the nine *Allium* species nuclear genome (Zhu *et al.*, 2017). The deeper and broader the genus phylogeny is examined, the more examples of the incongruence between nuclear and plastid phylogenies emerge, which indicate hybridogenic events in the earlier phases of the genus phylogeny (Li *et al.*, 2016; Han *et al.*, 2019; Xie *et al.*, 2019, 2020; Yusupov *et al.*, 2020; Friesen *et al.*, 2021).

1.4. Time of origin

Large discrepancies exist in estimated divergence times of the genus *Allium* – from 11 up to 52.2 mya, mainly due to differences in fossil placement, the dating methods used and the fact that no *Allium* fossil records exist. Published data is therefore based on a secondary calibration point estimated for other monocotyledons (Chen *et al.*, 2013; Li *et al.*, 2016; Hauenschild *et al.*, 2017; Costa *et al.*, 2020; Xie *et al.*, 2020). We trust that employment of nrITS substitution rates for herbaceous annual/perennial angiosperms is the most appropriate means to accomplish the task (Kay *et al.*, 2006). *Paleoallium*, a single fossil (*c.*49 mya) from Washington State, USA (Pigg *et al.*, 2018), of another genus of Amaryllidaceae is unlikely an *Allium*, yet it could be used for calibration. Such analyses with confidence in secondary calibration points often show that certain plant groups are younger than expected. The discrepancies in the origin and age of the angiosperms were confirmed (Coiro *et al.*, 2019; Li *et al.*, 2019). In my opinion, of the published datings, only Costa *et al.* (2020) with 52.2 mya (58.1–44.4 mya: 95% HPD) provided a close estimate of the actual genesis age of the genus *Allium*. The intercontinental disjunction hypothesis in the Allioideae tribes as the result of the vicariance after the Gondwana break-up is also discussed for the first time (Costa *et al.*, 2020). Using an extensive molecular clock analysis covering 800 monocots, Janssen and Bremer (2004) proposed an earlier genesis age for subfamily Allioideae (87 mya). Based on the analysis of the genus *Allium* subgenera distribution, the out-of-India hypothesis (Briggs, 2003; Bossuyt *et al.*, 2006; Datta-Roy and Karanth, 2009; Costa *et al.*, 2020) could also be validated. Most overlaps in the subgenera distribution are concentrated to the north of the Himalayas (Tian Shan, Karakorum, Tibet) and from there some migrated west (e.g. *Melanocrommyum*) and east (*Butomissa*, *Cyathophora*). Some monotypic subgenera grow only in Central Asia north of the Himalayas (e.g. *Vvedenskya*, *Porphyroprason*), others only in eastern Asia (e.g. *Caloscordum* and *Microscordum*) or only in west Asia and southeast Europe (*Nectaroscordum*). Some subgenera (*Cepa*, *Rhizirideum*, *Reticulatobulbosa*) are distributed in Europe and Asia, but exhibit the greatest diversity in Asia, north of the Himalayas. Others have disjunction centres in the Mediterranean, east Asia and North America (*Amerallium* and *Anguinum*). The migration to North America probably occurred several times from Asia via the Bering Land Bridge (Huang *et al.*, 2014; Herden *et al.*, 2016), as did *Allium schoenoprasum* (subgenus *Cepa*, section *Schoenoprasum*), the only native representative of the third lineage in North America (Friesen and Blattner, 2000) (Fig. 1.4).

2. Edible Alliums

2.1. Edible *Allium*

The genus *Allium*'s economic significance depends on several important vegetable crops (onion, garlic, leek, bunching onion, Chinese chive and others) and ornamental species mostly belong to the third evolutionary line (Galmarini, 2017) (Table 1.1), while man consumes wild-growing species of the three evolutionary lineages.

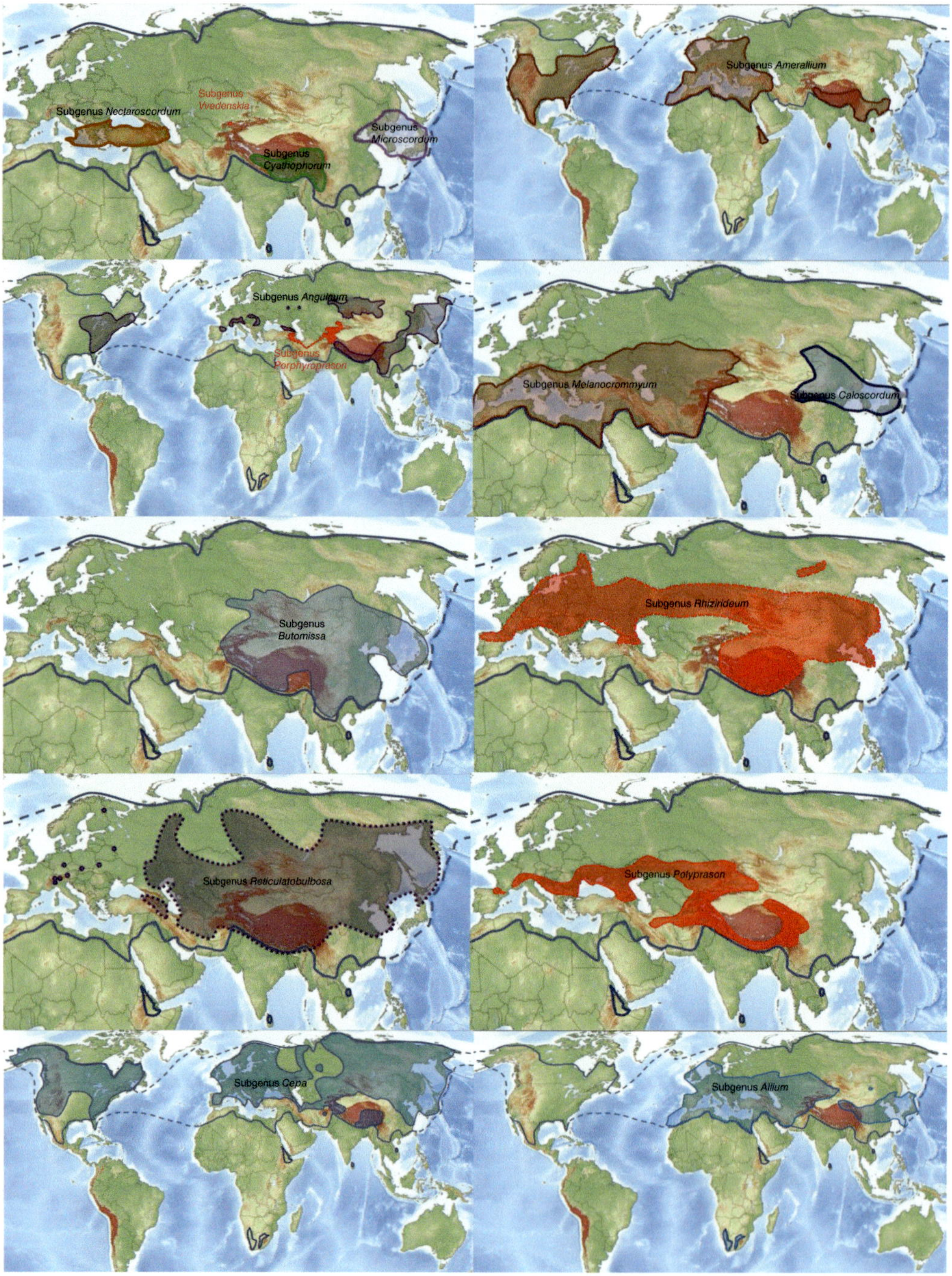

Fig. 1.4. Distribution of the subgenera of the genus *Allium*.

Generally, all parts of all *Allium* species may be consumed by humans. Many taste good, yet some are less edible or even unpalatable, e.g. members of subg. *Nectaroscordum* that taste and smell like burnt rubber.

Many wild species are consumed locally, depending on size, availability and taste. The following provides a shortlist of human-consumed wild species. Mongolia: *Allium mongolicum* Regel; Siberia: *A. microdictyon* Prokh.; Russian Far

Table 1.1. Allium crop species and their areas of cultivation.

Evolution lineage/ Subgenera/ Section	Botanical names of the crop groups	Other names used in the literature	Area of cultivation	English names
I/*Amerallium*/ *Bromatorrhiza*	*A. hookeri* Thw.		Butan, Yunnan, northwest Thailand	
I/*Amerallium*/ *Bromatorrhiza*	*A. wallichii* Kunth	*A. platyphyllum* Diels, *A. lancifolium* Stearn	East Tibet	
I/*Amerallium*/ *Rhopetoprason*	*A. kunthii* G.Don		Mexico	
I/*Amerallium*/ *Amerallium*	*A. canadense* L.		Cuba	
I/*Amerallium*/ *Molium*	*A. neapolitanum* Cyr.		Mexico	Naples garlic
III/*Allium*/*Allium*	*A. ampeloprasum* L. s.l.			
	Leek group	*A. porrum* L., *A. ampeloprasum* var. *porrum* (L.) J.Gay	Mainly Europe and North America	Leek
	Kurrat group	*A. kurrat* Schweinf. ex Krause	Egypt and adjacent areas	Kurrat, salad leek
	Great-headed garlic group - *A. ampeloprasum* var. ampeloprasum	*A. ampeloprasum* var. *holmense* (Mill.) Aschers. et Graebn.	Eastern Mediterranean, California	Great-headed garlic
	Pearl-onion group	*A. ampeloprasum* var. *sectivum* Lued.	Atlantic and temperate Europe	Pearl onion
	Tarée group		Iran	Tarée irani
III/*Allium*/*Allium*	*A. tuncelianum* (Kollmann) N.Özhatay, Mathew & Şiraneci		Turkey	Tunceli garlic***
III/*Allium*/*Allium*	*A. sativum* L.	*A. longicuspis* Regel	Worldwide	garlic
III/*Allium*/*Allium*	*A. rotundum* L.	*A. scorodoprasum* ssp. *rotundum* (L.) Stearn	Turkey	
III/*Allium*/ *Scorodon*	*A. macrostemon* Bunge	*A. uratense* Franch., *A. grayi* Regel	China, Korea, Japan	Chinese garlic, Japanese garlic
III/*Butomissa*/ *Butomissa*	*A. tuberosum* Rottl. ex Sprengel*		China, Japan, Central Asia, worldwide now	Chinese chive
III/*Cepa*/ *Schoenoprasum*	*A. schoenoprasum* L.		Worldwide in temperate areas	Chive

Continued

Table 1.1. Continued.

Evolution lineage/ Subgenera/ Section	Botanical names of the crop groups	Other names used in the literature	Area of cultivation	English names
III/*Cepa*/*Cepa*	*A. cepa* L. Common onion group Ever-ready onions Aggregatum group	*A. cepa*. ssp. *cepa*/ var. *cepa* *A. cepa* ssp. *australe* Trofimez ex Kazakova *A. cepa* var. *perutile* Stearn *A. ascalonicum* auct. hort. *A. cepa* var. *aggregatum* G.Don, var. *ascalonicum* Backer *A. cepa* ssp. *orientalis* Kazakova	Worldwide/Great Britain Nearly worldwide	onion, common onion ever-ready onion shallot, potato onion, multipler onion
III/*Cepa*/*Cepa*	*A. fistulosum* L.		East Asia, temperate Europe and North America	bunching onion, Welsh onion
	A. x *cornutum* Clem. ex Vis.**	*A. cepa* var. *viviparum* auct.**	Locally in south Asia, Europe, Canada, Antilles, China, Korea, Japan, southeast Asia	
III/*Cepa*/*Cepa*	*A. oschaninii* O.Fedtsch.		France, Italy	French shallot**
III/*Cepa*/*Cepa*	*A . proliferum* (Moench.) Schrader east Asian group Eurasian group	*A. aobanum* Araki, *A. wakegi* Araki, *A. cepa* var. *viviparum* (Metzg.) Alef., *A. cepa* var. *proliferum* (Moench)Alef.	China, Japan, south-east Asia North America, Europe, northeast Asia	Wakegi onion, top onion, tree onion, Egyptian onion, Catawissa onion
III/*Cepa*/*Cepa*	*A. pskemense* B.Fedtsch.		Uzbekistan, Kyrgyzstan, Kazakhstan	
III/*Cepa*/ *Sacculiferum*	*A. chinense* G.Don	*A. bakeri* Regel	China, Korea, Japan, Southeast Asia	Rakkyo, Japanese scallions

*see Blattner and Friesen (2006); Oyuntsetseg *et al.* (2012); **see Friesen and Klaas (1998); ***see Ipek *et al.* (2008); Kizil and Khawar (2017).

East: *A. ochotense* Prokh.; USA and east Canada: *A. tricoccum*; China: *A. prattii*; south Siberia and Mongolia, south Ural and Altai: *A. altaicum* Pall., *A. microdictyon* Prokh., *A. obliquum* L.; east Siberia and Mongolia: *A. ramosum* L.; Europe: *A. ursinum* L.; Yunnan and Sichuan, China: *A. omeiense* Z.Y. Zhu; India (Kashmir), Afghanistan and Pakistan: *A. humile* Kunth; Uzbekistan: *A. suworowii* Regel and *A. tschimganicum* B. Fedtsch.; Tajikistan: *A. giganteum* Regel; Pakistan and Afghanistan: *A. roylei* Stearn; west Siberia and east Kazakhstan: *A. nutans* L.; south Ural and north Kazakhstan: *A. angulosum* L.; northwest India: *A. consanguineum* Kunth (Keusgen *et al.*, 2008; Ozturk *et al.*, 2012; Fritsch and Abbasi, 2013; Ju *et al.*, 2013; Kang *et al.*, 2013). These natural resources are often improperly managed, and over-collected with consequent severe population decline.

2.2. Domestication

Domestication probably started by both protection and the rational use of wild plants, followed by transplanting into gardens (Hanelt, 1990). Human selection and natural events resulted in the development of variation now common in several cultivated species.

The cultivation of domesticated *Allium* species (onion, garlic and others) during ancient times is well covered (Helm, 1956; Jones and Mann, 1963; Havey, 1995; Ekşi *et al.*, 2020). Here I will discuss another aspect: the location of the crops in the genus' phylogenetic system, and whether its progenitor exists in nature. Were the crop plants' domestication a single or multiple event?

2.3. Onions (section *Cepa*)

Allium section *Cepa* comprises ten wild species (*A. altaicum, A. asarense, A. farctum, A. galanthum, A. oschaninii, A. praemixtum, A. pskemense, A. rhabdotum, A. roylei, A. vavilovii* and the cultivated *A. cepa* (bulb onion) and *A. fistulosum* (bunching onion), all of which are consumed by man as condiment vegetables.

The wild taxa inhabit the Irano-Turanian floristic region, mainly in the Tian-Shan and Pamiro-Alai regions. Occurrences in neighbouring floristic provinces are only marginal extensions. The exceptions are *A. altaicum* and *A. rhabdotum*, which inhabit southern Siberia and the Mongolian mountains and the eastern Himalayas, respectively (Stearn, 1960; Fritsch and Friesen, 2002; Gurushidze *et al.*, 2007).

2.3.1 Common onion

The wild progenitor and origin of the common onion are not clear. Nuclear ITS sequences analyses revealed that *Allium vavilovii* that crosses freely with *A. cepa* is its genetically closest related species (Gurushidze *et al.*, 2007). These clear results confirm earlier studies on chloroplast data (Havey, 1992), yet morphological characterization contradicts this conclusion. *Allium cepa* has thick, fistulous, slightly bent leaves, and a somewhat biconical inflated scape, resembling the species of the other clades of section *Cepa*. *Allium vavilovii* and *A. asarense*, however, have flat, sickle-shaped leaves and bubble-like inflation of the scape. These discrepancies between molecular and morphological characterizations and the similarity between *A. cepa* and *A. oschaninii* might imply a sense of a hybrid origin of the common onion.

Strong crossing barriers exist between *A. oschaninii* and both *A. vavilovii* and *A. cepa*, while *A. vavilovii* crosses to some extent with *A. galanthum* and *A. fistulosum* (van Raamsdonk *et al.*, 2003). Leaves and scape morphologies of common onion are similar to *A. galanthum*, but the morphologically distinct *A. fistulosum* crosses better with *A. cepa*. Apart from these morphological inconsistencies the high variation of ITS sequences within *A. cepa* cannot result only from multiple domestication events but indicate an extended history of hybridization (Gurushidze *et al.*, 2007). Complete CP genome sequencing in seven section *Cepa* species, unfortunately without *A. vavilovii*, showed *A. galanthum* as the next related species to *A. cepa* (Yusupov *et al.*, 2020). This confirms the possible hybridogenic origin of *A. cepa*.

To finally clarify the *A. cepa* origin, genome sequencing of several accessions of all section *Cepa* species from different distribution areas is required.

2.3.2 Bunching onion (Allium fistulosum *L.*)

RAPD and PCR RFLP data confirmed the monophyletic origin of *A. fistulosum* from *A. altaicum* (Friesen *et al.*, 1999). This variable vegetable is common in China, Japan and Korea (Inden and Asahira, 1990), where the slender bulbs and basal parts of the pseudostem are much esteemed as fresh or cooked vegetables. In the west, it is consumed mostly as fresh green leaves forced in the wintertime.

2.3.3 Allium × proliferum *(Moench) Schrad. (top onion, tree onion, Egyptian onion, Catawissa onion, Wakegi onion)*

These *A. fistulosum* × *A. cepa* hybrids (Schubert *et al.*, 1983; Havey, 1991; Maaß, 1997a; Friesen and Klaas, 1998) mostly fail to develop flowers. The few buds that reach anthesis are sterile and some topsets develop on the receptacle. The crops are popular in North America, Europe and northeastern Asia home gardens for their topsets and young leaves. Hanelt (1990) suggested a Chinese origin, but morphological differences and genetic diversity in the top onion support a polytopic origin (Maaß, 1997a). Indeed, *A. cepa* and *A. fistulosum* are often cultivated side-by-side, hence multiple hybridizations possibly occurred.

2.3.4 *Triploid onion* (Allium × cornutum *Clementi ex Visiani*)

Another sterile viviparous onion with a slender stature and pinkish-flushed flowers is cultivated in Tibet, Jammu, Croatia, central and western Europe, Canada and the Antilles. *Allium cepa* is accepted as one donor of these triploids (Havey, 1991; Maaß, 1997b; Friesen and Klaas, 1998), and *Allium roylei* was proposed as another parent (Puizina and Papeŝ, 1996). Fredotović *et al.* (2014) confirmed the above and proposed *A. pskemense* as the third parent, and that Jammu, Afghanistan or Pakistan are the monophyletic origin of the triploid species.

2.3.5 *French grey shallot*

French grey shallot cv. Grise de la Drôme has long been cultivated in southern France and Italy (Messiaen *et al.*, 1993; D'Antuono, 1998; Rabinovich and Kamenetsky, 2002. Messiaen *et al.* (1993) described the plants and, based on scape and umbel morphologies, proposed relationships with *A. oschaninii*. Maaß (1996) ruled out *A. cepa* relationship but proposed *A. oschaninii* or *A. vavilovii* as origins. Friesen and Klaas (1998) used both genomic *in situ* hybridization and fingerprint method RAPD and concluded that most grey shallot chromosomes originated from *A. oschaninii* and one-and-a-half chromosome arms from either *A. cepa* or *A. vavilovii*. All analysed grey shallot clones by fingerprinting were identical, hence monophyletic origin is very likely, but no information is available on where and when domestication occurred.

2.4. Garlic (*A. sativum*)

Like cultivated garlic, wild populations of *Allium longicuspis* Regel are sterile. The two plants are practically indistinguishable and are thus considered synonyms (Maaß and Klaas, 1995; Ipek *et al.*, 2003; Shemesh-Mayer and Kamenetsky-Goldstein, 2019; POWO, 2020). Supposedly, the transition to vegetative reproduction resulted from selections for earliness and large bulbs (Kamenetsky *et al.*, 2005; Shemesh-Mayer and Kamenetsky-Goldstein, 2019), and fertility was restored by physiological manipulations, experimentally (Pooler and Simon, 1994) and in open fields (Etoh and Simon, 2002; Kamenetsky *et al.*, 2004a; Kamenetsky *et al.*, 2007; Shemesh-Mayer and Kamenetsky-Goldstein, 2019).

It was suggested that *A. longicuspis* is the direct wild-growing ancestor of *A. sativum* (Vvedensky, 1944; Etoh and Ogura, 1984; Pooler and Simon, 1993; Maaß and Klaas, 1995; Mathew, 1996; Hong, 1999) and also that it is feral cultivated garlic since it grows along roads and in places of abandoned settlements in Central Asia (Etoh and Simon, 2002; Fritsch and Friesen, 2002; Klaas and Friesen, 2002; Kamenetsky *et al.*, 2004b).

The Turkish wild species *A. tuncelianum* (Kollmann) Özhatay, B.Mathew & Siraneci, assumed the wild progenitor of garlic (Mathew, 1996), is only distantly related and cannot be the progenitor (Ipek *et al.*, 2008).

2.5. Leek and relatives

Allium ampeloprasum L. *sensu lato* represents extremely variable polyploid complexes of several wild and cultivated taxa commonly grown across the Mediterranean basin through the Middle East into middle Asia and China (Kollmann, 1971; Wendelbo, 1971; Mathew, 1996; Hanelt, 2001). Bothmer (1970) introduced the concept of *A. ampeloprasum* complex, a group of closely related species including *A. ampeloprasum*, *A. bourgeaui* Rech. fil. and *A. commutatum* Guss, and later on added *A. atroviolaceum* Boiss to the list (Bothmer, 1974), and *A. polyanthum* should also be included in the complex (Guern *et al.*, 1991).

Cloning nrITS fragments of the hexaploid great-headed garlic and tetraploid leek revealed that both are allopolyploids with different parent species in their genomes (Hirschegger *et al.*, 2010), while the tetraploid minor crop taxa (Kurrat, Taree Irani and pearl onion) are related to leek. *Allium ampeloprasum* L. lectotype is from Island Flat Holm, UK (BM. Herb. Sloan 152, folio 153). Plants of the Flat Holm vegetable cultivar (Dr T. Rich, London, 2021, personal communication) are sterile hexaploids 2n=48 (De Sarker *et al.*, 1997). Hence, the name *A. ampeloprasum* L. s. str. should only apply to great-headed garlic and other *A. ampeloprasum* hexaploid plants, but leek should be named *Allium porrum* L. Nevertheless, this does not clarify the nomenclatural, taxonomic and phylogenetic issues of the *A. ampeloprasum* polyploid complex. Further genomic studies are needed to elucidate the subtleties in the origin of polyploid species in this complex followed by a thorough nomenclature and morphological processing including all relevant taxa. Hence, the questions of where and how crop species leek and great-headed garlic came about remain open.

2.6. Chinese chive (*Allium tuberosum* Rottl. ex Sprengel)

Domestication of Chinese chive occurred in northern China more than 3000 years ago (Hanelt, 2001) and currently it is the second-most economically important *Allium* crop in eastern Asia. *A. ramosum* L. is considered to be ancestral to the crop species (Hanelt, 1988). Since both are tetraploids (2n=32) that have similar morphology, Hanelt (2001) merged them into *A. ramosum*. Phenetic, cladistic and multivariate analyses of RAPD data revealed an unexpected relationship between wild *A. ramosum* and domesticated plants (Blattner and Friesen, 2006). Yet molecular data separated wild and crop populations as sister species.

Molecular data points to northern China as the most suitable site for the domestication of the diploid progenitor population but not the diploid *A. ramosum* from east Mongolia (Oyuntsetseg *et al.*, 2012).

The high *A. tuberosum* genetic variability is attributed to either multiple parallel domestications – namely, different populations contributed to the crop's gene pool – or post-domestication hybridization events with advanced wild populations (Blattner and Friesen, 2006).

References

Blattner, F.R. and Friesen, N. (2006) Chapter 10. Relationship between Chinese chive (*Allium tuberosum*) and its putative progenitor *A. ramosum* as assessed by random amplified polymorphic DNA (RAPD). In: Zeder, M.A., Decker-Walters, D., Bradley, D. and Smith, B.D. (eds) *Documenting Domestication: New Genetic and Archaeological Paradigms*. University of California Press, pp. 134–142.

Block, E. (2010) *Garlic and other Alliums. The Lore and the Science*. RSC Publishing.

Bothmer, R. (1970) Cytological studies in *Allium* I. Chromosome numbers and morphology in *Allium* Sect. *Allium* from Greece. *Botaniska Notiser* 123, 518–550.

Bothmer, R. (1974) Studies in the Aegean Flora XXI. Biosystematic studies in the *Allium ampeloprasum* complex. *Opera Botanica (Lund)* 34, 1–104.

Bossuyt, F., Brown, R.M., Hillis, D.M., Cannatella, D.C. and Milinkovitch, M.C. (2006) Phylogeny and biogeography of a cosmopolitan frog radiation: late cretaceous diversification resulted in continent-scale endemism in the family Ranidae. *Systematic Biology* 55, 579–594. DOI: 10.1080/10635150600812551.

Bradeen, J.M. and Havey, M.J. (1995) Randomly amplified polymorphic DNA in bulb onion and its use to assess inbred integrity. *Journal of the American Society for Horticultural Science* 120, 752–758.

Briggs, J.C. (2003) The biogeographic and tectonic history of India. *Journal of Biogeography* 30, 381–388. DOI: 10.1046/j.1365-2699.2003.00809.x.

Chase, M.W., Christenhusz, M.J.M., Fay, M.F., Byng, J.W., Judd, W.S. *et al.* (Angiosperm Phylogeny Group) (2016) An update of the Angiosperm Phylogeny Group classification for the orders and families of flowering plants: APG IV. *Botanical Journal of the Linnean Society* 181, 1–20. DOI: 10.1111/boj.12385.

Chen, S., Kim, D.K., Chase, M.W. and Kim, J.H. (2013) Networks in a large-scale phylogenetic analysis: reconstructing evolutionary history of asparagales (Lilianae) based on four plastid genes. *PLOS ONE* 8, e59472. DOI: 10.1371/journal.pone.0059472.

Choi, H.J. and Oh, B.U. (2011) A partial revision of *Allium* (Amaryllidaceae) in Korea and north-eastern China. *Botanical Journal of the Linnean Society* 167, 153–211.

Coiro, M., Doyle, J.A. and Hilton, J. (2019) How deep is the conflict between molecular and fossil evidence on the age of angiosperms? *New Phytologist* 223, 83–99. DOI: 10.1111/nph.15708.

Costa, L., Jimenez, H., Carvalho, R., Carvalho-Sobrinho, J., Escobar, I. *et al.* (2020) Divide to conquer: evolutionary history of Allioideae tribes (Amaryllidaceae) is linked to distinct trends of Karyotype evolution. *Frontiers in Plant Science* 11, 320. DOI: 10.3389/fpls.2020.00320.

D'Antuono, L.F. (1998) A new taxon among vegetable crops? *Allium Improvement Newsletter* 8, 1–3.

Datta-Roy, A. and Karanth, K.P. (2009) The out-of-India hypothesis: What do molecules suggest? *Journal of Biosciences* 34, 687–697. DOI: 10.1007/s12038-009-0057-8.

De Sarker, D., Johnson, M.A.T., Reynolds, A. and Brandham, P.E. (1997) Cytology of the highly polyploid disjunct species, Allium dregeanum (Alliaceae), and of some Eurasian relatives. *Botanical Journal of the Linnean Society* 124, 361–373.

De Wilde-Duyfjes, B.E.E. (1976) A revision of the genus *Allium* L. (Liliaceae) in Africa. *Mededelingen Landbouwhogeschool Wageningen* 76, 1–237. Available at: https://edepot.wur.nl/200358 (accessed 28 June 2022).

Ekşi, G., Özkan, A.M.G. and Koyunzu, M. (2020) Garlic and onions: an eastern tale. *Journal of Ethnopharmacology* 253, 112675. DOI: 10.1016/j.jep.2020.112675.

Etoh, T. and Ogura, H. (1984) Comparison between *Allium longicuspis* Regel and fertile garlic clone and their hybrid seeds. In: *194 Abstract of the Japanese Society for Horticultural Science, Spring Meeting.* Japanese Society for Horticultural Science, Tokyo, pp. 170–171.

Etoh, T. and Simon, P.W. (2002) Diversity, fertility and seed production of garlic. In: Rabinowitch, H.D. and Currah, L. (eds). *Advances in Allium Science*. CABI Publishing, New York, New York, pp. 101–118.

Filyushin, M.A., Beletsky, A.V., Mazur, A.M. and Kochieva, E.Z. (2016) The complete plastid genome sequence of garlic *Allium sativum* L. *Mitochondrial DNA Part B Resources* 1, 831–832. DOI: 10.1080/23802359.2016.1247669.

Filyushin, M.A., Beletsky, A.V., Mazur, A.M. and Kochieva, E.Z. (2018) Characterization of the complete plastid genome of lop-sided onion *Allium obliquum* L. (Amaryllidaceae). *Mitochondrial DNA Part B Resources* 3, 393–394. DOI: 10.1080/23802359.2018.1456369.

Fredotović, Ž., Šamanić, I., Weiss-Schneeweiss, H., Kamenjarin, J., Jang, T.S. *et al.* (2014) Triparental origin of triploid onion, *Allium × cornutum* (Clementi ex Visiani, 1842), as evidenced by molecular, phylogenetic and cytogenetic analyses. *BMC Plant Biology* 14, 24.

Friesen, N. (2007) Die Gattung *Allium* – Taxonomischer Überblick und wissenschaftliche Sammlung im Botanischen Garten der Universität Osnabrück. *Osnabrücker Naturwissenschaftliche Mitteilungen* 33/34, 95.

Friesen, N. and Blattner, F.R. (2000) Geographical isolation predominates over ecological differentiation in the phylogeny of *Allium* section *Schoenoprasum* (Alliaceae). *Plant Biology* 2, 297–305.

Friesen, N. and Klaas, M. (1998) Origin of some minor vegetatively propagated *Allium* crops studied with RAPD and GISH. *Genetic Resources and Crop Evolution* 45, 511–523.

Friesen, N., Zuev, V.V. and Aljanskaja, N. (1986) Krasivoluk nereidozwetnyi – *Calloscordum neriniflorum* Herbert. In: Sobolevskaja, K. (ed.) *Biologicheskie osobennosti rasteni Sibiri, nuzhdajushikhsja v okhrane*. Nauka, Sibirskoje Otdelenie, Novosibirsk, Russia, pp. 83–92. [In Russian].

Friesen, N., Pollner, S., Bachmann, K. and Blattner, F. (1999) RAPDs and non-coding chloroplast DNA reveal a single origin of the cultivated *Allium fistulosum* from *A. altaicum*. *American Journal of Botany* 86, 554–562.

Friesen, N., Fritsch, R.M. and Blattner, F.R. (2006) Phylogeny and new intrageneric classification of *Allium* L. (Alliaceae) based on nuclear rDNA ITS sequences. *Aliso* 22, 372–395. DOI: 10.5642/aliso.20062201.31.

Friesen, N., Smirnov, S., Herden, T., Oyuntsetseg, B., Shmakov, A. *et al.* (2020) *Allium* species of section *Rhizomatosa*, early members of the Central Asian steppe vegetation. *Flora* 263. DOI: 10.1016/j.flora.2019.151536.

Friesen, N., Smirnov, S.V., Leweke, M., Seregin, A.P. and Fritsch, R.M. (2021) Taxonomy and phylogeny of *Allium* section *Decipientia* (Amaryllidaceae): morphological characters do not reflect the evolutionary history verified by molecular markers. *Botanical Journal of the Linnean Society.* DOI: 10.1093/botlinnean/boab023.

Fritsch, R.M. (1988) Anatomische Untersuchungen an der Blattspreite bei *Allium* L. (Alliaceae) – Arten mit einer einfachen Leitbündelreihe. *Flora* 181, 83–100.

Fritsch, R.M. (2009) New *Allium* (Alliaceae) species from Tajikistan, Kyrgyzstan, and Uzbekistan. *Botanischer Jahrbücher für Systematic, Pflanzengeschichte und Pflanzengeographie* 127, 459–471. DOI: 10.1127/0006-8152/2009/0127-0459.

Fritsch R.M. (2012) Illustrated key to the sections and subsections and brief general circumscription of *Allium* subg. *Melanocrommyum*. *Phyton (Horn, Austria)* 52, 1–37.

Fritsch, R.M. (2016) A preliminary review of *Allium* subg. *Melanocrommyum* in Central Asia. [Key also in Russian] DOI: 10.5447/IPK/2016/60.

Fritsch, R.M. and Abbasi, M. (2013) In: Fritsch, R.M. (ed.) *A Taxonomic Review of Allium subg. Melanocrommyum in Iran*. Gatersleben.

Fritsch, R.M. and Friesen, N. (2002) Evolution, domestication, and taxonomy. In: Rabinowitch, H.D. and Currah, L. (eds) *Allium Crop Science: Recent Advances*. CAB International, Wallingford, UK, pp. 5–30.

Fritsch, R.M., Khasanov, F.O. and Friesen, N. (1998) New taxa, new combinations, and taxonomic remarks on *Allium* L. from Fergan depression, *Middle Asia. Linzer Biol. Beitr.* 30, 281–292.

Fritsch, R.M., Blattner, F.R. and Gurushidze, M. (2010) New classification of *Allium* L. subg. *Melanocrommyum* (Webb & Berth.) Rouy (Alliaceae), based on molecular and morphological characters. *Phyton (Horn)* 49, 145–220.

Galmarini, C.R. (2017) Chapter 1: Economic and academic importance. In: Shigyo, M., Khar, A., Abdelrahmam, M. and Kole, C. (eds) *The Allium Genomes. Compedium of Plant Genomes*. Springer Nature Switzerland AG, pp. 1–9. DOI: 10.1007/978-3-319-95825-5_1.

Germishuizen, G. and Meyer, N.L. (eds) (2003) *Plants of Southern Africa: An Annotated Checklist. Strelitzia*, vol. 14. National Botanical Institute, Pretoria, South Africa, pp. i–vi, 1–1231.

Govaerts, R., Kington, S., Friesen, N., Fritsch, R.M., Snijman, D.A. *et al.* (2005–2020) World checklist of Amaryllidaceae. Available at: http://apps.kew.org/wcsp/ (accessed 20 June 2020).

Guern, M., Lecorff, J. and Boscher, J. (1991) Comparative karyology of the *Allium ampeloprasum* complex in France. *Bulletin de la Societe Botanique de France – Lettres Botaniques* 138, 303–313.

Gurushidze, M., Mashayekhi, S., Blattner, F.R., Friesen, N. and Fritsch R.M. (2007) Phylogenetic relationships of wild and cultivated species of *Allium* section *Cepa* inferred by nuclear rDNA ITS sequence analysis. *Plant Systematic and Evolution* 269, 259–269.

Gurushidze, M., Fritsch, R.M. and Blattner, F.R. (2010) Species-level phylogeny of *Allium* subgenus *Melanocrommyum*: incomplete lineage sorting, intraspecific polymorphism, hybridization and trnF gene duplication. *Taxon* 59, 829–840.

Han, T.S., Zheng, Q.J., Onstein, R.E., Rojas-Andries, B.M., Hauenschild, F. *et al.* (2019) Polyploidy promotes species diversification of *Allium* through ecological shifts. *New Phytologist.* DOI: 10.1111/nph.16098.

Hanelt, P. (1988) Taxonomy as a tool for studying plant genetic resources. *Die Kulturpflanze* 36, 169–187.

Hanelt, P. (1990) Taxonomy, evolution, and history. In: Rabinovich, H.D. and Brewster, J.L. (eds) *Onions and Allied Crops, Vol. 1. Botany, Physiology, and Genetics*. CRC Press, Boca Raton, Florida, pp. 1–26.

Hanelt, P. (2001) Alliaceae. In: Hanelt, P. (ed.) *Mansfeld's Encyclopedia of Agricultural and Horticultural Crops*, vol. 4, 3rd edn. Springer Verlag, Vienna, pp. 2250–2269.

Hauenschild, F., Favre, A., Schnitzler, J., Michalak, I., Freiberg, M. *et al.* (2017) Spatio-temporal evolution of *Allium* L. in the Qinghai–Tibet–Plateau region: immigration and in situ radiation. *Plant Diversity* 39, 167–179. DOI: 10.1016/j.pld.2017.05.010.

Havey, M.J. (1991) Molecular characterization of the interspecific origin of viviparous onion. *Journal of Heredity* 82, 501–502.

Havey, M.J. (1992) Restriction enzyme analysis of the chloroplast and nuclear 45s ribosomal DNA of *Allium* sections *Cepa* and *Phyllodolon* (Alliaceae). *Plant Systematics and Evolution* 183, 17–31.

Havey, M.J. (1995) Onions and other cultivated alliums. In: Smartt, J. and Simmonds, N.W. (eds) *Evolution of Crop Plants*, 2nd edn. Longman Scientific and Technical, Burnt Mill, UK, pp. 344–350.

Helm, J. (1956) Die zu Würz- und Speisezwecken kultivierten Arten der Gattung *Allium* L. *Kulturpflanze* 4, 130–180.

Herden, T., Hanelt, P. and Friesen, N. (2016) Phylogeny of *Allium* L. subgenus *Anguinum* (G.Don ex W.D. J.Koch) N. Friesen (Amaryllidaceae). *Molecular Phylogenetics and Evolution* 95, 79–83. DOI: 10.1016/j.ympev.2015.11.004.

Hirschegger, P., Jakše, J., Trontelj, P. and Bohanec, B. (2010) Origins of *Allium ampeloprasum* horticultural groups and a molecular phylogeny of the section *Allium* (*Allium*: Alliaceae), *Molecular Phylogenetics and Evolution* 54, 488–497. DOI: 10.1016/j.ympev.2009.08.030.

Hong, C.J. (1999) *Fundamental studies on crossbreeding in garlic,* Allium sativum. PhD thesis, Kagoshima University, Kagoshima, Japan.

Huang, D.Q., Yang, J.T., Zhou, C.J., Zhou, S.D. and He, X.J. (2014) Phylogenetic reappraisal of *Allium* subgenus *Cyathophora* (Amaryllidaceae) and related taxa, with a proposal of two new sections. *Journal of Plant Research* 127, 275–286.

Inden, H. and Asahira, T. (1990) Japanese bunching onion (*Allium fistulosum* L.). In: Brewster, J.I. and Rabinowich, H.D. (eds) *Onion and Allied Crops. Vol. III. Biochemistry, Food Science, and Minor Crops*. CRC Press, Boca Raton, Florida, pp. 159–178.

Ipek, M., Ipek, A. and Simon, P.W. (2003) Comparison of AFLPs, RAPD markers, and isozymes for diversity assessment of garlic and detection of putative duplicates in germplasm collections. *Journal of American Society for Horticultural Science* 128, 246–252.

Ipek, M., Ipek, A. and Simon, P.W. (2008) Genetic characterisation of *Allium tuncelianum*: an endemic edible *Allium* species with garlic odor. *Scientia Horticuturae* 115, 409–415. DOI: 10.1016/j.scienta.2007.11.002.

Janssen, T. and Bremer, K. (2004) The age of major monocot groups inferred from 800+ rbcL sequences. *Botanical Journal of the Linnean Society* 146, 385–398. DOI: 10.1111/j.1095-8339.2004.00345.x.

Jimenez, H.J., da Silva, A.D.F.L., Martins, S.S., de Carvalho, R. and de Moraes Filho, R.M. (2020) Comparative genomics plastomes of the Amaryllidaceae family species. *Scientia Plena* 16, 8. DOI: 10.14808/sci.plena.2020.060202.

Jones, H.A. and Mann, L.K. (1963) *Onions and their Allies: Botany, Cultivation and Utilization*. Leonard Hill, London and Interscience, New York.

Ju, Y., Zhuo, J., Bo Liu, B. and Long, G. (2013) Eating from the wild: diversity of wild edible plants used by Tibetans in Shangri-la region, Yunnan, *China. Journal of Ethnobiology and Ethnomedicine* 9, 28.

Kamenetsky, R., Shafir, I.L., Baizeman, M., Khassanov, F., Kik, C. *et al.* (2004a) Garlic (*Allium sativum* L.) and its wild relatives from Central Asia: evaluation for fertility potential. *Acta Horticulturae* 637, 83–91.

Kamenetsky, R., Shafir, I.L., Zemah, H., Barzilay, A. and Rabinowitch, H.D. (2004b) Environmental control of garlic growth and florogenesis. *Journal of the American Society for Horticultural Science* 129, 144–151.

Kamenetsky, R., London Shafir, I., Khassanov, F., Kik, C., van Heusden, A.W. *et al.* (2005) Diversity in fertility potential and organo-sulphur compounds among garlics from Central Asia. *Biodiversity and Conservation* 14, 281–295.

Kamenetsky, E., Khassanov, F., Rabinovich, H.D., Auger, J. and Kik, C. (2007) Garlic biodiversity and genetic resources. *Medicinal and Aromatic Plant Science and Biotechnology*. *Global Science Books* 1, 1–5.

Kang, Y., Łuczaj, Ł., Kang, J. and Zhang, S. (2013) Wild food plants and wild edible fungi in two valleys of the Qinling Mountains (Shaanxi, central China). *Journal of Ethnobotany and Ethnomedicine* 9, 26.

Kay, K.M., Whittall, J.B. and Hodges, S.A. (2006) A survey of nuclear ribosomal internal transcribed spacer substitution rates across angiosperms: an approximate molecular clock with life history effects. *BMC Evolutionary Biology* 6, 36. DOI: 10.1186/1471-2148-6-36.

Kazakova, A.A. (1978) Luk, *Kul'turnaja Flora SSSR, X*, Kolos. Leningrad, USSR.

Keusgen, M., Jedelská, J. and Fritsch, R.M. (2008) Phytochemical analysis of *Allium* species from Central Asia. In: Keusgen, M. and Fritsch, R.M (eds) *Proceedings, First Kazbegi Workshop on 'Botany, Taxonomy and Phytochemistry of Wild Allium L. Species of the Caucasus and Central Asia'*, 4–8 June 2007. Marburg & Gatersleben, Kazbegi, Caucasus, Georgia, pp. 103–130.

Khassanov, F.O. (1996) Conspectus of the wild growing *Allium* species of Middle Asia. In: ÖztÜrk, M., Sec,men, Ö. and Görk, G. (eds) *Plant Life in Southwest and Central Asia*. EGE University Press, Izmir, Turkey, pp. 141–59.

Khassanov, F.O. (2000) Subinfrageneric grouping in genus *Allium* subgenus *Allium*. In: Ashurmetov, O., Khassanov, F. and Salieva, Y. (eds) *Plant Life in Southwest and Central Asia. Proceedings of the 5th International Symposium, 18–22 May 1998, Tashkent, Uzbekistan*, pp. 107–112.

Khassanov, F.O. (2018) Taxonomical and ethnobotanical aspects of *Allium* species from Middle Asia with particular reference to subgenus *Allium*. In: Shigyo, M., Khar, A., Abdelrahmam, M. and Kole, C. (eds) *The Allium Genomes. Compedium of Plant Genomes*. Springer Nature Switzerland AG, pp. 11–21. Available at: https://doi.org/10.1007/978-3-319-95825-5_2 (accessed 28 June 2022).

Khassanov, F.O. and Fritsch, R.M. (1994) New taxa in *Allium* L. subg. *Melanocrommyum* (Webb & Berth.) Rouy from Central Asia. *Linzer biologische Beiträge* 26, 965–990.

Khassanov, F.O., Shomuradov, H.F. and Kadyrov, G.U. (2011) Taxonomic revision of *Allium* L. sect. *Allium* s. l. in Central Asia. *Stapfia* 95, 171–174.

Khassanov, F.O., Karimov, F. and Tirkasheva, B. (2013) Taxonomic revision and lectotypification of *Allium* L. sect. *Coerulea* (Omelcz.) F.O. *Khass. Stapfia* 99, 208–234.

Kik, C. (2008) Allium genetic resources with particular reference to onion. *Acta Horticulturae* 770. DOI: 10.17660/ActaHortic.2008.770.14.

Kizil, S. and Khawar, K.M. (2017) Introduction of endemic *Allium tuncelianum* Kollman from hot and temperate climate to semi-arid climatic conditions. *Acta Scientarum Polonorum Hororum Cultus* 16, 117–124. DOI: 10.24326/asphc.2017.5.12.

Klaas, M. and Friesen, N. (2002) Chapter 4: molecular markers in *Allium*. In: Rabinovich, H.D. and Currah, L. (eds) *Allium Crop Science – Recent Advances*. CAB International, Wallingford, UK, pp. 159–185.

Kollmann, F. (1971) *Allium ampeloprasum* – a polyploid complex 1. *Ploidy levels. Israel Journal of Botany* 20, 13–20.

Lee, J., Chon, J.K., Lim, J.S., Kim, E.K. and Nah, G. (2017) Characterization of complete chloroplast genome of *Allium victorialis* and its application for barcode markers. *Plant Breeding and Biotechnology* 5, 221–227. DOI: 10.9787/PBB.2017.5.3.221.

Li, H.T., Yi, T.S., Gao, L.M., Ma, P.F., Zhang, T. *et al.* (2019) Origin of angiosperms and the puzzle of the Jurassic gap. *Nature Plants* 5, 461–470.

Li, Q.Q., Zhou, S.D., He, X.J., Yu, Y., Zhang, Y.C. *et al.* (2010) Phylogeny and biogeography of *Allium* (Amaryllidaceae: Allieae) based on nuclear ribosomal transcribed spacer and chloroplast rps16 sequences focusing on the inclusion of species endemic to China. *Annals of Botany* 106, 709–733.

Li, Q.Q., Zhou, S.D., Huang, D.Q., He, Y.J. and Wei, X.Q. (2016) Molecular phylogeny divergence time estimates and historical biogeography within one of the world's largest monocot genera. *AoB Plants* 8, plw041.

Linnaeus, C. von (1753) *Species plantarum, exhibentes plantas rite cognitas ad genera relatas Tomus I*. Holmiae, Impensis Laurentii Salvii.

Maaß, H. (1996) About the origin of the French grey shallot. *Genetic Resources and Crop Evolution* 43, 291–292.

Maaß, H. (1997a) Genetic diversity in the top onion, *Allium ×proliferum* (Alliaceae), analysed by isozymes. *Plant Systematics and Evolution* 208, 35–44.

Maaß, H. (1997b) Studies on triploid viviparous onions and their origin. *Genetic Resources and Crop Evolution* 44, 95–99.

Maaß, H. and Klaas, M. (1995) Intraspecific differentiation of garlic (*Allium sativum* L.) by isozyme and RAPD markers. *Theoretical and Applied Genetics* 91, 89–97.

Mashayekhi, S. and Columbus, J.T. (2014) Evolution of leaf blade anatomy in *Allium* (Amaryllidaceae) subgenus *Amerallium*, with a focus on the North American species. *American Journal of Botany* 101, 63–85. DOI: 10.3732/ajb.1300053.

Mathew, B. (1996) *A Review of Allium section Allium*. The International Board for Plant Genetic Resources IBPGR. Royal Botanical Gardens, Kew, London.

McNeal, D.W. and Jacobsen, T.D. (2002) *Allium* in flora of North America. Flora of North America Editorial Committee. *New York Oxford, Oxford University Press*, vol. 26, pp. 224–276.

Messiaen, C.M., Cohat, J., Leroux, I.P., Pichon, M. and Beyris, A. (1993) *Les Allium alimentaires reproduits par voie vegetative*. INRA, Paris.

Namgung, J., Do, H.D.K., Kim, C., Choi, H.J. and Kim, J.H. (2021) Complete chloroplast genomes shed light on phylogenetic relationships, divergence time, and biogeography of Allioideae (Amaryllidaceae). *Scientific Reports* 11, 3262.

Nguyen, N.H., Driscoll, H.E. and Specht, C.D. (2008) A molecular phylogeny of the wild onions (*Allium*; Alliaceae) with a focus on the western North American centre of diversity. *Molecular Phylogenetics and Evolution* 47, 1157–1172.

Omelczuk, T.J. (1962) Sistematicheskij sklad cibul' Ukraïni (rid *Allium* L.). *Ukrainskii Botanichnii Zhurnal* 19, 66–73. [In Ukrainian].

Oyuntsetseg, B., Blattner, F.R. and Friesen. N. (2012) Diploid *Allium ramosum* from East Mongolia: a missing link for the origin of the crop species *A. tuberosum? Erforschung biologisches Ressources Mongolei* (Halle/Saale) 12, 415–424.

Ozturk, M., Guzel, S., Altay, V. and Ernaz, A. (2012) Alliums, an underutilized genetic resource in the East Mediterranean. *Acta Horticulturae* 969. DOI: 10.17660/ActaHortic.2012.969.39.

Pigg, B.K., Bryan, F.A. and DeVore, M.L. (2018) *Paleoallium billgenseli* gen. et sp. nov.: fossil monocot remains from the latest early Eocene Republic Flora, Northeastern Washington State, USA. *International Journal of Plant Sciences* 179, 477–486.

Pooler, M.R. and Simon, P.W. (1993) Characterization and classification of isozyme and morphological variation in a diverse collection of garlic clones. *Euphytica* 68, 121–130.

Pooler, M.R. and Simon, P.W. (1994) True seed production in garlic. *Sexual Plant Reproduction* 7, 282–286. Available at: https://doi.org/10.1007/BF00227710 (accessed 28 June 2022).

POWO (2020) *Plants of the World online. Allium longicuspis Regel*. Available at: http://www.plantsoftheworldonline.org/taxon/urn:lsid:ipni.org:names:528333-1 (accessed 20 January 2021).

Puizina, J. and Papeŝ, D. (1996) Further cytogenetic analyses of the Croatian triploid shallot 'Ljutika' (*Allium cepa* var. *viviparum*, Alliaceae) and its comparison with the Indian triploid 'Pran'. *Plant Systematics and Evolution* 208, 11–23.

Quang, B.H., Bach, T.T., Eum, S., Hai, D.V., Khang, N.S. *et al.* (2020) Towards a floristic inventory of Bat Xat Nature Reserve, Vietnam: thirteen new national records of vascular plants. *Wulfenia* 27, 233–250.

Rabinovich, H.D. and Kamenetsky, R. (2002) Shallot (*Allium cepa*, *Aggregatum* Group). In: Rabinowitch, H.D. and Currah, L. (eds) *Allium Crop Science: Recent Advances*. CAB International, Wallingford, UK, pp. 409–430.

Saghir, A.R.B., Mann, L.K., Ownbey, M. and Berg, R.Y. (1966) Composition of volatiles in relation to taxonomy of American Alliums. *American Journal of Botany* 53, 477–484.

Schubert, I., Ohle, H. and Hanelt, P. (1983) Phylogenetic conclusion from Giemsa banding and NOR staining in top onions (Liliaceae). *Plant Systematics and Evolution* 143, 245–256.

Seregin, A., Anackov, G. and Friesen, N. (2015) Molecular and morphological revision of the *Allium saxatile* group (Amaryllidaceae): geographical isolation as the driving force of underestimated speciation. *Botanical Journal of the Linnean Society* 178, 67–101.

Shemesh-Mayer, E. and Kamenetsky-Goldstein, R. (2019) Recent advances in sexual propagation and breeding of garlic. In: Warrington, I. (ed.) *Horticultural Reviews 46*. John Wiley & Sons, Inc.

Sinitsyna, T., Herden, T. and Friesen, N. (2016) Dated phylogeny and biogeography of the Eurasian *Allium* section *Rhizirideum* (Amaryllidaceae). *Plant Systematics and Evolution* 302, 1311–1328. DOI: 10.1007/s00606-016-1333-3.

Stearn, W.T. (1960) *Allium* und *Milula* in the central and eastern Himalaya. *Bulletin of the British Museum (Natural History). Botany* 2, 161–187.

van Raamsdonk, L.W.D., Ensink, W., van Heusden, A.W. *et al.* (2003) Biodiversity assessment based on cpDNA and crossability analysis in selected species of *Allium* subgenus *Rhizirideum*. *Theoretical and Applied Genetics* 107, 1048–1058.

Vvedensky, A.I. (1944) The genus *Allium* in the USSR. *Herbertia* 11, 65–219.

Wendelbo, P. (1971) Alliaceae. In: Rechinger, K.H. (ed.) *Flora Iranica 76*. Graz, Austria, pp. 1–100.

Wheeler, E.J., Mashayekhi, S., McNeal, D.W., Columbus, J.T. and Pires, J.C. (2013) Molecular systematics of *Allium* subgenus *Amerallium* (Amaryllidaceae) in North America. *American Journal of Botany* 100, 701–711.

Xie, D.F., Yu, H.X., Xie, C., Yi-Qi, Deng, Y.D. *et al.* (2019) Phylogeny of Chinese *Allium* species in Section *Daghestanica* and adaptive evolution of *Allium* (Amaryllidaceae, Allioideae) species revealed by the chloroplast complete genome. *Frontiers in Plant Science* 10, 460. DOI: 10.3389/fpls.2019.00460.

Xie, D.F., Tan, J.B., Yu, Y., Gui, L.J., Su, D.M. *et al.* (2020) Insights into phylogeny, age and evolution of *Allium* (Amaryllidaceae) based on the whole plastome sequences. *Annals of Botany* 125, 1039–1055. DOI: 10.1093/aob/mcaa024.

Yang, J., Zhou, S., Huang, D. and He, X.J. (2017) Phylogeography of two closely related species of *Allium* endemic to East Asia: population evolution in response to climate oscillations. *Ecology and Evolution* 8, 7986–7999. DOI: 10.1002/ece3.4338.

Yang, X., Xie, D.F., Zhou, S.D. and He, X.J. (2019) Characterization of the complete chloroplast genome of *Allium kingdonii*. *Mitochondrial DNA B: Resources* 4, 868–869.

Yusupov, Z., Deng, T., Volis, S., Khassanov, F., Makhmuradov, D. *et al.* (2020) Phylogenomics of *Allium* section *Cepa* (Amaryllidaceae) provides new insights on the domestication of onion. *Plant Diversity* 7, 8. DOI: 10.1016/j.pld.2020.07.008.

Zhu, S., Tang, S., Tan, Z., Yu, Y., Dai Q. *et al.* (2017) Comparative transcriptomics provide insight into the morphogenesis and evolution of fistular leaves in *Allium*. *BMC Genomics* 18, 60. DOI: 10.1186/s12864-016-3474-8.

Appendix 1. Currently Accepted Infrageneric Groups in *Allium* L.

First evolutionary lineage

1. Subgenus *Nectaroscordum* (Lindl.) Asch. et Graebn., Type: *A. siculum* Ucria
Sect. *Nectaroscordum* (Lindl.) Gren. et Godr., Type: *A. siculum* Ucria
2. Subgenus *Microscordum* (Maxim.) N. Friesen, Type: *A. monanthum* Maxim
Sect. *Microscordum* Maxim., Type: *A. monanthum* Maxim
3. Subgenus *Amerallium* Traub., Type *A. canadense* L.
American sections:*
Sect. *Amerallium* Traub, Type: *A. canadense* L.
Sect. *Rhophetoprason* Traub, Type: *A. glandulosum* Link et Otto
Sect. *Lophioprason* Traub, Type: *A. sanbornii* Wood.
Sect. *Caulorhizideum* Traub, Type: *A. validum* S.Wats
East Asian sections
Sect. *Bromatorrhiza* Type: *A. wallichii* Kunth
Sect. *Kingdonia* X.J. He et D.Q. Huang, Type: *A. kingdonii* Stearn
Mediterranean sections
Sect. *Ophioscorodon* (Wallr.)Endl., Type: *A. ursinum* L.
Sect. *Narkissoprason* Type: *A. insubricum* Boiss. et Reut.
Sect. *Molium* G. Don, Type: *A. roseum* L.
Sect. *Briseis* (Salisb.) Stearn, Type: *A. triquetrum* L.
Sect. *Chamaeprason* F. Hermann, Type: *A. chamaemoly* L.
Sect. *Rhynchocarpum* Brullo, Type: *A. ruhmerianum* Asch. ex E.A. Durand et Barratte

*Modern authors divide American species from the subgenus *Amerallium* not into sections but into alliances (Saghir *et al.*, 1960; McNeal and Jacobsen, 2002; Nguyen *et al.*, 2008; Wheeler *et al.*, 2013).

Second evolutionary lineage

4. Subgenus *Caloscordum* (Herb.) R.M. Fritsch, Type: *A. neriniflorum* (Herb.) Baker
Sect. *Caloscordum* (Herb.) Baker, Type: *A. neriniflorum* (Herb.) Baker
5. Subgenus *Anguinum* (G.Don ex Koch) N. Friesen, Type: *A. victorialis* L.
Sect. *Anguinum* G. Don ex Koch, Type: *A. victorialis* L.
6. Subgenus *Porphyroprason* (Ekberg) R.M. Fritsch, Type: *A. oreophilum* C.A.Mey.
Sect. *Porphyroprason* Ekberg, Type: *A. oreophilum* C.A.Mey.
7. Subgenus *Vvedenskya* (Kamelin) R.M. Fritsch, Type: *A. kujukense* Vved.
Sect. *Vvedenskya* Kamelin, Type: *A. kujukense* Vved.
8. Subgenus *Melanocrommyum* (Webb et Berth.) Rouy. Type: *A. nigrum* L.
Sect. *Melanocrommyum* Webb et Berth. Type: *A. nigrum* L.
Sect. *Acmopetala* R.M. Fritsch, Type: *A. backhousianum* Regel
Sect. *Megaloprason* Wendelbo, Type: *A. rosenbachianum* Regel
Sect. *Regeloprason* Wendelbo, Type: *A. regelii* Trautv.
Sect. *Kaloprason* Koch, Type: *A. caspium* (Pall.) M. Bieb.
Sect. *Acanthoprason* Wendelbo, Type: *A. akaka* S.G.Gmel. ex Schult. et Schult. f.
Sect. *Compactoprason* R.M. Fritsch, Type: *A. giganteum* Regel
Sect. *Pseudoprason* (Wendelbo) K. Perss. et Wendelbo, Type: *A. koelzii* (Wendelbo) K. Perss. et Wendelbo
Sect. *Miniprason* R.M. Fritsch, Type: *A. karataviense* Regel
Sect. *Brevicaule* R.M.Fritsch, Type: *A. sergii* Vved.

Sect. *Thaumasioprason* Wendelbo, Type: *A. mirum* Wendelbo
Sect. *Verticillata* Kamelin, Type: *A. verticillatum* (Regel) Regel
Sect. *Acaule* R.M. Fritsch, Type: *A. hexaceras* Vved.
Sect. *Aroidea* F.O.Khass. et R.M. Fritsch, Type: *A. aroides* Popov & Vved.
Sect. *Popovia* F.O. Khass. et R.M. Fritsch, Type: *A. gypsaceum* Popov & Vved.
Sect. *Longibidentata* (R.M. Fritsch) R.M. Fritsch, Type: A. *fetissowii* Regel
Sect. *Asteroprason* R.M. Fritsch, Type: *A. elburzense* Wendelbo
Sect. *Procerallium* R.M. Fritsch, Type: *A. stipitatum* Regel
Sect. *Stellata* (F.O. Khass. et R.M. Fritsch) R.M. Fritsch, Type: *A. taeniopetalum* Popov & Vved
Sect. *Decipientia* (Omelczuk) R.M. Fritsch, Type: *A. decipiens* Fisch. ex Schult. & Schult.f.
Sect. *Tulipifolia* R.M. Fritsch & N. Friesen, Type: *A. tulipifolium* Ledeb.

Third evolutionary lineage

9. Subgenus *Butomissa* (Salisb.) N. Friesen, Type: *A. ramosum* L.
Sect: *Butomissa* (Salisb.) Kamelin, Type: *A. ramosum* L.
Sect. *Trifurcatum* X.J. He et D.Q.Huang, Type: A. *trifurcatum* (F.T. Wang & Tang) J.M. Xu
Sect. *Austromontana* N. Friesen, Type: *A. oreoprasum* Schrenk
10. Subgenus *Cyathophora* (R.M. Fritsch) R.M. Fritsch, Type: *A. cyathophorum* Bur. et Franch.
Sect. *Cyathophora* R.M. Fritsch, Type: *A. cyathophorum* Bur. et Franch.
Sect. *Milula* (Prain) N. Friesen, Type: *A. spicatum* (Prain) N. Friesen
Sect. *Coleoblastus* Ekberg, Type: *A. mairei* H.Lev.
11. Subgenus *Rhizirideum* (G. Don ex Koch) Wendelbo, Type: *A. senescens* L.
Sect. *Rhizirideum* G. Don ex Koch, Type: *A. senescens* L.
Sect. *Tenuissima* (Tsagolova) Hanelt, Type: *A. tenuissimum* L.
Sect. *Rhizomatosa* Egorova emend N. Friesen (incl. syn. sect. *Caespitosoprasum* N. Friesen), Type: *A. caespitosum* Sievers ex Bong. et C.A.Mey
Sect. *Eduardia* N. Friesen, Type: *A. eduardii* Stearn
12. Subgenus *Allium* Type: *A. sativum* L.
Sect. *Allium* Type: *A. sativum* L.
Sect. *Codonoprasum* Koch, Type: *A. oleraceum* L.
Sect. *Avulsea* F.O. Khass., Type: *A. griffithianum* Boiss.
Sect. *Brevidentia* F.O. Khass. et Yengal., Type: *A. brevidens* Vved.
Sect. *Coerulea* (Omelczuk) F.O. Khass., Type: *A. coeruleum* Pall.
Sect. *Costulatae* F.O. Khass. et Yengal., Type: *A. filidens* Regel.
Sect. *Haneltia* F.O. Khass., Type: *A. haneltii* F.O. Khass. & R.M. Fritsch.
Sect. *Crystallina* F.O. Khass. et Yengal., Type: *A. crystallinum* Vved.
Sect. *Eremoprasum* (Kamelin) F.O. Khass., R.M. Fritsch et N. Friesen, Type: *A. sabulosum* Stev. ex Bunge
Sect. *Kopetdagia* F.O. Khass., Type: *A. kopetdagense* Vved.
Sect. *Longivaginata* (Kamelin) F.O. Khass., R.M. Fritsch et N. Friesen, Type: *A. longivaginatum* Wendelbo
Sect. *Mediasia* F.O. Khass., Yengal. et N. Friesen, Type: *A. turkestanicum* Regel
Sect. *Minuta* F.O. Khass., Type: *Allium minutum* Vved.
Sect. *Multicaulia* F.O. Khass. et Yengal., Type: *A. borszczowii* Regel
Sect. *Pallasia* (Tzag.) F.O. Khass., R.M. Fritsch et N. Friesen, Type: *A. pallasii* Murray
Sect. *Unicaulia* F.O. Khass., Type: *Allium kotschyi* Boiss.
Sect. *Rechingeria* F.O. Khass. & Tirkasheva, Type: *A. rechingeri* Wendelbo

13. Subgenus *Cepa* (Mill.) Radić, Type: *A. cepa* L.
- Sect. *Cepa* (Mill.) Prokh., Type: *A. cepa* L.
- Sect. *Schoenoprasum* Dumort., Type: *A. schoenoprasum* L.
- Sect. *Annuloprason* Egorova, Type: *A. fedschenkoanum* Regel
- Sect. *Sacculiferum* P.P. Gritz., Type: *A. sacculiferum* Maxim.
- Sect. *Flavovirens* Q.Q. Li & X.J. He, Type: *A. flavovirens* Regel (nomen nudum)
- Sect. *Condensatum* N. Friesen, Type: *A. condensatum* Turch.

14. Subgenus *Polyprason* Radić, Type: *A. moschatum* L.
- Sect. *Scorodon* Koch, Type: *A. moschatum* L.
- Sect. *Oreiprason* F. Hermann, Type: *A. saxatile* M. Bieb.
- Sect. *Falcatifolia* N. Friesen, Type: *A. carolinianum* Redouté
- Sect. *Daghestanica* (Tscholok.) N. Friesen, Type: *A. daghestanicum* Grossg.

15. Subgenus *Reticulatobulbosa* (Kamelin) N. Friesen, Type: *A. lineare* L.
- Sect. *Reticulatobulbosa* Kamelin, Type: *A. lineare* L.
- Sect. *Campanulata* Kamelin, Type: *A. xiphopetalum* Aitch.
- Sect. *Scabriscapa* (Tscholok.) N. Friesen, Type: *A. scabriscapum* Boiss.
- Sect. *Nigrimontana* N. Friesen, Type: *A. drobovii* Vved.
- Sect. *Sikkimensia* (Traub) N. Friesen, Type: *A. sikkimense* Baker

2 A Review on *Allium* Biodiversity

Andrés Bermúdez[1], Jesús Martín[1], Turgay Unver[2], Pilar Hernández[3] and Gabriel Dorado[4*]

[1]Dep. Microbiología, Universidad de Córdoba, Córdoba, Spain; [2]Ficus Biotechnology, Ankara, Turkey; [3]Instituto de Agricultura Sostenible (IAS-CSIC), Consejo Superior de Investigaciones Científicas (CSIC), Córdoba, Spain; [4]Dep. Bioquímica y Biología Molecular, Universidad de Córdoba, Córdoba, Spain.

1. Introduction

The genus *Allium* is a monocotyledon that belongs to the order Asparagales, family Amaryllidaceae and tribe Allieae (Byng *et al.*, 2016). It comprises 2140 species, as compiled by Kew Science of the Royal Botanic Gardens, Kew, as of 01/01/2017, at https://wcsp.science.kew.org/incfamilies.do and https://wcsp.science.kew.org/reviewers.do (accessed September 2022) (Deniz *et al.*, 2015). About 70 species are currently cultivated, as vegetables, spices or ornamentals. This genus originated in the Indian subcontinent, 52 million years (myr) ago, during the early Eocene. It spread throughout Eurasia and North America 10–20 myr ago, during the middle Miocene. Both periods were significantly warm (Costa *et al.*, 2020). *Allium* is a monophyletic genus comprising three major clades (I, II, III). A total of 90% of the species of clades I and II belong to subgenus *Amerallium* and *Melanocrommyum*, respectively. On the other hand, clade III includes the subgenus *Allium*, being the most diversified and comprising all the cultivated species (Li *et al.*, 2010; Costa *et al.*, 2020).

In accord with the needs and requirements of the exponentially growing population and in an attempt to lower costs, agriculture has been industrialized over the years. That was followed by large-scale trading worldwide, as well as product usage diversification. Originally, many varieties of agricultural plants were cultivated. However, the traditional farming model has evolved into a new one, where only the most productive crops and varieties are farmed (Lanz *et al.*, 2018). For example, in Spain, which is the main garlic (*Allium sativum*) producer in the European Union, mainly four garlic varieties are currently marketed: (i) Spanish White ('Blanco'), mainly from Valladolid province; (ii) Spanish Purple or Red ('Morado' or 'Rojo'), mainly from Las Pedroñeras; it excels in bioactive compounds, and thus is labelled with Protected Geographical Indication (PGI); (iii) Chinese Spring White; and (iv) Chinese Spring Violet. Market demands have promoted the cultivation of the two highly productive Chinese varieties, due to increased profitability. This has resulted in a substantial loss of landraces and heirloom varieties, which were quite common in the market until

*Address for correspondence: bb1dopeg@uco.es

 Edible Alliums: Botany, Production and Uses (eds H.D. Rabinowitch and B. Thomas)
DOI: 10.1079/9781789249996.0002

very recently. As a consequence, consumers get cheaper garlic, albeit with lower quality than the Spanish White or the Spanish Purple ones.

The loss of crop biodiversity increases their vulnerability to diseases and pests. Such effect is particularly relevant in the current trend of global warming, drought, climate change and other environmental problems. Besides, such genetic erosion and loss of biodiversity may lead to the extinction of some varieties and species.

In fact, depending on the crop, from 17% to 30% of the world's annual harvest is lost due to diseases and pests, mostly in developing countries (Savary *et al.*, 2019). An illustrative example is a recent and abrupt irruption of the condition known as waxy breakdown in Spain. That physiopathy mainly affects the native Spanish White garlic, which is highly appreciated in central European markets. Consequently, the cultivation of this variety has declined, and the already short spectrum of local Spanish varieties cultivated is getting even narrower, while the white genotype is being replaced by Chinese Spring White garlic. Other garlic White varieties, like the French one from Drôme, have not been taken on, mainly because of their lower productivity compared with the Chinese one.

2. Status of *Allium* Biodiversity

Biodiversity represents both a global and national heritage for each country. They are invaluable, not only *per se*, but also because they may possess variability with regard to the nutraceutical, medical-pharmacological, cosmetics and related industries, as well as genetics useful for molecular biology and enzymology (Egea *et al.*, 2017). It is therefore of paramount importance to identify, rescue, preserve, manage and study Plant (PGR) and Animal (AGR) Genetic Resources. Thus, the World Database on Protected Areas (WDPA) has been created, including terrestrial and marine ecosystems (https://www.protectedplanet.net; accessed September 2022). The collaboration between the United Nations Environment Programme (UNEP) and the International Union for Conservation of Nature (IUCN) provides useful information on protected areas, and Other Effective area-based Conservation Measures (OECM) (https://www.protectedplanet.net). IUCN shows the status of the natural world, including safeguarding strategies (https://www.iucn.org; accessed September 2022). They are managed by UNEP World Conservation Monitoring Centre (UNEP-WCMC), a rich resource of biodiversity knowledge (https://www.unep-wcmc.org; accessed September 2022). Likewise, the European Cooperative Programme for Plant Genetic Resources (ECPGR; https://www.ecpgr.cgiar.org; accessed September 2022) promotes the conservation and usage of plant genetic resources in Europe. It includes the ECPGR *Allium* Working Group (https://www.ecpgr.cgiar.org/working-groups/allium).

As with animals, natural biodiversity reserves have been established mainly for wild plants (Fig. 2.1), and genebanks, mainly for crop plants (Ebert and Engels, 2020; Panis *et al.*, 2020; Ruta *et al.*, 2020). However, many low-income countries cannot afford the costs of PGR management, thus the significant difference in the conservation-protection levels between them. Most members of the genus *Allium* grow wild in Eurasia and North America. This implies the need for international co-operation to collect and manage PGR. Moreover, the level of vulnerability reduces with back-up copies of all PGR, thus increasing the significance of international co-operation.

Seed banks, the first kind of genebanks, were followed by the initiation of *in situ* field banks by Nikolai Vavilov. This Russian botanist collected seeds or propagation organs from 200,000 cultivated plants around the world and conserved them in the Vavilov Institute (Cohen, 1991). In recent decades, tissue and cryobanks were constructed to further improve the safekeeping of some specifically vulnerable genotypes. Indeed, there are more than 1700 genebanks all around the world, stocking over 7 million plant accessions (Singh *et al.*, 2012).

The main genebanks that conserve *Allium* species are in Russia, Europe, USA, Asia (including India, Taiwan and Japan) and Africa (Table 2.1).

Several online databases of plant resources provide farmers, plant breeders and scientists with access to plant genetic materials; for instance, the World Information and Early Warning System (WIEWS) (http://www.fao.org/wiews; accessed September 2022) on Plant Genetic Resources for Food and Agriculture (PGRFA), also known as the International Treaty on Plant

Fig. 2.1. Plant genetic resources. The world's protected areas are shown in green (terrestrial) and blue (marine) colour gradients. (Copyright © 2021 WDPA, OECM, IUCN, UNEP-WCMC, IUCN, Mapbox 'How we built it: mapping the world's 250,000 conservation areas' – https://www.mapbox.com/blog/how-we-built-it-mapping-the-worlds-250-000-conservation-areas; OpenStreetMap – https://www.openstreetmap.org; and Protected Planet – https://www.protectedplanet.net/en/thematic-areas/wdpa?tab=WDPA)

Genetic Resources for Food and Agriculture (ITPGRFA), or with the shorter names of International Seed Treaty (IST) or Plant Treaty (PT) (http://www.fao.org/plant-treaty). WIEWS has registered 30,346 *Allium* accessions at the time of writing.

Likewise, Genesys (https://www.genesys-pgr.org; accessed September 2022) is an internet platform containing information about PGRFA conserved in genebanks worldwide. Genesys collects data from three main networks: (i) the European Search Catalogue for Plant Genetic Resources (EURISCO) (https://eurisco.ipk-gatersleben.de); (ii) the Germplasm Resources Information Network (GRIN) (https://www.ars-grin.gov) of the United States Department of Agriculture (USDA) (https://www.usda.gov); and (iii) the System-wide Information Network on GEnetic Resources (SINGER) (http://www.singer.cgiar.org) of the Consultative Group on International Agricultural Research (CGIAR) (https://www.cgiar.org) and the Asian Vegetable Research and Development Centre (AVRDC), now World Vegetable Center (WorldVeg)

Table 2.1. Main *Allium* germplasm banks worldwide. Total *Allium* accessions are shown, as well as onion, garlic or other (when specified in last column). (Genesys and WIEWS)

	Genera		Species	
Germplasm bank	*Allium*	*A. cepa*	*A. sativum (other)*	References
Vavilov Research Institute of Plant Industry (VRIPI), now Vavilov Institute of Plant Genetic Resources (IPGR; Russia)	2616	1335	439 (*A. ampeloprasum*) 430	http://vir.nw.ru/index.htm http://www.vir.nw.ru/en
Leibniz Institute of Plant Genetics and Crop Plant Research (IPK; Germany)	2615	486	466	https://www.ipk-gatersleben.de/en/institute/about-us
National Plant Germplasm System (NPGS; USA)	2123	1877	528	https://data.nal.usda.gov/dataset/national-plant-germplasm-system
United Kingdom Vegetable Genebank (UKVGB; UK)	1900	1394	4	https://warwick.ac.uk/fac/sci/lifesci/wcc/gru/genebank
Directorate of Onion and Garlic Research (DOGR) and Indian Council of Agricultural Research – National Bureau of Plant Genetic Resources (ICAR-NBPGR; India)	1518	989		https://dogr.icar.gov.in/index.php?lang=en
National Institute for Agricultural Research and Experimentation (NIARE; Spain)	1511	588	754	http://www.inia.es/IniaPortal/goUrlDinamica.action?url=http://wwwsp.inia.es/en-us/Investigacion/centros/crf
Asian Vegetable Research and Development Center (AVRDC), now World Vegetable Center (WorldVeg; Asia & Africa)	1231	578	505	https://avrdc.org
Centre for Genetic Resources (CGN; The Netherlands)	431	227	42 (*A. fistulosum*) 24 (*A. ampeloprasum*)	https://www.wur.nl/en/Research-Results/Statutory-research-tasks/Centre-for-Genetic-Resources-the-Netherlands-1.htm
National Institute of Agrobiological Sciences (NIAS), now National Agriculture and food Research Organization (NARO; Japan)	182	78	104 (*A. fistulosum*)	http://www.naro.affrc.go.jp/english

(https://avrdc.org) (all websites accessed September 2022). Genesys has registered 22,352 *Allium* accessions at the time of writing, distributed in germplasm banks all around the world, most of them from Europe (Fig. 2.2).

Private initiatives are also involved in conservation and development of economically profitable crop plant varieties. Thus, the association of companies Bejo Zaden (https://www.bejo.com) and De Groot en Slot (https://www.degrootenslot.nl/en) (websites accessed September 2022) is leading the breeding, production and sales of *Allium*. They have patented preservation and improvement methods, as shown by

Fig. 2.2. Origin of *Allium* accessions in genebanks. (Sources are shown with green colour; © 2021 Genesys)

the following patents: (i) 'cytoplasmic male sterile leek plants, methods for the preparation and use thereof, and a cytoplasmic male sterile garlic plant' (international publication number: WO/2010/007059; international application number: PCT/EP2009/058980 (https://patentscope.wipo.int/search/en/detail.jsf?docId=WO2010007059); and (ii) 'interspecific hybrid plant resulting from a cross between an *Allium ampeloprasum* plant with an *Allium sativum* plant' (international publication number: WO/2018/134236; international application number: PCT/EP2018/051085) (https://patentscope.wipo.int/search/en/detail.jsf?docId=WO2018134236). Other companies like Nickerson-Zwaan (https://www.menaco.nl/our-brands/nickerson-zwaan), part of Menaco (https://www.menaco.nl) have registered the patent 'resistance to downy mildew of onion caused by the fungus *Peronospora destructor*' (international publication number: WO/ 2006/061256; international application number: PCT/EP2005/013537) (https://patentscope.wipo.int/search/en/detail.jsf?docId=WO2006061256).

3. *Ex Vitro* and *In Vitro* Cost-effectiveness of Methodologies to Preserve Biodiversity of Fertile and Infertile *Allium*

As indicated above, it is important to maintain and manage biodiversity in genebanks. Additionally, *Allium* species, like onion (*Allium cepa*) and garlic, are very heterozygous, showing a high polymorphism and large amounts of repetitive sequences (King *et al.*, 1998; Sun *et al.*, 2020). That coupled with their giant genomes requires high-throughput genotyping-by-sequencing (GBS) strategies, to properly identify their variability. That is now possible thanks to the latest developments in nucleic-acid sequencing, as we have reviewed (Dorado *et al.*, 2019b). Thus, we have demonstrated such approach, genotyping a large garlic germplasm bank, with high-throughput genotyping-by-sequencing DArT-seq (Egea *et al.*, 2017). It combines Diversity Arrays Technology (DArT) complexity reduction with Second-Generation Sequencing (SGS) or Third-Generation Sequencing (TGS). Such sequencing platforms are sometimes called the ambiguous name of 'Next'-Generation Sequencing (NGS). Besides, the most relevant commercial garlics are also sterile. They reproduce vegetatively only and therefore do not generate true botanical seeds through sexual reproduction. That increases the chances of losing biodiversity if varieties are not properly identified, protected and managed in germplasm banks (Fig. 2.3).

The most favourable scenario to maintain plant germplasms is when the accessions produce sexual seeds. Currently, many *Allium* species and varieties, including ornamental ones, are conserved in seed banks. Partial drying and maintenance of these seeds at low temperature extends their natural period of viability. The average annual cost for maintenance of one of these accessions ranges between 14 and 43 USD (Virchow, 2003), taking into account the inflation rate to date.

Yet some species produce recalcitrant seeds, being very sensitive to cold and desiccation.

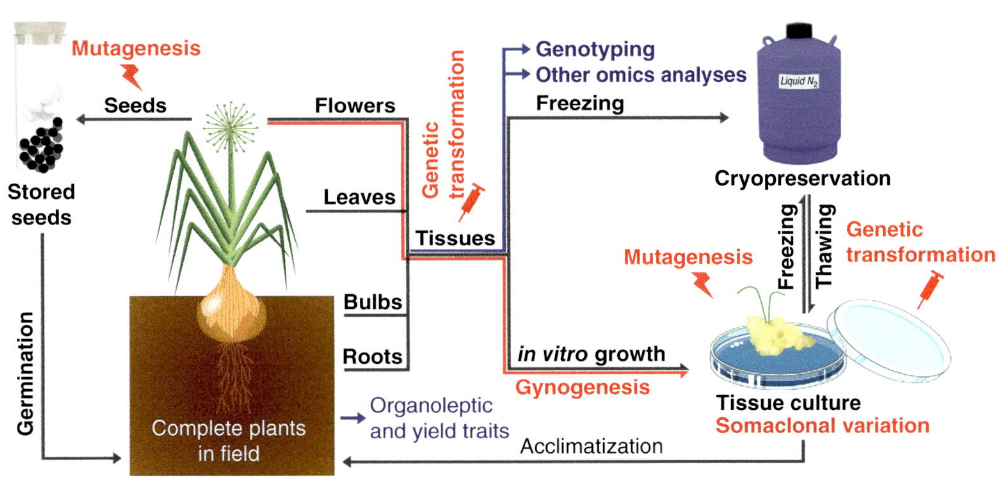

Fig. 2.3. Methodologies to preserve, increase and manage *Allium* biodiversity, including preserving germplasm (black), mutagenesis and genetic engineering (red), and biodiversity analyses (blue).

Therefore, their conservation in seed banks is not appropriate. Additionally, some species like most commercial garlics are sterile, vegetatively reproducing through bulbs. The traditional alternative to maintain these species would be field genebanks. In these, plants are grown, showing their phenotypes while being exposed to different environmental stresses, diseases and pests. Therefore, special care has to be taken to avoid damage to such accessions. As a consequence, such genebanks are more expensive than the seed ones, mainly due to the greater need for agricultural work. Thus, the average annual cost for one of this kind of accession ranges between 36 and 370 USD (Virchow, 2003), considering the inflation rate to date.

An alternative to overcome the abiotic and biotic stresses of field genebanks is tissue genebanks. Thus, plant tissues are grown *in vitro*, away from climatic shocks, free of pathogens and at a high multiplication rate (Normah *et al.*, 2013). However, this *in vitro* conservation method is the most labour-intensive and time-plus-capital-consuming. The annual average unit costs for the accessions ranges between 140 and 995 USD (Virchow, 2003), applying the inflation rate to date. Yet, interestingly, regenerating whole plants from non-vascularized tissues, such as meristems, calli, or somatic embryos, allows sanitation of the plant tissue used as explants. Indeed, these procedures can eliminate pathogens like viruses, since they are generally transported through the vascular system. Consequently, the virus-free plants so obtained are healthier and more productive. *In vitro* tissue culture also allows a rapid plantlet micropropagation (i.e. cloning) in a minimum space. The peculiar anatomy of *Allium*, with a stem reduced to a basal disc and its seated leaves that evolve to give bulbs, conditions the meristems to be hidden within the bulbs. This facilitates the chemical sterilization of the samples used as explants, without damaging the meristems. This allows their extraction under sterile conditions, but also implies a very time-consuming and costly extraction.

It is also possible to obtain organs or complete plants by direct regeneration from explants, such as bulbs, shoot tips, basal plates, flower stalks, peduncles-pedicels and roots, or, even better, from the meristems of the explants. For this purpose, explants are cultured on a basal media, MS (Murashige and Skoog, 1962), B5 (Gamborg *et al.*, 1968) or LS (Linsmaier and Skoog, 1965), supplemented with plant growth regulators (PGR). More specifically, they include a synergistic mixture of auxins and cytokinins. Garlic plants have been already regenerated from meristems (Messiaen *et al.*, 1970). Organs or whole plants can also be obtained by indirect regeneration from the explants already mentioned or from others, but passing through an intermediate stage of regenerative callus. A basal media MS, BDS (Dunstan and Short, 1977), B5 or N6 (Nitsch and Nitsch, 1969) is used for such purpose, supplemented with the synthetic auxin 2,4-dichlorophenoxyacetic acid (2,4-D).

Moreover, complete plants can also be obtained by indirect somatic embryogenesis, where one or several somatic cells are induced to undergo differentiation to a zygote-like state. Thereafter they develop into an embryo and a developed plant (George *et al.*, 2008a). Basal media and PGR are used for regeneration, albeit at different concentrations (George *et al.*, 2008c). This process implies virus elimination (pathogenic or not) from the regenerants (Keresa *et al.*, 2021). That is also the case in most botanical or true seeds derived from fertilization (Cobos *et al.*, 2019). Also, callogenesis is an initial step in the protocol. Therefore, another advantage of this procedure is micropropagation, which is greater than through meristems (George *et al.*, 2008c). Even more, each embryo evolves from only one cell and follows the developmental programme of the species. This guarantees that true-to-type plants are obtained (George *et al.*, 2008b). Thus, somaclonal variation (i.e. gene mutation or epigenetic marks in plants derived from cytogenetic failures during cell division) should be minimized. Somaclonal variation occurs during regenerative callogenesis (Larkin and Scowcroft, 1981). But it does not happen when callogenesis is not accompanied by cell de-differentiation until the somatic zygote stage, subsequently followed by an embryo developmental programme. All those advantages lead to the conclusion that indirect somatic embryogenesis is the method of choice, whenever an economically affordable protocol is available. That is particularly relevant, at least at the beginning of the conservation programme. Protocols have been mainly developed for somatic embryogenesis in garlic (Sata *et al.*, 2000), yet they are very labour- and time-consuming. Direct somatic embryogenesis, not passing through embryogenic callus, is also possible (Gantait *et al.*, 2010).

Yet another alternative for germplasm bank conservation is freezing at very low (ultra-low) temperatures (cryobanks). It offers some interesting valuable advantages, like genetic stability and cost-effectiveness. There are several cryopreservation methods described for garlic, including vitrification (Niwata, 1995; Makowska *et al.*, 1999; Keller, 2005; Ellis *et al.*, 2006), droplet vitrification (Volk *et al.*, 2004; Ellis *et al.*, 2006; Kim *et al.*, 2006; Liu *et al.*, 2017) and encapsulation-dehydration (Lynch *et al.*, 2016). Vitrification involves rapidly freezing in liquid nitrogen. Thus, explants (meristems, embryos, pollen etc.) are re-suspended in a mixture of cryoprotective substances, such as Plant Vitrification Solution 2 (PVS2), made of glycerol, ethylene glycol, Di-Methyl SulfOxide (DMSO) and sucrose. Thus, the water solidifies in the form of amorphous ice instead of crystal needles. That prevents breaking cells and tissues during freezing-thawing. The droplet vitrification approach involves dripping the explants (immersed in cryoprotectant liquid medium) inside strips of aluminium foil, into liquid nitrogen (–196°C). The encapsulation-dehydration technology consists of gelling into beads the media containing the explants, previously supplemented with alginate and Ca^{2+}, then drying them in a desiccator, before freezing in liquid nitrogen. As some practical examples, the Japanese shallot (*A. wakegi* Araki; synonym of *A. fistulosum* L.) has been cryopreserved by droplet vitrification (Kohmura *et al.*, 1994). Other cryopreservation methods have recently been used with other shallots (*A. cepa* var. *aggregatum*) (Wang *et al.*, 2019, 2020). In summary, cryobanks are the most cost-effective long-term solution for conservation. The estimated cost of cryopreservation ranges from 78 to 116 USD per accession, for initial transfer to storage (labour costs not included), plus a small annual upkeep cost of 1.6 USD for liquid nitrogen consumption (Hummer and Reed, 1996; Escobar *et al.*, 2000; Roca *et al.*, 2000), correcting for the inflation rate to date.

4. Mutations and Bioengineering to Increase Biodiversity

As we have already said, *in vitro* tissue culture is useful to conserve biodiversity, but it can also generate somaclonal variation. This is a problem for conservation, but an advantage for breeding programmes of obligate vegetative reproduction species. Mutagenesis using chemicals or radiation (like X or gamma rays) has been traditionally used to produce plant varieties. Yet it is a random process and, as such, it may also generate unwanted effects. Among them are reduced germination rate, plant and bulb sizes, as well as other desirable phenotypes. As an example, a dose of gamma rays in the range of 5–10 Gy reduces garlic germination by 50%, depending on the variety (Benke *et al.*, 2019). On the other hand, the most used methodology for the transformation of plants of the genus *Allium* has been the use of *Agrobacterium tumefaciens* in *A. cepa* (Eady *et al.*, 2000; Xu *et al.*, 2014) for resistance to insects (Zheng *et al.*, 2000, 2005). Likewise, Agrobacterium-mediated transformation has been demonstrated in *A. sativum* (Kondo *et al.*, 2000), for tolerance to fungi (Lagunes-Fortiz *et al.*, 2013) and *Allium ampeloprasum* (Eady *et al.*, 2000, 2005). Additionally, herbicide resistance was achieved in *A. cepa* (Eady *et al.*, 2003) and insect resistance was obtained in *A. sativum* (Zheng *et al.*, 2004). The particle bombardment method has also been used in *A. cepa* (Xu and Cui, 2007) and *A. sativum* (Sawahel, 2002; Robledo-Paz *et al.*, 2004) to obtain resistance to a herbicide (Park *et al.*, 2002).

5. Cost-effectiveness of Genotyping Biodiversity

The cost of maintaining accessions in germplasm banks makes it essential to detect and discard duplicate accessions and somaclonal variation. Whenever possible, the generation of doubled haploids aids genotyping, due to their genetic uniformity. The most used method for producing haploids in many species of *Allium* is gynogenesis, obtained by culturing unfertilized flowers, ovaries or ovules. This is useful to get complete homozygosity in double-haploid lines, reduced DNA methylation, elimination of deleterious alleles, and amenability to genetic analyses (Khan *et al.*, 2020). For instance, it has been reported that calluses obtained from non-pollinated shallot flowers regenerate in haploid, diploid and mixoploid plants (Sulistyaningsih *et al.*, 2006; Panahandeh and Farhadi, 2019). In any case, several strategies have been followed to ascertain the genetic biodiversity. They include morpho-physiological, biochemical, cytological and DNA-based molecular marker approaches (Krishna *et al.*, 2016). Different molecular markers that we have reviewed (Dorado *et al.*, 2017) have been used to genotype *Allium* species (Khosa *et al.*, 2016; Peska *et al.*, 2019; Malik *et al.*, 2020). They include pre-Polymerase Chain Reaction (PCR) reviewed by (Dorado *et al.*, 2019a), such as Restriction Fragment Length Polymorphisms (RFLP) (King *et al.*, 1998; Mccallum *et al.*, 2001), as well as post-PCR ones, such as microsatellites or Simple Sequence Repeats (SSR) (Baldwin *et al.*, 2012) and Amplified Fragment Length Polymorphisms (AFLP) (Ohara *et al.*, 2005; Wang *et al.*, 2016). Likewise, garlics have also been genotyped using Random Amplified Polymorphic DNA (RAPD) (Ipek *et al.*, 2003), SSR (Ma *et al.*, 2009; Cunha *et al.*, 2012; Wang *et al.*, 2016), AFLP (Morales *et al.*, 2013), Sequence Related Amplified Polymorphisms (SRAP) (Chen *et al.*, 2013), Inter-Simple Sequence Repeats (ISSR) (Jabbes *et al.*, 2011) and Insertions-Deletions (InDels) (Wang *et al.*, 2016). Recently, a DNA Melting peak Profile-based High Resolution Melting (MP-HRM) approach has been developed for germplasm genotyping (Lee *et al.*, 2020).

However, classical molecular markers (Dorado *et al.*, 2017) do not have the resolution power required for genotyping and generating core collections of germplasm, where they have giant and complex genomes, such as in *Allium* species like garlic (Egea *et al.*, 2017). High-throughput platforms based on massively parallel sequencing (SGS and TGS) have higher resolution, being more convenient. Reduced Representation Sequencing (RRS), Restriction-site Associated DNA sequencing (RAD-seq), double-digested Restriction-site Associated DNA sequencing (ddRAD-seq) and Specific Length Amplified Fragment (SLAF) sequencing (Fang *et al.*, 2020) to allow genotyping of Single Nucleotide Polymorphisms (SNP). They have several advantages, such as high abundance and even distribution across genomes (Villano *et al.*, 2019; Labate *et al.*, 2020). Additionally, they may be associated to traits of interest (Hayward *et al.*, 2012), including Quantitative Trait loci (QTL) and expression Quantitative Trait loci (eQTL) (Baldwin *et al.* 2014; Zhu *et al.*, 2019; Choi *et al.*, 2020), and used to construct linkage maps (Choi *et al.*, 2020; Cho *et al.*, 2021) and for germplasm genotyping (Lee *et al.*, 2018a, b).

On the other hand, RRS have similar costs per marker data point as Whole Genome Resequencing (WGR) but differ in their suitability for various tasks. Although all methods can use previous genomic information, several WGR approaches are dependent on these. This rules them out for characterizing the genomes of previously neglected non-model species. These RRS approaches are generally more suitable for *de novo* applications. They are more cost-effective when genotyping populations with large genomes, or high heterozygosity (Scheben *et al.*, 2017) as is the case of *Allium* with 10-30 Gbp (Ricroch *et al.*, 2005). For example, *A. cepa* has a genome of 16 Gbp (18 and 100 times larger than *Solanum lycopersicum* and *Arabidopsis*, respectively), with high heterozygosity (King *et al.*, 1998). Different sequencing platforms (Lario *et al.*, 1997; Dorado *et al.*, 2019b) have been used to generate SNP markers in *A. cepa* via transcriptomic and selected genomic region bases (Duangjit *et al.*, 2013; Jo *et al.*, 2017; Choi *et al.*, 2020). In *A. sativum*, studies of genomic biodiversity have been made (Egea *et al.*, 2017), and of the transcriptome related to organic sulphur biosynthesis (Chen *et al.*, 2018) or yield traits (Zhu *et al.*, 2019).

6. Future Prospects to Protect, Increase and Manage *Allium* Biodiversity

Regarding micropropagation, over 60% of its cost is for labour. The robotization of some steps would make this work cheaper. Cryopreservation is still only viable for a few species of *Allium*. Improving techniques to make it work in more species would be essential for long-term conservation. Of course, the restoration of fertility through hybridization, transgenesis or genome editing would allow improvement in sterile species without sexuality such as *A. sativum*. In relation to that, the Clustered Regularly Interspaced Short Palindromic Repeats (CRISPR) technology holds a great promise. Regarding genotyping, although RRS approaches will remain the most cost-effective for some time, it is expected that *de novo* sequencing and WGR will become more widespread as sequencing costs continue to decline (Scheben *et al.*, 2017), as we have reviewed (Dorado *et al.*, 2019b). That is promising not only for nuclear genomes, but also for organelles like mitochondria (Kim *et al.*, 2016; Tsujimura *et al.*, 2019) and chloroplasts (Huo *et al.*, 2019; Xie *et al.*, 2020).

Furthermore, *Allium* phenological studies (Lopez-Bellido *et al.*, 2016), together with omics ones, like genomics, metabolomics, transcriptomics, proteomics and metagenomics, will elucidate characteristic genes and metabolites of interest (Kim *et al.*, 2015; Havey, 2018; Khandagale *et al.*, 2020). That will further shed light on biosynthesis pathways, as well as regulatory-related genes and networks of different accessions, landraces and cultivars (Abdelrahman, 2020). The study of the rhizosphere microbiome and its effects on different *Allium* species should be also taken into consideration. Shockingly, the point of no return of climate change was predicted in 2000 (King *et al.*, 2020). Whatever the situation, it is clear that temperatures are increasing, with new records each year. Therefore, adaptation to such global warming is critical for survival. Biotechnology can help in this situation. Yet, the European Union and some countries worldwide do not allow marketing transgenics for human consumption. Fortunately, new genetic engineering strategies like CRISPR do not need transgenesis, but just genome-editing. Both factors may promote rethinking the prohibition of such bioengineered varieties.

Much work remains to be done to collect and to catalogue the biodiversity of *Allium* (Rabinowitch and Brewster, 1990; Rabinowitch and Currah, 2002; Shigyo *et al.*, 2018). New expeditions must be made to search for wild and cultivated varieties, some of which may be better adapted to the current global warming and climate change; likewise, for better tolerance to diseases and pests. That is especially relevant for countries with difficulties in protecting their natural heritage. Although much progress has been made in their development in recent decades, conservation technologies like cryopreservation have yet to be further optimized to be applied to more species. Likewise, genotyping techniques should be fine-tuned to properly identify variability and generate germplasm core collections. Lowering their maintenance costs would also be a significant accomplishment, mostly for countries with lower resources. Further improvement programmes must be carried out in order to obtain varieties with desirable traits, including higher and earlier production, increase in bioactive compounds and tolerance to both abiotic and biotic stresses. Last, but not least, the restoration of fertility is essential for proper breeding of various species of *Allium*, like the most relevant commercial varieties of garlic.

Acknowledgments

Supported by Ministerio de Economía y Competitividad (MINECO) and Instituto Nacional de Investigación y Tecnología Agraria y Alimentaria (MINECO and INIA RF2012-00002-C02-02); Consejería de Agricultura y Pesca (041/C/2007, 75/C/2009 and 56/C/2010), Consejería de Economía, Innovación y Ciencia (P1-AGR-7322) and Grupo PAI (AGR-248) of Junta de Andalucía; and Universidad de Córdoba (Ayuda a Grupos), Spain. We thank Cristina Mallor Giménez, of Centro de Investigación y Tecnología Agroalimentaria de Aragón (CITA) for critical reading of the manuscript.

References

Abdelrahman, M. (2020) *Allium* functional genomic development for future climatic changes. In: Kole, C. (ed.) *Genomic Designing of Climate-Smart Vegetable Crops*. Springer International Publishing, Cham, Switzerland. 978-3-319-97415-6.

Baldwin, S., Pither-Joyce, M., Wright, K., Chen, L.S. and Mccallum, J. (2012) Development of robust genomic simple sequence repeat markers for estimation of genetic diversity within and among bulb onion (*Allium cepa* L.) populations. *Molecular Breeding* 30, 1401–1411. DOI: 10.1007/s11032-012-9727-6.

Baldwin, S., Revanna, R., Pither-Joyce, M., Shaw, M., Wright, K. *et al.* (2014) Genetic analyses of bolting in bulb onion (*Allium cepa* L.). *Theoretical and Applied Genetics* 127, 535–547. DOI: 10.1007/s00122-013-2232-4.

Benke, A.P., Dukare, S., Jayaswall, K., Yadav, V.K. and Singh, M. (2019) Determination of proper gamma radiation dose for creating variation in Indian garlic varieties. *Indian Journal of Traditional Knowledge* 18, 547–552.

Byng, J.W., Chase, M.W., Christenhusz, M.J.M., Fay, M.F., Judd, W.S. *et al.* (2016) An update of the Angiosperm Phylogeny Group classification for the orders and families of flowering plants: APG IV. *Botanical Journal of the Linnean Society* 181, 1–20. DOI: 10.1111/boj.12385.

Chen, S.X., Zhou, J., Chen, Q., Chang, Y.X., Du, J.N. and Meng, H.W. (2013) Analysis of the genetic diversity of garlic (*Allium sativum* L.) germplasm by SRAP. *Biochemical Systematics and Ecology* 50, 139–146. DOI: 10.1016/j.bse.2013.03.004.

Chen, X.J., Liu, X., Zhu, S.Y., Tang, S.W., Mei, S.Y. *et al.* (2018) Transcriptome-referenced association study of clove shape traits in garlic. *DNA Research* 25, 587–596. DOI: 10.1093/dnares/dsy027.

Cho, Y., Kim, B., Lee, J. and Kim, S. (2021) Construction of a high-resolution linkage map and chromosomal localization of the loci determining major qualitative traits in onion *(Allium cepa* L.*). Euphytica* 217. DOI: 10.1007/s10681-020-02746-z.

Choi, Y., Kim, S. and Lee, J.D. (2020) Construction of an onion (*Allium cepa* L.) genetic linkage map using genotyping-by-sequencing analysis with a reference gene set and identification of QTLs controlling anthocyanin synthesis and content. *Plants-Basel* 9. DOI: 10.3390/plants9050616.

Cobos, A., Montes, N., Lopez-Herranz, M., Gil-Valle, M. and Pagan, I. (2019) Within-host multiplication and speed of colonization as infection traits associated with plant virus vertical transmission. *Journal of Virology* 93. DOI: 10.1128/JVI.01078-19.

Cohen, B.M. (1991) Vavilov, Nikolai, Ivanovich – the explorer and plant collector. *Economic Botany* 45, 38–46. DOI: 10.1007/Bf02860048.

Costa, L., Jimenez, H., Carvalho, R., Carvalho-Sobrinho, J., Escobar, I. and Souza, G. (2020) Divide to conquer: evolutionary history of Allioideae tribes (Amaryllidaceae) is linked to distinct trends of karyotype evolution. *Frontiers in Plant Science* 11. DOI: 10.3389/fpls.2020.00320.

Cunha, C.P., Hoogerheide, E.S.S., Zucchi, M.I., Monteiro, M. and Pinheiro, J.B. (2012) New microsatellite markers for garlic, *Allium sativum* (Alliaceae). *American Journal of Botany* 99, E17–E19. DOI: 10.3732/ajb.1100278.

Deniz, I.G., Genc, I. and Sari, D. (2015) Morphological and molecular data reveal a new species of *Allium* (Amaryllidaceae) from SW Anatolia, Turkey. *Phytotaxa* 212, 283–292. DOI: 10.11646/phytotaxa.212.4.4

Dorado, G., Unver, T., Budak, H. and Hernández, P. (2017) Molecular markers. In: Caplan, M. (ed.) *Reference Module in Biomedical Sciences. Biochemistry, Cell Biology and Molecular Biology*. Elsevier, Amsterdam.

Dorado, G., Besnard, G., Unver, T. and Hernández, P. (2019a) Polymerase chain reaction (PCR). In: Caplan, M. (ed.) *Reference Module in Biomedical Sciences. Biochemistry, Cell Biology and Molecular Biology*. Elsevier, Amsterdam.

Dorado, G., Gálvez, S., Budak, H., Unver, T. and Hernández, P. (2019b) Nucleic-acid sequencing. In: Caplan, M. (ed.) *Reference Module in Biomedical Sciences. Biochemistry, Cell Biology and Molecular Biology*. Elsevier, Amsterdam.

Duangjit, J., Bohanec, B., Chan, A.P., Town, C.D. and Havey, M.J. (2013) Transcriptome sequencing to produce SNP-based genetic maps of onion. *Theoretical and Applied Genetics* 126, 2093–2101. DOI: 10.1007/s00122-013-2121-x.

Dunstan, D.I. and Short, K.C. (1977) Improved growth of tissue cultures of the onion, *Allium cepa*. *Physiologia Plantarum* 41, 70–72. DOI: 10.1111/j.1399-3054.1977.tb01525.x.

Eady, C., Weld, R.J. and Lister, C.E. (2000) *Agrobacterium tumefaciens*-mediated transformation and transgenic-plant regeneration of onion (*Allium cepa* L.) *Plant Cell Reports* 19, 376–381. DOI: 10.1007/s002990050743.

Eady, C., Davis, S., Farrant, J., Reader, J. and Kenel, F. (2003) *Agrobacterium tumefaciens*-mediated transformation and regeneration of herbicide-resistant onion (*Allium cepa*) plants. *Annals of Applied Biology* 142, 213–217. DOI: 10.1111/j.1744-7348.2003.tb00243.x.

Eady, C., Davis, S., Catanach, A., Kenel, F. and Hunger, S. (2005) *Agrobacterium tumefaciens*-mediated transformation of leek (*Allium porrum*) and garlic (*Allium sativum*). *Plant Cell Reports* 24, 209–215. DOI: 10.1007/s00299-005-0926-z.

Ebert, A.W. and Engels, J.M.M. (2020) Plant biodiversity and genetic resources matter! *Plants-Basel* 9. DOI: 10.3390/plants9121706.

Egea, L.A., Merida-Garcia, R., Kilian, A., Hernandez, P. and Dorado, G. (2017) Assessment of genetic diversity and structure of large garlic (*Allium sativum*) germplasm bank, by diversity arrays technology 'genotyping-by-sequencing' platform (DArTseq). *Frontiers in Genetics* 8. DOI: 10.3389/fgene.2017.00098.

Ellis, D., Skogerboe, D., Andre, C., Hellier, B. and Volk, G. (2006) Implementation of garlic cryopreservation techniques in the national plant germplasm system. *Cryoletters* 27, 99–106.

Escobar, R.H., Debouck, D. and Roca, W.M. (2000) Development of cassava cryopreservation. In: Engelmann, F. and Takagi, H. (eds) *Cryopreservation of Tropical Plant Germplasm: Current Research Progress and Applications*. International Plant Genetic Resources Institute, Rome.

Fang, H.T., Liu, H.Y., Ma, R.S., Liu, Y.X., Li, J.N. *et al.* (2020) Genome-wide assessment of population structure and genetic diversity of Chinese Lou onion using specific length amplified fragment (SLAF) sequencing. *PLOS ONE* 15. DOI: 10.1371/journal.pone.0231753.

Gamborg, O.L., Miller, R.A. and Ojima, K. (1968) Nutrient requirements of suspension cultures of soybean root cells. *Experimental Cell Research* 50, 151. DOI: 10.1016/0014-4827(68)90403-5.

Gantait, S., Mandal, N. and Das, P. (2010) An overview on *in vitro* culture of genus *Allium*. *American Journal of Plant Physiology* 5, 325–337. DOI: 10.3923/ajpp.2010.325.337.

George, E.F., Hall, M.A. and Klerk, G.J.D. (2008a) Adventitious regeneration. In: George, E.F., Hall, M.A. and Klerk, G.J.D. (eds) *Plant Propagation by Tissue Culture*, 3rd edn. Springer, Dordrecht, The Netherlands. 978-1-4020-5005-3.

George, E.F., Hall, M.A. and Klerk, G.J.D. (2008b) Micropropagation: uses and methods. In: George, E.F., Hall, M.A. and Klerk, G.J.D. (eds) *Plant Propagation by Tissue Culture*, 3rd edn. Springer, Dordrecht, The Netherlands. 978-1-4020-5005-3.

George, E.F., Hall, M.A. and Klerk, G.J.D. (2008c) Somatic embryogenesis. In: George, E.F., Hall, M.A. and Klerk, G.J.D. (eds) *Plant Propagation by Tissue Culture*, 3rd edn. Springer, Dordrecht, The Netherlands. 978-1-4020-5005-3.

Havey, M.J. (2018) Impact of the *Allium* genomes on plant breeding. *Compendium of Plant Genomes* 197–214. DOI: 10.1007/978-3-319-95825-5_14.

Hayward, M., Hayward, A., Mason, A., Dalton-Morgan, J., Zander, M. *et al.* (2012) SNP discovery and applications in *Brassica napus*. *Journal of Plant Biotechnology* 39. DOI: 10.5010/jpb.2012.39.1.049.

Hummer, K. and Reed, B.M. (1996) Establishment and operation of a temperate clonal field genebank. In: Engelmann, F. (ed.) *Management of Field and In Vitro Germplasm Collections, Proceedings of a Consultation Meeting*. CIAT, Cali, Colombia; International Plant Genetic Resources Institute, Rome.

Huo, Y.M., Gao, L.M., Liu, B.J., Yang, Y.Y., Kong, S.P. *et al.* (2019) Complete chloroplast genome sequences of four *Allium* species: comparative and phylogenetic analyses. *Scientific Reports* 9. DOI: 10.1038/s41598-019-48708-x.

Ipek, M., Ipek, A. and Simon, P.W. (2003) Comparison of AFLPs, RAPD markers, and isozymes for diversity assessment of garlic and detection of putative duplicates in germplasm collections. *Journal of the American Society for Horticultural Science* 128, 246–252. DOI: 10.21273/Jashs.128.2.0246.

Jabbes, N., Geoffriau, E., Le Clerc, V., Dridi, B. and Hannechi, C. (2011) Inter simple sequence repeat fingerprints for assessing genetic diversity of Tunisian garlic populations. *The Journal of Agricultural Science* 3. DOI: 10.5539/jas.v3n4p77.

Jo, J., Purushotham, P.M., Han, K., Lee, H.R., Nah, G. and Kang, B.C. (2017) Development of a genetic map for onion (*Allium cepa* L.) using reference-free genotyping-by-sequencing and SNP assays. *Frontiers in Plant Science* 8. DOI: 10.3389/fpls.2017.01606.

Keller, E.R.J. (2005) Improvement of cryopreservation results in garlic using low-temperature preculture and high-quality *in vitro* plantlets. *Cryoletters* 26, 357–366.

Keresa, S., Kurtovic, K., Ban, S.G., Voncina, D., Jercic, I.H. *et al.* (2021) Production of virus-free garlic plants through somatic embryogenesis. *Agronomy-Basel* 11. DOI: 10.3390/agronomy11050876.

Khan, P.S.S.V., Vijayalakshmi, G., Raja, M.M., Naik, M.L., Germana, M.A. *et al.* (2020) Doubled haploid production in onion (*Allium cepa* L.): from gynogenesis to chromosome doubling. *Plant Cell, Tissue and Organ Culture (PCTOC)* 142, 1–22. DOI: 10.1007/s11240-020-01831-4.

Khandagale, K., Krishna, R., Roylawar, P., Ade, A.B., Benke, A. *et al.* (2020) Omics approaches in *Allium* research: progress and way ahead. *PeerJ* 8. DOI: 10.7717/peerj.9824.

Khosa, J.S., Mccallum, J., Dhatt, A.S. and Macknight, R.C. (2016) Enhancing onion breeding using molecular tools. *Plant Breeding* 135, 9–20. DOI: 10.1111/pbr.12330.

Kim, B., Kim, K., Yang, T.J. and Kim, S. (2016) Completion of the mitochondrial genome sequence of onion (*Allium cepa* L.) containing the CMS-S male-sterile cytoplasm and identification of an independent event of the ccmF (N) gene split. *Current Genetics* 62, 873–885. DOI: 10.1007/s00294-016-0595-1.

Kim, H.H., Lee, J.K., Yoon, J.W., Ji, J.J., Nam, S.S. *et al.* (2006) Cryopreservation of garlic bulbil primordia by the droplet-vitrification procedure. *Cryoletters* 27, 143–153.

Kim, S., Kim, M.S., Kim, Y.M., Yeom, S.I., Cheong, K. *et al.* (2015) Integrative structural annotation of de novo RNA-Seq provides an accurate reference gene set of the enormous genome of the onion (*Allium cepa* L.) *DNA Research* 22, 19–27. DOI: 10.1093/dnares/dsu035.

King, J.J., Bradeen, J.M., Bark, O., Mccallum, J.A. and Havey, M.J. (1998) A low-density genetic map of onion reveals a role for tandem duplication in the evolution of an extremely large diploid genome. *Theoretical and Applied Genetics* 96, 52–62. DOI: 10.1007/s001220050708.

King, M.D., Howat, I.M., Candela, S.G., Noh, M.J., Jeong, S., Noël, B.P.Y., Van Den Broeke, M.R., Wouters, B. and Negrete, A. (2020) Dynamic ice loss from the Greenland Ice Sheet driven by sustained glacier retreat. *Communications Earth & Environment* 1, 1. 10.1038/s43247-020-0001-2.

Kohmura, H., Ikeda, Y. and Sakai, A. (1994) Cryopreservation of apical meristems of Japanese shallot (*Allium-Wakegi* a) by vitrification and subsequent high plant-regeneration. *Cryoletters* 15, 289–298.

Kondo, T., Hasegawa, H. and Suzuki, M. (2000) Transformation and regeneration of garlic (*Allium sativum* L.) by *Agrobacterium*-mediated gene transfer. *Plant Cell Reports* 19, 989–993. DOI: 10.1007/s002990000222.

Krishna, H., Alizadeh, M., Singh, D., Singh, U., Chauhan, N. *et al.* (2016) Somaclonal variations and their applications in horticultural crops improvement. *3 Biotech* 6. DOI: 10.1007/s13205-016-0389-7.

Labate, J.A., Glaubitz, J.C. and Havey, M.J. (2020) Genotyping by sequencing for SNP marker development in onion. *Genome* 63, 607–613. DOI: 10.1139/gen-2020-0011.

Lagunes-Fortiz, E., Robledo-Paz, A., Gutiérrez-Espinosa, M.A., Mascorro-Gallardo, J.O. and Espitia-Rangel, E. (2013) Genetic transformation of garlic (*Allium sativum* L.) with tobacco chitinase and glucanase genes for tolerance to the fungus *Sclerotium cepivorum*. *African Journal of Biotechnology* 12, 3482–3492.

Lanz, B., Dietz, S. and Swanson, T. (2018) The expansion of modern agriculture and global biodiversity decline: an integrated assessment. *Ecological Economics* 144, 260–277. DOI: 10.1016/j.ecolecon.2017.07.018.

Lario, A., Gonzalez, A. and Dorado, G. (1997) Automated laser-induced fluorescence DNA sequencing: Equalizing signal-to-noise ratios significantly enhances overall performance. *Analytical Biochemistry* 247, 30–33. DOI: 10.1006/abio.1996.9933.

Larkin, P.J. and Scowcroft, W.R. (1981) Somaclonal variation: a novel source of variability from cell-cultures for plant improvement. *Theoretical and Applied Genetics* 60, 197–214. DOI: 10.1007/Bf02342540.

Lee, G.H., Lee, H.M., Kim, S.J. and Park, Y.D. (2020) Development of DNA melting peak profile-based high-resolution melting (MP-HRM) analysis for genotyping germplasms in onion breeding. *Horticulture, Environment, and Biotechnology* 61, 139–152. DOI: 10.1007/s13580-019-00204-x.

Lee, J.H., Natarajan, S., Biswas, M.K., Shirasawa, K., Isobe, S.K. *et al.* (2018a) SNP discovery of Korean short day onion inbred lines using double digest restriction site-associated DNA sequencing. *PLOS ONE* 13. DOI: 10.1371/journal.pone.0201229.

Lee, J.H., Robin, A.H.K., Natarajan, S., Jung, H.J. and Nou, I.S. (2018b) Varietal identification of open-pollinated onion cultivars using a nanofluidic array of single nucleotide polymorphism (SNP) markers. *Agronomy-Basel* 8. DOI: 10.3390/agronomy8090179.

Li, Q.Q., Zhou, S.D., He, X.J., Yu, Y., Zhang, Y.C. *et al.* (2010) Phylogeny and biogeography of *Allium* (Amaryllidaceae: Allieae) based on nuclear ribosomal internal transcribed spacer and chloroplast rps16 sequences, focusing on the inclusion of species endemic to China. *Annals of Botany* 106, 709–733. DOI: 10.1093/aob/mcq177.

Linsmaier, E.M. and Skoog, F. (1965) Organic growth factor requirements of tobacco tissue cultures. *Physiologia Plantarum* 18, 100. DOI: 10.1111/j.1399-3054.1965.tb06874.x.

Liu, X.X., Wen, Y.B., Cheng, Z.H. and Mou, S.W. (2017) Establishment of a garlic cryopreservation protocol for shoot apices from adventitious buds *in vitro*. *Scientia Horticulturae* 226, 10–18. DOI: 10.1016/j.scienta.2017.08.016.

Lopez-Bellido, F.J., Lopez-Bellido, R.J., Munoz-Romero, V., Fernandez-Garcia, P. and Lopez-Bellido, L. (2016) New phenological growth stages of garlic (*Allium sativum*) *Annals of Applied Biology* 169, 423–439. DOI: 10.1111/aab.12312.

Lynch, P.T., Souch, G.R., Zamecnik, J. and Harding, K. (2016) Optimization of water content for the cryopreservation of *Allium sativum in vitro* cultures by encapsulation-dehydration. *Cryoletters* 37, 308–317.

Ma, K.H., Kwag, J.G., Zhao, W., Dixit, A., Lee, G.A. *et al.* (2009) Isolation and characteristics of eight novel polymorphic microsatellite loci from the genome of garlic (*Allium sativum* L.) *Scientia Horticulturae* 122, 355–361. DOI: 10.1016/j.scienta.2009.06.010.

Makowska, Z., Keller, J. and Engelmann, F. (1999) Cryopreservation of apices isolated from garlic (*Allium sativum* L.) bulbils and cloves. *Cryo-Letters* 20, 175–182.

Malik, G., Dhatt, A.S. and Malik, A.A. (2020) A review of genetic understanding and amelioration of edible *Allium* species. *Food Reviews International.* DOI: 10.1080/87559129.2019.1709202.

Mccallum, J., Leite, D., Pither-Joyce, M. and Havey, M.J. (2001) Expressed sequence markers for genetic analysis of bulb onion (*Allium cepa* L.). *Theoretical and Applied Genetics* 103, 979–991. DOI: 10.1007/s001220100630.

Messiaen, C.M., Marrov, J., Quiot, J.B., Leclant, F. and Leroux, J.P. (1970) Study in France south east of a sanitary selection scheme of garlic and shallot. *Rendered Accounts of 7th Conferance of Plant Pathology*. C.N.R.A. Montfavet, France, pp. 101–103.

Morales, R.G.F., Resende, J.T.V., Resende, F.V., Delatorre, C.A., Figueiredo, A.S.T. and Da-Silva, P.R. (2013) Genetic divergence among Brazilian garlic cultivars based on morphological characters and AFLP markers. *Genetics and Molecular Research* 12, 270–281. DOI: 10.4238/2013.February.4.1.

Murashige, T. and Skoog, F. (1962) A revised medium for rapid growth and bio assays with tobacco tissue cultures. *Physiologia Plantarum* 15, 473–497. DOI: 10.1111/j.1399-3054.1962.tb08052.x.

Nitsch, J.P. and Nitsch, C. (1969) Haploid plants from pollen grains. *Science* 163, 85. DOI: 10.1126/science.163.3862.85.

Niwata, E. (1995) Cryopreservation of apical meristems of garlic (*Allium-sativum* L) and high subsequent plant-regeneration. *Cryo-Letters* 16, 102–107.

Normah, M.N., Chin, H.F. and Reed, B.M. (2013) *Conservation of Tropical Plant Species*. Springer, New York. 978-1-4614-3776-5, 978-1-4614-3775-8, 978-1-4899-8724-2.

Ohara, T., Song, Y.S., Tsukazaki, H., Wako, T., Nunome, T. *et al.* (2005) Genetic mapping of AFLP markers in Japanese bunching onion (*Allium fistulosum*). *Euphytica* 144, 255–263. DOI: 10.1007/s10681-005-6768-5.

Panahandeh, J. and Farhadi, N. (2019) Haploid induction via *in vitro* gynogenesis in Persian shallot (*Allium hirtifolium*). *Journal of Horticultural Sciences* 27, 91–98. DOI: 10.2478/johr-2019-0017.

Panis, B., Nagel, M. and Van Den Houwe, I. (2020) Challenges and prospects for the conservation of crop genetic resources in field genebanks in *in vitro* collections and/or in liquid nitrogen. *Plants-Basel* 9. DOI: 10.3390/plants9121634.

Park, M.Y., Yi, N.R., Lee, H.Y., Kim, S.T., Kim, M. *et al.* (2002) Generation of chlorsulfuron-resistant transgenic garlic plants (*Allium sativum* L.) by particle bombardment. *Molecular Breeding* 9, 171–181. DOI: 10.1023/A:1019702705996.

Peska, V., Mandakova, T., Ihradska, V. and Fajkus, J. (2019) Comparative dissection of three giant genomes: *Allium cepa*, *Allium sativum*, and *Allium ursinum*. *International Journal of Molecular Sciences* 20. DOI: 10.3390/ijms20030733.

Rabinowitch, H.D. and Brewster, J.L. (1990) *Onions and Allied Crops*. CRC Press, Boca Raton, Florida.

Rabinowitch, H.D. and Currah, L. (2002) *Allium Crop Science: Recent Advances*. CAB International, Wallingford, UK. 0851995101 (alk. paper).

Ricroch, A., Yockteng, R., Brown, S.C. and Nadot, S. (2005) Evolution of genome size across some cultivated *Allium* species. *Genome* 48, 511–520. DOI: 10.1139/G05-017.

Robledo-Paz, A., Cabrera-Ponce, J.L., Villalobos-Arambula, V.M., Herrera-Estrella, L. and Jofre-Garfias, A.E. (2004) Genetic transformation of garlic (*Allium sativum* L.) by particle bombardment. *Hortscience* 39, 1208–1211. DOI: 10.21273/Hortsci.39.6.1208.

Roca, W.M., Debouck, D., Escobar, R.H., Mafla, G. and Fregene, M. (2000) Cryopreservation and cassava germplasm conservation at CIAT. In: Engelmann, F. and Takagi, H. (eds) *Cryopreservation of Tropical Plant Germplasm: Current Research Progress and Applications*. International Plant Genetic Resources Institute, Rome.

Ruta, C., Lambardi, M. and Ozudogru, E.A. (2020) Biobanking of vegetable genetic resources by *in vitro* conservation and cryopreservation. *Biodiversity and Conservation* 29, 3495–3532. DOI: 10.1007/s10531-020-02051-0.

Sata, S.J., Bagatharia, S.B. and Thaker, V.S. (2000) Induction of direct somatic embryogenesis in garlic (*Allium sativum*). *Methods in Cell Science* 22, 299–304. DOI: 10.1023/a:1017541500318.

Savary, S., Willocquet, L., Pethybridge, S.J., Esker, P., Mcroberts, N. *et al.* (2019) The global burden of pathogens and pests on major food crops. *Nature Ecology & Evolution* 3, 430. DOI: 10.1038/s41559-018-0793-y.

Sawahel, W.A. (2002) Stable genetic transformation of garlic plants using particle bombardment. *Cellular & Molecular Biology Letters* 7, 49–59.

Scheben, A., Batley, J. and Edwards, D. (2017) Genotyping-by-sequencing approaches to characterize crop genomes: choosing the right tool for the right application. *Plant Biotechnology Journal* 15, 149–161. DOI: 10.1111/pbi.12645.

Shigyo, M., Khar, A. and Abdelrahman, M. (2018) The *Allium* genomes. *Compendium of Plant Genomes*. Springer International Publishing, Cham, Switzerland.

Singh, A.K., Varaprasad, K.S. and Venkateswaran, K. (2012) Conservation Costs of Plant Genetic Resources for Food and Agriculture: Seed Genebanks. *Agricultural Research*, 1, 223–239. 10.1007/s40003-012-0029-3

Sulistyaningsih, E., Aoyagi, Y. and Tashiro, Y. (2006) Flower bud culture of shallot (*Allium cepa* L. *Aggregatum* group) with cytogenetic analysis of resulting gynogenic plants and somaclones. *Plant Cell, Tissue and Organ Culture (PCTOC)* 86, 249–255. DOI: 10.1007/s11240-006-9114-2.

Sun, X.D., Zhu, S.Y., Li, N.Y., Cheng, Y., Zhao, J. *et al.* (2020) A chromosome-level genome assembly of garlic (*Allium sativum*) provides insights into genome evolution and allicin biosynthesis. *Molecular Plant* 13, 1328–1339. DOI: 10.1016/j.molp.2020.07.019.

Tsujimura, M., Kaneko, T., Sakamoto, T., Kimura, S., Shigyo, M. *et al.* (2019) Multichromosomal structure of the onion mitochondrial genome and a transcript analysis. *Mitochondrion* 46, 179–186. DOI: 10.1016/j.mito.2018.05.001.

Villano, C., Esposito, S., Carucci, F., Iorizzo, M., Frusciante, L. *et al.* (2019) High-throughput genotyping in onion reveals structure of genetic diversity and informative SNPs useful for molecular breeding. *Molecular Breeding* 39, 5. DOI: 10.1007/s11032-018-0912-0.

Virchow, D. (2003) Efficient conservation of crop genetic diversity: the costs and implications. In: Virchow, D. (ed.) *Efficient Conservation of Crop Genetic Diversity: Theoretical Approaches and Empirical Studies*. Springer, Berlin and Heidelberg. 978-3-662-05212-9.

Volk, G.M., Maness, N. and Rotindo, K. (2004) Cryopreservation of garlic (*Allium sativum* L.) using plant vitrification solution 2. *Cryoletters* 25, 219–226.

Wang, H.P., Li, X.X., Liu, X.H., Oiu, Y., Song, J.P. *et al.* (2016) Genetic diversity of garlic (*Allium sativum* L.) germplasm from China by fluorescent-based AFLP, SSR and InDel markers. *Plant Breeding* 135, 743–750. DOI: 10.1111/pbr.12424.

Wang, M.R., Hamborg, Z., Haugslien, S., Sivertsen, A., Rasmussen, M. *et al.* (2019) Cryopreservation of shallot (*Allium cepa* var. *aggregatum*) shoot tips by droplet-vitrification. *Acta Horticulturae* 241–248. DOI: 10.17660/ActaHortic.2019.1234.32.

Wang, M.R., Zhang, Z.B., Zamecnik, J., Bilavcik, A., Blystad, D.R. *et al.* (2020) Droplet-vitrification for shoot tip cryopreservation of shallot (*Allium cepa* var. *aggregatum*): effects of PVS3 and PVS2 on shoot regrowth. *Plant Cell, Tissue and Organ Culture (PCTOC)* 140, 185–195. DOI: 10.1007/s11240-019-01721-4.

Xie, D.F., Tan, J.B., Yu, Y., Gui, L.J., Su, D.M. *et al.* (2020) Insights into phylogeny, age and evolution of *Allium* (Amaryllidaceae) based on the whole plastome sequences. *Annals of Botany* 125, 1039–1055. DOI: 10.1093/aob/mcaa024.

Xu, K.D., Huang, X.H., Wu, M.M., Wang, Y., Chang, Y.X. *et al.* (2014) A rapid, highly efficient and economical method of agrobacterium-mediated *in planta* transient transformation in living onion epidermis. *PLOS ONE* 9. DOI: 10.1371/journal.pone.0083556.

Xu, Q.J. and Cui, C.R. (2007) Genetic transformation of OSISAP1 gene to onion (*Allium cepa* L.) mediated by microprojectile bombardment. *Journal of Plant Physiology and Molecular Biology* 33, 188–96 (In Chinese).

Zheng, S.J., Henken, B., Wietsma, W., Sofiari, E., Jacobsen, E. *et al.* (2000) Development of bio-assays and screening for resistance to beet armyworm (*Spodoptera exigua* Hubner) in *Allium cepa* L. and its wild relatives. *Euphytica* 114, 77–85. DOI: 10.1023/A:1004089424419.

Zheng, S.J., Henken, B., Ahn, Y.K., Krens, F.A. and Kik, C. (2004) The development of a reproducible *Agrobacterium tumefaciens* transformation system for garlic (*Allium sativum* L.) and the production of transgenic garlic resistant to beet armyworm (*Spodoptera exigua* Hubner). *Molecular Breeding* 14, 293–307. DOI: 10.1023/B:MOLB.0000047775.83715.b5.

Zheng, S.J., Henken, B., De Maagd, R.A., Purwito, A., Krens, F.A. *et al.* (2005) Two different *Bacillus thuringiensis* toxin genes confer resistance to beet armyworm (*Spodoptera exigua* Hubner) in transgenic Bt-shallots (*Allium cepa* L.). *Transgenic Research* 14, 261–272. DOI: 10.1007/s11248-005-0109-2.

Zhu, S.Y., Chen, X.J., Liu, X., Zhao, J. and Liu, T.M. (2019) Transcriptome-wide association study and eQTL analysis to assess the genetic basis of bulb-yield traits in garlic (*Allium sativum*). *BMC Genomics* 20. DOI: 10.1186/s12864-019-6025-2.

3 Composition, Nutritional and Health Benefits

Abdul Karim Darkal[1], Muhammad Sarfraz[1], Muhammad Jawad Nasim[1], Yannick Ney[1], Rahman Shah Zaib Saleem[2] and Claus Jacob[1]*

[1]Division of Bioorganic Chemistry, School of Pharmacy, Saarland University, Saarbruecken, Germany; [2]Department of Chemistry, SBA School of Science and Engineering (SBASSE), Lahore University of Management Sciences (LUMS), Lahore, Pakistan

1. Introduction

Garlic and related plants of the *Amaryllidaceae*, such as onions and chives, have been part of folk medicine for millennia, the world over, and cherished for the culinary delights and supposed treatment of numerous human and animal disorders (Rivlin, 2001). Garlic juices, oils, extracts and powders have been, and still are, employed by many as a welcome natural and household remedy against minor ailments, such as mild infections and the common cold (Fig. 3.1). Then again, the question of whether garlic and its products may provide an alternative for modern medication in cases of complicated disorders, such as diabetes and cardiovascular complications, is debatable for a number of reasons, and indeed is often debated rather controversially and emotionally (Bayan *et al.*, 2014). Despite such questions, the unique palette of reactive and biologically active ingredients, especially reactive sulfur species (RSS), renders garlic rather special among medicinal plants (Gruhlke, 2019). It is therefore hardly surprising that today, one witnesses many garlic-based remedies in personal use and also in commercial circulation, some prepared *ad hoc* at home and others sold commercially in drug and department stores, in pharmacies and over the internet, including garlic juices, powders, tablets and capsules, as illustrated in Fig. 3.1 (Rajasekaran *et al.*, 2012). In parallel, modern ecological agriculture values garlic juices and garlic-laced granules as eco-friendly and effective alternatives to traditional nematicides, insecticides, fungicides and acaricides (Anwar *et al.*, 2017).

The garlic bulb contains a wide and rich spectrum of nutrients, minerals, flavonoids, chalcogens and other secondary metabolites with an established record of biological activity *in vitro* and *in vivo* (Choi *et al.*, 2007). Garlic scape is rich in phytochemicals such as allicin, polyphenols and flavonoids, and its extracts exhibit strong antioxidant activity (Nencini *et al.*, 2011). The scape is characterized by a distinct garlicky fragrance and a typical spicy flavour and is consumed in several Asian regions, especially in China, as an essential ingredient of national cuisine. Fresh garlic leaves provide another interesting source of nutraceuticals, such as cyclic organosulfur compounds (OSCs), foliogarlic disulfanes A1, A2, and A3, and foliogarlic trisulfanes

*Address for correspondence: c.jacob@mx.uni-saarland.de

DOI: 10.1079/9781789249996.0003

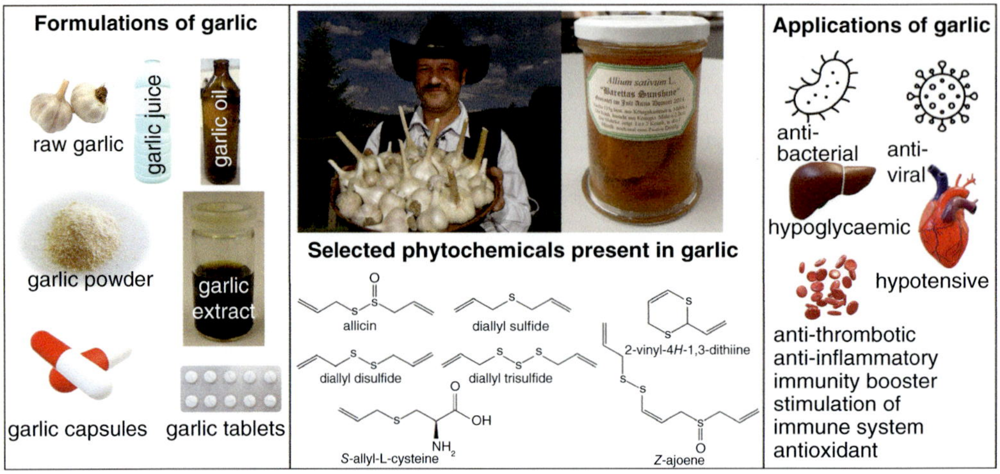

Fig. 3.1. Detlef Kittelmann with his 'Barettas sunshine' garlic, a selection of different dosage forms of garlic, and various possible health benefits associated with it.

A1 and A2 (Fukaya *et al.*, 2020). The presence of disulfane, trisulfanes or tetrahydro-2*H*-difuro [3,2-b:2′,3′-c]furan-5(5a*H*)-one motifs in these structures promises exciting biological activities.

In general, this interplay of basic nutritional and culinary values of garlic on the one hand and biological activities, such as possible antioxidant, anti-microbial, anti-hyperglycemic and anti-hypertensive actions, on the other distinguishes garlic from most other vegetables, herbs, spices and medicinal plants.

2. The Garlic Bulb as a Treasure Chest of Nutrients and Ancient Defence Systems

2.1. Composition of the garlic bulb and its nutritional value

Since the first isolation and characterization of biologically active ingredients of garlic, including allicin, by Cavallito and Bailey in 1944, a range of structurally diverse and often rather aggressive substances has been obtained and extensively studied for several beneficial biological activity (Cavallito and Bailey, 1944; Chen and Huang, 2019). These substances include nutrients, such as carbohydrates, some minerals, e.g. selenium, iron, zinc and copper, and flavonoids, such as allexin (Table 3.1; Fig. 3.2).

A typical garlic clove is composed of ~65% water, 28% carbohydrates, including ten different kinds of sugars, vitamins A and C, adenosine, 1.2% amino acids, a wide range of minerals, especially selenium in the form of L-selenomethionine, Se-methyl-L-selenocysteine, allyl-selenocysteine, and methylselenic acid, an aspect of chalcogen biology discussed in detail in Section 6 of this Chapter. The bulb is rich in some common flavonoids, such as catechin, epicatechin, myricetin, apigenin and quercetin, and also less common substances, such as the phytoalexin allixin (Fig. 3.2). These secondary metabolites are joined by saponins, such as eruboside-B, *β*-chlorogenin, sativoside B1 and procto-desgalactotigonin, a handful of rather special enzymes, such as alliinase, myrosinase and peroxidases, and, remarkably, 2.3% of sulfur compounds, the centre of our discussion here (Suleria *et al.*, 2015).

2.2. Reactive sulfur species (RSS)

Garlic is rich in many rather unusual sulfur compounds, ranging from the thiosulfinate allicin (diallylthiosulfinate) to several organic polysulfanes. The high and often (cyto-) toxic activity associated with RSS is part of intricate ancient defence systems against predators and microbial infections. Some of these phytoalexins and phytoanticipins are so aggressive that they cannot

Table 3.1. Typical composition and nutritional values of a garlic bulb (Nutrition Values, 2021).

Raw garlic	Nutritional values (per 100 g)	Vitamins	Nutritional values (per 100 g)	Minerals	Nutritional values (per 100 g)
Energy	149.0 kcal	Vitamin A	9.0 IU	Calcium	181.0 mg
Water	62–68 g	Vitamin C	31.2 mg	Iron	1.7 mg
Carbohydrates	26–30 g	Vitamin E	0.1 mg	Magnesium	25.0 mg
Dietary fibres	2.1 g	Vitamin K	1.7 mcg	Phosphorus	153.0 mg
Sugars	1.0 g	Thiamine	0.2 mg	Potassium	401.0 mg
Proteins	1.5–2.1 g	Riboflavin	0.1 mg	Sodium	17.0 mg
Amino acids	1.0–0.5 g	Niacin	0.7 mg	Zinc	1.2 mg
Fat	0.5 g	Vitamin B6	1.2 mg	Copper	0.3 mg
–	–	Folate	3.0 mcg	Manganese	1.7 mg
–	–	Pantothenic acid	0.6 mg	Selenium	14.2 mcg
–	–	Choline	23.2 mg	–	–

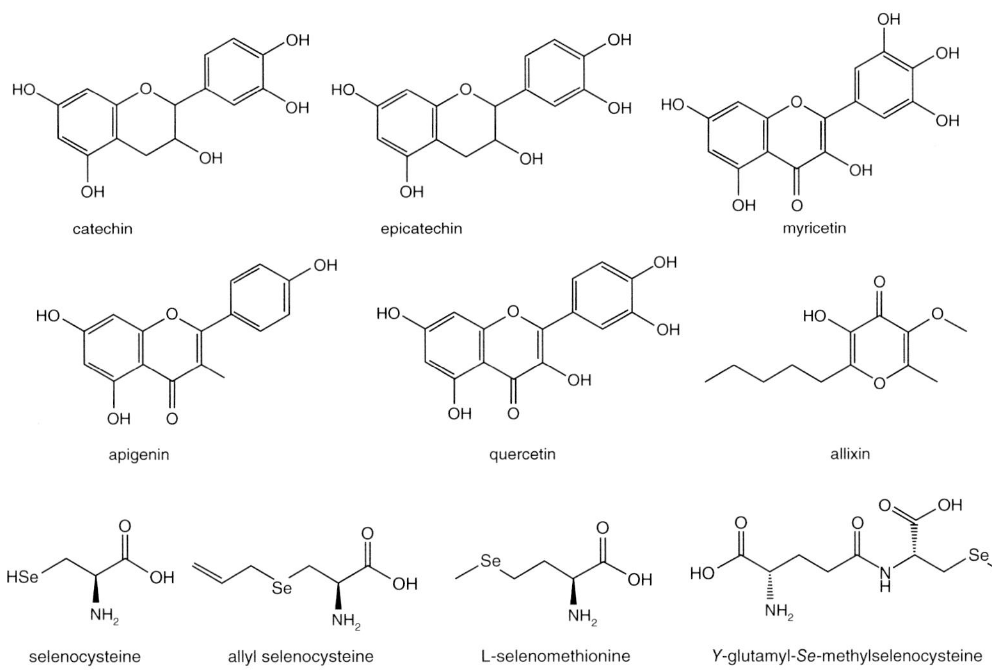

Fig. 3.2. Selection of biologically interesting non-sulfur-containing substances from garlic, mostly phytoalexins, phytoanticipins and trace elements.

be stored in the cell but need to be synthesized rapidly *in situ*, if and when the plant is under attack. Allicin, for instance, is a highly reactive compound synthesized enzymatically from the less aggressive alliin (*S*-allylcysteine sulfoxide) in the presence of the *C-S*-lyase enzyme alliinase (EC 4.4.1.4), upon injury to the bulb or by the physical mixing of the two components of this binary defence system (Singh and Singh, 2008). Allicin and some of its decomposition products, e.g. diallyldisulfide (DADS), diallyltrisulfide (DATS) and diallyltetrasulfide (DATTS), and various reaction products with cellular thiols, e.g. *S*-allyl-cysteine, also exhibit a pronounced, often (cyto-) toxic or antioxidant biological activity (Fig. 3.3).

A similar, albeit less popular, defence system in garlic is based on allixin, a non-sulfur

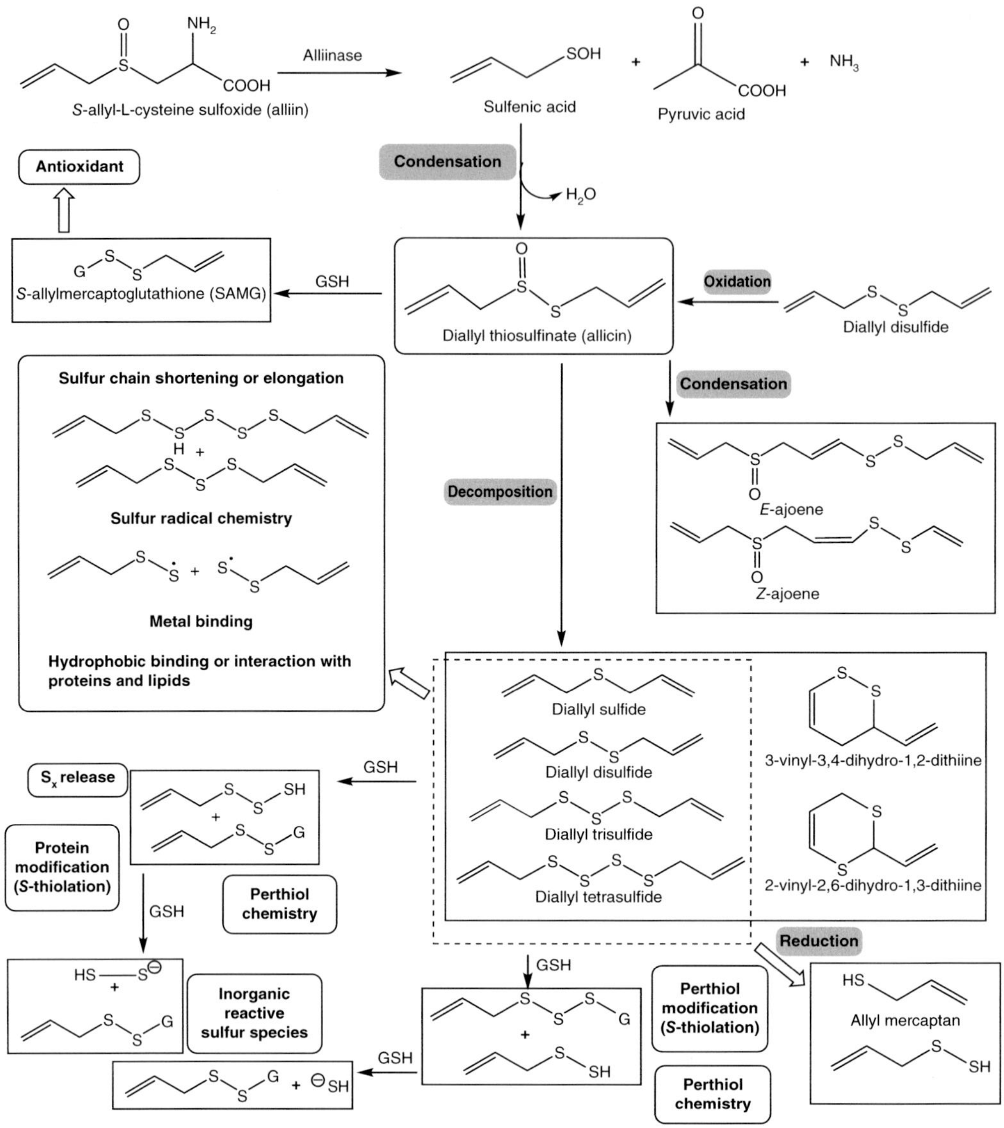

Fig. 3.3. Reactive sulfur species present in garlic, in particular allicin and polysulfanes, their biosynthesis, organic synthesis, reactivity and mode of action.

redox metabolite which exhibits various antioxidant and anti-microbial activities (Kodera *et al.*, 2002; Balqis *et al.*, 2018). Interestingly, some *Brassica* plants, including broccoli and wasabi, employ a similar binary defence system based on glucosinolates and the enzyme myrosinase, which, once in contact, produce a set of reactive electrophilic isothiocyanates (Kołodziejski *et al.*, 2019).

This strategy of binary defence systems is of considerable interest today, particularly in medicine and agriculture, as it provides the blueprints for innovative binary drugs or pro-drugs activated by enzymes in or at the target site, and efforts to shuttle enzymes such as alliinase to target sites with the help of antibodies and subsequently flooding the organism with alliin have been explored (Miron *et al.*, 2003).

3. Antioxidant and Immunomodulatory Activities

3.1. Direct and indirect antioxidant activities of garlic

Since garlic secondary metabolites serve as phytoalexins or phytoanticipins against microbes, they may be useful in the context of anti-microbial research. The pronounced reactivity of garlic RSS against thiol groups affirms the notion that such compounds react fast, efficiently and fairly selectively, with redox-sensitive cysteine residues in proteins and enzymes which either serve as cellular redox sensors or are required for cell survival (Banjac *et al.*, 2008).

The concept of the cellular thiolstat, for instance, suggests that redox modifications caused by RSS in mammalian cells either stimulate the cellular antioxidant defence systems or, if these modifications cannot be reversed, trigger apoptosis (Giles and Jacob, 2002). The reactivity of such sulfur secondary metabolites may therefore result in direct antioxidant action by simply reducing reactive oxygen species (ROS), as has been shown, for instance, for allyl mercaptan and many flavonoids found in garlic, namely catechins and quercetin. It may also unleash an indirect antioxidant action by upregulating *in situ* antioxidant defence mechanisms via redox sensing and response loops, as found for allicin itself. Other ingredients, such as zinc and L-selenomethionine, also upregulate antioxidant defences by stimulating the expression of zinc and selenium-containing antioxidant enzymes, such as superoxide dismutase (SODs) and glutathione peroxidase (GPx), respectively (Nasr, 2014). Hence, garlic is often considered as an antioxidant, albeit many of its actions are not, strictly speaking, reducing. On the contrary, most RSS found in garlic are highly oxidizing and their antioxidant action is therefore indirect, whereas their pro-oxidant, redox-modulating action on the cellular thiolstat, with often detrimental consequences for the cell affected, is the true, direct reactivity and action of these RSS.

3.2. Garlic and the immune system

Garlic and its ingredients trigger a mild form of oxidative stress (OS), hence provoke an antioxidant response in the human body. This may stimulate the body's antioxidant defences and also alert the immune system against external stressors and attacks (Schafer and Kaschula, 2014). This system is comparable to vaccination and acts to protect against the OS-related damages by upregulating the body's antioxidant defences with a minor oxidative stimulus (Yoshitake *et al.*, 2015). Garlic may indeed be considered as a health protective agent because of its own reactivity once inside the body and its indirect stimulating actions. These stimulating actions include the upregulation of antioxidant enzymes due to the supplementation of selenium and other trace elements, on one hand, and the positive impact caused by mild forms of OS, on the other.

It has previously been reported that garlic enhances the immune system in humans by increasing the white blood cell count (Nwabueze, 2012). It promotes the release of nitric oxide, which augments the liberation of interferons and reinforces the antiviral defences (Bhattacharyya *et al.*, 2007; Wilson and Demmig-Adams, 2007). Additional immune-stimulating activities of garlic relate to the activity of natural killer (NK) cells and lymphokine-activated killer (LAK) cells (Iciek *et al.*, 2009; Kim and Kwon, 2009).

These actions of garlic ingredients are not the result of classic direct antioxidant or defensive actions, rather they are the consequences of indirect events triggering the body's antioxidant and immune defences. Hence the amount of garlic or its components required to trigger such larger and more concerted actions of the human body may be small, compared to the concentrations required for direct antioxidant or anti-microbial actions. This aspect of the protective roles often associated with garlic needs to be emphasized when discussing controversial issues such as concentrations and bioavailability of active ingredients in animals and humans.

3.3. Apoptosis in mammalian cells

Although garlic and the various RSS contained therein, e.g. allicin, DADS, DATS and DATTS, can act as cellular redox modulators, none of them is – strictly redox-action speaking – an antioxidant. By generating mild oxidative events they tweak the cellular thiolstat and thereby trigger various processes, including antioxidant defences.

Nonetheless, higher concentrations of garlic and its ingredients are achievable via injection. Then, with the more aggressive oxidative onslaught, the redox modulatory activity associated with RSS manifests itself in a wide range of pro-oxidant, redox modulatory, cytotoxic and pro-apoptotic actions, including anti-cancer and anti-microbial activities observed *in vitro* and also in animal models. Hence, the activity of 100 mg ml^{-1} raw garlic extract (RGE) against elliptocytosis type 4 (EL4) in mice cancer cells confirmed the efficacy of injected RGE as compared to RGE provided orally (Li *et al.*, 2018).

These protective and therapeutic aspects of garlic are extensive yet controversial. Marut *et al.* have reported a beneficial activity of dipropyltetrasulfide (DPTTS) in systemic sclerosis (SSc) *in vitro* and in mouse model (Marut *et al.*, 2013). Despite the rather modest concentration, the intravenous administration yielded impressive results, probably due to cellular thiol modifications, to such an extent that the cell can no longer provide antioxidative protection. The unfavourable redox balance pushed the cell over a critical redox threshold and induced apoptosis (Saidu *et al.*, 2013). Since healthy cells with a more stable redox balance and defence are less affected, this activity of DPTTS is somewhat selective as it depends on the predisposition of the target cell, namely its redox signature.

Similarly, cancer cells are naturally under OS and hence close to their redox threshold, and apoptosis has been reported in a range of cancer cells treated with allicin, DADS, DATS and DATTS, e.g. gastric cancer SGC-7901 adenocarcinoma cells and hepatocellular carcinoma (HCC) cells (Sun and Wang, 2003; Chu *et al.*, 2013). Together with certain preventive actions ascribed to garlic, these anti-cancer activities are as amazing as they are controversial and should be evaluated further. Cells naturally close to the redox threshold also include redox-sensitive microorganisms. *Plasmodia*, for instance, are affected by such an effective and selective activity of garlic phytoalexins and anticipins, as are pathogenic bacteria and fungi, including *Aspergillus niger* (Kodera *et al.*, 2002).

Since the sulfur agents are highly reactive, they are often captured by their surroundings, such as other food ingredients, or sequestered by cellular components, such as the redox buffer glutathione (gamma-glutamyl-cysteinylglycine). When ingested, most of the unstable sulfur-containing substances are metabolized readily or decompose before reaching their targets in sufficient amounts to exert any significant activity. Furthermore, normal garlic consumption is not sufficient to achieve the levels required for the above actions, and injections are therefore often employed as the preferred route of administration. Consequently, the expected actions of garlic and its components are indirect, and the most likely targets for any direct action, if any, include the skin, the gastrointestinal and respiratory tract. We therefore shall focus on some of the more realistic applications of garlic.

4. Garlic and the Microbiome

4.1. Anti-bacterial actions

Since most bacteria, plasmodia and viruses are sensitive to redox changes, especially towards oxidation, it is hardly surprising that ROS, reactive nitrogen species (RNS) and RSS kill bacteria efficiently, similar to the strong disinfectants H_2O_2, hypochlorite (ClO^-) and nitric oxide $^{\cdot}NO$ (Mannick *et al.*, 1996; Allen and Stephens, 2011).

Certain RSS, such as the thiosulfinate allicin, exhibit pronounced bactericidal activity against a wide range of pathogenic Gram-negative and Gram-positive bacteria, including some of the most hazardous and on occasion also resistant organisms (Table 3.2). Raw garlic, for instance, is active against *Staphylococcus aureus*, *Escherichia coli*, *Proteus mirabilis*, *Klebsiella pneumoniae* and *Pseudomonas aeruginosa* at 37.5 mg ml^{-1} (Tessema *et al.*, 2006). Different studies have also reported a slightly higher activity against *S. aureus* with a Minimum Inhibitory Concentration (MIC) of 11.25 mg ml^{-1}. Garlic juice exhibits promising anti-microbial activity against *S. aureus* and *E. coli* at concentrations of 4–8% v/v (Bruno *et al.*, 2019).

More eloquent and concentrated aqueous, methanolic and ethanolic extracts exhibit considerably higher activity, for instance against *S. aureus*, *E. coli*, *P. aeruginosa* and *Salmonella typhi*, with MIC values ranging from 0.05 to 1.0 mg ml^{-1} (Gull *et al.*, 2012). Assuming an average molecular weight of ~200 g mol^{-1}, these

Table 3.2. Selection of anti-microbial activities associated with garlic in different formulations.

Garlic formulations	MIC values of different formulations of garlic against selected bacteria					References
	Staphylococcus aureus	*Proteus mirabilis*	*Klebsiella pneumoniae*	*Escherichia coli*	*Pseudomonas aeruginosa*	
Raw	37.5 mg ml^{-1}	37.5 mg ml^{-1}	37.5 mg ml^{-1}	37.5 mg ml^{-1}	37.5 mg ml^{-1}	(Tessema *et al.*, 2006)
Juice	4–8% *v/v*	4–8% *v/v*	4–8% v/v	4–8% v/v	4–16% v/v	(Yadav *et al.*, 2015)
Aqueous extract	1.6 μg ml^{-1}	–	–	3.2 μg ml^{-1}	1.6 μg ml^{-1}	(Elsom, 2000)
Ethanolic extract	78.9 μg ml^{-1}	89.0 μg ml^{-1}	160.2 μg ml^{-1}	65.5 μg ml^{-1}	58.5 μg ml^{-1}	(Karuppiah and Rajaram, 2012)
Methanolic extract	100 mg ml^{-1}	–	–	50 mg ml^{-1}	200 mg ml^{-1}	(Garba *et al.*, 2013)
Garlic and honey extracts	6.25 % *v/v*	–	–	6.25 % *v/v*	–	(Andualem, 2013)

concentrations correspond to around 250 μM to 5 mM, and hence are in the range required for some standard antibiotics. Nonetheless, one needs to handle these numbers carefully, as other studies have required ethanolic extracts of 100 mg ml^{-1} for activity against *S. aureus* (Khashan, 2014), and an MIC of 0.14 μg ml^{-1} for a methanolic extract against *S. aureus* and *P. aeruginosa* (Mauti *et al.*, n.d.). Such MIC values formally corresponding to sub-micromolar concentrations of active ingredients and easily reached locations of possible applications on the skin imply that applications of garlic are not entirely speculative. Intriguingly, a combination of garlic extract with honey also exhibits strong anti-microbial activities, especially against *S. aureus* and *E. coli* at a concentration of 6.25% v/v (Andualem, 2013).

Additional activities associated with garlic extracts and realistic modes of application include antioxidant and anti-fungal activities of the ethanolic extract against *Meyerozyma guilliermondii* and *Rhodotorula mucilaginosa* (Pârvu *et al.*, 2019).

4.2. Garlic and the gut microbiota

The gastrointestinal tract (GIT) may benefit from the ingested garlic. Besides the immediate oral antibacterial actions against *E. coli*, garlic may provide a more subtle beneficial activity, by fine-tuning the gut microbiota (Ried *et al.*, 2018) and processes associated with ageing and obesity (Magne *et al.*, 2020), and reducing the abundance of Gram-negative bacteria *Bacteroidetes* and *Prevotella* (Chen *et al.*, 2019). It is therefore likely that garlic impacts humans' gut microbiome without any need for excessive consumption, as demonstrated in human adults whose diets were supplemented for 12 weeks with 1.2 g of aged garlic powder (AGP), 1.2 mg of *S*-allyl cysteine, or cellulose as a placebo. PCR analysis of faeces confirmed an increase in *Lactobacillus* and *Clostridia* in the AGP consumers, in contrast to an increase of *Faecalibacterium prausnitzii* in the controls (Ried *et al.*, 2018). These gut microbiome changes are not necessarily of direct benefit to human health. Yet the benefits of garlic consumed in acceptable amounts are often indirect and subtle, and, as mentioned already, need to be taken with a pinch of salt.

More complex and possibly also more tasty alternatives to the rather mildly aggressive AGP include freshly mixed garlic and honey preparations. This mixture at the concetration of 6.25% to 12.5% release their hydrophobic ingredients and show synergistic anti-bacterial activity against *S. aureus*, *Streptococcus pneumonia*, *Salmonella* and *Listeria monocytogenes* (Andualem, 2013). Similarly, a combination of garlic, ginger, honey and lemon extracts is active against *Streptococcus mutants* (Mathai *et al.*, 2017). Indeed, garlic and ginger look like a perfect combination to optimize the mutual biological activity and also culinary delight, and, not surprisingly, such a refined relish was at the centre of the notorious raw garlic-eating challenge in 2015 between the authors and Detlef Kittelmann in the town of Zwota, Saxony, Germany.

4.3. Garlic breath

The respiratory tract is an obvious target for applications of garlic products, due to two rather unusual physical and metabolic aspects of garlic. Firstly, many of the garlic OSCs, including allicin, are rather volatile and can be inhaled. Secondly, many of the garlic metabolic products are exhaled via the lungs, hence also present and active in this organ.

It is therefore especially interesting to discover significant activities of garlic against bacteria responsible for the infection of the respiratory tract, such as *P. aeruginosa*, *Streptococcus pneumoniae*, *Klebsiella pneumoniae*, *Haemophilus influenza* and *S. aureus*. Indeed, vaporized allicin at 40–110 mM attacks multi-drug resistant strains of *P. aeruginosa*, *S. pyogenes*, *Streptococcus agalactiae*, *Streptococcus dysgalactiae*, *Streptococcus pneumoniae* and *S. aureus* rather efficiently (Reiter *et al.*, 2017).

Amazingly, garlic breath and breathing are not a one-way road. The sulfur-containing volatile products exhaled through the lungs include allyl mercaptan (AM), allyl methyl sulfide (AMS) and DADS (Minami *et al.*, 1989; Rosen *et al.*, 2000), which are rather active in a biological setting. AM inhibits the growth of *Giardia duodenalis*, a parasitic micro-organism which inhabits and replicates in the small intestine and provokes giardiasis (Argüello-García *et al.*, 2018).

Notably, simulation studies based on first-principles density functional theory (DFT) confirm that AMS interacts with the crucial phospholipids and amino acids of viruses which are involved in their attachment to host cells, thereby possibly providing a protective effect against viral infections (Sinha *et al.*, 2020).

AMS, DADS and other garlic ingredients are excreted also via the skin. Hypothetically, activity of these substances may also be found there (Sato *et al.*, 2020). Compounds such as allicin are active against a range of pathogens affecting the skin and the respiratory tract, including methicillin-resistant *Staphylococcus epidermidis* (MRSE) and, notably, biofilms formed by these Gram-positive bacteria (Pérez-Giraldo *et al.*, 2003). If and how excretion and exhalation of the garlic ingredients and its metabolized products may be turned into a health benefit in lungs and skin is a topic that demands further investigation.

5. Antiviral Activity

It is important to elucidate if viruses are affected by garlic active ingredients as oxidation of the viral coat, for instance, may impair its infectious potential. Studies on garlic and the Sars-CoV-2 virus report an indirect positive impact of garlic on the stimulation of the human immune system (Oso *et al.*, 2020; Tufan *et al.*, 2020). A direct antiviral attack of garlic essential oil on the viral spike glycoprotein and the angiotensin-converting enzyme 2 receptor (ACE-2 receptor), which is a crucial step for the entry of the virus into the human cell, has been reported (Asif *et al.*, 2020). On closer inspection, 17 OSCs in garlic essential oil have been identified in docking studies as possibly preventing the actions on the ACE-2 receptor, including DADS and DATS (Thuy *et al.*, 2020). Additionally, some of these sulfur compounds also inhibit the main protease PDV6LU7 of the Sars-CoV-2 virus (Shekh *et al.*, 2020).

Quercetin in combination with ascorbic acid has also been considered as supportive in the treatment or prophylaxis of Covid-19, and a quercetin-3-beta-galactoside has been identified as an important protease inhibitor against this virus in molecular docking and surface plasmon resonance/fluorescence resonance energy transfer (SPR/FRET)-based bioassays (Chen *et al.*, 2006).

The application of many of these promising antiviral findings remains vague as many of these studies have been conducted *in vitro*. Furthermore, dietary garlic cannot be compared to highly active pure compounds administered in high doses. As for quercetin, this compound is not specific to garlic and may be more leisurely consumed in the form of fruits (Sultana and Anwar, 2008).

Garlic is not an antiviral remedy on a par with modern medications, yet literature suggests that certain secondary metabolites of garlic may provide a remedy. Aqueous garlic extract exhibits activity against the influenza virus at 100 $\mu g\ ml^{-1}$ in an MTT assay exploiting vero cells (Mehrbod *et al.*, 2009; Rasool *et al.*, 2017). These fairly high concentrations are possibly not pharmaceutically relevant. The mixture of honey, ginger and garlic extract shows a synergistic activity against different types of influenza A virus isolated from human nasal swab, the lungs of chicken, and wild bird droppings, with concentrations from 0.005 to 5% v/v required in the *in vitro* assay for antiviral action (Vahed and Batool Jafri, 2016).

Specific dosage forms of garlic have been prepared to treat acute respiratory diseases (Sobenin *et al.*, 2010). A microparticle, for instance, for the treatment of severe lung infections is based on the original binary defence system of garlic and contains both alliin and alliinase, which produces its allicin activity once inhaled and slowly dissolving in the moisture of the lungs (Strehlow *et al.*, 2016).

Aqueous garlic extracts have been evaluated against poultry Newcastle virus, the human cytomegalovirus (HCMV), the herpes simplex virus (HSV) types 1 and 2, and the human parainfluenza virus type 3, with concentrations required for activity depending on the virus and the assay conditions and ranging from 10 mg ml^{-1} in the case of the Newcastle virus to a staggering 1 g ml^{-1} required against the parainfluenza virus (Weber *et al.*, 1992; Guo *et al.*, 1993).

Some pure ingredients of garlic have also been tested. Here, the diallyl polysulfanes, and especially DATS, have shown good activity, for instance against HCMV at 10 μg ml^{-1}. Alliin, DAS and DADS reduced inflammation associated with Dengue virus infection at doses ranging from 0.1 μg to 50 μg for alliin and 10 μg to 250 μg for DAS and DADS (Hall *et al.*, 2017). It should be emphasized once again that these concentrations often depend on the assay conditions employed and the activity of garlic is not limited to its sulfur agents. Quercetin acts against the MERS-CoV virus by inhibiting 3C-like protease at a concentration of 20 μM (Jo *et al.*, 2019). Moreover, garlic extracts produce gold nanoparticles by reduction of Au^{3+} solutions. These gold nanoparticles have subsequently shown some activity against the measles virus (MeV) with EC_{50} values in vero cells of around 10 μg ml^{-1} (Meléndez-Villanueva *et al.*, 2019).

As for animal studies, DATS provided antiviral activity against the murine cytomegalovirus (MCMV) in mice at doses of 25 and 75 mg kg^{-1} administered via intraperitoneal injections (Liu *et al.*, 2004). DATS suppresses the expression of the human cytomegalovirus (HCMV) genome at 9.6 μg ml^{-1} in human embryo lung fibroblast (HEL) cells (Zhang *et al.*, 2013).

6. Enrichment of Garlic with Trace Elements

The garlic bulb accumulates high levels of many important micronutrients, such as calcium and potassium. Its natural high content of sulfur lends itself to biofortification with the related element selenium and maybe also with other trace metals showing a high affinity to chalcogens, such as zinc and copper. The individual bulb's composition, however, varies with genotype and environment. Various strategies have been employed to enrich garlic with trace elements, especially selenium, to increase its nutritional value (Ip *et al.*, 2000). Hence, garlic cultivated on selenium-enriched soil accumulated a staggering amount of 1600, 1000 and 800 μg of Se per leaf, bulb and root, respectively (Pérez *et al.*, 2019). Such biofortified garlic a day may therefore keep the food supplement at bay. Notably, the leaves rather than the bulb are particularly rich in selenium, therefore shifting the green parts of the enriched plant into the focus, also for culinary uses.

The inorganic selenite (SeO_3^{2-}), is metabolized in garlic to produce a palette of organic compounds (Fig. 3.2). Analyses of selenium-enriched garlic reveal that selenium is found mostly in amino acids selenocysteine (SeCys), allyl selenocysteine, L-selenomethionine (SeMet) and the dipeptide gamma-glutamyl-Se-methylselenocysteine and as inorganic ions SeO_3^{2-} and SeO_4^{2-} (Ge *et al.*, 1996). Of the total selenium concentration (~300 μg g^{-1} garlic), around 73% is stored as gamma-glutamyl-Se-methylselenocysteine (Ip *et al.*, 2000). Subsequently, it has been reported that Se-methyl-selenocysteine is the chief bioactive selenium compound in garlic and confirmed that freeze-drying is the preferred method for the stability of selenium compounds on long preservation (Yang, 2002).

As food supplements, selenium compounds are less controversial compared to SeO_3^{2-} and SeMet is most commonly employed in selenium supplementation, for instance in selenium-enriched yeast (Wang *et al.*, 2017; Wu *et al.*, 2019). Once consumed, SeMet often finds its way into animal and humans proteins as a substitute for its sulfur analogue methionine, therefore endowing traditional sulfur proteins with extra selenium redox and antioxidant activity (Schrauzer, 2000).

General health benefits associated with selenium include a better antioxidant defence, often by upregulation of GPx. An increase in two liver selenoenzymes, GPx and Type 1 5'-deiodinase, has been observed in rats fed with a diet containing selenium-enriched garlic (0.1–0.5 ppm) and this activity is comparable to the traditional Se supplement selenite (SeO_3^{2-}) (Ip and Lisk, 1993). A similar increase of GPx activity has also been obtained in chickens fed with 18.5 mg Se kg^{-1} body weight. The total selenium concentration in the liver and plasma increased from 0.514 to 0.627 mg Se kg^{-1}, thereby also resulting in increased GPx activity from 617.6 to 724.3 units L^{-1} in the liver and from 111.6 to 253.4 units L^{-1} in plasma (Seo *et al.*, 2008). Selenium-enriched garlic has also been reported to inhibit the development and progression of chronic gastritis and early-stage mammary carcinogenesis in rodent models (Ip *et al.*, 1996; Gu *et al.*, 2007).

In short, garlic fortified with selenium provides an attractive alternative to the consumption of supplements containing the rather controversial SeO_3^{2-} or selenium-enriched yeast. Since garlic is already part of several cuisines and the usual amounts of garlic consumption would be sufficient to supply the recommended daily amounts of selenium, the biofortified garlic combines the benefits of standard garlic with the ones associated with selenium supplementation yet without the need for compliance. It is therefore not surprising that various products containing selenium-enriched garlic can be obtained commercially, such as SelenoForce® sold in New Zealand (Sabsina, n.d.).

Garlic is also amenable to enrichment with other minerals and trace metals, such as magnesium and iron, although biofortification with these ions often decreases during selenium enrichment, possibly because of the mutual exclusiveness of these trace elements (Pérez *et al.*, 2019). Here other nutritional sources of these minerals may be more realistic and easier to achieve in practice.

7. Social and Cultural Ingredients of Garlic

Garlic is mainly consumed as a condiment herb worldwide. Is it, therefore, possible to improve the health of individuals and entire populations by promoting the regular intake of garlic and/or diets rich in garlic products? This question is highly controversial and the responses are complex, as RSS concentrations are moderate and

Fig. 3.4. Selenium enrichment of soil yields selenium-enriched plants for human food and animal feed. The insert details the 'CJ Special' healthy garlic cocktail.

their bioavailability is problematic. The strategy of providing active substances via food, however, is an attractive alternative, especially in the elderly, who are confronted with multimedication and compliance, often taking six to ten different types of medicines per day (Hajjar *et al.*, 2007). Nonetheless, caution against the replacement of proven drugs for garlic and other herbs should be considered, especially in the case of serious illnesses.

So does garlic have a proven track record as a medication? No. Is garlic a worthy substitute for traditional drugs in persons suffering from illnesses such as high blood pressure or infections? Certainly not. In such cases, there is no alternative to consulting a qualified medical doctor and to follow their advice and prescriptions. Then again, does garlic contain small amounts of biologically active compounds? Yes. Is garlic harmless if consumed in moderate amounts? Yes. Can garlic be enriched with selenium and other minerals? Yes. Is garlic, as a spice, a tasty and worthy alternative to table salt (NaCl) and therefore worth advocating in your cuisine? Certainly yes.

Although garlic itself should therefore not be considered as an alternative to modern medications, its consumption as a culinary delight is another matter and certainly not debateable. We shall therefore conclude on a positive note with the recipe for a garlic shake, which in many ways spices up the nutritional and health benefits of garlic and also exemplifies the added value of this plant. Free of any claims of being particularly healthy, the 'CJ Special' garlic cocktail in Fig. 3.4 is a worthy substitute for not-so-healthy soft or alcoholic drinks. Enjoy.

Dedication

We would like to dedicate this chapter to Detlef Kittelmann, for 38 years of cultivating the most exotic types of garlic on his farm in Zwota, Saxony, Germany, and studying different aspects of garlic and nutrition.

References

Alemseged, M., Adugna, S. and Bayu, E. (2018) Antibacterial properties of mixture honey and garlic (*Allium sativum*) extracts against respiratory tract infection-causing bacteria. *Journal of Pharmacovigilance* 6, 1–4. DOI: 10.4172/2329-6887.1000267.

Allen, R.C. and Stephens, J.T. (2011) Myeloperoxidase selectively binds and selectively kills microbes. *Infection and Immunity* 79, 474–485. DOI: 10.1128/IAI.00910-09.

Andualem, B. (2013) Combined antibacterial activity of stingless bee (*Apis mellipodae*) honey and garlic (*Allium sativum*) extracts against standard and clinical pathogenic bacteria. *Asian Pacific Journal of Tropical Biomedicine* 3, 725–731. DOI: 10.1016/S2221-1691(13)60146-X.

Anwar, A., Gould, E., Tinson, R., Groom, M. and Hamilton, J.C. (2017) Think yellow and keep green: role of sulfanes from garlic in agriculture. *Antioxidants* 6, 3. DOI: 10.3390/antiox6010003.

Argüello-García, R., de la Vega-Arnaud, M., Loredo-Rodríguez, I.J., Mejía-Corona, A.M., Melgarejo-Trejo, E. *et al.* (2018) Activity of thioallyl compounds from garlic against *Giardia duodenalis* trophozoites and in experimental giardiasis. *Frontiers in Cellular and Infection Microbiology* 8. DOI: 10.3389/fcimb.2018.00353.

Asif, M., Saleem, M., Saadullah, M., Yaseen, H.S. and Al Zarzour, R. (2020) COVID-19 and therapy with essential oils having antiviral, anti-inflammatory, and immunomodulatory properties. *Inflammopharmacology* 28, 1153–1161. DOI: 10.1007/s10787-020-00744-0.

Balqis, W., Lukiati, B. and Amin, M. (2018) Active compounds with antioxidant potential in boiled local Papua-Indonesian garlic. In: *The 9th International Conference on Global Resource Conservation (ICGRC) and AJI*. Ritsumeikan University, Malang City, Indonesia, 020012. DOI: 10.1063/1.5061848.

Banjac, A., Perisic, T., Sato, H., Seiler, A., Bannai, S. *et al.* (2008) The cystine/cysteine cycle: a redox cycle regulating susceptibility versus resistance to cell death. *Oncogene* 27, 1618–1628. DOI: 10.1038/sj.onc.1210796.

Bayan, L., Koulivand, P.H. and Gorji, A. (2014) Garlic: a review of potential therapeutic effects. *Avicenna Journal of Phytomedicine* 4, 1–14.

Bhattacharyya, M., Girish, G.V., Karmohapatra, S.K., Samad, S.A. and Sinha, A.K. (2007) Systemic production of IFN-alpha by garlic (*Allium sativum*) in humans. *Journal of Interferon & Cytokine Research* 27, 377–382. DOI: 10.1089/jir.2006.0124.

Bruno, M.J., Landry, M., Destiny, M.B., Bertin, M.J. and Antoine, A.A. (2019) Phytochemical screening and antibacterial and antifungal activity of *Allium sativum* juice on multi-resistant strains. *Advances in Analytical Chemistry* 9, 34–37.

Cavallito, C.J. and Bailey, J.H. (1944) Allicin, the antibacterial principle of *Allium sativum*. I. Isolation, physical properties and antibacterial action. *Journal of the American Chemical Society* 66, 1950–1951. DOI: 10.1021/ja01239a048.

Chen, J. and Huang, G. (2019) Antioxidant activities of garlic polysaccharide and its phosphorylated derivative. *International Journal of Biological Macromolecules* 125, 432–435. DOI: 10.1016/j.ijbiomac.2018.12.073.

Chen, K., Xie, K., Liu, Z., Nakasone, Y., Sakao, K. *et al.* (2019) Preventive effects and mechanisms of garlic on dyslipidemia and gut microbiome dysbiosis. *Nutrients* 11, 1225. DOI: 10.3390/nu11061225.

Chen, L., Li, J., Luo, C., Liu, H., Xu, W. *et al.* (2006) Binding interaction of quercetin-3-beta-galactoside and its synthetic derivatives with SARS-CoV 3CL(pro): structure-activity relationship studies reveal salient pharmacophore features. *Bioorganic & Medicinal Chemistry* 14, 8295–8306. DOI: 10.1016/j.bmc.2006.09.014.

Choi, M.K., Chae, K.Y., Lee, J.Y. and Kyung, K.H. (2007) Antimicrobial activity of chemical substances derived from S-Alk(en)yl-L-Cysteine Sulfoxide (Alliin) in garlic, *Allium sativum* L. *Food Science and Biotechnology* 16, 1–7.

Chu, Y.L., Ho, C.T., Chung, J.G., Raghu, R., Lo, Y.C. *et al.* (2013) Allicin induces anti-human liver cancer cells through the p53 gene modulating apoptosis and autophagy. *Journal of Agricultural and Food Chemistry* 61, 9839–9848. DOI: 10.1021/jf403241s.

Elsom, G.K. (2000) An antibacterial assay of aqueous extract of garlic against anaerobic/microaerophilic and aerobic bacteria. *Microbial Ecology in Health and Disease* 12, 81–84. DOI: 10.1080/089106000435464.

Fukaya, M., Nakamura, S., Hayashida, H., Noguchi, D., Nakashima, S. *et al.* (2020) Structures of cyclic organosulfur compounds from garlic (*Allium sativum* L.) leaves. *Frontiers in Chemistry* 8. DOI: 10.3389/fchem.2020.00282.

Garba, I., Umar, A.I., Abdulrahman, A.B., Tijjani, M.B., Aliyu, M.S. *et al.* (2013) Phytochemical and antibacterial properties of garlic extracts. *Bayero Journal of Pure and Applied Sciences* 6, 45–48. DOI: 10.4314/bajopas.v6i2.10.

Ge, H., Cai, X.J., Tyson, J.F., Uden, P.C., Denoyer, E.R. *et al.* (1996) Identification of selenium species in selenium-enriched garlic, onion and broccoli using high-performance ion chromatography with inductively coupled plasma mass spectrometry detection. *Analytical Communications* 33, 279. DOI: 10.1039/ac9963300279.

Giles, G.I. and Jacob, C. (2002) Reactive sulfur species: an emerging concept in oxidative stress. *Biological Chemistry* 383, 375–388. DOI: 10.1515/BC.2002.042.

Gruhlke, M.C. (2019) Reactive sulfur species. In: Hasanuzzaman, M., Fotopoulos, V., Nahar, K. and Fujita, M. (eds) *Reactive Oxygen, Nitrogen and Sulfur Species in Plants*. John Wiley & Sons, Inc., Hoboken, New Jersey, pp. 715–728. DOI:10.1002/9781119468677.ch31.

Gu, L.K., Zhou, P., Zhou, J., Wang, R.M., Yang, W.J. *et al.* (2007) Effect of selenium-enriched garlic on chronic gastritis of the glandular stomach of Mongolian gerbils induced by *H. pylori*. *Zhonghua Yu Fang Yi Xue Za Zhi* 41(Suppl), 104–107.

Gull, I., Saeed, M., Shaukat, H., Aslam, S.M., Samra, Z.Q. *et al.* (2012) Inhibitory effect of *Allium sativum* and *Zingiber officinale* extracts on clinically important drug resistant pathogenic bacteria. *Annals of Clinical Microbiology and Antimicrobials* 11, 8. DOI: 10.1186/1476-0711-11-8.

Guo, N.L., Lu, D.P., Woods, G.L., Reed, E., Zhou, G.Z. *et al.* (1993) Demonstration of the anti-viral activity of garlic extract against human cytomegalovirus *in vitro*. *Chinese Medical Journal* 106, 93–96.

Hajjar, E.R., Cafiero, A.C. and Hanlon, J.T. (2007) Polypharmacy in elderly patients. *The American Journal of Geriatric Pharmacotherapy* 5, 345–351. DOI: 0.1016/j.amjopharm.2007.12.002.

Hall, A., Troupin, A., Londono-Renteria, B. and Colpitts, T.M. (2017) Garlic organosulfur compounds reduce inflammation and oxidative stress during dengue virus infection. *Viruses* 9. DOI: 10.3390/v9070159.

Iciek, M., Kwiecień, I. and Włodek, L. (2009) Biological properties of garlic and garlic-derived organosulfur compounds. *Environmental and Molecular Mutagenesis* 50, 247–265. DOI: 10.1002/em.20474.

Ip, C. and Lisk, D.J. (1993) Bioavailability of selenium from selenium-enriched garlic. *Nutrition and Cancer* 20, 129–137. DOI: 10.1080/01635589309514279.

Ip, C., Lisk, D.J. and Thompson, H.J. (1996) Selenium-enriched garlic inhibits the early stage but not the late stage of mammary carcinogenesis. *Carcinogenesis* 17, 1979–1982. DOI: 10.1093/carcin/17.9.1979.

Ip, C., Birringer, M., Block, E., Kotrebai, M., Tyson, J.F. *et al.* (2000) Chemical speciation influences comparative activity of selenium-enriched garlic and yeast in mammary cancer prevention. *Journal of Agricultural and Food Chemistry* 48, 2062–2070. DOI: 10.1021/jf000051f.

Jo, S., Kim, H., Kim, S., Shin, D.H. and Kim, M.-S. (2019) Characteristics of flavonoids as potent MERS-CoV 3C-like protease inhibitors. *Chemical Biology & Drug Design* 94, 2023–2030. DOI: 10.1111/cbdd.13604.

Karuppiah, P. and Rajaram, S. (2012) Antibacterial effect of *Allium sativum* cloves and *Zingiber officinale* rhizomes against multiple-drug resistant clinical pathogens. *Asian Pacific Journal of Tropical Biomedicine* 2, 597–601. DOI: 10.1016/S2221-1691(12)60104-X.

Khashan, A. (2014) Antibacterial activity of garlic extract (*Allium sativum*) against *Staphylococcus aureus in vitro*. *Global Journal of Bio-Science and Biotechnology* 3, 346–348.

Kim, J.Y. and Kwon, O. (2009) Garlic intake and cancer risk: an analysis using the Food and Drug Administration's evidence-based review system for the scientific evaluation of health claims. *The American Journal of Clinical Nutrition* 89, 257–264. DOI: 10.3945/ajcn.2008.26142.

Kodera, Y., Ayabe, M., Ogasawara, K., Yoshida, S., Hayashi, N. *et al.* (2002) Allixin accumulation with long-term storage of garlic. *Chemical and Pharmaceutical Bulletin* 50, 405–407. DOI: 10.1248/cpb.50.405.

Kołodziejski, D., Koss-Mikołajczyk, I., Abdin, A.Y., Jacob, C. and Bartoszek, A. (2019) Chemical aspects of biological activity of isothiocyanates and indoles, the products of glucosinolate decomposition. *Current Pharmaceutical Design* 25, 1717–1728. DOI: 10.2174/1381612825666190701151644.

Li, Z. Le, W. and Cui, Z. (2018) A novel therapeutic anticancer property of raw garlic extract via injection but not ingestion. *Cell Death Discovery* 4, 1–10. DOI: 10.1038/s41420-018-0122-x.

Liu, Z.F., Fang, F., Dong, Y.S., Li, G. and Zhen, H. (2004) Experimental study on the prevention and treatment of murine cytomegalovirus hepatitis by using allitridin. *Antiviral Research* 61, 125–128. DOI: 10.1016/s0166-3542(03)00087-1.

Magne, F., Gotteland, M., Gauthier, L., Zazueta, A., Pesoa, S. *et al.* (2020) The firmicutes/bacteroidetes ratio: a relevant marker of gut dysbiosis in obese patients? *Nutrients* 12. DOI: 10.3390/nu12051474.

Mannick, E.E., Bravo, L.E., Zarama, G., Realpe, J.L., Zhang, X.-J. *et al.* (1996) Inducible nitric oxide synthase, nitrotyrosine, and apoptosis in helicobacter pylori gastritis: effect of antibiotics and antioxidants. *Cancer Research* 56, 3238–3243.

Marut, W., Jamier, V., Kavian, N., Servettaz, A., Winyard, P.G. *et al.* (2013) The natural organosulfur compound dipropyltetrasulfide prevents HOCl-induced systemic sclerosis in the mouse. *Arthritis Research & Therapy* 15, R167. DOI: 10.1186/ar4351.

Mathai, K., Anand, S., Aravind, A., Dinatius, P., Krishnan, A.V. *et al.* (2017) Antimicrobial effect of ginger, garlic, honey, and lemon extracts on *Streptococcus mutans*. *The Journal of Contemporary Dental Practice* 18, 1004–1008. DOI: 10.5005/jp-journals-10024-2165.

Mauti, G.O., Mauti, E.M., Ouno, G.A. and Maronga, B. (2015) Antibacterial activity of garlic, tulsi, bitter guard and cinnamon extracts against wound pathogens. *Journal of Scientific and Innovative Research* 4, 178–181.

Mehrbod, P., Amini, E. and Tavassoti-Kheiri, M. (2009) Antiviral activity of garlic extract on Influenza virus. *Iranian Journal of Virology* 3, 19–23. DOI: 10.21859/isv.3.1.19.

Meléndez-Villanueva, M.A., Morán-Santibañez, K., Martínez-Sanmiguel, J.J., Rangel-López, R., Garza-Navarro, M.A. *et al.* (2019) Virucidal activity of gold nanoparticles synthesized by green chemistry using garlic extract. *Viruses* 11. DOI: 10.3390/v11121111.

Minami, T., Boku, T., Inada, K., Morita, M. and Okazaki, Y. (1989) Odor components of human breath after the ingestion of grated raw garlic. *Journal of Food Science* 54, 763–763. DOI: 10.1111/j.1365-2621.1989.tb04703.x.

Miron, T., Mironchik, M., Mirelman, D., Wilchek, M. and Rabinkov, A. (2003) Inhibition of tumor growth by a novel approach: in situ allicin generation using targeted alliinase delivery. *Molecular Cancer Therapeutics* 2, 1295–1301.

Nasr, A.Y. (2014) Protective effect of aged garlic extract against the oxidative stress induced by cisplatin on blood cells parameters and hepatic antioxidant enzymes in rats. *Toxicology Reports* 1, 682–691. DOI: 10.1016/j.toxrep.2014.09.003.

Nencini, C., Menchiari, A., Franchi, G.G. and Micheli, L. (2011) In vitro antioxidant activity of aged extracts of some Italian *Allium* species. *Plant Foods for Human Nutrition* 66, 11–16. DOI: 10.1007/s11130-010-0204-2.

Nutrition Values (2021) Nutrition facts for garlic, raw, recommended daily values and analysis. Available at: https://www.nutritionvalue.org/Garlic%2C_raw_nutritional_value.html (accessed 2 March 2021).

Nwabueze, A.A. (2012) The effect of garlic (*Allium sativum*) on growth and haematological parameters of *Clarias gariepinus* (Burchell, 1822). *Sustainable Agriculture Research*. 1, 222–228. DOI: 10.22004/ag.econ.231382.

Oso, B.J., Adeoye, A.O. and Olaoye, I.F. (2020) Pharmacoinformatics and hypothetical studies on allicin, curcumin, and gingerol as potential candidates against COVID-19-associated proteases. *Journal of Biomolecular Structure and Dynamics* 40, 1–12. DOI: 10.1080/07391102.2020.1813630.

Pârvu, M., Moţ, C.A., Pârvu, A.E., Mircea, C., Stoeber, L. *et al.* (2019) Allium sativum extract chemical composition, antioxidant activity and antifungal effect against *Meyerozyma guilliermondii* and *Rhodotorula mucilaginosa* causing onychomycosis. *Molecules* 24, 3958. DOI: 10.3390/molecules24213958.

Pérez, M.B., Lipinski, V.M., Filippini, M.F., Chacón-Madrid, K., Arruda, M.A.Z. *et al.* (2019) Selenium biofortification on garlic growth and other nutrients accumulation. *Horticultura Brasileira* 37, 294–301. DOI: 10.1590/s0102-053620190307.

Pérez-Giraldo, C., Cruz-Villalón, G., Sánchez-Silos, R., Martínez-Rubio, R., Blanco, M.T. *et al.* (2003) *In vitro* activity of allicin against *Staphylococcus epidermidis* and influence of subinhibitory concentrations on biofilm formation. *Journal of Applied Microbiology* 95, 709–711. DOI: 10.1046/j.1365-2672.2003.02030.x.

Rajasekaran, R., Lu, K.H. and Sheen, L.Y. (2012) Recent research progress on garlic (dà suàn) as a potential anticarcinogenic agent against major digestive cancers. *Journal of Traditional and Complementary Medicine* 2, 192–201. DOI: 10.1016/S2225-4110(16)30099-2.

Rasool, A., Khan, M.U.R., Ali, M.A., Anjum, A.A., Ahmed, I. *et al.* (2017) Anti-avian influenza virus H9N2 activity of aqueous extracts of *Zingiber officinalis* (Ginger) and *Allium sativum* (Garlic) in chick embryos. *Pakistan Journal of Pharmaceutical Sciences* 30, 1341–1344.

Reiter, J., Levina, N., Van der Linden, M., Gruhlke, M., Martin, C. *et al.* (2017) Diallylthiosulfinate (Allicin), a volatile antimicrobial from garlic (*Allium sativum*), kills human lung pathogenic bacteria, including MDR strains, as a vapor. *Molecules* 22, 1711. DOI: 10.3390/molecules22101711.

Ried, K., Travica, N. and Sali, A. (2018) The effect of kyolic aged garlic extract on gut microbiota, inflammation, and cardiovascular markers in hypertensives: the GarGIC trial. *Frontiers in Nutrition* 5. DOI: 10.3389/fnut.2018.00122.

Rivlin, R.S. (2001) Historical perspective on the use of garlic. *The Journal of Nutrition* 131, 951S–954S. DOI: 10.1093/jn/131.3.951S.

Rosen, R.T., Hiserodt, R.D., Fukuda, E.K., Ruiz, R.J., Zhou, Z. *et al.* (2000) The determination of metabolites of garlic preparations in breath and human plasma. *Biofactors* 13, 241–249. DOI: 10.1002/biof.5520130137.

Sabsina (n.d.) Repositioning: SelenoForce®. Available at: https://selenoforce.com/jan2008v81.html (accessed 17 November 2020).

Saidu, N.E.B., Abu Asali, I., Czepukojc, B., Seitz, B., Jacob, C. *et al.* (2013) Comparison between the effects of diallyl tetrasulfide on human retina pigment epithelial cells (ARPE-19) and HCT116 cells. *Biochimica et Biophysica Acta (BBA)-General Subjects* 1830, 5267–5276. DOI: 10.1016/j.bbagen.2013.08.004.

Sato, S., Sekine, Y., Kakumu, Y. and Hiramoto, T. (2020) Measurement of diallyl disulfide and allyl methyl sulfide emanating from human skin surface and influence of ingestion of grilled garlic. *Scientific Reports* 10, 465. DOI: 10.1038/s41598-019-57258-1.

Schafer, G.H. and Kaschula, C. (2014) The immunomodulation and anti-inflammatory effects of garlic organosulfur compounds in cancer chemoprevention. *Anti-Cancer Agents in Medicinal Chemistry (Formerly Current Medicinal Chemistry-Anti-Cancer Agents)* 14, 233–240.

Schrauzer, G.N. (2000) Selenomethionine: a review of its nutritional significance, metabolism and toxicity. *Journal of Nutrition* 130, 1653–1656. DOI: 10.1093/jn/130.7.1653.

Seo, T.C., Spallholz, J.E., Yun, H.K. and Kim, S.W. (2008) Selenium-enriched garlic and cabbage as a dietary selenium source for broilers. *Journal of Medicinal Food* 11, 687–692. DOI: 10.1089/jmf.2007.0053.

Shekh, S., Reddy, K.K.A. and Gowd, K.H. (2020) *In silico* allicin induced S-thioallylation of SARS-CoV-2 main protease. *Journal of Sulfur Chemistry* 42, 1–12. DOI: 10.1080/17415993.2020.1817457.

Singh, V. and Singh, D. (2008) Pharmacological effects of garlic (*Allium sativum* L.). *Annual Review of Biomedical Data Science* 10. DOI: 10.5016/1806-8774.2008.v10p6.

Sinha, A.K., Farooqui, S.A., Sharma, A., Mishra, A. and Verma, V. (2020) Reactivity of allyl methyl sulphide, the in-vitro metabolite of garlic, with some amino acids and with phospholipid involved in viral infections. *Journal of Biomolecular Structure and Dynamics* 1–7. DOI: 10.1080/07391102.2020.1811154.

Sobenin, I.A., Pryanishnikov, V.V., Kunnova, L.M., Rabinovich, Y.A., Martirosyan, D.M. *et al.* (2010) The effects of time-released garlic powder tablets on multifunctional cardiovascular risk in patients with coronary artery disease. *Lipids in Health and Disease* 9, 119. DOI: 10.1186/1476-511X-9-119.

Strehlow, B., Bakowsky, U., Pinnapireddy, S.R., Kusterer, J., Mielke, G. *et al.* (2016) A novel microparticulate formulation with Allicin *in situ* synthesis. *Journal of Pharmaceutics & Drug Delivery Research* 2016. DOI: 10.4172/2325-9604.1000143.

Suleria, H.A.R., Butt, M.S., Khalid, N., Sultan, S., Raza, A. *et al.* (2015) Garlic (*Allium sativum*): diet based therapy of 21st century: a review. *Asian Pacific Journal of Tropical Disease* 5, 271–278. DOI: 10.1016/S2222-1808(14)60782-9.

Sultana, B. and Anwar, F. (2008) Flavonols (kaempeferol, quercetin, myricetin) contents of selected fruits, vegetables and medicinal plants. *Food Chemistry* 108, 879–884. DOI: 10.1016/j.foodchem.2007.11.053.

Sun, L. and Wang, X. (2003) Effects of allicin on both telomerase activity and apoptosis in gastric cancer SGC-7901 cells. *World Journal of Gastroenterology* 9, 1930–1934. DOI: 10.3748/wjg.v9.i9.1930.

Tessema, B., Mulu, A., Kassu, A. and Yismaw, G. (2006) An *in vitro* assessment of the antibacterial effect of garlic (*Allium sativum*) on bacterial isolates from wound infections. *Ethiopian Medical Journal* 44, 385–389.

Thuy, B.T.P., My, T.T.A., Hai, N.T.T., Hieu, L.T., Hoa, T.T. *et al.* (2020). Correction to investigation into SARS-CoV-2 resistance of compounds in garlic essential oil. *ACS Omega* 5, 16315–16315. DOI: 10.1021/acsomega.0c02641.

Tufan, A., Avanoğlu Güler, A. and Matucci-Cerinic, M. (2020) COVID-19, immune system response, hyperinflammation and repurposing antirheumatic drugs. *Turkish Journal of Medical Sciences* 50, 620–632. DOI: 10.3906/sag-2004-168.

Vahed, H. and Batool Jafri, S. (2016) Propagation of influenza virus in lymphocytes determine by antiviral effects of honey, ginger and garlic decoction. *Journal of Antivirals & Antiretrovirals* 8. DOI: 10.4172/jaa.1000129.

Wang, X., Yang, Y., Zhang, H. and Liu, J. (2017) Safety assessment and comparison of sodium selenite and bioselenium obtained from yeast in mice. *BioMed Research International* 2017. DOI: 10.1155/2017/3980972.

Weber, N.D., Andersen, D.O., North, J.A., Murray, B.K., Lawson, L.D. *et al.* (1992) *In vitro* virucidal effects of *Allium sativum* (garlic) extract and compounds. *Planta Medica* 58, 417–423. DOI: 10.1055/s-2006-961504.

Wilson, E.A. and Demmig-Adams, B. (2007) Antioxidant, anti-inflammatory, and antimicrobial properties of garlic and onions. *Nutrition & Food Science* 37, 178–183. DOI: 10.1108/00346650710749071.

Wu, G., Liu, F., Sun, X., Lin, X., Zhan, F. *et al.* (2019) Preparation of selenium-enriched yeast by re-using discarded *Saccharomyces cerevisiae* from the beer industry for Se-supplemented fodder applications. *Applied Sciences* 9, 3777. DOI: 10.3390/app9183777.

Yadav, S., Trivedi, N.A. and Bhatt, J.D. (2015) Antimicrobial activity of fresh garlic juice: an *in vitro* study. *Ayu* 36, 203–207. DOI: 10.4103/0974-8520.175548.

Yoshitake, Y., Fukuma, D., Yuno, A., Hirayama, M., Nakayama, H. *et al.* (2015) Phase II clinical trial of multiple peptide vaccination for advanced head and neck cancer patients revealed induction of immune responses and improved OS. *Clinical Cancer Research* 21, 312–321. DOI: 10.1158/1078-0432.CCR-14-0202.

Zhang, J., Wang, H., Xiang, Z.D., Shu, S.N. and Fang, F. (2013) Allitridin inhibits human cytomegalovirus replication *in vitro*. *Molecular Medicine Reports* 7, 1343–1349. DOI: 10.3892/mmr.2013.1328.

4 Crop Development – Onion

Brian Thomas*
School of Life Sciences, University of Warwick, Coventry, UK

1. Introduction

This chapter describes the stages of development of the onion (*Allium cepa* L.) and the physiological mechanisms involved, as informed by recent research and publications. To gain a full picture of the development of onions at the crop level, it should be read in conjunction with the chapters on shallots, which share many aspects of development with onions (see Chapter 5) and Agronomy and Production (Chapter 11), which focuses on production aspects of the crop cycle along with the review by Khokhar (2017). The Common Onion group is characterized by forming large single bulbs, although some varieties are grown as salad onions or small bulbs and harvested when still quite young. They are thus distinct from the *Aggregatum* group, which have active lateral buds that develop into clusters of bulbs as in, e.g., shallots. Onions represent a major worldwide crop with global production in 2019 of approximately 100 million tonnes from approx. 5.2 million hectares compared with 4.5 tonnes from 220 hectares for salad onions and shallots. Onions are the third-largest vegetable crop by weight of production, behind potatoes and tomatoes. The largest producers of onions are China and India who, between them, grow over 45% of the world's supply of bulb onions (FAOSTAT, 2019).

Onions are grown as an annual crop, but it is a biennial species, which normally flowers in the second year after the bulb has reached the post-juvenile age and gone through a dormant or quiescent phase. This is critical for seed production (see Chapter 13). The key stages of onion development can be defined as seed germination and establishment, foliar production and photosynthesis, bulb initiation and development, bulb dormancy, floral initiation and development, and seed production. Onions are a global crop, being grown in over 140 countries across the world. They are thus adapted to a wide range of environmental conditions such as temperature, rainfall and daylength. Selection for the different climates and daylengths means that, across the wide range of varieties grown throughout the world, the basic developmental pattern is modified in a variety of ways. However, a common feature is that, under suitable temperatures, bulb formation is responsive to daylength, and onions are routinely classified by the permissive daylength for bulb formation. The main groups are long-day (LD), intermediate day (ID) and short-day (SD). The minimum daylength that allows bulbing is referred to as the

*Address for correspondence: Brian.thomas@warwick.ac.uk

 Edible Alliums: Botany, Production and Uses (eds H.D. Rabinowitch and B. Thomas)
DOI: 10.1079/9781789249996.0004

critical daylength and these values are typically >14–16 h (LD), 12–14 h (ID) and 10–12 h (SD). LD varieties are grown in high latitudes such as northern Europe while SD varieties are grown in low-latitude tropical countries and ID in mid-latitude countries. All onion plants need a period of leaf growth and photosynthetic assimilation to support bulb formation. LD varieties will not form bulbs if grown at low latitudes and, conversely, SD varieties will form bulbs far too early in the season when grown in high latitudes, resulting in small bulbs. Sometimes this is done intentionally to produce small bulbs as sets for reproduction or to be pickled as pearl onions. In such cases, plant density is up to 50–60 times higher than standard production.

2. Seed Germination and Plant Establishment

Onion is a small-seeded crop. Seeds can be planted either in the autumn for winter and spring growth and spring/early summer harvest, or in the spring for mid- or late-summer harvest. Germination involves the extension of the primary root followed by a single cotyledon that emerges as a loop, which unbends to form a hook or whip-like structure. Germination requires sufficient moisture and a temperature range between 5°C and 35°C with an optimum of about 25°C for germination rate and overall percentage (Harrington, 1962; Brewster, 1979). Germination can be inhibited by water or salt stress (Singh and Gopal, 2019) or mechanical impedance of the soil. The establishment of seedlings requires rapid germination and growth under both favourable and unfavourable conditions, a combination sometimes called seed vigour. Genetic variation in seed vigour was demonstrated in an onion diversity set (Taylor *et al.*, 2019), and the authors were able to identify putative markers for the trait. Other approaches are being developed for improving seed germination and establishment. Osmo-priming, or treating seeds with solutions that allow germination to begin but not allow radicle or shoot extension, has long been used to improve the uniformity of germination and establishment. Several variations of this approach have been used, e.g. using gibberellic acid (GA_3) and *Azotobacter* as biofertilizer improved germination in seeds stored at ambient temperature that were losing viability (Brar *et al.*, 2020). Another of these is the use of gold or silver nanoparticles for seed priming (Acharya *et al.*, 2019). Primed onion seeds were reported to show an increase in the emergence and an average yield increase of 23.9%. Beneficial micro-organisms such as *Trichoderma* or *Azotobacter* have also been used to improve seedling performance. Rivera-Mendez *et al.* (2018) showed that culture filtrates of *Trichoderma asperellum*, which give protection against pathogens, also improved the elongation and dry weight of onion seedlings. In an experiment comparing the effects of mineral fertilizers and microbial inoculants on onion growth and yield, it was shown that a combination of *Azotobacter*, *Sphingobacterium* and *Burkholderia* spp improved bulb yield and quality (Tinna *et al.*, 2020). In the presence of these microbial inoculants, mineral fertilizer could be reduced by 25% with no yield penalty.

In high latitudes, the growing season may be too short for the seed-sown crops to reach commercial-sized maturity in the growing season. In such regions, onions may be grown from transplants or sets. Sets are small bulbs, approximately 2–3 g, produced by growing plants at high density, e.g. 1000–2000 plants m^{-2}. To ensure that sets do not bolt after being held over the winter, they are either produced from bolting-resistant varieties, e.g. Sturon, which can be held at 3–5°C, or, if not bolting-resistant, after being stored at high temperature, e.g. 28°C (Brewster, 2008).

3. Growth and Photosynthesis

Once established, onion plants go through a phase of leaf growth and resource accumulation through photosynthesis. The pattern of leaf development and environmental impacts, including the timing of planting, spacing, temperature and rainfall, has been described in detail (e.g. Brewster, 2008). In general, increasing planting density resulted in earlier maturity dates, due in part to the reduction in red/far-red ratios in the crop (Brewster *et al.*, 1986; Mondal *et al.*, 1986a, b). Abou Khadrah *et al.* (2017) found that bulb yield and quality were greatest at a density of 45 plants m^{-2} as compared to

30 or 60 m^{-2}, although the overall yields in this study were low compared with countries such as Israel or The Netherlands (see also Chapter 11). However, in terms of overall profitability, it was reported that the highest densities were recommended, even though the size of the bulbs was reduced (De Menezes and Neto, 2012).

Onions are sensitive to reduced water (Khokhar, 2017), limiting the application of deficit irrigation strategies to conserve water. It was reported, however, that certain plant bioregulators, e.g. KNO_3, thiourea, salicylic acid, or GA_3, could partially offset the reduction in yield caused by limiting irrigation (Wakchaure *et al.*, 2018). However, the major environmental influences on development are intercepted radiation (Brewster *et al.*, 1986) and temperature (Khokhar, 2008, 2017). A decrease in intercepted radiation results in a delay of maturity. This can be caused by natural year-on-year changes in radiation or by increased planting density and mutual shading, particularly in temperate regions where light levels may be limiting (Mondal *et al.*, 1986a). Temperate onions prefer cool conditions and an increase in average temperature during growth leads to a decrease in bulb yield, primarily because of accelerated maturity, and consequently less time for radiation interception (Brewster, 2008). Experiments using a thermogradient tunnel showed that yield reduced by half between the temperatures of 14° and 22°C (Daymond *et al.*, 1997), accompanied by a decrease in time to maturity from 117 to 55 days. In the context of climate change, this might be a concern. However, the same authors showed that the yield increase in response to an increased CO_2 level from 374 to 532 ppm more than compensated for the temperature effect (Daymond *et al.*, 1997).

4. Bulb Initiation and Development

4.1. Juvenility

The distinctive feature of the onion is the bulb, a storage organ formed from expanded leaf sheaths arranged in a series of layers, made of leaf sheaths and young leaf buds, forming a globular structure. Bulb formation does not usually begin immediately after germination, but the plant undergoes a period of vegetative growth to produce the vegetative photosynthetic structures that are needed to fix and metabolize the carbon and nutritional resources for bulb formation and development. This delay in bulb formation parallels the situation in many plants, where flower and seed production are delayed, ensuring that there are sufficient resources for seed filling. The period of delay is normally referred to as the juvenile phase.

The concept of juvenility in plants was originally developed with regard to flowering and sexual reproduction (Taylor *et al.*, 2010; Baldwin *et al.*, 2014; Rashid *et al.*, 2019). It could be argued that, as flowering in onion occurs in the second year, after bulb formation in the first, the minimum leaf number for bulbing may be a manifestation of a minimum leaf number for flowering. Juvenility in flowering is believed to be regulated via the miRNAs 256 and 172 (Poethig, 2009). The study of miRNAs in alliums is in its infancy, with few publications so far and those papers that have been published being very general and not yet aimed at establishing function by experimentation ((Panda *et al.*, 2014; Mishra *et al.*, 2019; Khandagale *et al.*, 2020).

Initiation of bulbing involves the lateral swelling of vacuolated mature cells in the central region at the base of the innermost emergent leaves. For initials that have not yet emerged, cell division and extension slow down and they swell to form the scale leaves of the bulb unit. Consequently, no new foliage leaf blades are produced. A role for gibberellins in bulb formation was proposed based on the observation that s-3307, an inhibitor of gibberellin biosynthesis, caused swelling in the bases of young seedlings accompanied by the reorientation of microtubules from a transverse to a longitudinal direction (Mita and Shibaoka, 1984a). These effects were reversed by gibberellins, which stabilized microtubules in the transverse direction, which would inhibit bulb swelling (Mita and Shibaoka, 1984b). However, measurements on endogenous gibberellins found that levels were higher in LD than SD (Nojiri *et al.*, 1993), suggesting the gibberellins do not have a role in the daylength regulation of bulbing. More recently, Khokhar (2017) reported on several studies that showed beneficial effects of gibberellins or ethylene, as ethrel, on bulb yield and quality, although the mechanisms behind these observations are not clear.

4.2. Response to daylength

In northerly latitudes, bulb initiation and development are dependent on long days. These are called long-day (LD) onions. In lower latitudes, these temperate cultivars do not form bulbs and intermediate-day (ID) or short-day (SD) varieties must be grown. Confusingly, all three are classed as long-day plants (LDP), with regard to their photoperiod mechanism, i.e. they respond when the daylength exceeds a critical value. The critical daylength is longer for the LD varieties than the ID ones, which, in turn, is longer than for SD varieties. Growing onions in photoperiods of less than the critical daylength prevents bulb formation, and leaves continue to develop. Conversely, growing plants in daylengths that greatly exceed the critical daylength results in rapid bulbing onset and decreased leaf photosynthetic capacity, due to sequential leaf senescence, and the formation of small bulbs (Austin, 1972).

There are direct parallels between the promotion of bulbing in LD and the promotion of flowering in LDP. In both cases, the response is promoted in light rich in far-red (FR) wavelengths. Early studies found that day extensions with tungsten filament light promoted bulb formation, whereas fluorescent light was ineffective (Butt, 1968). An action spectrum for bulb induction showed a single peak at 714 nm and no activity below 700 nm (Lercari, 1983). This is typical of a phytochrome high irradiance response (HIR) found in etiolated seedlings. Flowering in the LDP wheat has a similar action spectrum with a peak at 714 nm for the light given as a day extension (Carr-Smith *et al.*, 1989). The HIR in plants is mediated by Phytochrome A (PHYA), which has also been shown to mediate the promotive effect of FR on Arabidopsis flowering (Valverde *et al.*, 2004). In further experiments it was found that giving FR light near the middle of a 24 h period was much more effective than giving it early or late in the day (Lercari and Deitzer, 1987). Plants grown in the absence of FR light, e.g. in fluorescent light, do not bulb in response to LD. The perception of LD in onion, therefore, involves a time-dependent response to FR mediated by PHYA. In Arabidopsis, time-dependent sensitivity to FR/PHYA involved an internal circadian rhythm in CONSTANS (CO) expression mediated by the genes GIGANTEA (GI) and FKF1 and stabilization in LD by FR or blue light (Valverde *et al.*, 2004). Taylor *et al.* (2010) found that the expression of GI and FKF1 in onion showed a diurnal pattern, similar to that found in Arabidopsis, consistent with a common time-measurement mechanism. Subsequently, Rashid and Thomas (2020) identified a member of the onion CO gene family that showed a strong diurnal expression pattern that was consistent with a role in daylength perception. Thus, the main components of the photoperiodic time-measurement system found in Arabidopsis can also be found in onions.

A further parallel between the daylength-dependent responses of flowering and bulbing is the separation of the site of perception in leaves and response in a spatially distant meristematic region (Thomas and Vince-Prue, 1997). Bulbing in onions relies on the perception of LD by young, developing leaves, which are much more sensitive than older leaves to LP cues, even though older plants respond more strongly than young plants (Sobeih and Wright, 1986). This, in turn, indicates that a leaf-generated mobile signal is required to enable bulbing to occur. In flowering plants, this elusive signal was called florigen, which was shown in Arabidopsis to be a phloem mobile protein called FT (Corbesier *et al.*, 2007). In many flowering responses, the effect of permissive daylength is irreversible (Thomas and Vince-Prue, 1997). The bulbing response does show reversibility, although after sufficient LD cycles, onions become fully committed to form rested bulbs (Heath and Holdsworth, 1948). Intercalation of SD between LD prevented the bulbing response, suggesting an antagonistic regulation by LD promotion and SD inhibition (Kato, 1964). One reason why bulbing may be more plastic than flowering is that individual leaves respond autonomously to daylength, depending on their stage of development and bulbing requires several individual leaf initials to be committed to forming scales (Fig. 4.1), whereas flowering requires the reprogramming of a single inflorescence or meristem.

4.3. FT genes

A significant breakthrough in helping to understand how bulbing is regulated came from the discovery of a family of six FT genes in onion (*AcFT1-6*) (Lee *et al.*, 2013). They assigned

Fig. 4.1. Autonomy of onion leaves showing the reversal of a long-day bulbing signal. The pictures show sequential leaves dissected from plants grown in long days (top) or short days (middle) and plants placed in long days and transferred to short days after the initiation of bulbing (bottom). (From the author's laboratory)

putative functions to three of these genes based on their effects in transgenic onions and Arabidopsis, with *AcFT2* being a promoter of flowering, *AcFT1* being a mobile bulbing promoter, and *AcFT4* a mobile inhibitor of bulbing. They were unable to observe any action of *AcFT3*, *5* and *6* in the transgenic plants and concluded that they formed a separate group that were unlikely to be mobile signals. In their experiments, they observed opposite patterns of expression of *AcFT1* and *4*. *AcFT1* expression was inhibited in transgenic onions over-expressing *AcFT4* and it was proposed that *AcFT4* regulates the expression of *AcFT1*. Using RNA-seq, Manoharan *et al.* (2016) confirmed the presence of *AcFT1-6* as previously described by Lee *et al.* (2013) and suggested a further member of the family as *AcFT7*, which may be involved in differentiating between the responses of early- and late-maturing varieties to daylength. Lyngkhoi *et al.* (2019) were also able to confirm the presence of AxFT1-6 in LD and SD onions and were able to detect their presence at different stages of development in plants grown in natural short-days. Ikeda *et al.* (2020) also found that *AcFT1* was expressed across a range of cultivars according to their time to bulb formation and maturity.

In a subsequent paper, it was shown that *AcFT1* and *AcFT4* have diurnal patterns of expression, with *AcFT4* peaking in the morning and *AcFT1* in the afternoon/evening (Rashid and Thomas, 2020). *AcFT1* was expressed in LD but not SD in both Renate, a long-day, and Hojem, a short-day, onions (Rashid and Thomas, 2020). *AcFT4* was expressed only in SD in Renate but in both LD and SD in Hojem. This indicated that the expression of *AcFT4* in Hojem was not sufficient to suppress *AcFT1* expression and additional factors must be involved.

This was supported by experiments that looked at the diurnal expression of *AcFT1* and *AcFT4* in long- and short-day onions over a range of daylengths from 10 h to 14 h (Cheng *et al.*, 2021b). It was found that *AcFT1* expression level increased with daylength in both LD and SD onions. In contrast, *AcFT4* was expressed in all daylengths, suggesting it has a less important role in the daylength-dependent bulbing process (Fig. 4.2). The results indicated that *AcFT1* and *AcFT4* are negatively co-regulated, but *AcFT1* is the predominant regulator of bulb formation in response to daylength (Fig. 4.3). In addition to the formation of scale leaves, bulbing is associated with changes in existing foliar leaves, which stop elongating during bulbing and resume growth after returning to SD (Fig. 4.1). Unlike changes at the meristem, the response occurs in the same organ as perception. It is interesting that *AcFT1* and *AcFT4* were found to be expressed in older leaf tissue, which would be photosynthetically active and capable of exporting a mobile signal, whereas *AcFT5* and *AcFT6* were expressed throughout the leaf, including the basal tissues (Rashid *et al.*, 2019). It was also found that in another LDP, Marco, *AcFT5* was expressed in the leaf in LD whereas *AcFT6* was only expressed in the leaf in SD (Cheng *et al.*, 2021a). It was previously proposed that unlike *AcFT1* and *4*, *AcFT5* and *6* were unable to act as mobile signals in complementing ft-1 mutant in Arabidopsis, suggesting that they may act locally (Lee *et al.*, 2013). This would be consistent with them acting directly in the leaf, regulating its response to daylength.

4.4. Changes associated with bulbing

In an RNA-seq experiment resported in Rashid and Thomas (2019), out of 23,280 sequences, 68.34% of the daylength-sensitive genes were tissue-selective while only 4.29% of the tissue-specific genes were expressed differentially in LD and SD (Table 4.1). Thus, daylength-specific changes are mediated by a relatively small subset of the expressed genes. The transition from foliar leaf growth to the bulb initiation and development phases represents a radical change in morphology and also a corresponding reassignment of resources from leaf blades to sheaths (Lancaster *et al.*, 1986; Brewster, 2008; Mallor and Thomas, 2008). Co-ordination of resource management and bulb formation is supported by the observation that onion photosynthesis and stomatal conductance increased at bulb initiation (Bachie *et al.*, 2019). The resources that are redistributed are carbon, nitrogen and sulphur compounds.

The changing level of sulphur compounds in the form of flavour precursors was followed in leaf blades and scales during bulb formation in field-grown onions (Lancaster *et al.*, 1986). They found that flavour precursors, particularly S-1-propyl-L-cysteine sulphoxide (PCSO), initially accumulated in the foliage leaves and subsequently decreased along with a concomitant increase in the scales as bulbs were formed. This pattern was confirmed by Mallor and Thomas (2008), who also found that if bulbing was inhibited by keeping plants in SD, the flavour

(a)

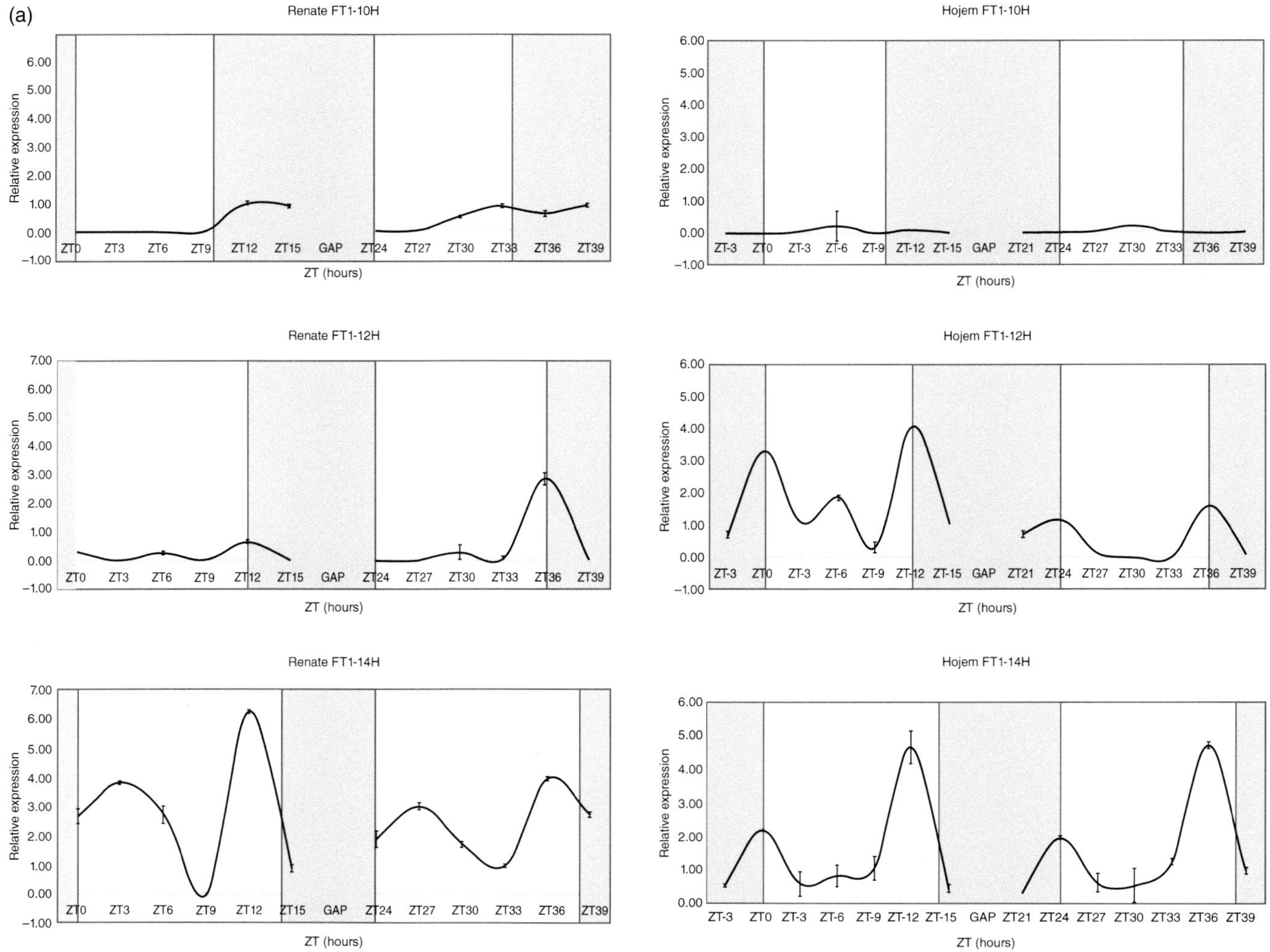

Fig. 4.2a. Expression of *AcFT1* in long-day (Renate) and short-day (Hojem) varieties of onion over a 48 h period in increasing daylengths using qRT-PCR. (After Cheng *et al.*, 2021b)

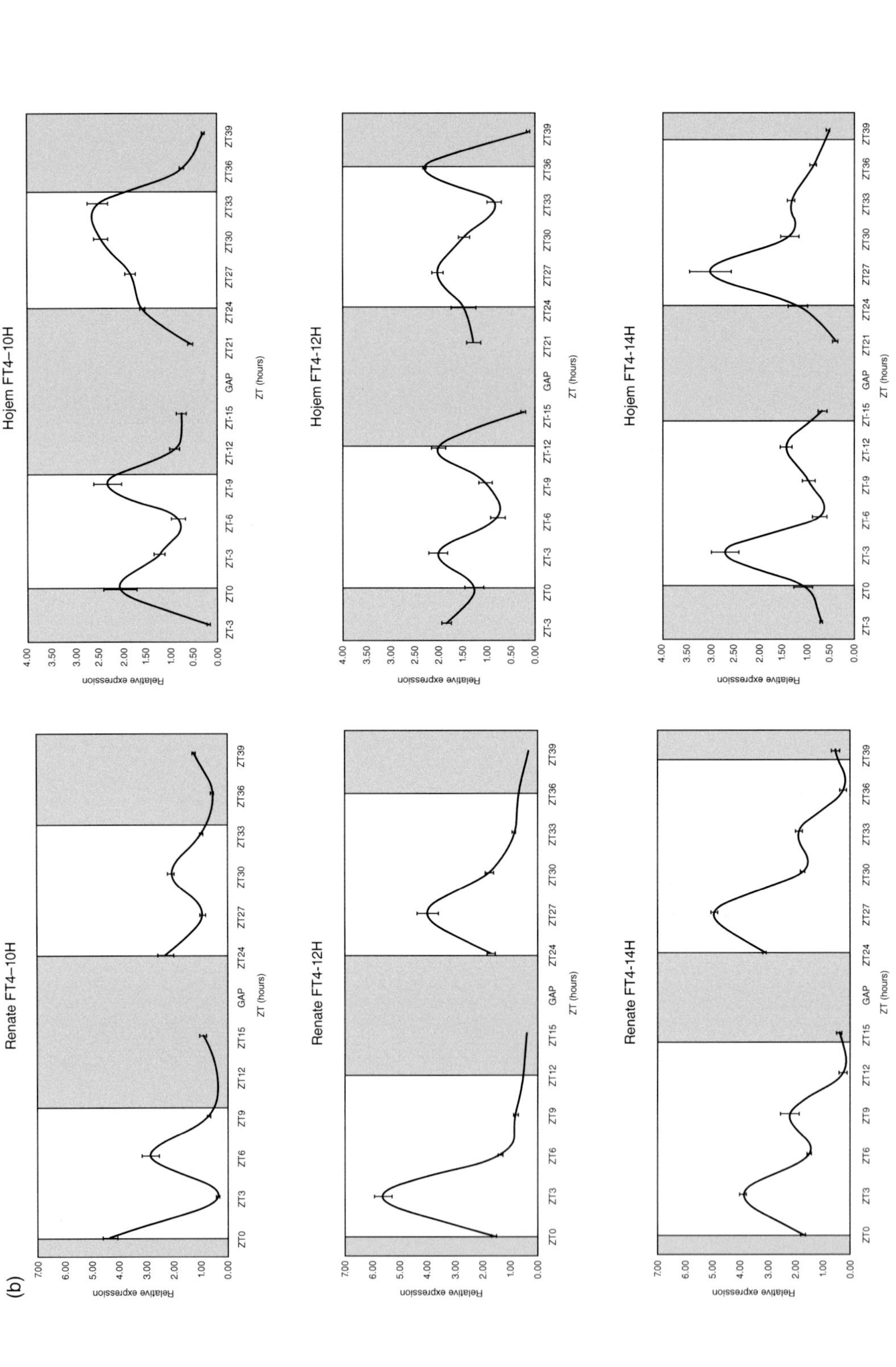

Fig. 4.2b. Expression of *AcFT1* in long-day (Renate) and short-day (Hojem) varieties of onion over a 48 h period in increasing daylengths using qRT-PCR. (After Cheng *et al*., 2021b)

precursors accumulated to high levels in the leaf blades. Both studies suggested that flavour precursors are synthesized in the foliage leaves and subsequently transported to the bulb. The pattern of precursor redistribution (sulphur) followed that of dry-weight (predominantly carbon) changes, whereas nitrogen was retained

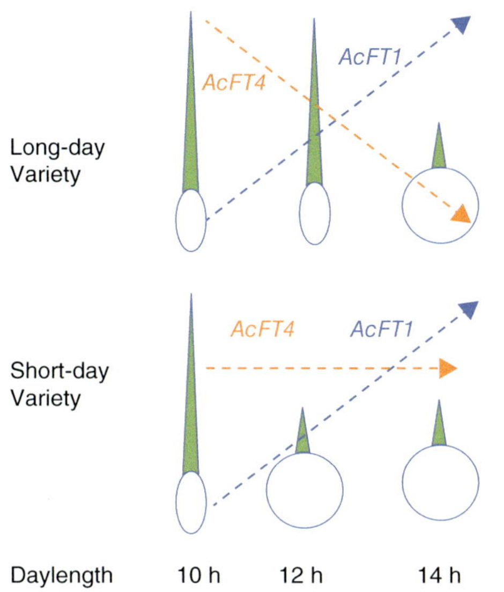

Fig. 4.3. Relative expression of *AcFT1* and *AcFT4* at different daylengths in long-day and short-day onion varieties. (Based on data from Cheng *et al.*, 2021b)

for longer in the foliage leaves and was redistributed as the leaves senesced (Mallor and Thomas, 2008), indicating the requirement to maintain enzymes for photosynthesis and other metabolic activity until leaf senescence.

The predominant form of carbon storage, i.e. non-structural carbohydrates, in onions is fructan oligosaccharides (Darbyshire and Henry, 1978). Fructan concentration is higher in onions bred for dehydration, but this is predominantly because the water content of the bulbs is lower (Darbyshire and Henry, 1979). Fructans are present in both the leaves and bulb scales. There is no evidence that fructans are themselves mobile in the plant, and carbohydrate reallocation from leaves to bulb tissues during bulb formation requires fructan breakdown and resynthesis. Fructans are synthesized as shown in the schematic below. Fructosyl residues, which may be derived from sucrose through the action of invertase, are added to sucrose to initially form the trisaccharide kestose through the action of sucrose: sucrose 1-fructosyltransferase (1-SST). Further fructosyl residues are added through the actions of 1-FFT, fructan: fructan 1-fructosyltransferase (1-FFT), and fructan: fructan 6G-fructosyltransferase (6G-FFT), which produces branched chains (Fig. 4.4). Fructan levels are high in leaf blades and leaf sheaths prior to bulbing. Upon bulb initiation, the levels of fructans decrease in these tissues, and they subsequently accumulate in sheath bases and

Table 4.1. Comparison of numbers of the genes with tissue specificity or daylength sensitivity as found by RNA-seq (Rashid and Thomas, 2019). Tissue-selective genes are those which show different expression between leaf and bulb. Daylength-responsive genes are those that show differential expression in LD and SD. (From Cheng *et al.*, 2021a)

	Tissue-selective genes		
	Leaf: LD-enhanced	Leaf: SD-enhanced	Leaf: daylength-insensitive
Bulb LD-enhanced	8	3	315
Bulb SD-enhanced	3	6	334
Bulb-daylength insensitive	181	148	**22,282**
	Daylength-responsive genes		
	SD-bulb	SD-leaf	SD-tissue-selective
LD-bulb	2375	17	1666
LD-leaf	29	5002	2136
LD-tissue-selective	2422	2262	**7371**
Total: 23,280			

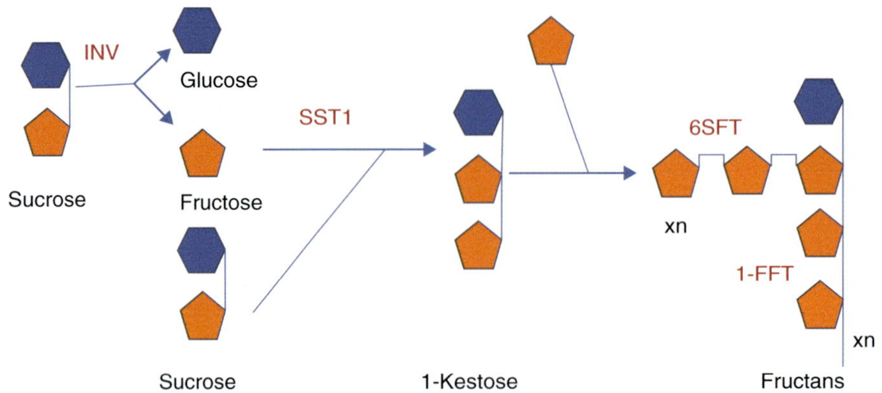

Fig. 4.4. Enzyme steps involved in the synthesis of fructans in onion. INV = invertase; SST1 (1-SST) = 1-fructosyltransferase; 1-FFT = fructan: fructan 1-fructosyltransferase; 6-SFT (6G-FFT) = fructan 6G-fructosyltransferase. Further fructosyl residues are added through the actions of 1-FFT, fructan: fructan 1-fructosyltransferase (1-FFT) and fructan: fructan 6G-fructosyltransferase (6G-FFT).

bulb tissues but not in aerial leaves (Oku *et al.*, 2019). Activities of the 1-SST, 1-FFT and 6G-FFT are initially high in leaf bases during development and subsequently continue to be present at a lower but relatively steady level. In their study with two varieties that accumulated fructans to a different extent, the low-fructan variety had a higher level of invertase and they speculated that this prevented the tissues from maintaining a high-enough level of sucrose to support high-fructan production (Oku *et al.*, 2019). Cheng *et al.* (2021a) compared the expression of different members of the 1-SST gene family in leaves and bulbs under LD and SD conditions. Two of the genes designated 1-SST B1 and 1-SST B2 were expressed at higher levels in leaf-base tissues than in leaf-blade tissue. In contrast, 1-SST L was expressed only in leaf blades. The management of carbohydrate resources during bulb formation, therefore, requires the tissue-specific deployment of members of the SST gene family.

5. Dormancy and Storage

5.1. Dormancy and sprouting

At the end of the bulb development and growth phase, plants are harvested after the tops have gone over and the bulbs are then dried. This can be either in the field in dry climates, or in temperate regions that are wetter, bulbs are mechanically harvested and dried with artificial ventilation and heating. During this drying process, referred to as curing, the neck of the bulb dries out and seals, which prevents pathogen entry and helps with skin development (Brewster, 2008). The outer layers form a dry skin and within the fleshy storage leaves (aka true scales), leaf initials are formed. They do not elongate immediately but remain quiescent during a period of dormancy. Dormancy is sometimes referred to as being in two phases: endo-dormancy, where cell division and growth of leaf initials cannot take place, and eco-dormancy, where leaf growth can occur, but environmental conditions determine when and how quickly (Chope *et al.*, 2012b). Carter *et al.* (1999) looked at the expression of Histone 2A, which would be linked to cell division, as a marker for the end of dormancy. Expression of Histone 2A tended to occur at the same time irrespective of when sprouting began. They concluded that differences between varieties were not due to inherent differences in dormancy but in the inhibition of subsequent sprouting. Chope *et al.* (2012a) looked at transcriptional changes during the period from end-dormancy to eco-dormancy. Significant changes took place in the relatively short period after curing and there were greater changes before sprouting than after, again suggesting a distinction between true end-dormancy and post-dormancy sprout suppression.

The relationship between storage temperature and time to sprouting is not straightforward.

In general, low (e.g. <5°C) or high (e.g. 30°C) temperatures inhibit sprouting and extend storage life. This could be viewed as the consequence of the optimum temperature for vernalization, which was shown to be between 8°C and 12°C with minima and maxima at 3°C and 15°C (Brewster, 1987). Modified-atmosphere treatments have been used in an attempt to extend storage. Chavez-Mendoza *et al.* (2016) evaluated the storage of onion in controlled atmosphere regimes (CA) at 1% O_2 and 1% or 5% CO_2, at 2.5°C. Control treated onions sprouted at 20% after three months and up to 100% at the end of storage at seven months. The CA treatments showed no sprouting and dry matter remained constant over a six-month period. Cho *et al.* (2010) reported that forced air pre-drying for 15–20 days is an effective procedure, allowing bulbs to be stored at 0°C for up to six months. Also, browning, due to oxidation of phenolics by polyphenol oxidase, can be controlled by applying citric acid or a controlled atmosphere with nitrogen gas (Cho *et al.*, 2010).

Pak *et al.* (1995) looked at dry storage at 16°C. Sprouts were initiated after two weeks from harvest indicating that real dormancy was minimal at this intermediate temperature. Fructan content of the scales decreased with an associated increase in fructose. Sprouts were well-supplied with carbohydrates to support growth, including 30% of the carbohydrate as fructans. Sucrose synthase also increased as the sprouts grew in a linear way. Yasin and Bufler (2007) also looked at carbohydrate metabolism during storage. Sucrose synthase activity increased at the time of sprouting, again consistent with a separation of dormancy and sprouting. Benkeblia *et al.* (2005) studied variation in fructans during long-term storage. They found that both low- and high-polymerization fructans decreased during the storage period. Enzymes that hydrolyse fructans and kestose peaked after about 16 weeks in storage and were suggested as potential markers for the end of dormancy. Ohanenye *et al.* (2019) monitored fructan distribution in bulbs during storage. They found that fructans with DP 7–8 were initially distributed in the top and bottom sections of the storage leaf scales but prior to sprouting they accumulated in the baseplate, indicating that mobilization of the fructans may be a marker for the onset of sprouting.

5.2. Growth substances

Ethylene has been found to be an effective treatment for inhibiting sprout growth and hence extending storage life. Bufler (2009) found that ethylene inhibited sprout growth in dormant bulbs but also in bulbs that were already sprouting. Ethylene did not extend the period of dormancy but the application of MCP was able to break dormancy and give earlier sprouting. A similar conclusion had been reached by Chope *et al.* (2007) who found that the MCP had no effect on the levels of ABA in the bulbs. Alamar *et al.* (2020) used a transcriptomic approach to look at the expression of genes related to ethylene and other growth substances during storage. They found that ABA increased under ethylene during a long dormancy period but cytokinin increases were independent of ethylene. Ethylene modified the expression of genes in the ethylene, ABA, and cytokinin synthesis pathways. ABA metabolism has also been linked to dormancy and storage. Onions were stored in controlled atmospheres and the level of ABA followed in long-, medium-, and short-storage varieties. In all cases, the ABA levels decreased exponentially, being initially highest in the long-storing varieties. The authors suggested that sprouting took place when the level of ABA was reduced to a critical threshold (Chope *et al.*, 2006). Murkute (2015), however, analysed ABA levels in 22 varieties of SD onions and found two varieties with low ABA levels that had not sprouted, even after five months, indicating that ABA levels alone could not account for storage time.

6. Flowering

6.1. Environmental effects

Onions are biennial crops, normally flowering in the second year after overwintering as a bulb, having received a sufficient vernalization treatment, or cold exposure in the autumn-planted fields. Flowering can be induced by low temperature in some spring-sown varieties. The prerequisite conditions to be met for flowering are, firstly, sufficient numbers of leaves produced to move beyond the juvenile phase. This could be prior to bulb formation, or in seedlings in spring-sown crops. Secondly, the bulb or seedling size needs to be sufficient, although this can vary for different varieties and crop types. Thirdly, the plants may require to be exposed to an appropriate temperature for vernalization.

D'angelo and Goldman (2018) conducted experiments over four years to investigate the effects of vernalization on flowering in three onion varieties. They found an endodormancy period of 8–10 weeks and the optimum for scape formation was 10°C for 14 weeks. After ten weeks of chilling, each additional week decreased the time to flower by about 4%. For one of the varieties, CUDH2107, in one of two experiments, many of the bulbs proceeded to flower without storage vernalization, suggesting that rarely flowering may be facultative. They concluded that for the tested genotypes, vernalization is not the only path to flowering, but its impact on flowering time is substantial. This situation is similar to that seen in some varieties of spring-sown wheat, which have no vernalization requirement for flowering (Jedel *et al.*, 1986). Models have been developed to predict bolting rates of overwintered varieties and suggest that such models can be based on air temperature (Ikeda *et al.*, 2019).

Brewster (1982) looked at the relationship between mean temperatures and daylengths on the rates of inflorescence development in Senshyu Semi-globe Yellow. The time from inflorescence appearance and seed ripening was largely accounted for by mean temperature, with about 900 day-degrees C required for progression from inflorescence appearance and seed ripening.

In autumn-sown fields, bolting-resistant bulbs require overwintering before flowering; this places a restriction on the time to flowering and hence the speed at which breeding programmes can take place (Brewster and Butler, 1989). Spring-sown cultivars can, with appropriate management, produce flowers in the first year after summer plantings. In this paper, the authors looked at environmental factors that could improve flowering in such material. They found that there were considerable differences between varieties in their responses but overall daylength extensions, supplementary lighting, low nitrogen and/or gibberellins could enhance the flowering response. Naamni *et al.* (1980) and Mushtaq *et al.* (2018) also reported that foliar application of gibberellins to Grano and to locally available cultivars Pulkara and Dark Red, respectively, could increase seed yield and also the quality as represented by seedling vigour.

6.2. Genetic regulation

In their analysis of FT genes in onion, the authors proposed that the *AcFT2* gene was responsible for flowering (Lee *et al.*, 2013). This was based on the observation that *AcFT2* expression was the only FT gene that increased upon vernalization of the bulbs when that treatment promoted flowering. It was not expressed in Renate, an LD variety during development in either LD or SD but was expressed very late in Hojem, an SD variety in short-days (Rashid *et al.*, 2019). Also, unlike *ACFT1* and *ACFT4*, which mediate the bulbing response to daylength, *AcFT2* had no obvious diurnal expression rhythm. Interestingly, Lee *et al.* (2013) found that *AcFT2* had only a small effect on promoting flowering in Arabidopsis. The small effect was only seen using the CaMV35s promoter but it had no effect under the SUC2 promoter, which would be expected for a mobile signal. This contrasted with *AcFT1*, which was able to substitute for Arabidopsis FT in the *ft-1* mutant under the SUC2 promoter to promote flowering. This is consistent with the vernalization signal being detected at the apex, as is usually the case, rather than in the leaves, which would require a mobile signal.

Genetic analysis of bolting was carried out to identify QTL and a low-density linkage map (Baldwin *et al.*, 2014). Across seven open-pollinated populations grown in LD and SD, strong differentiation was shown at the chromosomal regions around *AcFT2* and AcSOC1. However, no QTLs for bolting were found in the region of *AcFT2* on chromosome 5, although a QTL found on chromosome 1, designated AcBlt1, was consistently linked to bolting susceptibility in the mapping population.

6.3. Flower and seed development

The flowering process is favoured by higher temperatures. It takes about 400 day-degree days

>10°C from spathe appearance to peak flowering and 900 day degrees >10°C to the shedding of ripe seeds from capsules (Brewster, 1982). Occasionally, small bulbs or topsets form instead of flowers in the inflorescence, which may occur in response to scarring the spathe. This can be a consequence of high-temperature bulb storage, which diverts development from reproductive to vegetative through a devernalization process (Brewster, 2008).

Scapes are commonly 1–2 m in length and lateralized mother-bulbs can produce several inflorescences. The scape carries the flower bud or umbel, surrounded by a spathe. On anthesis, the spathe breaks and consequently the umbel with 200–600 flowers becomes exposed. The flower structure supports insect pollination, comprised of a whorl of petals, which opens and exposes the nectaries that secrete nectar, two whorls of anthers that dehisce sequentially, followed by style elongation and the receptive, sticky, stigma. Pollination results in the fertilization of the ovules and subsequent seed development. Maximum seed weight takes about 45 days after fertilization and the potential for full germination takes up to 80 days (Gray and Ward, 1987; Chapter 12, this volume).

7. Conclusions

In conclusion, this chapter describes the key stages in onion development from seed germination through growth, bulb formation, dormancy, storage and flowering. Our understanding of how these stages are regulated at the physiological and genetic level has made progress over recent years, but there are still many questions to be asked.

Onions are one of the major global crops, grown throughout the world and key components of the diets of billions of people. They are also fascinating subjects for biological study, demonstrating morphologically distinct phases of growth and exhibiting classic physiological processes such as dormancy, photoperiodism and vernalization with a biennial growth habit. With this in mind, it is surprising how little is known in detail about the genetic and physiological mechanisms regulating the progression through its key physiological stages. This is in contrast to some crops such as rice, wheat, maize, potatoes, tomatoes and others, where the physiological and genetic characteristics are extensively studied. Contrast also with Arabidopsis, a small weed that has no commercial or nutritive value whatsoever but there is extensive knowledge of its functional genetics.

A list of areas, about which we know very little, include the genetic factors that determine whether a leaf initial becomes a photosynthetic foliage leaf or a fleshy storage organ. The conversion of leaves into different organs during development is an example of homeosis. Much is known, for example, of how modified leaves form petals, anthers and stigmas in flowers. These generally involve the action of homeotic genes, particularly of the MADS box family. Although bulb scales and flower parts appear very different, the ability of inflorescences to form topsets suggests that the regulatory gene differences may be small. However, almost nothing is known about the regulatory genes that establish organ identity in onions.

A second area where little is known about a potentially important regulatory system concerns the role of microRNAs in plant maturity, determining, for example, the minimum leaf numbers for bulb initiation and/or flowering. Some labs are beginning to look at this but the work is in its infancy.

A third area concerns the response to daylength, where the detail of how photoreceptors, endogenous clocks and mobile signals link to form a robust but flexible mechanism is only just beginning to emerge. Assuming *FT* genes and mobile FT proteins are involved, experimental proof that GI/CO regulate *FTs* has not been forthcoming and neither are there any alternative working hypotheses. At the site of response, the reaction partner for FT in the target tissue (equivalent of *FD*) has not been identified. Similarly, what is the molecular mechanism involved in the temperature requirement for emergence from dormancy and vernalization requirement for flowering? Are there genes equivalent to Arabidopsis *Flc* and *Vrn* involved? One thing is for sure: there is no shortage of questions for molecular physiologists to address in future research.

References

Abou Khadrah, S.A., El-Sayed, A.A., Geries, L.S.M. and Abdelmasieh, W.K.L. (2017) Response of onion yield and quality to different planting date, methods and density. *Egyptian Journal of Agronomy* 39, 203–219. DOI: 10.21608/agro.2017.1203.1065.

Acharya, P., Jayaprakasha, G.K., Crosby, K.M., Jifon, J.L. and Patil, B.S. (2019) Green-synthesized nanoparticles enhanced seedling growth, yield, and quality of onion (*Allium cepa* L.). *ACS Sustainable Chemistry & Engineering* 7, 14580–14590. DOI: 10.1021/acssuschemeng.9b02180.

Alamar, M.C., Anastasiadi, M., Lopez-Cobollo, R., Bennett, M.H., Thompson, A.J. *et al.* (2020) Transcriptome and phytohormone changes associated with ethylene-induced onion bulb dormancy. *Postharvest Biology and Technology* 168. DOI: 10.1016/j.postharvbio.2020.111267.

Austin, R.B. (1972) Bulb formation in onions as affected by photoperiod and spectral quality of light. *Journal of Horticultural Science* 47, 493–504.

Bachie, O.G., Santiago, L.S. and Mcgiffen, M.E. (2019) Physiological responses of onion varieties to varying photoperiod and temperature regimes. *Agriculture-Basel* 9. DOI: 10.3390/agriculture9100214.

Baldwin, S., Revanna, R., Pither-Joyce, M., Shaw, M., Wright, K. *et al.* (2014) Genetic analyses of bolting in bulb onion (*Allium cepa* L.). *Theoretical and Applied Genetics* 127, 535–547. DOI: 10.1007/s00122-013-2232-4.

Benkeblia, N., Ueno, K., Onodera, S. and Shiomi, N. (2005) Variation of fructooligosaccharides and their metabolizing enzymes in onion bulb (*Allium cepa* L. cv. Tenshin) during long-term storage. *Journal of Food Science* 70, S208–S214.

Brar, N.S., Kaushik, P. and Dudi, B.S. (2020) Effect of seed priming treatment on the physiological quality of naturally aged onion (*Allium cepa* L.) seeds. *Applied Ecology and Environmental Research* 18, 849–862. DOI: 10.15666/aeer/1801_849862.

Brewster, J.L. (1979) Response of growth-rate to temperature in seedlings of several allium crop species. *Annals of Applied Biology* 93, 351–357. DOI: 10.1111/j.1744-7348.1979.tb06551.x.

Brewster, J.L. (1982) Flowering and seed production in overwintered cultivars of bulb onions. 2. Quantitative relationships between mean temperatures and daylengths and the rate of inflorescence development. *Journal of Horticultural Sciences* 57, 103–108. DOI: 10.1080/00221589.1982.11515029.

Brewster, J.L. (1987) The effect of temperature on the rate of sprout growth and development within stored onion bulbs. *Annals of Applied Biology* 111, 463–467. DOI: 10.1111/j.1744-7348.1987.tb01475.x.

Brewster, J.L. (2008) *Onions and Other Vegetable Alliums*. CAB International, Wallingford, UK.

Brewster, J.L. and Butler, H.A. (1989) Inducing flowering in growing plants of overwintered onions: effects of supplementary irradiation, photoperiod, nitrogen, growing medium and gibberellins. *Journal of Horticultural Sciences* 64, 301–312. DOI: 10.1080/14620316.1989.11515958.

Brewster, J.L., Mondal, F.M. and Morris, G.E.L. (1986) Bulb development in onion (*Allium cepa* L.). 4. Influence on yield of radiation interception, its efficiency of conversion, the duration of growth and dry-matter partitioning. *Annals of Botany* 58, 221–233. DOI: 10.1093/oxfordjournals.aob.a087200.

Bufler, G. (2009) Exogenous ethylene inhibits sprout growth in onion bulbs. *Annals of Botany* 103, 23–28. DOI: 10.1093/aob/mcn203.

Butt, A.M. (1968) Vegetative growth, morphogenesis and carbohydrate content of the onion plant as a function of light and temperature under field and controlled conditions. PhD thesis, University of Wageningen, The Netherlands. Available at: https://research.wur.nl/en/publications/vegetative-growth-morphogenesis-and-carbohydrate-content-of-the-o (accessed 31 July 2022).

Carr-Smith, H.D., Johnson, C.B. and Thomas, B. (1989) Action spectrum for the effect of day-extensions on flowering and apex elongation in green, light-grown wheat (*Triticum aestivum* L). *Planta* 179, 428–432. DOI: 10.1007/bf00397581.

Carter, C.E., Partis, M.D. and Thomas, B. (1999) The expression of histone 2A in onion (*Allium cepa*) during the onset of dormancy, storage and emergence from dormancy. *New Phytologist* 143, 461–470. DOI: 10.1046/j.1469-8137.1999.00484.x.

Chavez-Mendoza, C., Vega-Garcia, M.O., Guevara-Aguilar, A., Sanchez, E., Alvarado-Gonzalez, M. *et al.* (2016) Effect of prolonged storage in controlled atmospheres on the conservation of the onion (*Allium cepa* L.) quality. *Emirates Journal of Food and Agriculture* 28, 842–852. DOI: 10.9755/ejfa.2016-06-643.

Cheng, W., Rashid, H.A., Stark, R. and Thomas, B. (2021a) The role of organ- and daylength-specific gene expression in bulb development and resource management in onion (*Allium cepa* L.). *Scientia Horticulturae* 286. DOI: 10.1016/j.scienta.2021.110223.

Cheng, W., Rashid, M.H.A. and Thomas, B. (2021b) Changes in the expression of photoperiodic bulbing genes in response to increasing daylength in long-day and short-day onion varieties. *The Journal of Horticultural Science and Biotechnology* 96, 718–727. DOI: 10.1080/14620316.2021.1928556.

Cho, J., Bae, R.N. and Lee, S.K. (2010) Current research status of postharvest technology of onion (*Allium cepa* L.). *Korean Journal of Horticultural Science & Technology* 28, 522–527.

Chope, G.A., Terry, L.A. and White, P.J. (2006) Effect of controlled atmosphere storage on abscisic acid concentration and other biochemical attributes of onion bulbs. *Postharvest Biology and Technology* 39, 233–242. DOI: 10.1016/j.postharvbio.2005.10.010.

Chope, G.A., Terry, L.A. and White, P.J. (2007) The effect of 1-methylcyclopropene (1-MCP) on the physical and biochemical characteristics of onion cv. SS1 bulbs during storage. *Postharvest Biology and Technology* 44, 131–140. DOI: 10.1016/j.postharvbio.2006.11.012.

Chope, G.A., Cools, K., Hammond, J.P., Thompson, A.J. and Terry, L.A. (2012a) Physiological, biochemical and transcriptional analysis of onion bulbs during storage. *Annals of Botany* 109, 819–831. DOI: 10.1093/aob/mcr318.

Chope, G.A., Cools, K., Terry, L.A., Hammond, J.P. and Thompson, A.J. (2012b) Association of gene expression data with dormancy and sprout suppression in onion bulbs using a newly developed onion microarray. *Acta Horticulturae* 969, 169–174.

Corbesier, L., Vincent, C., Jang, S., Fornara, F., Fan, Q. *et al.* (2007) FT protein movement contributes to long-distance signaling in floral induction of Arabidopsis. *Science* 316, 1030–1033. DOI: 10.1126/science.1141752.

D'angelo, C.J. and Goldman, I.L. (2018) Temporal aspects of vernalization and flowering in long-day storage onion. *Journal of the American Society for Horticultural Science* 143, 446–453. DOI: 10.21273/jashs04495-18.

Darbyshire, B. and Henry, R.J. (1978) Distribution of fructans in onions. *New Phytologist* 81, 29. DOI: 10.1111/j.1469-8137.1978.tb01600.x.

Darbyshire, B. and Henry, R.J. (1979) Association of fructans with high percentage dry-weight in onion cultivars suitable for dehydrating. *Journal of the Science of Food and Agriculture* 30, 1035–1038. DOI: 10.1002/jsfa.2740301103.

Daymond, A., Wheeler, T., Hadley, P., Ellis, R. and Morison, J. (1997) The growth, development and yield of onion (*Allium cepa* L.) in response to temperature and CO_2. *Journal of Horticultural Sciences* 72, 135–145.

De Menezes, F.O.G. and Neto, J.V. (2012) Onion production depending on planting density. *Horticultura Brasileira* 30, 733–739. DOI: 10.1590/s0102-05362012000400028.

FAOSTAT (2019) http://www.fao.org/faostat/en/#data/QC (accessed 5 July 2022).

Gray, D. and Ward, J.A. (1987) A comparison of leek (*Allium porrum*) and onion (*Allium cepa*) seed development. *Annals of Botany* 60, 181–187. DOI: 10.1093/oxfordjournals.aob.a087435.

Harrington, J.F. (1962) The effect of temperature on the germination of several kinds of vegetable seeds. *Proceedings of the XVI International Horticultural Congress* Vol. II., pp. 435–441.

Heath, O.V.S. and Holdsworth, M. (1948) Morphogenetic factors as exemplified by the onion plant growth. In: Daneellie, J.F. and Brown, R. (eds) *Symposia of the Society of Experimental Biology*, Vol. 2. Cambridge University Press, Cambridge, UK, pp. 326–350.

Ikeda, H., Kinoshita, T., Yamamoto, T. and Yamasaki, A. (2019) Sowing time and temperature influence bulb development in spring-sown onion (*Allium cepa* L.). *Scientia Horticulturae* 244, 242–248. DOI: 10.1016/j.scienta.2018.09.050.

Ikeda, H., Yamamoto, T., Kinoshita, T., Tsukazaki, H. and Yamasaki, A. (2020) A comparative study on the growth and bulb development of several onion (*Allium cepa* L.) cultivars sown in spring in the Northeast Region of Japan. *The Horticulture Journal* 89, 586–592. DOI: 10.2503/hortj.UTD-200.

Jedel, L.E., Evans, L.E. and Scarth, R. (1986) Vernalization responses of a selected group of spring wheat (*Triticum aestivum* L.). *Canadian Journal of Plant Science* 66, 1–9. DOI: 10.4141/cjps86-001.

Kato, T. (1964) Physiological studies on the bulbing and dormancy of onion plants. III. Effects of external factors on the bulb formation and development. *Journal of the Japanese Society of Horticultural Science* 33, 53–61.

Khandagale, K., Krishna, R., Roylawar, P., Ade, A.B., Benke, A. *et al.* (2020) Omics approaches in Allium research: progress and way ahead. *PeerJ* 8. DOI: 10.7717/peerj.9824.

Khokhar, K.M. (2008) Effect of temperature and photoperiod on the incidence of bulbing and bolting in seedlings of onion cultivars of diverse origin. *The Journal of Horticultural Science and Biotechnology* 83, 488–496. DOI: 10.1080/14620316.2008.11512412.

Khokhar, K.M. (2017) Environmental and genotypic effects on bulb development in onion – a review. *The Journal of Horticultural Science and Biotechnology* 92, 448–454. DOI: 10.1080/14620316.2017.1314199.

Lancaster, J.E., McCallion, B.J. and Shaw, M.L. (1986) The dynamics of the flavor precursors, the s-alk(en)yl-l-cysteine sulfoxides, during leaf blade and scale development in the onion (*Allium cepa*). *Physiologia Plantarum* 66, 293–297. DOI: 10.1111/j.1399-3054.1986.tb02423.x.

Lee, R., Baldwin, S., Kenel, F., McCallum, J. and MacKnight, R. (2013) FLOWERING LOCUS T genes control onion bulb formation and flowering. *Nature Communications* 4. DOI: 10.1038/ncomms3884.

Lercari, B. (1983) Action spectrum for the photoperiodic induction of bulb formation in *Allium cepa* L. *Photochemistry and Photobiology* 38, 219–222. DOI: 10.1111/j.1751-1097.1983.tb03865.x.

Lercari, B. and Deitzer, G.F. (1987) Time-dependent effectiveness of far-red light on the photoperiodic induction of bulb formation in *Allium cepa* L. *Photochemistry and Photobiology* 45, 831–835. DOI: 10.1111/j.1751-1097.1987.tb07891.x.

Lyngkhoi, F., Khar, A., Mangal, M., Gaikwad, A.B. and Thirunavukkarasu, N. (2019) Expression analysis and association of bulbing to FLOWERING LOCUS T (FT) gene in short day onion (*Allium cepa* L.). *Indian Journal of Genetics and Plant Breeding* 79, 77–81. DOI: 10.31742/ijgpb.79.1.10.

Mallor, C. and Thomas, B. (2008) Resource allocation and the origin of flavour precursors in onion bullbs. *The Journal of Horticultural Science and Biotechnology* 83, 191–198. DOI: 10.1080/14620316.2008.11512369.

Manoharan, R.K., Han, J.S.H., Vijayakumar, H., Subramani, B., Thamilarasan, S.K. *et al.* (2016) Molecular and functional characterization of FLOWERING LOCUS T homologs in *Allium cepa*. *Molecules* 21. DOI: 10.3390/molecules21020217.

Mishra, R., Mohapatra, R., Mahanty, B. and Joshi, R.K. (2019) Analysis of microRNAs and their targets from onion (*Allium cepa*) using genome survey sequences (GSS) and expressed sequence tags (ESTs). *Bioinformation* 15, 907–916. DOI: 10.6026/97320630015907.

Mita, T. and Shibaoka, H. (1984a) Effects of s-3307, an inhibitor of gibberellin biosynthesis, on swelling of leaf sheath-cells and on the arrangement of cortical microtubules in onion seedlings. *Plant & Cell Physiology* 25, 1531–1539. DOI: 10.1093/oxfordjournals.pcp.a076866.

Mita, T. and Shibaoka, H. (1984b) Gibberellin stabilizes microtubules in onion leaf sheath-cells. *Protoplasma* 119, 100–109. DOI: 10.1007/bf01287822.

Mondal, M.F., Brewster, J.L., Morris, G.E.L. and Butler, H.A. (1986a) Bulb development in onion (*Allium cepa* L.). 1. Effects of plant-density and sowing date in field conditions. *Annals of Botany* 58, 187–195. DOI: 10.1093/oxfordjournals.aob.a087197.

Mondal, M.F., Brewster, J.L., Morris, G.E.L. and Butler, H.A. (1986b) Bulb development in onion (*Allium cepa* L.). 2. The influence of red far-red spectral ratio and of photon flux-density. *Annals of Botany* 58, 197–206. DOI: 10.1093/oxfordjournals.aob.a087198.

Murkute, A.A. (2015) Abscisic acid treatment for delaying bulb sprouting in short day onion. *Indian Journal of Horticulture* 72, 310–312. DOI: 10.5958/0974-0112.2015.00060.2.

Mushtaq, S., Amjad, M., Ziaf, K. and Afzal, I. (2018) Gibberellins application timing modulates growth, physiology, and quality characteristics of two onion (*Allium cepa* L.) cultivars. *Environmental Science and Pollution Research* 25, 25155–25161. DOI: 10.1007/s11356-018-2542-9.

Naamni, F., Rabinowitch, H.D. and Kedar, N. (1980) The effect of GA_3 application on flowering and seed production in onion (*Allium cepa* L.). *Journal of the American Society for Horticultural Science* 105, 164–167.

Nojiri, H., Toyomasu, T., Yamane, H., Shibaoka, H. and Murofushi, N. (1993) Qualitative and quantitative-analysis of endogenous gibberellins in onion plants and their effects on bulb development. *Bioscience, Biotechnology, and Biochemistry* 57, 2031–2035. DOI: 10.1271/bbb.57.2031.

Ohanenye, I.C., Alamar, M.C., Thompson, A.J. and Terry, L.A. (2019) Fructans redistribution prior to sprouting in stored onion bulbs is a potential marker for dormancy break. *Postharvest Biology and Technology* 149, 221–234. DOI: 10.1016/j.postharvbio.2018.12.002.

Oku, S., Ueno, K., Tsuruta, Y., Jitsuyama, Y., Suzuki, T. *et al.* (2019) Sugar accumulation and activities of enzymes involved in fructan dynamics from seedling to bulb formation in onion (*Allium cepa* L.). *Scientia Horticulturae* 247, 147–155. DOI: 10.1016/j.scienta.2018.12.013.

Pak, C., Vanderplas, L.H.W. and Deboer, A.D. (1995) Importance of dormancy and sink strength in sprouting of onions (*Allium cepa*) during storage. *Physiologia Plantarum* 94, 277–283.

Panda, D., Dehury, B., Sahu, J., Barooah, M., Sen, P. *et al.* (2014) Computational identification and characterization of conserved miRNAs and their target genes in garlic (*Allium sativum* L.) expressed sequence tags. *Gene* 537, 333–342. DOI: 10.1016/j.gene.2014.01.010.

Poethig, R. (2009) Small RNAs and developmental timing in plants. *Current Opinion in Genetics & Development* 19, 374–378. DOI: 10.1016/j.gde.2009.06.001.

Rashid, M.H.A. and Thomas, B. (2020) Diurnal expression of Arabidopsis gene homologs during daylength-regulated bulb formation in onion (*Allium cepa* L.). *Scientia Horticulturae* 261. DOI: 10.1016/j.scienta.2019.108946.

Rashid, M.H.A., Cheng, W. and Thomas, B. (2019) Temporal and spatial expression of arabidopsis gene homologs control daylength adaptation and bulb formation in onion (*Allium cepa* L.). *Scientific Reports* 9. DOI: 10.1038/s41598-019-51262-1.

Rivera-Mendez, W., Brenes-Madriz, J. and Zuniga-Vega, C. (2018) Effects of the application of *Trichoderma asperellum* and its culture filtrate on the growth of onion seedlings. *Tecnología en Marcha* 31, 98–105. DOI: 10.18845/tm.v31i2.3627.

Singh, P. and Gopal, J. (2019) Effect of water and salinity stress on germination and seedling characters in onion. *Indian Journal of Horticulture* 76, 368–372. DOI: 10.5958/0974-0112.2019.00059.8.

Sobeih, W.Y. and Wright, C.J. (1986) The photoperiodic regulation of bulbing in onions (*Allium cepa* L.): effects of plant age and size. *Journal of Horticultural Sciences* 61, 337–341. DOI: 10.1080/14620316.1986.11515710.

Taylor, A., Massiah, A. and Thomas, B. (2010) Conservation of *Arabidopsis thaliana* photoperiodic flowering time genes in onion (*Allium cepa* L.). *Plant & Cell Physiology* 51, 1638–1647. DOI: 10.1093/pcp/pcq120.

Taylor, A., Teakle, G.R., Walley, P.G., Finch-Savage, W.E., Jackson, A.C. *et al.* (2019) Assembly and characterisation of a unique onion diversity set identifies resistance to Fusarium basal rot and improved seedling vigour. *Theoretical and Applied Genetics* 132, 3245–3264. DOI: 10.1007/s00122-019-03422-0.

Thomas, B. and Vince-Prue, D. (1997) *Photoperiodism in Plants*, 2nd edn. Academic Press, London.

Tinna, D., Garg, N., Sharma, S., Pandove, G. and Chawla, N. (2020) Utilization of plant growth promoting rhizobacteria as root dipping of seedlings for improving bulb yield and curtailing mineral fertilizer use in onion under field conditions. *Scientia Horticulturae* 270. DOI: 10.1016/j.scienta.2020.109432.

Valverde, F., Mouradov, A., Soppe, W., Ravenscroft, D., Samach, A. *et al.* (2004) Photoreceptor regulation of CONSTANS protein in photoperiodic flowering. *Science* 303, 1003–1006. DOI: 10.1126/science.1091761.

Wakchaure, G.C., Minhas, P.S., Meena, K.K., Singh, N.P., Hegade, P.M. *et al.* (2018) Growth, bulb yield, water productivity and quality of onion (*Allium cepa* L.) as affected by deficit irrigation regimes and exogenous application of plant bio-regulators. *Agricultural Water Management* 199, 1–10. DOI: 10.1016/j.agwat.2017.11.026.

Yasin, H.J. and Bufler, G. (2007) Dormancy and sprouting in onion (*Allium cepa* L.) bulbs. I. Changes in carbohydrate metabolism. *The Journal of Horticultural Science and Biotechnology* 82, 89–96. DOI: 10.1080/14620316.2007.11512203.

5 Crop Development – Shallot

Haim D. Rabinowitch*
The RH Smith Institute of Plant Science and Genetics in Agriculture, The Hebrew University of Jerusalem, Rehovot, Israel

1. Introduction

This chapter complements previous reviews (Rabinowitch and Kamenetsky, 2002; Shabir *et al.*, 2020; Rabinowitch, 2021) with revised, updated information on shallot. The biennial herbaceous Jersey shallot, a member of the *Allium cepa* taxon, shares cytological and morphological traits (Tashiro *et al.*, 1982) and intercrosses freely with bulb onion (Moshin *et al.*, 2017). They differ, however, in morphology, size, lateralization (Krontal *et al.*, 1998; Shabir *et al.*, 2020; Figs 5.1, 5.2) and flavour constituents (Ariyanti *et al.*, 2018).

Like doubled onions (Rabinowitch, 1979), shallots' apical meristem intensely divides (Fig. 5.2) throughout the growing phase (Fig. 5.1a, b). The developing divisions remain attached to the original basal plate, thus forming a cluster of doubled irregular bulbs that sprout when dormancy/rest breaks (Fig. 5.1).

2. Taxonomy and Centre of Origin

The genus *Allium* (Amaryllidaceae) consists of ~1000 species (Govaerts *et al.*, 2021), of which dozens are farmed/gathered as vegetables, condiments, herbs, folk medicine and/or ornamentals (Fritsch and Friesen, 2002; Swamy and Veere Gowda, 2006).

Allium cepa L. assumed ancestors (e.g. *A. vavilovii* M. Popov et Vved.; Gurushidze *et al.*, 2007; *A. galanthum*; Yusupov *et al.*, 2020; Chapter 1, this book) grow wild in mountainous Central Asia (Fritsch and Friesen, 2002; Nguyen *et al.*, 2008), while the two major domesticated crops, bulb onion and shallots, are known only in cultivation.

From an unknown point of time and reason(s), selections of caespitose bulb onion were cloned (Bailey, 1949). Further selections yielded fast-dividing plants richer in dry matter, milder and more delicate than the original common onion (Legin *et al.*, 2004; Auger *et al.*, 2005).

Shallot and bulb onion share phytomorphology, cytological (x=8, 2n=16) (Khrustaleva, 2018), physiological and molecular traits (Jones and Mann, 1963; Messiaen *et al.* 1993; Maaß, 1996; Friesen and Klaas, 1998). They produce fertile offspring (Moshin *et al.*, 2017), e.g. echalion: doubled, banana-like bulbs coated by thin pale-brown to grey or light-purple tunics (Fig. 5.3). It differs from onion bulbs in size and shape and has a sweeter, subtle milder flavour (UPOV, 2007).

*Address for correspondence: haim.rabinowitch@mail.huji.ac.il

DOI: 10.1079/9781789249996.0005

Fig. 5.1. (a) shallot seedlings; (b) shallot sprouts from sets; (c) developed plant from sets; (d, e) seed propagated F_1 hybrids; clustered divisions attached at the primary basal plate; (f, g) clustered divisions of clonally propagated plants.

Fig. 5.2. Onion and shallot bulb's architecture. Dichotomous division of (a) onion and (b) shallot bulbs, surrounded by false scales and tunics; (c) single-centred onion bulbs.

Bailey (1949) classified shallot as an onion cultigen; Jones and Mann (1963) assigned both to *A. cepa*. Based on the marked phenotypic differences, lateralization and flavour (Fritsch and Friesen, 2002; Prosea shallot, 2016; Saraswathi *et al.*, 2017; Potato onion, 2020; Shabir *et al.*, 2020), while Hanelt (1990) subgrouped them as common onion (synonyms: *A. cepa* L. var. *cepa*; *A. cepa* L. ssp. *cepa* and ssp. australe Trofim.), and Aggregatum (synonyms: *A. ascalonicum* Auct. non-Strand; *A. cepa* ssp. Orientale Kazak. *A. cepa* var. *ascalonicum* Backer; *Allium cepa* L. var. *aggregatum* Don.), respectively. Fritch and Friesen (2002) added a third group: the ever-ready onion.

The subordinate *A. cepa* L. Aggregatum group (Fritsch and Friesen, 2002) also includes apomictic potato onion (aka multiplier onion;

Table 5.1. Main shallot crops, other than *A. cepa* – summary (details in Rabinowitch, 2021).

Botanical name	Vernacular name	Country	Reference
Shallot x *A. fistulosum* amphidiploids	Shallot	Louisiana, USA[1]	Tashiro *et al.* (1982); Jones (1990); Messiaen *et al.* (1993); Kik (2002)
A. cepa var. *aggregatum*	Eschalote	Australia	https://www.newworldencyclopedia.org/entry/Shallot (accessed September 2022)
A. wakegi	Shallot	Japan	Tashiro *et al.* (1982)
A. chinense G. Don	Rakkyo/ eschalote	Japan	Tashiro *et al.* (1982); Hu (2005); Seidemann (2005)
A. angolense Baker	Shallot	Africa	Seidemann (2005)
A. x *cornutum* Clementi ex Vis	Shallot	Croatia	Puizina and Papfš (1996); Major *et al.* (2018)
A. altissimum[2]	Persian shallot; Mooseer	Iran, Iraq	Askari-Khorasgani and Pessarakli (2020)
A. hirtifolium Boiss. (syn. *A. stipitatum* Regel)[2]	Persian shallot; Mooseer	Iran, Iraq	Fritsch (1996); Ebrahimi *et al.* (2008); Asili *et al.* (2010); Ghahremani-Majd *et al.* (2012); Ghahremani-Majd and Dashti (2014); Khorasani *et al.* (2018); Askari-Khorasgani *et al.* (2019); Askari-Khorasgani and Pessarakli (2020)
A. oschaninii O. Fedtsch	Scalogno di Romagna	Ravenna, Italy; France	Maaß (1996); D'Antuono *et al.* (2002)
A. oschaninii O. Fedtsch	Grise de la Drôme	France, Argentina	Messiaen *et al.* (1993); Maaß (1996); Friesen and Klaas (1998); Klaas and Friesen (2002); Ravindran (2017)
A. oschaninii O. Fedtsch	Griselle; Grisombelle	France, Argentina	Messiaen *et al.* (1993); Rabinowitch and Kamenetsky (2002); Anon (n.d.)

[1]Details in Table 5.2.
[2]~140 *Allium* species in Iran, classified into seven subgenera and 32 sects (Fritsch and Gurushidze, 2009), are named Persian shallots (Askari-Khorasgani and Pessarakli, 2020)

Table 5.2. Shallot varieties from Louisiana, USA, selections from amphidiploid crosses: shallot x Welsh onion. (AgCenter, 2017; Fontenot *et al.*, 2017)

Name	PRR[1]	Evergreen	Propagules	Size	Edible parts
Bonheur	+	–	Sets	Medium	Sets
Delta giant	+	–	Sets	Large vigorous	Sets
Louisiana Evergreen	+	+	Divisions	Large	Divisions
Summergreen	+	+	Sets and seeds	Large	Divisions

[1]Pinkroot (*Phoma terrestris*) tolerance.

Stephens, 2018) and intermediate progenies (Pandey, 2006), namely, small onion, underground onion, nesting onion, ever-ready onion, and Egyptian ground onion (Permadi and Van der Meer, 1993); a Russian vegetatively propagated onion; cv. Utrechtse Sint Jansui (Netherlands: Shigyo and Kik, 2007); cv. Pran (India: Havey, 1991) and cv. Ljutika (Croatia: Puizina and Papfš, 1996).

Other alliaceous crops were also named shallot (Rabinowitch, 2021; Table 5.1). However, henceforth, the term relates solely to the Jersey shallot.

3. Economy, Distribution and Leading Types

In 2018, onion and garlic global production amounted to ~90 million and 28.55 million tons, respectively. FAO statistics for 2011–2018 report on ~5–5.5 million tons shallot/year

Fig. 5.3. Echalion bulbs, a bulb onion x shallot hybrid.

(including other minor alliums). France, the largest European shallot producer, farms ~2000 ha shallot, including grey shallots (*A. oschaninii* O. Fedtsch) (200–300 ha) and Jersey shallot, mainly in Finistère, Brittany. France markets ~40,000 tons, annually (Cohat *et al.*, 2001). Rabinowitch (2021) summarized world production and listed varieties from the USA (Blue Book Shallot, n.d.); Canada (Canada Factsheet Shallot, 2021); Europe (EU Plant Variety Catalogues, 2020) and Argentina (Burba and Galmarini, 1997).

Shallot types and production goals vary with location and markets (Alfu *et al.*, 2013). In the temperate zone, bulbs are consumed for their unique flavour, due to high dry matter and aroma (Dembele and Dubois, 1973; Cohat *et al.*, 2001), and young inflorescences as greens and garnish (New World Encyclopedia, 2019). Hence, within the environment-imposed variation, clonal propagation guarantees perpetuated quality performance of the heterozygous vegetable/condiment.

In warm-humid climates, onion seed production is difficult (Peters, 1990; Kamenetsky and Rabinowitch, 2002), growing seasons are short for profitable production, and bulbs' quality is poor (Currah and Proctor, 1990). Locally selected shallots store longer and tolerate heat, pests and diseases better than many popular short-day (SD) onions (Currah, 2002; Shabir *et al.*, 2020) and young inflorescences consumed as greens (Fig. 5.4). Hence, shallots substitute for bulb onion and scallions as vegetables, and in folk medicine (Brewster, 2008).

In the cold subtropics and temperate zones, shallot farming and bulbs' dormancy are long (e.g. Oregon State University, 1982, 1987, 2010), while the tropical shorter growing cycle (Getachew and Asfaw, 2004; Shimeles, 2014) and dormancy (Woldetsadik and Workneh, 2010; Rosliani *et al.*, 2016) enable 2–3 crops annually (Sinnadurai, 1973; Woldetsadik, 2003), thus supplying consumers' needs almost year-round. Shallot is therefore locally preferred over bulb onion (Tesfa *et al.*, 2015; Ravindran, 2017) in the tropics, and vegetative propagation guarantees self-sufficiency of propagules.

4. Developmental Morphology

Jones and Mann (1963), DeMason (1990), Brewster (2008), Ravindran (2017) and Khokhar (2017) reviewed the developmental morphology of *Allium cepa*. In summary, the black, wrinkled, crescent-shaped seed contains a curved monocot embryo embedded in an endosperm made of phytine, fat and mainly mannan (DeMason, 1990). It consists of a short radicle at the condensed stem's base and a cylindrical basal sheath with a blade at the top. The cotyledon envelops the apex and the embryonic first true leaf.

Fig. 5.4. Shallot reproduction: (a) edible inflorescence buds; (b) mature capsules and seed; (c) shallot seedlings.

The germinating *A. cepa* radicle emerges through the micropyle, while the sheath base grows upward. Thus the U-shaped cotyledon elongates, and crosses the soil surface (Fig. 5.4) before unbending, with the testa attached to its top. The apical dome forms an uneven ring around the apex (Electron microscopy pictures: Rabinowitch, 1990a; Rabinowitch and Kamenetsky Goldstein, 2020). It grows upward, thus forming leaves made of a cylindrical sheath and a concaved closed blade that turns hollow. The first true leaf emerges through a pore at the top of the cotyledonary sheath. The true stem expands slowly while its oldest tissues decay and the degenerating taproot is replaced by adventitious, rarely branching fibrous roots, which form a 15–30 cm-deep rhizosphere, thus shallots' vulnerability to drought.

Both main and lateral apices continually form new leaves, each emerging from a pore at the top inner side of the preceding sheath, thus forming prostrate, erect or intermediate distichous foliage (UPOV, 2007). The successive sheaths are densely locked one within the other to form a firm structure – the false stem (Fig. 5.1b).

When raised from sets, laterals remain attached to the original stem (Fig. 5.1), while shoots' physiological/developmental morphology resembles developing seedlings.

When bulbing commences, leaf initiation ceases, thus the pseudostem's core becomes hollow, and mechanically cannot further support the blades' weight. Hence, foliage collapses with bulb maturation. Foliage turgor weakens with assimilates' basipetal transport, and partition between sheaths' bases and leaf buds, where sugars polymerize into fructans. Cell swelling leads to transformation of organs into false and true fleshy scales, respectively (Brewster, 2008; Fig. 5.2). Concomitantly, slow metabolism-maintenance processes proceed (Galsurker *et al.*, 2017), and photosynthesis continues as long as green blades remain active, with the consequent increase in both bulb size and yield (Jones and Davis, 1944).

Sequential senescence proceeds from the oldest to the youngest leaf. The three outermost sheaths per bulb undergo programmed cell death, desiccation and pigment accumulation, thus forming protective tunics (Eshel *et al.*, 2014).

Each shallot plant forms a cluster (Rabinowitch and Kamenetsky, 2002; Brewster, 2008; Fig. 5.1) of 3–28 cloves (Messiaen *et al.*, 1993a; Krontal *et al.*, 1998), generally smaller than onion bulbs.

Bulbing index, the ratio between maximum bulb and minimum neck diameter, is the preferred non-destructive quantifier for onion bulb formation (Rabinowitch and Kamenetsky Goldstein, 2020). By definition, bulbing commences when the quotient is 2 (Brewster, 2008). The caespitose shallot, however, is difficult to measure, thus a roughly averaged quotient of individual divisions serves as an estimate for the bulb's formation.

The shallot's single, thin, erect, fistulose, inflated, 30–100 cm long scape produces a compact globose umbel made of a few dozen whitish flowers (Palupi *et al.*, 2017), aggregated as cymes, each with 5–10 perfect flowers, and occasionally topsets. The protandrous flowers or cyme open sequentially, while groups of cymes flower a synchronically over 2–4 weeks, thus improving chances of both sexual reproduction and self-pollination. Hence, as in bulb onion (Peters, 1990) there is non-uniform seed ripening.

The sub-campanulate to urceolate flower consists of five whorls (for schematic presentation, see Chapter 8, Fig. 8.4, this volume; Puizina, 2013), with three organs each: outer and inner erect to somewhat spreading tepals; outer and inner stamens; and a superior ovary with three united carpels (pistil). The trilocular ovary, attached by axile placentation, contains two ovules per locule, six per fruit. On ripening, the globular loculicidal capsules (fruits) split open to release the firm, black, wrinkled mature non-dormant seeds.

5. Physiology

Qualitatively, bulb onion and shallot respond similarly to environmental cues but quantitatively they vary in the physiological age for apical division, branching mode, responses to vernalization and photoperiod, and post-bloom development lateralization (Krontal *et al.*, 1998, 2000; Rabinowitch and Kamenetsky Goldstein, 2020).

5.1. Lateralization

In modern onion cultivars, seedlings form ≥13 leaves before doubling occurs. The inner leaf buds form groups of concentric rings (Rabinowitch, 1979; Fig. 5.2) of dormant lateral(s) sympodial branch(es) between the main apex and the youngest leaf (Brewster, 2008) with concomitant flowering (DeMason, 1990), but clusters form rarely. When grown from sets, multi-doubling is common in certain genotypes (Rabinowitch, 1979).

Shallot seedlings develop apical lateralization already after the third leaf-bud formation, and doubling commenced together with apical transition and inflorescence development (Krontal *et al.*, 1998, 2000), while the energy-rich sets support sprouting and lateralization (Fig. 5.1).

GA_3 application promoted tillering in Welsh onions (Yamazaki *et al.*, 2015), but no information on shallot response is available. It is worth trying both ways: promotion and retardation for sets size control.

5.2. Bulb formation

The obligate long photoperiod (LP) *Allium cepa* (Damte *et al.*, 2017; Ikeda *et al.*, 2019; Rabinowitch and Kamenetsky Goldstein, 2020), sense light cues by their youngest leaf (Brewster, 1990). Thermal period >600 degree-days (Lancaster *et al.*, 1996; Damte *et al.*, 2017) together with uninterrupted LP (Rabinowitch and Kamenetsky, 2002; Brewster, 2008) induce bulbing irrespective of age, while the process is accelerated with increased temperatures and age.

Horticulturally, cultivar classification is based on the approximate number of dark hours required for bulbing, as very long (VLD), long (LD), intermediate (ID) and short (SD) -day types, thus indicating the genotype's general adaptability to particular latitudes. Importantly, SD cultivars will *not* form bulbs under nights longer than the genotype requirement for LP.

Under LP, spectral quality and intensity control bulbing and quality (Rabinowitch and Kamenetsky, 2002; Rabinowitch and Kamenetsky Goldstein, 2020), via phytochrome (Lercari, 1984; Sobeih and Wright, 1987). High irradiance (Kedar *et al.*, 1975; Wright and Sobeih, 1986), and high far-red energy, accelerated bulbing (Lercari, 1982; Mondal. *et al.*, 1986a, b, c), while blue light is less efficient (Terabun, 1965). Hence, in dense versus thin stands or under weed competition, where the red:far-red ratio is low, early bulbing and fast maturation are common, thus affecting size and quality.

Neither an FR-free LP nor far-red supplemented SP induced bulbing, while LP interruptions with red or blue lights inhibited bulbing (Brewster, 2008). Interruption by SP resulted in reverting of fully developed non-dormant bulbs, and resprouting (Kedar *et al.*, 1975; Brewster, 2008).

Fertigation and other field factors also affect *A. cepa* bulbing (van Kampen, 1970;

Fita, 2009; Esuyawukal-Moges and Biruk-Masrie, 2019).

Compared with *A. cepa* bulb formation physiology, little is known about the genetics of the process (Rabinowitch, 2021). Recently, onion flowering and bulbing were attributed to the expression of vernalization-promoted FT protein (Lee *et al.*, 2013; Atif *et al.*, 2020; Chapter 4, this volume).

The role of hormones in *A. cepa* bulbing is unclear (Brewster, 2008; Atif *et al.*, 2020). Ethylene initiated bulbing (Levy *et al.*, 1979) yet under LP ethylene antagonists did not inhibit the process (Lercari, 1983). Gibberellins play an inhibitory role (Brewster, 2008) while IAA (Clark and Heath, 1962), and cytokinins' levels (Lercari and Micheli, 1981) increase under LP. The entire picture, however, is blurred.

5.3. Florogenesis

The biennial onion acquires flowering competence only when the juvenile phase of 10–12 leaves is crossed (Rabinowitch and Kamenetsky, 2002), thus, under favourable conditions, florogenesis in the first growing season occurs. Professional growers time the sowing date and optimize agro-management for minimum bolting and maximum quality yields (Rabinowitch and Kamenetsky Goldstein, 2020), while seed-to-seed producers take different actions (Chapter 12).

Shallot scapes serve as foods (Fig. 5.4), thus mid-season harvest provides income while benefitting bulb growth and quality yields by removing the competitive sink.

Optimum storage and/or field vernalization ranges between 5°C and 13°C for 7–90 days depending on genotype and plant age (Bertaud, 1988; De Bon and Rhino, 1988; Khokhar *et al.*, 2007). Older plants are more responsive to vernalization than younger ones (Khokhar, 2008a). Flowering/bolting is poor at high (15.5–30°C) and low (3–0°C) temperatures (Khokhar, 2008b; Rabinowitch and Kamenetsky Goldstein, 2020), while LP and intermediate-high temperatures enhance florogenesis in transitioned growing plants (Brewster, 1983).

Shallot cold requirements vary with genotype (Esnault *et al.*, 2005) and environment. In phytotron and field-grown seedlings, critical physiological ages for apical transition to the reproductive state were 13–16 (Krontal *et al.*, 2000) and six (Krontal *et al.*, 1998) true leaves (including leaf buds), respectively. The marked differences in juvenile age probably reflect the impact of ever-changing field environments compared with the phytotron-constant LP, temperature and humidity.

Most shallot sets are in a post-juvenile state, thus responsive to vernalization (Krontal *et al.*, 2000; Fita, 2004; Tabor *et al.*, 2006), and large sets (physiologically older) require shorter vernalization to bolt/flower than smaller ones. Specifically, 8–12°C (Krontal *et al.*, 2000; Tabor *et al.*, 2006), 6–10°C (Esnault *et al.*, 2005) or 4–10°C (Tendaj *et al.*, 2013) induce the apical transition of several divisions/cluster. Seedlings bolted at 17/9°C but not at 28/21°C day/night, respectively. Hence, genotype, storage conditions and set size/age can efficiently be applied to meet growers' needs, e.g. ambient and cold storage for bulbs for consumption; mild-cold storage of large sets for bolting (marketing inflorescence buds, and/or seed production).

Protandry (anthers dehiscence precedes stigma receptivity) forms a weak barrier to self-pollination within an umbel. The shallot's weeks-long asynchronous anthesis (for bulb onion: Rabinowitch and Kamenetsky, 2002; Brewster, 2008) improves both seed production and self-pollination between younger and older flowers (Brewster, 2008). Note: self-pollination is the rule within a clonally propagated population.

Pollination biology (Williams and Free, 1974; Rao and Suryanarayan, 1989; Currah and Proctor, 1990; Rasekh Adel *et al.*, 2013; Devi *et al.*, 2015), seed production (Rabinowitch, 1990b; Askari-Khorasgani and Pessarakli, 2019), including development, ripening, harvest, threshing and storage, are as detailed for onions (Nikus and Mulugeta, 2010; Chapters 4 and 12, this volume).

Spring-foraging wild and domesticated insects pollinate the sequentially maturing stigmas, with the consequent asynchronous seed-setting within and between umbels, cultivars and production environments. GA_3 treatment improves synchronization (Naamni *et al.*,1980; Mushtaq *et al.*, 2018).

Seed maturation in onions takes 4–7 weeks (Brewster, 1982; Pathak, 2000) and up to 11 weeks (Neal and Ellerbrock, 1986) before the capsules (simple fruits derived from compound ovaries) break open along their seams, releasing the black crinkled and triangular (in cross-section) seeds (Rasekh Adel *et al.*, 2013; Kamenetsky and Rabinowitch, 2017; Askari-Khorasgani and Pessarakli, 2019). Shallot seed development is similar (author's observation; Fig. 5.4), and 1000-seed weight amounts to 3–3.5 g (Tendaj *et al.*, 2014; Palupi *et al.*, 2017; Wassu *et al.*, 2018).

Shallot seeds do not enter dormancy. Under optimal conditions, they store for years, ready for sowing whenever needed (Tendaj *et al.*, 2013; Shimeles, 2014; Buda *et al.*, 2018), and pre-sowing GA_3 and KNO_3 treatments enhanced germination and seedling vigour (Agung and Diara, 2017).

5.4. Bulb harvest, storage and dormancy

On ripening, SD shallot bulbs enter days, weeks or months-long dormancy (Sinnadurai and Amuti, 1971; Sinnadurai, 1973; Currah and Proctor, 1990), while generally, LD shallot dormancy is longer. Dormant bulbs maintain slow life activities, e.g. cell division, differentiation and growth, compared with growing plants (Abdalla and Mann, 1963; Lang *et al.*, 1987). Long dormancy maintains the quality of stored bulbs, while significantly reducing flexibility in planting time for the following season.

Oregon State University (2010) recommends machine harvest, while manual picking is common in the tropics. Bulbs' storage life depends on genotype, size, maturation state at harvest, dormancy, health and intactness, curing, field and storage environments (Yasin and Bufler, 2007; Phillips, 2010; Eshel *et al.*, 2014; Calica *et al.*, 2018; Gulumbe *et al.*, 2018). Dry-season crops (Tekalign *et al.*, 2012) and high-nitrogen bulbs (Ruaysoongnern, 1994) store less well than low-nitrogen bulbs and those from rainfed/-irrigated fields. Occasionally, the sprout suppressor maleic hydrazide is applied to minimize sprouting during storage (Oregon State University, 2010).

Bulb curing is important for long-keeping (Lestari *et al.*, 2019), thus applied worldwide, with variations: in Indonesia, 7–10 days of open field curing is common (Fig. 5.5); in Ethiopia, 15 days of open-shed curing is practised; storage

Fig. 5.5. Shallot bulb harvest and curing in southeast Asia: (a) manual harvest; (b) curing; (c, d, e) protected storage from insolation and rain.

at ~30°C (Stow and Ward, 1978) in open sheds with passive or active ventilation is practised in many warm-climate countries, as excessive moisture removal minimizes storage diseases, delays sprouting and root growth.

In Europe, many LD shallots are stored for 8–10 months (Tendaj, 2005; Leino *et al.*, 2018) at 60–70% RH and 0–2°C (Tendaj *et al.*, 2013). In Poland, cold storage under a controlled atmosphere minimized dryweight and flavonoid losses of cured shallot bulbs (Pudzianowska *et al.*, 2012).

6. Quality, Flavour, Colour and Potential Health Benefits

USDA (Anon, 2016) and other agencies developed standards for shallot bulb quality grades (Thai Ministry of Agriculture and Cooperatives, 2008); ASEAN Standard (ASEAN, 2009); OECD (2014); South African Ministry of Agriculture (2015); FAO (2019), and more.

Shallot bulbs are richer in fats and contain 16–33% soluble solids compared with 7–15% in bulb onions (Currah and Proctor, 1990; Legin *et al.*, 2004). These, together with a distinct aroma and texture (Block *et al.*, 1992a, b), make shallot bulbs a favoured gourmet condiment.

Carbohydrates (mainly fructans, glucose, fructose and sucrose), together with cell-wall components (pectin, cellulose), make up ~70–85% of shallot bulbs' dry matter. Additionally, shallot accumulates some biologically active compounds, i.e. organosulphur molecules (e.g. alliin), selenium and other elements, vitamins A, B and C (Fattorusso *et al.*, 2002; Hedges and Lister, 2007; Anon, 2020), saponins, polyphenols (quercetin, kaempferol; Leelarungrayub *et al.*, 2006), flavonoids, phenolic acids, isorhamnetin, cyanidin derivatives (Joslym and Peterson, 1958) and their glycosides (Fattorusso *et al.*, 2002; Masuzaki *et al.*, 2006) in the red and yellow shallots, respectively. The richness in phenolics and diallyl-sulfides makes shallot antioxidative strength comparable to garlic (Leelarungrayub *et al.*, 2006). Wang and Ng (2002) isolated ascalin, an anti-fungal peptide, from shallot bulbs, and Amin and Kapadnis (2005) reported that fungi were relatively more susceptible than bacteria to the heat-stable shallot extracts.

7. Genetics and Breeding

The knowledge of shallot genetics is very poor (Rabinowitch, 2021). As in bulb onion (McCallum, 2007), the large genome size, biennial life cycle, cross-pollinated nature, high inbreeding-depression, the small research community, and low funding retarded the development and utilization of molecular tools for shallot studies.

D'Angelo and Goldman (2019) proposed optimization by the annualization of the LD onion breeding cycle. Factually, SD shallot seedlings complete their life cycle within a season, and a similar performance is achievable in LP shallot, thus breeders can easily adopt this methodology.

Recent advances in omics research (Khandagale *et al.*, 2020), next-generation sequencing technologies, molecular markers (e.g. Peška *et al.*, 2019), mapping populations (e.g. Jo *et al.*, 2017), saturated linkage maps (Shigyo *et al.*, 2018; Chapters 14–17, this volume), and completion of *A. cepa* sequencing (Finkers *et al.*, 2021) generated a wealth of genomic resources and shed light on the genetics of important traits, thus advancing breeding efficiency through combinations of classical breeding (e.g. double haploids (Khosa *et al.*, 2016)), with state-of-the-art molecular genetics, including, but not limited to, genetic engineering (e.g. Zheng *et al.*, 2005; Cardi *et al.*, 2017). Adapting genome-editing tools for *A. cepa* could further contribute to crop improvement.

8. Propagation

Shallots are mostly propagated from lateral divisions, but an assortment of cultivars propagated by true shallot seeds (TSS) is now available (Rabinowitch, 2021).

8.1. Clonal propagation

Successful production of solid, predictable high-quality heavy yields depends on a timely supply of healthy sets at the right physiological age. Professional propagules-producers should intimately know bulbs' ageing physiology, storage environmental effects on field performance, as well as storehouse and production-field pest control. Moreover, clear definitions of the production goals (earliness; bulbs/bulbs and inflorescence buds for garnish; divisions' size; fresh consumption; storage for food or propagation, TSS production etc.) are essential.

Clonal propagation suffers major weaknesses, including complete dependence on the previous season, thus a limited choice of genotypes and acreage size, while dormancy length dictates planting date. Low multiplication rates, voluminous transportation loads and storage space (passive/active ventilated, ambient/cold), evapotranspiration, decay, and/or sprouting losses are more costly than TSS (Woldetsadik and Workneh, 2010; Firmansyah *et al.*, 2017). Together with manual planting, these add up to ~40% of the total production costs in Ethiopia (Getahun and Zelleke, 2009); Indonesia (Adiyoga and Soetiarso, 1997) and elsewhere.

The most troubling factor, however, is skipping the sexual-cycle cleansing of pathogens (Rosliani, 2016). Shallot propagules' contamination increases geometrically with every growing cycle (Proctor, 1987; Hailekidan *et al.*, 2013), thus production fields become perpetuating repositories of pathogens and bear low yields/quality (Tendaj, 2005; Tabor *et al.*, 2006; Soegianto *et al.*, 2011; Katis *et al.*, 2012; Hailekidan *et al.*, 2013; Marais *et al.*, 2019).

Where specialized growers/companies produce healthy quality sets, meristem tip culture with and without thermotherapy, cryotherapy and/or antiviral agents, is practised, and strict certification systems apply (e.g. strawberries, sweet potato, garlic, potatoes and more). Similar techniques apply to shallots (Wang *et al.*, 2018, 2020; Marais, 2019).

Like virus-free garlic (Conci *et al.*, 2003, 2005; Pérez Moreno *et al.*, 2010), shallot clean-start results in vigorous growth and high quality yields (e.g. France: GEVES, n.d.; The Netherlands: NAK Tuinbouw, 2018; the USA's official organizations for variety evaluation and seed testing).

Some French provinces set their own propagules' quality standards (French Certification, n.d.; Traditional French Shallot (n.d.) and, similarly, Louisiana established a state certification system (La Admin Code, 2021).

Several niche sectors set their standards for shallot production and handling, e.g. the organic sector (Petitti *et al.*, 2021); special local brands, e.g. Protected Geographical Indication (PGI) (n.d.), and producers of traditional shallot, e.g. Protected Designation of Origin (PDO, n.d.).

Smallholders in developing countries grow >50% of the global production. There, abiding by strict regulations and the costs of meticulous propagation and maintaining routines are prohibitive (Taşkın *et al.*, 2013), especially among self-sufficient growers (Sinnadurai, 1973).

The scarcity of plant-protection experts, specialized tissue-culture laboratories, and the inadequate fields and storehouses for phytosanitation result in inferior propagules – a major reason for poor yields. Hence, mean yields of 3–8 mt ha^{-1} are common in Ethiopia and Indonesia (Grubben, 1994; Pathak, 1994; Shimeles, 2014; Tesfa *et al.*, 2015) compared with potential yields of 10–30 mt ha^{-1} (Woldetsadik, 2003; Woldetsadik and Workneh, 2010; Shimeles, 2014), and actual 35–40 mt ha^{-1} in France (Cohat *et al.*, 2001) and Oregon (Oregon State University, 2010).

Regional/governmental/international co-operation in establishing professional quality propagules production systems can make the difference, with minimum extra costs.

8.2. True shallot seed (TSS)

Hybrid seed production depends on rarely available male sterility, while male/female synchronous flowering (Currah and Proctor, 1990) requires physiological manipulations (Van Nocker, 2001).

Self-pollinated *A. cepa* (Jones and Davis, 1944; Smith and Crowther, 1995; Khan *et al.*, 2001; Khodadadi and Hassnpanah, 2010; Tendaj and Mysiak, 2013) and other alliaceous crops (e.g. Stevens and Bougourd, 1988; Smith and Crowther, 1995) suffer sever inbreeding depression generally (Cramer, 2001) yet educated propagation of OP shallot seeds (Tabor, 2018) results in plants more vigorous (Farid *et al.*, 2012), more adaptive, and resilient to changing field conditions than clones. The ever-going selection of best-performing TSS cultivars results

Table 5.3. The legal state of shallot variety IP protection programmes in European countries.

State	Year	Activity
France[1]	1990	France legislated that **only** vegetatively propagated shallot can be named shallots.
Netherlands[2]	1992	The first Dutch TSS varieties were released in 1992 (UNECE, 2002).
EU[3]	2006	EU court decided on acceptance of TSS but required labelling as 'shallots from seeds' while clonally propagated shallots are labelled 'traditional shallots'.
UK[4]		UK national list/Plant breeders rights technical protocol for the official examination of distinctness, uniformity and stability (DUS) shallot

[1]Article 19 in https://eur-lex.europa.eu/legal-content/EN/TXT/PDF/?uri=CELEX:62004CC0147&from=SL (accessed 9 July 2022) and Article 73 in UNECE (2002).
[2]Listed in the reference section, this chapter.
[3]https://www.stradalex.com/en/sl_src_publ_jur_int/document/cjeu2006_C-147_04_57 (accessed 9 July 2022).
[4]https://assets.publishing.service.gov.uk/government/uploads/system/uploads/attachment_data/file/425893/dus-protocol-shallot.pdf

in continuously improved performance, quality and yield; a broader varietal spectrum; optimal use of land and water; and targeting the local optimal growing season (Shimeles, 2014; Shimeles and Dessalenge, 2015; Buda *et al.*, 2018; Wassu *et al.*, 2018; Rahayu *et al.*, 2019).

From the 1970s/1980s onwards, public breeders in Ethiopia, France, Holland, Israel and more, and seed companies (e.g. Bejo Zaden, De Groot en Slot, Holland; Hazera, Israel; PlantiCo Zielonki, Poland; Sumbawang Superior, Indonesia, and more) initiated TSS OPs and F_1 hybrids breeding programmes. The legal situation is detailed in Table 5.3.

Compared with 1–4 mt sets/divisions ha^{-1} (Sinnadurai, 1973; Shimeles, 2014), only 1–5 kg TSS ha^{-1} are required for shallot production. Hence, growers' costs of healthy propagules for direct sowing, seedlings or sets are lower; the choice of genotypes, year-round, is wider; optimum seeding/planting time becomes a reality, and mechanization similar to onion agriculture is practical.

Significantly, the major TSS benefits are breaking the pest seasonal circulation, from one field to the other (Özer and Köycüm, 2004), and the pathogen-free start with consequent minimization of yield and quality losses.

TSS penetration rates in traditional shallot territories seem to be slow (Van den Brink and Basuki, 2012; Tendaj *et al.*, 2014; Shimeles and Dessalenge, 2015; Rabinowitch, 2021), since the shift requires conceptual and agro-management adaptation of agro-practices, e.g. sowing, germination, stand, fertigation, plant protection, etc.

8.2.1. Quality considerations

Vegetatively propagated plants produce many small laterals, compared with current TSS mostly single, twin or triplet divisions in both OP (Wassu *et al.*, 2018) and hybrid cultivars (Rabinowitch and Kamenetsky, 2002; Fig. 5.1). Other traits depend on the interaction between breeders' achievements (Rabinowitch, 2020) and field and storage environments. Consumers' perceptions and acceptance depend on quality and price compared with the alternatives. In our case, professionals ranked hybrids' performance as equal to the best European Jersey shallots.

Note: as common with all crops, healthy seedlings are vulnerable to infection, hence TSS introduction requires regional co-ordination for prevention/minimization of reinfestation.

As in bulb onion, production of quality TSS, storage and handling require intensive breeding work (Sulistyaningsih *et al.*, 2002; Rabinowitch, 2021), and a very high degree of seed production professionalism (Chapter 12).

9. Conclusions and Prospects

Shallot, a minor crop, plays important regional economic and nutritional roles as an onion substitute and a distinctive flavouring condiment. Yet only a little is invested in studying its physiology and genetics; improvement efforts lag behind science frontlines and, accordingly, its horticultural performance.

The two members of *A. cepa* taxon vary in tolerance to biotic and abiotic stress, doubling, size, storage, bolting, flavour and more (Currah and Proctor, 1990; Odeny and Narina, 2011). However, the wealth of information gathered on onion biology applies to shallots. Leading-edge breeding without or with minor modifications and TSS acceptance are thus expected. The two wide germplasm collections share a variability richness and should broaden the breeding horizon for better, more sustainable production of both crops.

We should bear in mind, however, that shallot selection is aimed at different needs than for onions, thus characterized by a unique and high diversity of important traits. These rich diversities should be preserved for generations to come.

References

Abdalla, A.A. and Mann, L.K. (1963) Bulb development in the onion (*Allium cepa* L.) and the effect of storage temperature on bulb rest. *Hilgardia* 35(5), 85–112. Available at: http://hilgardia.ucanr.edu/fileaccess.cfm?article=152605&p=TJUMJN (accessed 9 July 2022).

Adiyoga, W. and Soetiarso, T.A. (1997) The study of comparative advantage and economic incentives in shallot production. *Indonesian Journal of Horticulture* 7, 614–621.

AgCenter (2017) Available at: https://www.lsuagcenter.com/profiles/bneely/articles/page1496096519086 (accessed 25 April 2022).

Agung, I.G.A.M.S. and Diara, I.W. (2017) Pre-sowing treatment enhanced germination and vigour of true shallot (*Allium cepa* var. *aggregatum*) seeds. *International Journal of Agriculture Environment and Biotechnology* 2, 3262–3267. DOI: 10.22161/ijeab/2.6.62.

Alfu, L., Endang, S. and Arif, W. (2013) Morphogenetic variation of shallot (*Allium cepa* L. Aggregatum Group). *Agricultural Science* 16(2), 1–11. Available at: https://journal.ugm.ac.id/jip/issue/view/447/showToc (accessed 25 April 2022).

Amin, M. and Kapadnis, B.P. (2005) Heat stable anti-microbial activity of *Allium ascalonicum* against bacteria and fungi. *Indian Journal of Experimental Biology* 43, 751–754. Available at: http://nopr.niscair.res.in/bitstream/123456789/23231/1/IJEB%2043%288%29%20751-754.pdf (accessed 25 April 2022).

Anon (2016) *United States Standards for Grades of Bunched Shallots*. USDA, Washington, DC. Available at: https://www.ams.usda.gov/sites/default/files/media/BunchedShallotsStandard.pdf (accessed 25 April 2022).

Anon (2020) *USDA National Nutrient Database*. Nutrition facts for shallots: raw, recommended daily values and analysis. Available at: https://www.nutritionvalue.org/Shallots%2C_raw_nutritional_value.html (accessed 25 April 2022).

Anon (n.d.) *Allium oschaninii – O. Fedtsch. Plants for a future*. Available at: https://pfaf.org/user/Plant.aspx?LatinName=Allium+oschaninii (accessed 29 July 2022).

Ariyanti, N.A., Torikai, K., Kirana, R.P., Sho, H., Sulistyaningsih, E. *et al.* (2018) Comparative study on phytochemical variations in Japanese F_1 varieties of bulb onions and South-East Asian shallot landraces. *The Horticulture Journal* 87, 63–72. DOI: 10.2503/hortj.OKD-066.

ASEAN (2009) ASEAN Standard for Shallot. *ASEAN Stan* 14: 2009. Available at: https://www.asean.org/wp-content/uploads/images/archive/AMAF%2031%20asean%20standard%20shallot.pdf (accessed 25 April 2022).

Asili, A., Behravan, J., Naghavi, M.R. and Asili, J. (2010) Genetic diversity of Persian shallot (*Allium hirtifolium*) ecotypes based on morphological traits, allicin content and RAPD markers. *Journal of Medicinal and Aromatic Plants* 1, 1–6. Available at: https://www.researchgate.net/profile/Javad-Asili/publication/279464029_Genetic_diversity_of_Persian_shallot_Allium_hirtifolium_ecotypes_based_on_morphological_traits_allicin_content_and_RAPD_markers/links/55ea8cc308aeb6516265e6c7/Genetic-diversity-of-Persian-shallot-Allium-hirtifolium-ecotypes-based-on-morphological-traits-allicin-content-and-RAPD-markers.pdf?_sg%5B0%5D=started_experiment_milestone&origin=journalDetail (accessed 29 July 2022).

Askari-Khorasgani, O. and Pessarakli, M. (2019) Agricultural management and environmental requirements for production of true shallot seeds – a review. *Advances in Plants & Agriculture Research* 9, 318–322. DOI: 10.15406/apar.2019.09.00441.

Askari-Khorasgani, O. and Pessarakli, M. (2020) Evaluation of cultivation methods and sustainable agricultural practices for improving shallot bulb production – a review. *Journal of Plant Nutrition* 43, 148–163. DOI: 10.1080/01904167.2019.1659329.

Atif, M.J., Ahanger, M.A., Amin, B., Ghani, M.I., Ali, M. *et al.* (2020) Mechanism of *Allium* crops bulb enlargement in response to photoperiod: a review. *International Journal of Molecular Sciences* 21, 1325. DOI: 10.3390/ijms21041325.

Auger, J., Arnault, I., Legin, A., Rudnitskaya, A., Seleznev, B. *et al.* (2005) Comparison of gas chromatography-mass spectrometry and electronic tongue analysis for the classification of onions and shallots. *International Journal of Environmental Analytical Chemistry* 85, 12–13, 971–980. DOI: 10.1080/03067310500154452.

Bailey, L.H. (1949) *Manual of Cultivated Plants*. Macmillan, New York.

Bertaud, D.S. (1988) Effects of chilling duration, photoperiod and temperature on floral initiation and development in sprouted and unsprouted onion bulbs. *Proceedings of the 4th EUCARPIA Allium Symposium*. IHR, Wellesbourne, UK, pp. 245–261.

Block, E., Naganathan, S., Putman, D. and Zhao, S.H. (1992a) *Allium* chemistry: HPLC analysis of thiosulfinates from onion, garlic, wild garlic, leek, scallion, shallot, elephant (great-headed) garlic, chive and Chinese chive; uniquely high allyl-to-methyl ratios in some garlic samples. *Journal of Agricultural and Food Chemistry* 40, 2418–2430. DOI: 10.1021/jf00024a017.

Block, E., Putman, D. and Zhao, S.H. (1992b) *Allium* chemistry: GC-MS analysis of thiosulfinates and related compounds from onion leek scallion shallot chive and Chinese chive. *Journal of Agricultural and Food Chemistry* 40, 2431–2438. DOI: 10.1021/jf00024a018.

Blue Book Services (n.d.) Available at: https://www.producebluebook.com/know-your-commodity/shallots (accessed 9 July 2022).

Brewster, J.L. (1982) Flowering and seed production in overwintered cultivars of bulb onions: I. Effects of different raising environments, temperatures and day lengths. *The Journal of Horticultural Science and Biotechnology* 57, 93–101. DOI: 101080/00221589198211515028.

Brewster, J.L. (1983) Effects of photoperiod, nitrogen nutrition and temperature on inflorescence initiation and development in onion (*Allium cepa* L.). *Annals of Botany* 51, 429–440. Available at: https://www.jstor.org/stable/42756610?seq=1#metadata_info_tab_contents (accessed 9 July 2022).

Brewster, J.L. (1990) Cultural systems and agronomic practices in temperate climates. In: Rabinowitch, H.D. and Brewster, J.L. (eds) *Onions and Allied Crops II: Agronomy, Biotic Interactions*. CRC Press, Boca Raton, Florida, pp. 1–30. Available at: https://www.taylorfrancis.com/chapters/edit/10.1201/9781351075152-1/cultural-systems-agronomic-practices-temperate-climates-brewster?context=ubx&refId=9f6887ac-7711-4c98-a015-311806c05bbb (accessed 9 July 2022).

Brewster, J.L. (2008) *Onions and Other Vegetable Alliums*. CAB International, Wallingford, UK.

Buda, I.M., Agung, I.G.A.M.S. and Ardhana, I.P.G. (2018) Nitrogen fertilizer increased bulb diameter and yields of true seed and bulb – propagated shallot varieties. *International Journal of Scientific Engineering and Technology* 7, 80–86. DOI: 10.15680/IJIRSET.2018.0701007.

Burba, J.L. and Galmarini, C.R. (1997) *Allium* crop situation in Argentina. *Acta Horticulturae* 433, 35–52. DOI: 10.17660/ActaHortic.1997.433.2.

Calica, G.B., Mae, M. and Dulay, N. (2018) Assessment of the postharvest systems and losses of shallots in Ilocos, Philippines. *Asian Journal of Postharvest and Mechanization* 1, 81–92. Available at: https://www.philmech.gov.ph/assets/publication/Journal/2018_AJPM%20Vol.1%20No.1.pdf#page=86 (accessed 25 April 2022).

Canada Factsheet Shallot (2021) Available at: http://www.omafra.gov.on.ca/english/crops/facts/98-037.htm (accessed 25 April 2022).

Cardi, T., D'Agostini, N. and Tripodi, P. (2017) Genetic transformation and genomic resources for next-generation precise genome engineering in vegetable crops. *Frontiers in Plant Science* 8, 241. DOI: 10.3389/fpls.2017.00241.

Clark, J.E. and Heath, O.V.S. (1962) Studies in the physiology of the onion plant: V. An investigation into the growth substance content of bulbing onions. *Journal of Experimental Botany* 13, 227–249. DOI: 10.1093/jxb/13.2.227.

Cohat, J., Chauvin, J.E. and Le Nard, M. (2001) Shallot (*Allium cepa* var. *aggregatum*) production and breeding in France. *Acta Horticulturae* 555, 221–225. DOI: 10.17660/ActaHortic.2001.555.32.

Conci, V.C., Canavelli, A., Lunello, P., Di Rienzo, J., Nome, S.F. and Zumelzu, G. (2003) Yield losses associated with virus-infected garlic plants during five successive years. *Plant Disease* 87, 1411–1415. Available at: https://apsjournals.apsnet.org/doi/pdf/10.1094/PDIS.2003.87.12.1411 (accessed 9 July 2022).

Conci, V.C., Perotto, M.C., Cafrune, E.P. and Lunello, P. (2005) Program for the intensive production and protocols optimization for the production of virus-free garlic plants. *Acta Horticulturae* 688, 195–200. DOI: 10.17660/ActaHortic.2005.688.25.

Cramer, C.S. (2001) Comparison of open-pollinated and hybrid onion varieties for New Mexico. *HortTechnology* 11, 119–123. DOI: 10.21273/HORTTECH.11.1.119.

Currah, L. (2002) Onions in the tropics: cultivation and country reports. In: Rabinowitch, H.D. and Currah, L. (eds) *Allium Crop Science: Recent Advances*. CAB International, Wallingford, UK, pp. 379–407.

Currah, L. and Proctor, F.J. (1990) *Onions in Tropical Regions*. Natural Resources Institute Bulletin, Chatham, UK.

Damte, T., Tabor, G., Haile, M., Mitiku, G. and Lulseged, T. (2017) Determination of beginning of bulb enlargement time in shallot, *Allium cepa* var. aggregatum for managing onion thrips (*Thrips tabaci*). *Scientia Horticulturae* 220, 154–159. DOI: 10.1016/j.scienta.2017.03.035.

D'Angelo, C.J. and Goldman, I.L. (2019) Annualization of the long day onion breeding cycle through threshold vernalization and dormancy disruption. *Crop Breeding, Genetics and Genomics* 1, e190009. DOI: 10.20900/cbgg20190009.

Davis, G.N. and Jones, H.A. (1944) Experiments with the transplant onion crop in California. *Bulletin of the California Experimental Station* 682.

De Bon, H. and Rhino, B. (1988). Flowering in the various cultivars of onions (*Allium cepa* L.) brought to Martinique (French West Indies). In: *Proceedings of the 4th EUCARPIA Allium Symposium*. IHR, Wellesbourne, UK, pp. 262–266.

DeMason, D.A. (1990) Morphology and anatomy of *Allium*. In: Rabinowitch, H.D. and Brewster, J.L. (eds) *Onions and Allied Crops*. CRC Press, Boca Raton, Florida, pp. 27–51.

Dembele, S. and Dubois, P. (1973). Composition des essences d'échalote (*Allium cepa* var. aggregatum). *Annales de Technologie Agricole* 22, 121–129.

Devi, S., Gulati, R., Tehri, K. and Poonia A. (2015) The pollination biology of onion (*Allium cepa* L.) – a review. *Agricultural Reviews* 36, 1–13. DOI: 10.5958/0976-0741.2015.00001.X.

Dowker, B.D. and Fennell, J.F.M. (1981) The relative performance of inbreds and open-pollinated populations of spring-sown onions. *The Journal of Agricultural Science* 97, 25–30. DOI: 10.1017/S0021859600035802.

Ebrahimi, R., Zamani, Z.A., Kashi, A.A.K. and Jabari, A. (2008) Comparison of fatty acids, mineral elements of 17 Iranian shallot landraces (*Allium hirtifolium* Boiss.) *Iran Journal of Food Science Technology* 5, 61–68.

Eshel, D., Teper-Bamnolker, P., Vinokur, Y., Saad, I., Zutahy, Y. and Rodov, V. (2014) Fast curing: a method to improve postharvest quality of onions in hot climate harvest. *Postharvest Biology and Technology* 88, 34–39. DOI: 10.1016/j.postharvbio.2013.09.002.

Esnault, F., Tromeur, C., Kermarrec, M.P. and Pouliquen, R. (2005) Effect of temperature on growth and flowering ability of three Jersey shallot (*Allium cepa* L. group Aggregatum) cultivars. *The Journal of Horticultural Science and Biotechnology* 80, 413–420. DOI: 10.1080/14620316.2005.11511953.

Esuyawukal-Moges, K. and Biruk-Masrie, Z. (2019) Growth and yield response of shallot (*Allium cepa* var. aggregatum) varieties to intra-row spacing in Eastern Amhara, Ethiopia. *Current Research in Agricultural Science* 6, 157–168. DOI: 10.18488/journal.68.2019.62.157.168.

EU Plant Variety Catalogues (2020) Available at: https://ec.europa.eu/food/sites/food/files/plant/docs/plant_variety_catalogues_vegetable-species.pdf (accessed 9 July 2022).

Evoor, S., Veere Gowda, R., Gangappa, E. and Krisna Monohar, R. (2007) Heterosis for yield, yield components and quality traits in onion (*Allium cepa* L.). *Karnataka Journal of Agricultural Science* 20, 813–815.

FAO (2019) Joint FAO/WHO Food Standards Programme Codex Committee on Fresh Fruits and Vegetables 21st Session Proposed Draft Standard for Onions and Shallot. *CODEX Alimentarius Commission, July 2019*. Available at: http://www.fao.org/fao-who-codexalimentarius/sh-proxy/en/?lnk=1&url=https%253A%252F%252Fworkspace.fao.org%252Fsites%252Fcodex%252FMeetings%252F-CX-731-21%252FWorking%2Bdocuments%252Fffv21_09e.pdf (accessed 28 April 2022).

Farid, N., Sugiharto, A.N., Herison, C., Purwito, A. and Sutjahjo, S.H. (2012) Analysis of combining ability, heterosis effect and heritability estimate of yield-related characters in shallot (*Allium cepa* var. *ascalonicum* Baker). *Agrivita* 34, 36–43. Available at: https://agrivita.ub.ac.id/index.php/agrivita/article/view/142/592 (accessed 9 July 2022).

Fattorusso, E., Iorizzi, M., Lanzotti, V. and Taglialatela-Scafati, O. (2002) Chemical composition of shallot (*Allium ascalonicum* Hort.). *Journal of Agricultural and Food Chemistry* 50, 5686–5690. DOI: 10.1021/jf020396t.https://doi.org/10.1021/jf020396t

Finkers, R., van Kaauwen, M., Ament, K., Burger-Meijer, K., Egging, R. *et al.* (2021) Insights from the first genome assembly of onion (*Allium cepa*). *G3 Genes|Genomes|Genetics* 11. DOI: 10.1093/g3journal/jkab243.

Firmansyah, A.M., Liana. T. and Bhermana, A. (2017) The effect of age, form and seed cutting to the growth and production of shallot (*Allium ascalonicum* L.) at dry lands with low altitude. *Journal of Agricultural Science and Food Research* 8, 194. Available at: https://www.longdom.org/open-access/the-effect-of-age-form-and-seed-cutting-to-the-growth-and-productionof-shallot-allium-ascalonicum-l-at-dry-lands-with-low-altitude.pdf (accessed 9 July 2022).

Fita, G.T. (2004) Manipulation of flowering for seed production of shallot *(Allium cepa* var. *ascalonicum* Backer*)*. PhD thesis, Hannover University, Germany. Available at: https://www.repo.uni-hannover.de/counterlike-stats/esstatistics/123456789/6521?locale=en (accessed 28 April 2022).

Fontenot, K., Sexton, M., Singh, R. and Strahan R. (2017) *Allium Crops. Onions, Shallots, Garlic. Louisiana State University, Agricultural Center*. Available at: https://www.lsuagcenter.com/~/media/system/0/1/f/8/01f81e42f920be38f87b75f9d34f4fc4/pub2318%20allium%20crops%20rev%200417pdf.pdf (accessed 28 April 2022).

French Certification (n.d.) Available at: https://ail-echalote-certifie.org/en/monitoring-process-and-certification-of-french-garlic-and-shallot/ (accessed 28 April 2022).

Fritsch, R.M. (1996) The Iranian species of *Allium* subg. Melnnocrommyum sect. Megaloprason (Alliaceae). *Nordic Journal of Botany* 16, 9–17. Available at: https://onlinelibrary.wiley.com/doi/pdf/10.1111/j.1756-1051.1996.tb00211.x (accessed 29 July 2022).

Friesen, N. and Klaas, M. (1998) Origin of some minor vegetatively propagated *Allium* crops studied with RAPD and GISH. *Genetic Resources and Crop Evolution* 45, 511–523. DOI: 10.1023/A:1008647700251.

Fritsch, R.M. and Friesen, N. (2002) Evolution, domestication and taxonomy. In: Rabinowitch, H.D. and Currah, L. (eds) *Allium Crop Science: Recent Advances*. CAB International, Wallingford, UK, pp. 5–30.

Fritsch, R.M. and Gurushidze, M. (2009) Phylogenetic relationships of ornamental species in *Allium* L. subg. Melanocrommyum (Webb et Berthel.) Rouy (Alliaceae). *Israel Journal of Plant Sciences* 75, 287–295. DOI: 10.1560/IJPS.57.4.287.

Galsurker, O., Faigenboim, A.D., Teper-Bamnolker, P., Daus, A., Fridman, Y. *et al.* (2017) Cellular and molecular changes associated with onion skin formation suggest involvement of programmed cell death. *Frontiers in Plant Science* 7, 2031.09. DOI: 10.3389/fpls.2016.02031.

Getachew, T., Asfaw, Z. (2004) *Achievements in Shallot and Garlic Research*. Research Report No. 38. Ethiopian Agricultural Research Organization, Addis Ababa.

Ghahremani-Majd, H. and Dashti, F. (2014) Genetic diversity of Persian shallot (*Allium hirtifolium* Boiss.) populations based on morphological traits and RAPD markers. *Plant Systematics and Evolution* 300, 1021–1030. DOI: 10.1007/s00606-013-0940-5.

Ghahremani-Majd, H., Dashti, F., Dastan, D., Mumivand, H., Hadian, J. and Esna-Ashari, M. (2012) Antioxidant and antimicrobial activities of Iranian mooseer (*Allium hirtifolium* Boiss) populations. *Horticultural and Environmental Biotechnology* 53, 116–122. DOI: 10.1007/s13580-012-0131-2.

Getahun, D. and Zelleke, A. (2009) Identification of appropriate planting period for shallot (*Allium cepa* L. var. *ascalonicum* Baker) seed production at Debre Zeit Ethiopia. In: *The Conference of the Crop Science Society of Ethiopia, 11*, Addis Abeba, Ethiopia, 26–28 Apr 2004, pp. 169–175. Available at: https://agris.fao.org/agris-search/search.do?recordID=ET2007000298) accessed 9 July 2022).

Getahun, D., Zelleke, A., Derso, E. and Kiflu, E. (2003) Storability of shallot cultivars (*Allium cepa* L. var. *ascalonicum* Baker) at Debre Zeit, Ethiopia. *Acta Horticulturae* 604, 639–646. DOI: 10.17660/ActaHortic.2003.604.78.

GEVES (n.d.) Variety and seed expertise. Available at: https://www.geves.fr/variety-seed-expertise/vegetable/marketing-and-certification-of-seeds-and-propagating-material-of-vegetable-species/ (accessed 9 July 2022).

Govaerts, R., Kington, S., Friesen, N., Fritsch, R.M., Snijman, D.A. *et al.* (2005–2021) World Checklist of Selected Plant Families. Available at: http://apps.kew.org/wcsp (accessed 9 July 2022).

Grubben, G.J.H. (1994) Constraints for shallot garlic and Welsh onion in Indonesia: a case study on the evolution of *Allium* crops in the equatorial tropics. *Acta Horticulturae* 358, 333–339. DOI: 10.17660/ActaHortic.1994.358.55.

Gulumbe, A.A., Abubakar, L., Sokoto, B.M. and Aliero, A.A. (2018) Heterosis for enhanced shelf life and earliness in onions (*Allium cepa* L.) genotypes. *Asian Research Journal of Agriculture* 10, 1–9. DOI: 10.9734/ARJA/2018/33865.

Gurushidze, M., Mashayekhi, S., Blattner, F.R., Friesen, N. and Fritsch, R.M. (2007) Phylogenetic relationships of wild and cultivated species of *Allium* section Cepa inferred by nuclear rDNA ITS sequence analysis. *Plant Systematics and Evolution* 269, 259–269. DOI: 10.1007/s00606-007-0596-0.

Hailekidan, B., Andargie, M. and Assefa, K. (2013) *In vitro* plantlet regeneration from the bulbs of shallot (*Allium cepa* var. group Aggregatum). *Research in Plant Sciences* 1, 45–52. DOI: 10.12691/plant-1-2-7.

Hanelt, P. (1990) Taxonomy, evolution, and history. In: Rabinowitch, H.D. and Brewster, J.L. (eds) *Onions and Allied Crops*. CRC Press, Boca Raton, Florida, pp. 1–26.

Havey, M.J. (1991) Molecular characterization of the interspecific origin of viviparous onion. *Journal of Heredity* 82, 501–503. DOI: 10.1093/oxfordjournals.jhered.a111137.

Hedjes, L.J. and Lister, C.E. (2007) The nutritional attributes of *Allium* species. *Crop Food Research Confidential Report No.1814.* A Report Prepared for Horticulture NZ. Available at: https://www.vri.org.nz/research/nutritional-attributes-of-allium-species/ (accessed 28 April 2022).

Hu, S.Y. (2005) *Food Plants of China*. Chinese University Press, Hong Kong.

Ikeda, H., Kinoshita, T., Yamamoto, T. and Yamasaki, A. (2019) Sowing time and temperature influence bulb development in spring-sown onion (*Allium cepa* L.). *Scientia Horticulturae* 244, 242–248. DOI: 10.1016/j.scienta.2018.09.050.

Jo, J., Purushotham P.M., Koeun, H., Lee, H.R., Nah, G. *et al.* (2017) Development of a genetic map for onion (*Allium cepa* L.) using reference-free genotyping-by-sequencing and SNP assays. *Frontiers in Plant Science* 8, 1606. DOI: 10.3389/fpls.2017.01606.

Jones, R.N. (1990) Cytogenetics. In: Rabinowitch, H.D. and Brewster, J.L. (eds) *Onion and Allied Crops*, Vol. I. CRC Press, Boca Raton, Florida, pp. 199–212.

Jones, H.A. and Davis, G.N. (1944) Inbreeding and heterosis and their relation to the development of new varieties of onion. *Technical Bulletin of USDA* 874, pp. 1–28. Available at: https://ageconsearch.umn.edu/record/169717 (accessed 28 April 2022).

Jones, H.A. and Mann, L.K. (1963) *Onions and Their Allies: Botany Cultivation and Utilization*. Interscience Publishers, New York.

Joslym, M.A. and Peterson, R.G. (1958) Reddening of white onion bulb purees. *Journal of Agricultural and Food Chemistry* 6, 754–765. DOI: 10.1021/jf60092a006.

Kamenetsky, R. and Rabinowitch, H.D. (2002) Florogenesis. In: Rabinowitch H.D. and Currah, L. (eds) *Allium Crop Science: Recent Advances*. CAB International, Wallingford, UK, pp. 31–57.

Kamenetsky, R. and Rabinowitch, H.D. (2017) Physiology of domesticated Alliums: onions, garlic, leek, and minor crops. In: Thomas, B., Murphy, D. and Murray, B. (eds) *Encyclopedia of Applied Plant Sciences*. Elsevier Science, pp. 255–261. Available at: https://doi.org/10.1016/b978-0-12-394807-6.00064-2 (accessed 29 July 2022).

Katis, N.I., Maliogka, V.I. and Dovas, C.I. (2012) Ch. 5 – Viruses of the genus Allium in the Mediterranean region. In: Loebenstein, G. and Lecoq, H. (eds) *Viruses and Virus Diseases of Vegetable in the Mediterranean Basin. Advances in Virus Research* 84, 163–208. Available at: https://doi.org/10.1016/B978-0-12-394314-9.00005-1 (accessed 29 July 2022).

Kedar, N., Levy, D. and Goldschmidt, E.E. (1975) Photoperiodic regulation of bulbing and maturation of Bet alpha onions (*Allium cepa* L.) under decreasing daylength conditions. *Journal of Horticultural Sciences* 50, 373–380. DOI: 10.1080/00221589.1975.11514646.

Kik, C. (2002) Exploitation of wild relatives for the breeding of cultivated *Allium* species. In: Rabinowitch, H.D. and Currah, L. (eds) *Allium Crop Science: Recent Advances*. CAB International, Wallingford, UK, pp. 409–430.

Khan, S.A., Amjad, M. and Khan, A. (2001) The extent of inbreeding depression in seven cultivars of onion (*Allium cepa* L). *International Journal of Agriculture and Biology* 3, 498–500.

Khandagale, K., Krishna, R., Roylawar, P., Ade, A.B., Benke, A. *et al.* (2020) Omics approaches in *Allium* research: progress and way ahead. *PeerJ* 8, e9824. DOI: 10.7717/peerj.9824.

Khodadadi, M. and Hassanpanah, D. (2010) Iranian onion (*Allium cepa* L.) cultivars responses to inbreeding depression. *World Applied Sciences Journal* 11, 426–428. Available at: http://www.idosi.org/wasj/wasj11(4)/8.pdf (accessed 9 July 2022).

Khokhar, K.M. (2008a) Effect of set-size and planting time on the incidence of bolting, bulbing, and seed yield in two onion cultivars. *The Journal of Horticultural Science and Biotechnology* 83, 481–487. DOI: 10.1080/14620316.2008.11512411.

Khokhar, K.M. (2008b) Effect of temperature and photoperiod on the incidence of bulbing and bolting in seedlings of onion cultivars of diverse origin. *The Journal of Horticultural Science and Biotechnology* 83, 488–496. DOI: 10.1080/14620316.2008.11512412.

Khokhar, K.M. (2017) Environmental and genotypic effects on bulb development in onion – a review. *The Journal of Horticultural Science and Biotechnology* 92, 448–454. DOI: 10.1080/14620316.2017.1314199.

Khokhar, K.M., Hadley, P. and Pearson, S. (2007) Effect of cold temperature durations of onion sets in store on the incidence of bolting, bulbing and seed yield. *Scientia Horticulturae* 112, 16–22. DOI: 10.1016/j.scienta.2006.12.038.

Khorasani, M., Mehrvarz, S.S. and Zarre, S. (2018) Scape anatomy and its systematic importance in the *Allium stipitatum* species complex (Amaryllidaceae). *Nordic Journal of Botany* 36, e02008. Available at: https://doi.org/10.1111/njb.02008 (accessed 29 July 2022).

Khosa, J.S., McCallum, J., Dhatt, A.S. and Macknight, R.C. (2016) Enhancing onion breeding using molecular tools. *Plant Breeding* 135, 9–20. DOI: 10.1111/pbr.12330.

Khrustaleva, L. (2018) Cytological details of genome. In: Shigyo, M., Khar, A. and Abdelrahman, M. (eds) *The Allium Genomes Compendium of Plant Genomes*. Springer, Cham, Switzerland, pp. 67–87. Available at: https://link.springer.com/content/pdf/10.1007%2F978-3-319-95825-5_5.pdf (accessed 9 July 2022).

Klaas, M. and Friesen, N. (2002) Molecular markers in *Allium*. In: Rabinowitch, H.D. and Currah, L. (eds) *Allium Crop Science – Recent Advances*. CAB International, Wallingford, UK, pp. 159–185.

Krontal, Y., Kamenetsky, R. and Rabinowitch, H.D. (1998) Lateral development and florogenesis of a tropical shallot – a comparison with bulb onion. *International Journal of Plant Sciences* 159, 57–64. Available at: https://www.journals.uchicago.edu/doi/pdf/10.1086/297521 (accessed 9 July 2022).

Krontal, Y., Kamenetsky, R. and Rabinowitch, H.D. (2000) Flowering physiology and some vegetative traits of short-day shallot: a comparison with bulb onion. *The Journal of Horticultural Science and Biotechnology* 75, 35–41. DOI: 10.1080/14620316.2000.11511197.

La Admin Code (2021) Available at: https://casetext.com/regulation/louisiana-administrative-code/title-7-agriculture-and-animals/part-xiii-seeds/chapter-7-certification-of-specific-cropsvarieties/subchapter-c-fruits-and-vegetables/section-xiii-785-shallot-seed-certification-standards-formerly-section-197 (accessed 3 May 2022).

Lancaster, J.E., Triggs, C.M., De Ruiter, J.M. and Gandar, P. (1996) Bulbing in onions: photoperiod and temperature requirements and prediction of bulb size and maturity. *Annals of Botany* 78, 423–430. DOI: 10.1006/ANBO.1996.0138.

Lang, G.A., Early, J.D., Martin, G.C. and Darnell, R.L. (1987) Endo-, para- and eco-dormancy: physiological dormancy and clarification of dormancy research. *HortScience* 22, 371–377.

Lee, R., Baldwin, S., Kenel, F., McCallum, J. and Macknight, R. (2013) *FLOWERING LOCUS T* genes control onion bulb formation and flowering. *Nature Communications* 4, 2884. DOI: 10.1038/ncomms3884.

Leelarungrayub, N., Rattanapanone, V., Chanarat, N. and Gebicki, J.M. (2006) Quantitative evaluation of the antioxidant properties of garlic and shallot preparations. *Nutrition* 22, 266–274. DOI: 10.1016/j.nut.2005.05.010.

Legin, A., Rudnitskaya, A., Seleznev, B., Gebicki, J.M., Sparfel, G. *et al.* (2004) Electronic tongue distinguishes onions and shallots. *Acta Horticulturae* 634, 183–191. DOI: 10.17660/ActaHortic.2004.634.22.

Leino, M.W., Solberg, S.Ø., Tunset, H.M., Fogelholm, J., Karlsson Strese, E.M. *et al.* (2018) Patterns of exchange of multiplying onion (*Allium cepa* L Aggregatum-Group) in Fennoscandian home gardens. *Economic Botany* 72, 346–356. DOI: 10.1007/s12231-018-9426-2.

Lercari, B. (1982) The effect of far-red light on the photoperiodic regulation of carbohydrate accumulation in *Allium cepa* L. *Physiology of Plants* 54, 475–479. Available at: https://doi.org/10.1111/j.1399-3054.1982.tb00711.x (accessed 29 July 2022).

Lercari, B. (1983) The role of ethylene in photoperiodic control of bulbing in *Allium cepa*. *Physiologia Plantarum* 59, 647–650. DOI: 10.1111/j.1399-3054.1983.tb06294.x.

Lercari, B. (1984) Role of phytochrome in photoperiodic regulation of bulbing and growth in the long day plant *Allium cepa*. *Physiologia Plantarum* 60, 433–436. DOI: 10.1111/j.1399-3054.1984.tb06088.x.

Lercari, B. and Micheli, P. (1981) Photoperiodic regulation of cytokin levels in leaf blades of *Allium cepa* L. *Plant and Cell Physiology* 22, 501–505. DOI: 10.1093/oxfordjournals.pcp.a076192.

Lestari, R.H.S., Sulistyaningsih, E. and Purwantoro, A. (2019) The effect of drying and storage on the quality of shallot (*Allium cepa* L aggregatum group) bulbs. *Agricultural Science* 3(3), 117–126.

Levy, D., Goldschmidt, E. and Kedar, N. (1979) Bulbing of onions (*Allium cepa* L.): the role of endogenous ethylene. *Plant, Cell & Environment* 2, 155–158. DOI: 10.1111/j.1365-3040.1979.tb00787.x.

Maaß, H.I. (1996) About the origin of the French grey shallot. *Genetic Resources and Crop Evolution* 43, 291–292. Available at: https://fdocuments.in/document/about-the-origin-of-the-french-grey-shallot.html (accessed 9 July 2022).

Major, N., Ban, S.G., Urlić, B., Ban, B., Dumičić, G. and Perković, J. (2018) Morphological and biochemical diversity of shallot landraces preserved along the Croatian coast. *Frontiers in Plant Science*. Available at: https://doi.org/10.3389/fpls.2018.01749 (accessed 29 July 2022).

Marais, A., Faure, C., Theil, S. and Candresse, T. (2019) Characterization of the virome of shallots affected by the shallot mild yellow stripe disease in France. *PLOS ONE* 14(7), e0219024. DOI: 10.1371/journal.pone.0219024.

Masuzaki, S.-I., Shigyo, M. and Yamaguchi, N. (2006) Complete assignment of structural genes involved in flavonoids biosynthesis influencing bulb color to individual chromosomes of the shallot (*Allium cepa* L.). *Genes & Genetic Systems* 81, 255–263. DOI: 10.1266/ggs.81.255.

McCallum, J. (2007) Onion. In: Kole, C. (ed.) *Vegetables: Genome Mapping and Molecular Breeding in Plants*, Vol. 5. Springer, New York, pp. 331–347.

Messiaen, C.M., Cohat, J., Pichon, M., Leroux, J.P. and Beyries, A. (1993) *Les Allium Alimentaires Reproduits pare voie Vegetative*. INRA, Paris.

Mondal, M.F., Brewster, J.L., Morris, G.E.L. and Butler, H.A. (1986a) Bulb development in onion. I. Effects of plant density and sowing date in field conditions. *Annals of Botany* 58, 187–195.

Mondal, M.F., Brewster, J.L., Morris, G.E.L. and Butler, H.A. (1986b) Bulb development in onions. II. Effects of red:far red spectral ratio and photon flux density. *Annals of Botany* 58, 197–206.

Mondal, M.F., Brewster, J.L., Morris, G.E.L. and Butler H.A. (1986c) Bulb development in onion (*Allium cepa* L.). III. Effects of the size of adjacent plants, shading by neutral and leaf filters, irrigation and nitrogen regime and the relationship between the red:far-red spectral ratio in the canopy and leaf area index. *Annals of Botany* 58, 207–219. Available at: https://doi.org/10.1093/oxfordjournals.aob.a087199 (accessed 29 July 2022).

Moshin, G., Farruk, A., Rahman, S. and Islam, M.S. (2017) Use of male sterility and synthesis of maintainer line for hybrid seed production in onion (*Allium cepa* L.) in Bangladesh. *Bangladesh Journal of Plant Breeding and Genetics* 29, 31–38. DOI: 10.3329/bjpbg.v29i1.33705.

Mushtaq, S., Amjad, M., Ziaf, K. and Afzal, I. (2018) Gibberellins application timing modulates growth, physiology, and quality characteristics of two onion (*Allium cepa* L.) cultivars. *Environmental Science and Pollution Research* 25(25), 25155–25161. DOI: 10.1007/s11356-018-2542-9.

Naamni, F., Rabinowitch, H.D. and Kedar, N. (1980). The effect of GA3 application on flowering and seed production in onion (*Allium cepa* L.). *Journal of the American Society for Horticultural Science* 105, 164–167.

NAK Tuinbouw (2018) Available at: https://www.naktuinbouw.com/sites/default/files/Inspection%20Regulations%20Naktuinbouw_2018-EN-WEB_08-02-2018.pdf (accessed 9 July 2022).

Neal, C.A. and Ellerbrock, L.A. (1986) Optimizing time of harvest for seed of *Allium cepa* L. *Journal of Seed Technology* 10, 37–45. Available at: https://www.jstor.org/stable/23433004 (accessed 9 July 2022).

New World Encyclopedia (2019) Available at: https://www.newworldencyclopedia.org/entry/shallot (accessed 9 July 2022).

Nguyen, N.H., Driscoll, H.E. and Specht, C.D. (2008) A molecular phylogeny of the wild onions (*Allium*; Alliaceae) with a focus on the western North American center of diversity. *Molecular Phylogenetics and Evolution* 47, 1157–1172. DOI: 10.1016/j.ympev.2007.12.006.

Nikus, O. and Mulugeta, F. (2010) *Onion Seed Production Techniques: A Manual for Extension Agents and Seed Producers*. FAO-Crop Diversification and Marketing Development Project Asella, Ethiopia. Available at: https://fdocuments.net/document/onion-seed-production-techniques-fao-seed-production-techniques-a-manual-for-extension.html (accessed 29 July 2022).

Odeny, D.A. and Narina, S.S. (2011) *Alliums*. In: Kole, C. (ed.) *Wild Crops Relatives: Genomics and Breeding Resources Vegetables*. Springer, pp. 1–10. DOI: 101007/978-3-642-20450-0.

OECD (2014) *Shallots, International Standards for Fruit and Vegetables*. OECD Publishing, Paris. Available at: https://doi.org/10.1787/9789264213265-en-fr (accessed 9 July 2022).

Oregon State University (1982) Shallot planting density trial. Available at: https://agsci.oregonstate.edu/oregon-vegetables/shallot-planting-density-trial (accessed 9 July 2022).

Oregon State University (1987) Introduction to the Overwintering Onions Report (1987). Available at: https://agsci.oregonstate.edu/oregon-vegetables/introduction-overwintering-onions-report-1987 (accessed 9 July 2022).

Oregon State University (2010) Shallots. Available at: https://horticulture.oregonstate.edu/oregon-vegetables/shallots-0 (accessed 9 July 2022).

Özer, N. and Köycüm, N.D. (2004) Seed-borne fungal diseases of onion and their control. In: Mukerji, K.G. (ed.) *Fruit and Vegetable Diseases. Management of Fruits and Vegetables*, vol. 1. Springer, Dordrecht, The Netherlands, pp. 281–306.

Palupi, E.R., Manik, F. and Suhartanto, M.R. (2017) Can we produce true seed of shallot (TSS) from small size shallot sets? *Journal of Tropical Crop Science* 4, 26–31. DOI: 10.29244/jtcs.4.1.26-31.

Pandey, U.B. (2006) Potato onion (Multiplier onion). In: Peter, K.V. (ed.) *Handbook of Herbs and Spices*, vol. 3. Woodhead Publishing, Cambridge, UK, pp. 495–501. Available at: https://vdoc.pub/documents/handbook-of-herbs-and-spices-vol-1-291gch3bjv30 (accessed 9 July 2022).

Pathak, C.S. (1994) *Allium* improvement for the tropics: problems and AVRDC strategy. *Acta Horticulturae* 358, 23–32. DOI: 10.17660/ActaHortic.1994.358.2.

Pathak, C.S. (2000) Hybrid seed production in onion. *Journal of New Seeds* 1, 3–4, 89–108. DOI: 101300/J153v01n03_04.

PDO (n.d.) Available at: https://ec.europa.eu/info/food-farming-fisheries/food-safety-and-quality/certification/quality-labels/quality-schemes-explained_en (accessed 9 July 2022).

PGI (n.d.) Available at: https://www.inao.gouv.fr/eng/Official-signs-identifying-quality-and-origin/Protected-Geographical-Indications (accessed 9 July 2022).

Pérez-Moreno, L., Navarro-León, J.N., Ramírez-Malagón, R. *et al.* (2010) Impacto e identificación de virus fitopatógenos sobre rendimiento y calidad de ajo en el estado de Guanajuato México. *Revista Mexicana de Fitopatología* 28, 75–85. Available at: http://www.scielo.org.mx/scielo.php?script=sci_arttext&pid=S0185-33092010000200003 (accessed 9 July 2022).

Permadi, A.H. and Van der Meer, Q.P. (1993) *Allium cepa* L cv. group Aggregatum. In: Siemonsma, J.S. and Piluek, K. (eds) *Plant Resources of South-East Asia. 8. Vegetables.* Pudoc Scientific Publishers, Wageningen, The Netherlands, pp. 64–68.

Peška, V., Mandáková, T., Ihradská, V. and Fajkus, J. (2019) Comparative dissection of three giant genomes: *Allium cepa, Allium sativum,* and *Allium ursinum. International Journal of Molecular Sciences* 20, 733. DOI: 10.3390/ijms20030733.

Peters, R. (1990) Seed production in onions and some other *Allium* species. In: Rabinowitch, H.D. and Brewster, J.L. (eds) *Onions and Allied Crops. I. Botany, Physiology and Genetics*. CRC, Boca Raton, Florida, pp 161–176. Available at: https://www.taylorfrancis.com/chapters/edit/10.1201/9781351075169-8/seed-production-onions-allium-species-ross-peters?context=ubx&refId=99c92945-6e5b-4d0b-9bac-cf12ebbbb4e8 (accessed 9 July 2022).

Petitti, M., Ortolani, L., Schäfer, F. and Messmer, M. (2021) *The state of organic seed in Europe*. EU LIVESEED Project. Available at: https://www.liveseed.eu/wp-content/uploads/2021/03/Booklet2-LIVESEED_2021_web.pdf (accessed 9 July 2022).

Phillips, N. (2010) Seed and bulb dormancy characteristics in New World *Allium* L (Amaryllidaceae): a review. *International Journal of Botany* 6, 228–234. Available at: https://docsdrive.com/pdfs/ansinet/ijb/2010/228-234.pdf (accessed 9 July 2022).

Potato Onion (2020) Available at: https://www.cultivariable.com/instructions/root-crops/how-to-grow-potato-onion (accessed 9 July 2022).

Proctor, F.J. (1987) *Report on Visit to Ethiopia to Discuss Post-Harvest Storage and Handling of Allium Species 20 March–1 April, 1987*. Tropical Development and Research Institute, Overseas Development Institute, London.

Prosea Shallot (2016) Available at: https://uses.plantnet-project.org/en/Allium_cepa_Aggregatum_(PROSEA) (accessed 9 July 2022).

Pudzianowska, M., Gajewski, M., Przybył, J.L., Buraczyńska, A., Gaczkowska, O. *et al.* (2012) Influence of storage conditions on flavonoids content and antioxidant activity of selected shallot (*Allium cepa* var. *ascalonicum* Backer) hybrid cultivars. *Vegetable Crops Research Bulletin* 77, 101–111. DOI: 10.2478/v10032-012-0019-9.

Puizina, J. (2013) Shallots in Croatia – genetics, morphology, and nomenclature. *Acta Botanica Croatica* 72, 387–398. DOI: 10.2478/botcro-2013-0016.

Puizina, J. and Papfš, D. (1996) Cytogenetical evidence for hybrid structure and origin of diploid and triploid shallots (*Allium cepa* var. *viviparum* Liliaceae) from Dalmatia (Croatia). *Plant Systematics and Evolution* 199, 203–215. DOI: 10.1007/BF00984905.

Rabinowitch, H.D. (1979) Doubling of onion bulbs as affected by size and planting date of sets. *Annals of Applied Biology* 93, 63–65. DOI: 10.1111/j.1744-7348.1979.tb04727.x.

Rabinowitch, H.D. (1990a) Physiology of flowering. In: Rabinowitch, H.D. and Brewster, J.L. (eds) *Onions and Allied Crops. I. Botany Physiology and Genetics*. CRC, Boca Raton, Florida, pp. 113–134. Available at: https://www.taylorfrancis.com/chapters/edit/10.1201/9781351075169-5/physiology-flowering-haim-rabinowitch?context=ubx&refId=bfe2fcb9-6c74-4249-93d7-62d3b989e5ed (accessed 9 July 2022).

Rabinowitch, H.D. (1990b) Seed development. In: Rabinowitch, H.D. and Brewster, J.L. (eds) *Onions and Allied Crops. I. Botany Physiology and Genetics*. CRC, Boca Raton, Florida, pp. 151–159. Available at: https://www.taylorfrancis.com/chapters/edit/10.1201/9781351075169-7/seed-development-haim-rabinowitch?context=ubx&refId=07c81d22-aadd-4d38-ac1e-0564eea580f9 (accessed 9 July 2022).

Rabinowitch, H.D. (2021) Shallot (*Allium cepa* L. Aggregatum group) breeding. In: Al-Khayri, J.M., Mohan, J.S. and Johnson, D.V. (eds) *Advances in Plant Breeding Strategies: Vegetable Crops. 8. Bulbs, Roots, Tubers, Flower Heads and Mushrooms*. Springer, pp. 99–154. Available at: https://doi.org/10.1007/978-3-030-66965-2 (accessed 31 July 2022).

Rabinowitch, H.D. and Kamenetsky, R. (2002) Shallot (*Allium cepa* Aggregatum group) In: Rabinowitch, H.D. and Currah, L. (eds) *Allium Crop Science: Recent Advances*. CAB International, Wallingford, UK, pp. 409–430.

Rabinowitch, H.D. and Kamenetsky Goldstein, R. (2020) *Allium* crops. In: Wien, H.C. and Stützel, H. (eds) *Physiology of Vegetable Crops*, 2nd edn. CAB International, Wallingford, UK, pp. 421–456.

Rahayu, H.S.P., Muchtar, S. and Saidah, S. (2019) The feasibility and farmer perception of true shallot seed technology in Sigi District, Central Sulawesi, Indonesia. *Asian Journal of Agriculture* 3, 16–21. DOI: 10.13057/asianjagric/g030103.

Rao, G.M. and Suryanarayan, M.C. (1989) Effect of honeybee pollination on seed yield in onion (*Allium cepa* L). *Indian Bee Journal* 51, 9–11.

Rasekh Adel, M., Sadeghi Namaghi, H. and Hosseini, M. (2013) Effects of insect pollinators on onion seed production quality and quantity. *Journal of Crop Protection* 2, 395–402. Available at: https://www.sid.ir/en/journal/ViewPaper.aspx?id=431706 (accessed 9 July 2022).

Ravindran, P.N. (2017) Shallot *Allium cepa* Aggregatum group. In: *The Encyclopedia of Herbs and Spices*. CAB International, Wallingford, UK, pp. 886–890.

Rosliani, R., Hidayat, I.M., Sulastrini, I. and Hilman Y. (2016) Dissemination of technology for shallot (*Allium ascalonicum* L.) seed production using true shallot seed (TSS) in Indonesia. *Acta Horticulturae* 1143, 345–351. DOI: 10.17660/ActaHortic.2016.1143.49.

Ruaysoongnern, S. (1994) Management factors affecting keeping quality of shallot in Sisaket, North Eastern Thailand. *Acta Horticulturae* 358, 375–381. DOI: 10.17660/ActaHortic.1994.358.62.

Saraswathi, T., Sathiyamurthy, V.A., Tamilaselvi, N.A. and Harish, S. (2017) Review on aggregatum onion (*Allium cepa* L var. *aggregatum* Don). *International Journal of Current Microbiology and Applied Sciences* 6, 1649–1667. DOI: 10.20546/ijcmas.2017.604.201.

Seidemann, J. (2005) *World Spice Plants: Economic Usage, Botany, Taxonomy*. Springer Science & Business Media.

Shabir, S., Sharma, G. and Irfan Ali, M. (2020) A review: *Allium cepa aggregatum*, characteristics, and properties. *Journal of Environmental Science and Technology* 6, 1–6. Available at: https://www.gyanvihar.org/journals/index.php/2020/03/13/a-review-allium-cepa-aggregatum-characteristics-and-properties (accessed 9 July 2022).

Shigyo, M. and Kik, C. (2007) Onions. In: Prohens, J. and Nuezm F. (eds) *Vegetables II: Fabaceae Liliaceae Solanaceae and Umbelliferae.* Springer, pp. 121–159.

Shigyo, M., Khar, A. and Abdelrahman, L. (2018) *The Allium Genomes*. Springer Nature, Switzerland. Available at: https://books.google.co.il/books?id=__1uDwAAQBAJ&pg=PA168&lpg=PA168&dq=SHALLOT+GENETICS&source=bl&ots=KaDgoZ7ciJ&sig=ACfU3U2ogJ7y9Np_628lDVbasaSOD4mikA&hl=en&sa=X&ved=2ahUKEwi-tsOwi-fkAhWMasAKHeVpAPk4ChDoATAPegQICBAB#v=onepage&q=SHALLOT%20GENETICS&f=true (accessed 9 July 2022).

Shimeles, A. (2014) The performance of true seed shallot lines under different environments of Ethiopia. *The Journal of Agricultural Science* 59, 129–139. DOI: 10.2298/JAS1402129S.

Shimeles, A. and Dessalenge, L. (2015) The performance of true seed shallot lines under two methods of planting at different environments of Ethiopia. *Research Journal of Agricultural and Environmental Management* 4, 174–179.

Sinnadurai, S. (1973) Shallot farming in Ghana. *Economic Botany* 27, 438–441. DOI: 10.1007/BF02860697.

Sinnadurai, S. and Amuti, S.K. (1971) Dormancy of shallots in Ghana. *Experimental Agriculture* 7, 17–20. DOI: 10.1017/S0014479700004737.

Smith, B.M. and Crowther, T.C. (1995) Inbreeding depression and single cross hybrids in leeks (*Allium ampeloprasum* ssp. *porrum*). *Euphytica* 86, 87–94. DOI: 10.1017/S0014479700004737.

Sobeih, W.Y. and Wright, C.J. (1987) The photoperiodic regulation of bulbing in onions (*Allium cepa* L.). 3. Response to red-far-red ratio and cyclic lighting. *The Journal of Horticultural Science and Biotechnology* 62, 379–389. DOI: 10.1080/14620316.1987.11515795.

Soegianto, A., Sugiharto, A.N. and Windiastika, G. (2011) Molecular identification of shallot progenitors generated from true seeds by PCR based techniques. *Journal of Agriculture and Food Technology* 1,

145–148. Available at: https://www.textroad.com/pdf/JAFT/J.%20Agric.%20Food.%20Tech.,%201(8)%20145-148,%202011.pdf (accessed 9 July 2022).

South African Ministry of Agriculture (2015) Available at: https://www.ecolex.org/es/details/legislation/regulations-relating-to-the-grading-packing-and-marking-of-onions-and-shallots-intended-for-sale-in-the-republic-of-south-africa-no-r-192-of-2015-lex-faoc143821/ (accessed 9 July 2022).

Stephens, J.M. (2018) *Onion, Potato – Allium cepa L. (Aggregatum group)*. University of Florida, IFAS Extension. Available at: http://edis.ifas.ufl.edu/pdffiles/MV/MV10000.pdf (accessed 9 July 2022).

Stevens, J. and Bougourd, S. (1988) Inbreeding depression and the outcrossing rate in natural populations of *Allium schoenoprasum* L. (wild chives). *Heredity* 60, 257–261. DOI: 10.1038/hdy.1988.40.

Stow, J.R. and Ward, C.M. (1978) The effect of transfer from low to high temperatures on losses of stored bulb onions (*Allium cepa*). *Annals of Applied Biology* 88, 193–197. DOI: 10.1111/j.1744-7348.1987.tb01475.x.

Sulistyaningsih, E., Yamashita, K. and Tashiro, Y. (2002) Haploid induction from F_1 hybrids between CMS shallot with *Allium galanthum* cytoplasm and common onion by unpollinated flower culture. *Euphytica* 125, 139–144. DOI: 10.1023/A:1015756729401.

Swamy, K.R.M. and Veere Gowda, R. (2006) *Leek and Shallot Handbook of Herbs and Spices*, vol. 3. Woodhead Publishing Series in Food Science Technology and Nutrition, pp. 365–389. Available at: https://doi.org/10.1533/9781845691717.3.365 (accessed 9 July 2022).

Tabor, G. and Zeleke, A. (2004) *Achievements in Shallot and Garlic Research*. Research report No. 36. Ethiopian Agricultural Research Organization, Addis Ababa, Ethiopia. Available at: http://webcache.googleusercontent.com/search?q=cache:YjgBIXYC3FIJ:publication.eiar.gov.et:8080/xmlui/bitstream/handle/123456789/2361/RESEARCH%2520REPORT%2520NO%252036.pdfabbyyy.pdf%3Fsequence%3D1%26isAllowed%3Dy+&cd=1&hl=en&ct=clnk&gl=il (accessed 9 July 2022).

Tashiro, Y., Miyazaki, S. and Kanazawa, K. (1982) On the shallot cultivated in the countries of southeastern Asia. *Bulletin of the Faculty of Agriculture-Saga University (Japan)* 53, 65–73.

Taşkın, H., Baktemur, G., Kurul, M. and Büsyükalaca, S. (2013) Use of tissue culture techniques for producing virus-free plant in garlic and their identification through real-time PCR. *The Scientific World Journal 2013*, Article ID 781282. DOI: 10.1155/2013/781282.

Tekalign, T., Abdissa, Y. and Pant, L.M. (2012) Growth, bulb yield and quality of onion (*Allium cepa* L.) as influenced by nitrogen and phosphorus fertilization on vertisol. II. Bulb quality and storability. *African Journal of Agricultural Research* 7, 5980–5985. Available at: https://doi.org/10.5897/ajar10.1025 (accessed 29 July 2022).

Tendaj, M. (2005) Shallot production and research in Poland. In: *Vegetable Crops Research Bulletin 62*. Research Institute for Vegetable Crops, Skierniewice, Poland, pp. 55–60.

Tendaj, M. and Mysiak, B. (2013) The effect of summer seedling planting dates on the development of seed stalks in shallot (*Allium cepa* L. var. *ascalonicum* Backer). *Hortum Cultusm Acta Scientiarum Polonorum* 12(6), 57–66. Available at: https://czasopisma.up.lublin.pl/index.php/asphc/article/view/2966 (accessed 9 July 2022).

Tendaj, M., Mysiak, B. and Krawiec, M. (2013) The effect of storage temperature of steckling bulbs on seed stalk development and seed yield of shallot (*Allium cepa* L var. *ascalonicum* Backer). *Acta Agrobotanica* 66(3), 41–48. DOI: 10.5586/aa.2013.036.

Tendaj, M., Krawiec, M., Palonka, S. and Mysiak, B. (2014) The effect of cultivation method on selected traits related to the sowing value of shallot (*Allium cepa* L var. *ascalonicum* Backer) seed. *Acta Scientiarum Polonorum Hortorum Cultus* 13, 107–115. Available at: https://czasopisma.up.lublin.pl/index.php/asphc/article/view/2795 (accessed 9 July 2022).

Terabun, M. (1965) Studies on the bulb formation in onion plants. I. Effects of light quality on the bulb formation and growth. *Journal of the Japanese Society for Horticultural Science* 34, 52–60. DOI: 10.2503/jjshs.34.196.

Tesfa, T., Woldestsadik, K. and Bayu, W. (2015) Shallot yield, quality and shelf-life as affected by nitrogen fertilizer. *International Journal of Vegetable Science* 21, 454–466. DOI: 10.1080/19315260.2014.895790.

Thai Ministry of Agriculture and Cooperatives (2008) *Thai Agricultural Standard. TAS* 1509–2008.

Traditional French Shallot (n.d.) Available at: http://plant-echalote.fr/en/plant-echalote.html (accessed 9 July 2022).

UNECE (2002) Seed shallots and the proposed French standards for shallots. Available at: https://unece.org/fileadmin/DAM/trade/agr/meetings/ge.01/document/2002_inf/2002_i15.pdf (accessed 9 July 2022).

UPOV (2007) Onion, echalion and shallot. Available at: https://www.upov.int/edocs/mdocs/upov/en/tc/43/tg_46_7_proj_3.pdf (accessed 9 July 2022).

Van den Brink, L. and Basuki, R.S. (2012) Production of true seed shallots in Indonesia. *Acta Horticulturae* 958, 115–120. DOI: 10.17660/ActaHortic.2012.958.12.

Van Kampen, J. (1970) Shortening the breeding cycle in onions. PhD thesis, Wageningen Agricultural University, The Netherlands.

Van Nocker, S. (2001) The molecular biology of flowering. *Horticultural Reviews* 27, 1–39.

Wang, H.X. and Ng, T.B. (2002) Ascalin, a new anti-fungal peptide with human immunodeficiency virus type 1 reverse transcriptase-inhibiting activity from shallot bulb. *Peptides* 23, 1025–1029. DOI: 10.1016/s0196-9781(02)00032-3.

Wang, M.R., Cui, Z.H., Li, J.W., Hao, X.Y., Zhao, L. *et al.* (2018) *In vitro* thermotherapy-based methods for plant virus eradication. *Plant Methods* 14, 87. DOI: 10.1186/s13007-018-0355-y.

Wang, M.R., Hamborg, Z., Blystad, D.R. and Wang, Q.C. (2020) Combining thermotherapy with meristem culture for improved eradication of onion yellow dwarf virus and shallot latent virus from infected *in vitro*-cultured shallot shoots. *Annals of Applied Biology* 178, 442–449. DOI: 10.1111/aab.12646.

Wassu, M., Woldetsadik, K. and Bekele, K. (2018) Registration of a new “Improved Huruta” shallot variety with true seed production potential. *East African Journal of Sciences* 12, 77–82. Available at: https://www.ajol.info/index.php/eajsci/article/view/181544 (accessed 9 July 2022).

Williams, I.H. and Free, J.B. (1974) The pollination of onion (*Allium cepa* L.) to produce hybrid seed. *Journal of Applied Ecology* 11, 409–417. DOI: 10.2307/2402194.

Woldetsadik, S.K. (2003) Shallot *(Allium cepa* var. *ascalonicum)* responses to plant nutrients and soil moisture in a sub-humid tropical climate. Doctoral thesis, Swedish University of Agricultural Sciences Alnarp, Sweden.

Woldetsadik, S.K. and Workneh, T. (2010) Effects of nitrogen levels harvesting time and curing on quality of shallot bulb. *African Journal of Agricultural Research* 5, 3342–3353. Available at: https://citeseerx.ist.psu.edu/viewdoc/download?doi=10.1.1.1014.7779&rep=rep1&type=pdf (accessed 9 July 2022).

Wright, C.J. and Sobeih, W.Y. (1986) The photoperiodic regulation of bulbing in onions (*Allium cepa* L.). I. Effects of irradiance. *Journal of Horticultural Science* 61, 331–335. DOI: 10.1080/14620316.1986.11515709.

Yamazaki, H., Shiraiwa, N., Itai, A. and Honda, I. (2015) Involvement of gibberellins in the regulation of tillering in Welsh onion (*Allium fistulosum* L.). *The Horticulture Journal* 84, 334–341. DOI: 10.2503/hortj.MI-050.

Yasin, H.J. and Bufler, G. (2007) Dormancy and sprouting in onion (*Allium cepa* L.) bulbs. I. Changes in carbohydrate metabolism. *Journal of Horticultural Science and Biotechnology* 82, 89–96. DOI: 10.1080/14620316.2007.11512203.

Yusupov, Z., Deng, T., Volis, S., Khassanov, F., Makhmuradov, D. *et al.* (2020) Phylogenomics of *Allium* section Cepa (Amaryllidaceae) provides new insights on the domestication of onion. *Plant Diversity* 43, 102–110. Available at: https://doi.org/10.1016/j.pld.2020.07.008 (accessed 29 July 2022).

Zheng, S.J., Henken, B., de Maagd, R.A., Purwito, A., Krens, F.A. *et al.* (2005) Two different *Bacillus thuringiensis* toxin genes confer resistance to beet armyworm (*Spodoptera exigua* Hübner) in transgenic Bt-shallots (*Allium cepa* L.) *Transgenic Research* 14, 261–272. DOI: 10.1007/s11248-005-0109-2.

6 Crop Development – Garlic

Rina Kamenetsky Goldstein* and Einat Shemesh-Mayer*
Agricultural Research Organization, The Volcani Institute, Rishon-LeZion, Israel

1. Introduction

Garlic, *Allium sativum* L. (Amaryllidaceae), is one of the world's most popular vegetable and medicinal crops. Annual garlic farming area of 1.6 million ha stretches from the equator to about 50° north and south latitudes, and global production amounts to ~30 million tons. China is by far the largest producer (23 MT/year), followed by India (2.9 MT) (FAOSTAT, 2021).

Garlic development, production, ripening and harvest benefit from mild winters and sunny dry summers, thus local environments greatly affect the plants physiology and morphological variability, their flavour, pungency, bulb characteristics, earliness and reproductive traits (Meredith, 2008).

Among Alliums, garlic generates the highest concentration of organosulphur compounds such as non-protein cysteine and glutathione derivatives. The major precursor of these compounds is sulphur-alk(en)yl cysteine sulfoxide (alliin), which, upon tissue-crushing, is catalyzed by alliinase to pyruvate, ammonia and thiosulfinate allicin, thus releasing the typical aroma and health-beneficial compounds (Block, 2010; Ramirez *et al.*, 2017; Putnik *et al.*, 2019; see Chapter 3, this volume). Flavour precursors generally amount to 1–5% of the garlic's dry weight (Martins *et al.*, 2016; Liu *et al.*, 2020) and their levels are affected by genotype, growth and storage conditions.

A. sativum is only known in cultivation, while its wild relatives commonly grow in Central Asia. After garlic domestication ~10,000 years ago, it was introduced to the Mediterranean basin, India and China and therefrom to other world regions (Engeland, 1991; see Chapter 1, this volume).

Presumably, native garlic produced only small bubs, flowers and seeds, similar to wild *Allium* species. Following domestication, local selections for large bulbs and earliness led to assimilates' diversion into the storage organs rather than the reproductive parts. The consequently limited resources for blooming resulted in sterility (Etoh and Simon, 2002). Hence, all garlic commercial varieties are clonally propagated and improvements are limited to selections of random mutations (Takagi, 1990) or stable epigenetic effects (Gimenez and Lampasona, 2018).

Collections of fertile genotypes (Etoh and Simon, 2002) and advances in fertility restoration by physiological and agro-manipulations resulted in sexual hybridization, recombination and increased variation (Jenderek and Zewdie, 2005; Kamenetsky, 2007), thus facilitating the breeding of seed-propagated garlic (Shemesh-Mayer and Kamenetsky Goldstein, 2021).

*Addresses for correspondence: vhrkamen@volcani.agri.gov.il; shemeshe@volcani.agri.gov.il

 Edible Alliums: Botany, Production and Uses (eds H.D. Rabinowitch and B. Thomas)
DOI: 10.1079/9781789249996.0006

The diploid garlic has a huge nuclear genome, and therefore functional genomic data was based mainly on transcriptome assembly and molecular analyses. However, in 2020, garlic became a first *Allium* species with a sequenced genome (Sun *et al.*, 2020). Analysis of garlic genome reveals an eruption of transposable elements, explaining the substantial expansion of the garlic genome.

In recent decades, major advances in garlic science follow three main directions:

1. Garlic physiology, flowering and dormancy mechanisms, genomic data, and traits' specific molecular markers paved innovative directions to the crop improvement.
2. Fertility restoration and true-seed production opened new horizons for classical breeding.
3. Improved agro-technologies and storage practices resulted in increased garlic production, marketing and customers' awareness of its functional benefits.

Crop production and storage, disease control, nutrition and biochemistry were thoroughly reviewed (Tripathi *et al.*, 2017; Lakshmi and Sekhar, 2018; Madhu *et al.*, 2019; El-Saber Batiha *et al.*, 2020). Therefore, we focus on recent advances in garlic development, clonal and sexual propagation, and physiological and genetic research.

2. Genetic Resources and Horticultural Diversity

A. sativum belongs to section *Allium* that consists of 114 species, ~25 of which are closely related to garlic, *e.g. A. longicuspis* L. in Central Asia, *A. tuncelianum* (Kollman) Özhatay, Matthews, Siraneci in Turkey, and *A. moschatum* L. in the Caucasus (Mathew, 1996; see Chapter 1, this volume).

Millennia of human selection and strong interaction with the environment yielded a variety of phenotypic expressions (Portela, 2001; Meredith, 2008; Fig. 6.1). Several garlic classifications are based on bio-geographical, morphological and horticultural traits, namely:

1. Bio-geographical division points at *Longicuspis* type from Central Asia, as the progenitor of four types: *Sativum* from the Mediterranean area, *Ophioscorodon* from central and eastern Europe, and *Subtropical* and *Pekinense* subtypes from south, southeast, and east Asia (Maaß and Klaas, 1995; Fritsch and Friesen, 2002).
2. Morphological traits based on bolting ability categorize varieties as *complete bolters* with a long, thick stalk, many topsets and umbels with many flowers or topsets; *semi-bolters* with a thin, short stalk and a few large topsets, and *non-bolters*, which produce only cloves (Takagi, 1990, Figs 6.1, 6.2). Accordingly, garlic commercial cultivars are classified as bolting *hardneck* and non-bolting *softneck* types (Engeland, 1991).
3. A wider horticultural division is based on plant phenotype, including bulb structure, bolting ability and production region (Meredith, 2008; Woodward, 2014; Fig. 6.2). Presumably, hardneck purple stripe is genetically closest to wild garlic. Marbled purple stripe, glazed purple stripe, porcelain and rocambole are strong bolters; Asiatic, Creole, Middle Eastern, and turban are semi-bolters, while artichoke, silverskin and subtropical types usually do not bolt.

In Europe, varieties are classified as autumn-planted bolters, autumn-planted non-bolters and spring-planted non-bolters (France, http://plant-certifie-ail.org, 2019; Russia, Belarus and Ukraine, Popkov, 2012).

Most of the 600 garlic world cultivars (Shemesh-Mayer and Kamenetsky Goldstein, 2021) were introduced from one region to another. For instance, the US-dominant cultivars 'California Early' and 'California Late' were probably brought from Italy and France. 'California Early' exhibits better tolerance to climatic conditions, while 'California Late' is less adapted to hot weather, but keeps better in storage. 'Music', a porcelain-type cultivar from Italy, is adapted to Canadian cold climate (Meredith, 2008). The Israeli short-day 'Shani', which was introduced from South America in the 1960s, is well adapted to the Mediterranean climate. The Ukrainian 'Lubasha', Central Asia 'Dungansky' and the Czech 'Lukan' and 'Havran' are popular in local markets for their high quality. It is possible, however, that the same varieties grown in various regions carry different names.

Clonal germplasm collections are maintained in different countries as a source of genetic variability. Large, clonally propagated collections are kept in Leibniz Institut für Pflanzengenetik und Kulturpflanzenforschung (IPK), Gatersleben, Germany (https://www.ipk-gatersleben.de/); USDA National Plant Germplasm System (NPGS) in Pullman, Washington (https://www.ars.usda.gov/pacific-west-area/pullman-wa/plant-germplasm-

Fig. 6.1. Garlic germplasm variability: A. plant and bulb structure of bolting, semi-bolting and non-bolting garlic types (drawings by Fiona Murray, https://www.garlicana.com). B. garlic horticultural types produced in a single site, Garlicana, Oregon (https://www.garlicana.com; photos and names provided by Avram Drucker: 1 – 'Lithuanian Purple'; 2 – 'Krasnodar White'; 3 – 'Vekan'; 4 – 'Sakura'; 5 – 'Creole Red'; 6 – 'Basque Turban'; 7 – 'Pyongyang'; 8 – 'Corsican Red'; 9 – 'Nootka Rose'; 10 – 'Polish White'; 11 – 'Silver Rose'; 12 – 'Bolivian. C'. seed-propagated bolting and semi-bolting genotypes in Israel (Photo: Shemesh-Mayer and Ben Michael).

introduction-and-testing-research) and CRI Olomouc, Czech Republic (http://cr-hana.eu/en) (websites accessed September 2022). The *Allium* Database, managed by the *Allium* Working Group of the European Cooperative Programme for Plant Genetic Resources (ECPGR) (http://www.ecpgr.cgiar.org/working-groups/allium), serves as a tool for the management of national collections in Europe, such as rationalization and safety duplication.

In addition to living collections, *in vitro* conservation and cryopreservation are used to revive outstanding accessions under disease-free conditions (Volk *et al.*, 2004b; Keller *et al.*, 2013; Liu *et al.*, 2017).

Germplasm collections served numerous studies in cultivar classification and molecular diversity. A large array of techniques has been used, e.g. isozymes, RAPD (random amplification of polymorphic DNA), AFLP (amplified fragment length polymorphism), SSRs (simple sequence repeat), ISSR (inter simple sequence repeat), EST (expressed sequence tags), miRNAs (microRNAs), SNPs (single nucleotide polymorphism), insertion-deletion (indel) and high-throughput

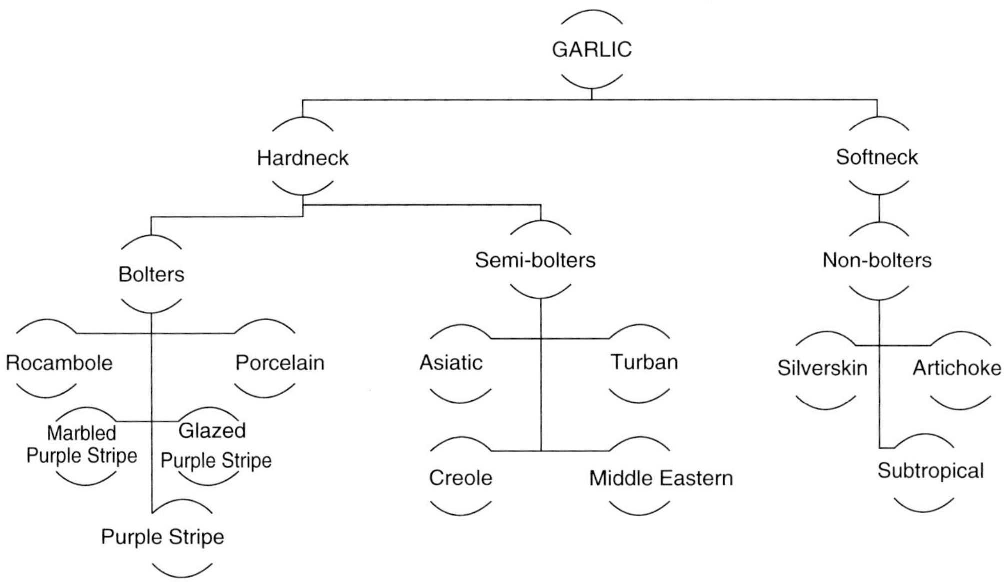

Fig. 6.2. Horticultural garlic types categorized by bolting ability and bulb structure. (Adapted from Woodward, 2014)

genotyping using genotyping-by-sequencing (GBS) (Maaß and Klaas, 1995; Fernandez Pappu *et al.*, 2003; Ipek *et al.*, 2003, 2015; Volk *et al.*, 2004a; Kim *et al.*, 2009; Ma *et al.*, 2009; Cunha *et al.*, 2012; Garcia Lampasona *et al.*, 2012; Panda *et al.*, 2014; Liu *et al.*, 2015; Havey and Ahn, 2016; Egea *et al.*, 2017; Anwar *et al.*, 2020; Shen *et al.*, 2021). The results allowed assessments of the genetic diversity of large germplasms, created core collections and enabled the management of garlic diversity in living collections.

Clonally propagated garlic cannot be improved by classic and molecular breeding tools. Hence, broadening garlic variability by hybridization and seed propagation is very appealing.

It is widely accepted that primitive Central Asian garlic landraces can flower and produce seeds (Etoh and Simon, 2002; Simon and Jenderek, 2004). In the early 1980s, Japanese expeditions collected several fertile garlic accessions in Central Asia (Etoh *et al.*, 1988), Armenia, Georgia, and Xinjiang in China (Etoh *et al.*, 1992). Later, screening of existing clonal collections in the USA identified fertile genotypes with seed germination of 67–93% (Pooler and Simon, 1994; Jenderek and Hannan, 2000, 2004), and international missions gathered over 300 Central Asian landraces (Baitulin *et al.*, 2000; Kamenetsky *et al.*, 2004b). Consequently, new garlic germplasm generated from sexual reproduction and breeding programmes were initiated in Israel and other climatic regions (Kamenetsky *et al.*, 2005). Selections from seedlings populations vary in morphological and quality traits (Fig. 6.1C). Most offspring are hardneck, yet only around 15% of progeny produce the next generation of true seeds (Shemesh *et al.*, 2008). Currently, the variability between commercial clones is larger than that available in sexually propagated populations, supposedly due to the long-term selection under a variety of environments.

3. Plant Development and Life Cycle

3.1. Seed and seedling development

Garlic seed shape and black colour are typical to the subgenus *Allium* (Shemesh *et al.*, 2008). Seed 1000 weight is 1.5–2 g, and germination lasts several weeks up to one year (Etoh and Simon, 2002; Simon and Jenderek, 2004). Germination rates of various seed populations range between 10% and 40% (Simon and Jenderek, 2004; Shemesh *et al.*, 2008) to 90% (Kamenetsky *et al.*,

2004b). Scarification, stratification and chilling promote germination, while phytohormone treatments had only little effect (Etoh and Simon, 2002; Ben Michael *et al.*, 2020b). Weak seedling performance and vigour, abnormal development, and low survivability were often observed (Simon and Jenderek, 2004).

Garlic seed germination begins with the elongation of a loop-shaped cotyledon, followed by the development of 5–15 true leaves (Simon and Jenderek, 2004; Shemesh *et al.*, 2008) that produce a cylindrical stem-like structure termed the 'false' stem. Plant development proceeds through the morphologically distinct successive juvenile, adult vegetative and reproductive phases, linked by the regulatory networks (Ben Michael *et al.*, 2020b). In the annual cycle of post-juvenile garlic, vernalization is required for meristem transition, bulbing and flowering (Simon and Jenderek, 2004; Ben Michael *et al.*, 2020a,b). The garlic bulb is made of a cluster of lateral cloves, which arise in the leaves' axils (Fig. 6.1A). At the end of the growth period, the foliage dries up, and following bulb maturation, the cloves enter physiological rest/dormancy.

3.2. Annual life cycle, florogenesis, seed production and bulbing

The annual cycle of the adult plant includes several distinct stages (Lopez-Bellido *et al.*, 2016; Fig. 6.3). Following planting, adventitious roots arise from the clove base, and leaf primordia of the apical bud sprout to produce flat leaves. Under suitable environmental conditions, some varieties bolt and develop inflorescences with flower buds and/or topsets (Takagi, 1990; Lopez-Bellido *et al.*, 2016).

In vernalized post-juvenile plants, SAM transition from a vegetative to a reproductive state occurs during the growth stage (Kamenetsky, 2007; Fig. 6.4A). In bolting genotypes, the inflorescence subdivides into clusters (cymes), each differentiates 5–6 individual flower meristems (Kamenetsky, 2007; Fig. 6.4B, C). Florogenesis progresses to floral development and anthesis, seed development and ripening are associated with the flower position within the cyme and inflorescence (Shemesh-Mayer *et al.*, 2013).

Most bolting garlic genotypes develop topsets of different sizes and colours (Kamenetsky, 2007; Shemesh *et al.*, 2008; Bhusal, 2021; Fig. 6.4 F, G), while in semi-bolters flower-forming meristems deteriorate at the very early stages of development, thus only large bulbils develop on the scape top (Fig. 6.4 H, I). Pseudovivipary (topset formation) is common in species growing in extreme or marginal environments as further insurance of reproduction (Tooke *et al.*, 2005). The damage to the flowers can be minimized by perpetual removal of bulbils, thus resulting in normal anthesis of some flowers and seed production (Etoh *et al.*, 1988; Simon and Jenderek, 2004).

3.3. Individual flowers and fertility barriers

Garlic flowers have six tepals, six stamens and a carpel (Kamenetsky, 2007). In fertile genotypes, when spathes break, the green closed tepals gradually change colour to pink and purple, unfold and reach a length of 3–4

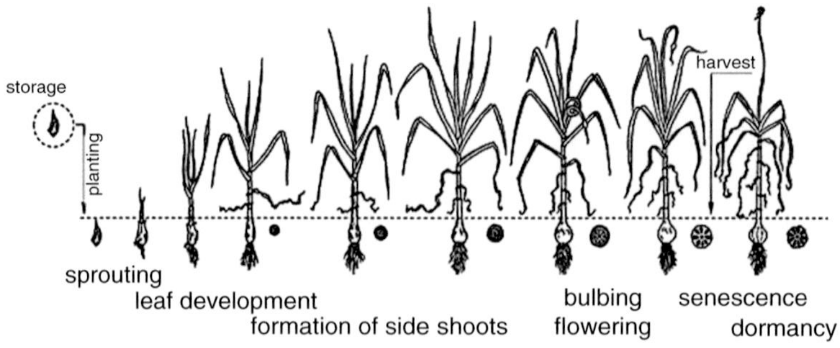

Fig. 6.3. Phenological stages of bolting garlic from planting to harvest. (Adapted from Lopez-Bellido *et al.*, 2016)

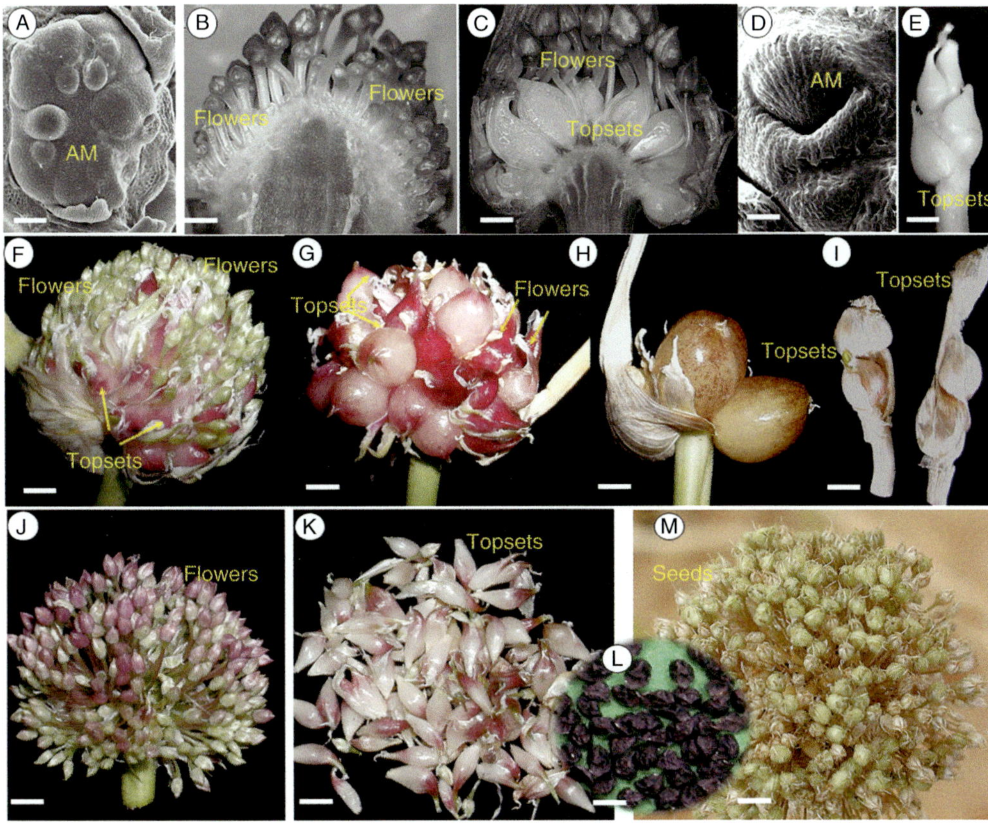

Fig. 6.4. Inflorescence, flowers and topsets development in bolting and semi-bolting garlic genotypes (Photo: Kamenetsky Goldstein and Yarmus). A. apical meristem (AM) transition to reproductive stage in bolting genotypes, nine weeks after planting. Meristem is dividing into several cymes. Bar = 0.2 mm. B. numerous flowers differentiated in bolting genotype inflorescence, 20 weeks after planting. Bar = 3.5 mm. C. fast-developing bulbils in topset-producing plant squeeze existing flowers, 24 days after planting. Bar = 3.5 mm. D. apical meristem (AM) of semi-bolting 'Shani' does not transform into floral meristems. Bar = 0.05 mm. E. semi-bolting genotypes produce only a few large topsets. Bar = 1 mm. F. fully developed inflorescence. Bar = 4 mm. G. inflorescence of topset-producing bolter with a few aborted flowers and many large topsets. Bar = 3.5 mm. H. semi-bolter's scape bearing a few large topsets. Bar = 5 mm. I. semi-bolting 'Shani' produces 2–3 topsets, solely. Bar = 10 mm. J. fully developed dense inflorescence of fertile genotype at anthesis. Bar = 5 mm. K. fertile inflorescence produces numerous small topsets. Bar = 5 mm. L. matured garlic seeds. Bar = 3 mm. M. seed capsules. Bar = 4.5 mm.

mm at anthesis. Concomitantly, the anther filaments extend above the tepals, the stomia open and pollen sheds for 3–4 days. Pollen viability in fertile garlic amounts to 80–100% (Simon and Jenderek, 2004; Shemesh-Mayer *et al.*, 2013).

The ovary of the protandrial flower changes colour to dark green or purple when pollen shedding ends. The style elongates above the anthers' level and reaches its final length when the individual flower pollen sacs are empty. The stigma becomes receptive for *c.*3 days and the surface's papillae turn wrinkly. The tepals wither about a week after anthesis (Shemesh-Mayer *et al.*, 2013). Fertility impairment occurs in both male and female organs (Fig. 6.5A).

Complete male sterility is caused by irregular chromosome pairing and/or deletions with the consequent post-meiotic abortion (Etoh and Simon, 2002). Metabolic interferences result in

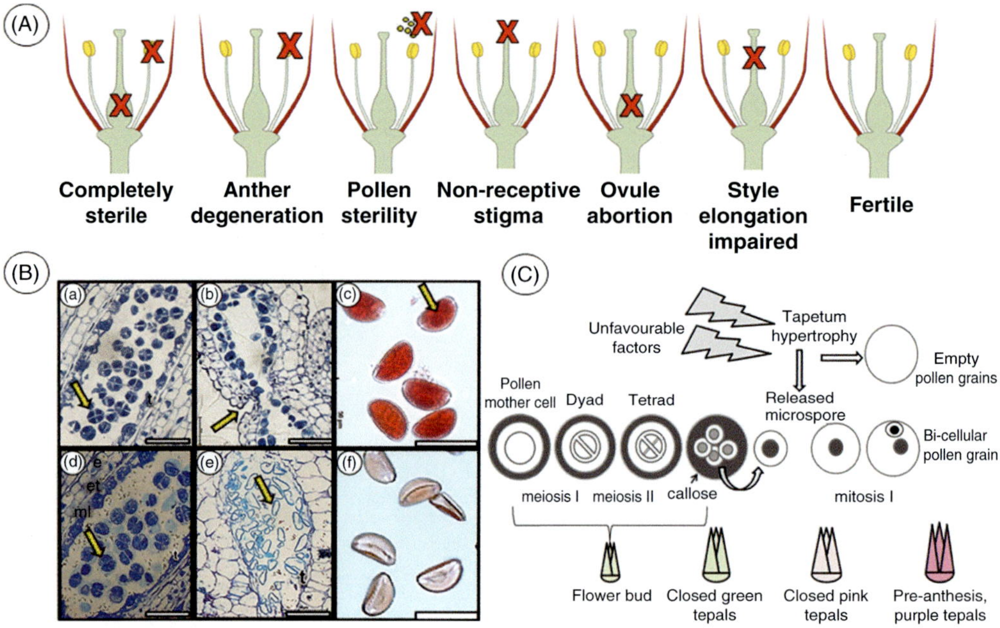

Fig. 6.5. Development barriers in garlic fertility. (Adapted from Shemesh-Mayer *et al.*, 2013, 2015a) A. scheme of fertile and complete male- and female-sterile flowers. B. anther and pollen development in fertile and male-sterile garlic genotypes: (a–c) fertile genotype; (d–f) male-sterile genotype. (a) pollen sac longitudinal section at tetrad (arrow) stage. Bar = 50 µm; (b) cross-section of an anther with mature microspores consisting of vegetative and generative cells; the beginning of the stomium opening is marked by an arrow. Bar = 80 µm; (c) acetocarmine-stained mature pollen grains containing vegetative and generative cells (arrow). Bar = 25 µm; (d) longitudinal section of the pollen sac at the tetrad (arrow) stage (e; epidermis, et; endothecium, ml; middle layer, t; tapetum.) Bar = 50 µm; (e) tapetum remnants (t) in the pollen sac; aborted microspores (arrow). Bar = 50 µm; (f) acetocarmine-stained empty pollen grains; no visible vegetative or generative cells. Bar = 25 µm. C. a schematic presentation of normal micro-sporogenesis in garlic and anomalies caused by genetic or environmental factors. The pollen mother cell undergoes two meiotic divisions, with the consequent release of four microspores, each of which divides mitotically to produce vegetative and generative cells in the mature bicellular pollen grain.

tetrads' callose walls degradation, the absence of a normal cortical cytoskeleton, and cytoplasm degeneration in the pollen mother cells (Winiarczyk *et al.*, 2012; Tchórzewska *et al.*, 2015, 2017; Fig. 6.5B, C). High levels of protease and acid phosphatase activity, as well as a low level of esterase activity in anthers' locules, are also associated with microspores degeneration (Winiarczyk and Gębura, 2016).

The second type of male sterility is characterized by anthers' degeneration after the microspores' release from the tetrads, enlargement of the tapetal cells and arrest of pollen differentiation (Shemesh-Mayer *et al.*, 2013). Finally, even visually normal pollen might be sterile for unknown reasons (Shemesh-Mayer *et al.*, 2013; Fig. 6.5B).

Transcriptome analyses of fertile and male-sterile garlic genotypes revealed approx. 16,000 differentially expressed genes. Gene activity in the former genotypes was associated with the development of reproductive tissues and higher-expression fertility-related genes, while in the male-sterile plant, gene activity was associated with modifications in the mitochondrial genome that affect pollen functions. Hence, male sterility probably results from interruption of mitochondrial functions, respiratory restrictions and non-regulated programmed tapetum cell death. Energy deficiency led to pollen abortion (Shemesh-Mayer *et al.*, 2015b).

Phenotypes with female sterility might possess abnormal embryo sacs, asymmetric development of the integuments or failure of a micropylar channel to facilitate pollen tube penetration. Absence of style elongation or non-receptive stigma might also lead to ovule abortion (Winiarczyk and Kosmala, 2009).

4. Environmental and Genetic Control of Development

Environment plays a major role in garlic development, quality bulb formation and bolting, and therefore a better understanding of garlic physiology and genetics contributes to improved agro-management in various climatic regions (Kamenetsky, 2007; Yamashita *et al.*, 2010; Shemesh-Mayer *et al.*, 2015a; Wu *et al.*, 2015, 2016; Atif *et al.*, 2019; Hsiao *et al.*, 2019; Ben Michael *et al.*, 2020a,b). The diploid garlic (2n=2x=16) has a huge nuclear genome, hence genome sequencing was challenging, and mainly transcriptome assembly and molecular analyses were employed in generating functional genomic data (Kamenetsky *et al.*, 2015; Liu *et al.*, 2015; Chen *et al.*, 2018; Yang *et al.*, 2021). Recently, garlic became the first *Allium* species with a sequenced genome (Sun *et al.*, 2020). A chromosome-level genome assembly has a total size of 16.24 Gb, and the annotation of 57 561 predicted protein-coding genes. Analysis of this garlic genome assembly revealed transposable elements that might explain the substantial expansion of the garlic genome. The reference genome of garlic and large-scale transcriptome data generated in several studies provide valuable resources for research on garlic biology and breeding.

Similar to other geophytes, garlic physiology is strongly regulated by external signals, such as temperature and photoperiod. Storage and/or field vernalization satisfies cold requirements for the life cycle completion. Garlic seeds require long stratification for germination and seedling development (Simon and Jenderek, 2004; Ben Michael *et al.*, 2020b). It further affects leaf morphology, bolting and bulb formation, while plants raised from non-stratified seeds, which are not exposed to sufficient vernalization, do not form flowers or bulbs (Ben Michael *et al.*, 2020b). Stratification induces molecular alterations in the embryonic meristem. Functional annotation analyses of 3000 differentially expressed genes showed that stratification reactivates the embryonic cell cycle, initiates the metabolism, and primes garlic seed germination. An eight-week stratification enhanced activities associated with cell cycle, sugar metabolism, cell division and proliferation, as well as expression of meristem identity and flowering genes in garlic seedlings. Phase transitions from seed germination to flowering and bulbing are tightly connected and are possibly associated with downregulation of the specific flowering repressor(s). Thus, embryonic SAM plays an important role not only in seed germination but also in the entire life cycle, providing the foundation for the genetic regulation of major functional shifts in metabolism and development (Ben Michael *et al.*, 2020b).

In many bulbous species from temperate climates, flowering and bulbing depend on vernalization. In garlic, it suffices the need for a transition of the apical meristem to the reproductive stage, for lateral bud development, and dormancy initiation (del Pozo and González, 2005; Ben Michael *et al.*, 2018, 2020a; Fig. 6.6A). Storage vernalization and/or field chilling induce meristem transition and augment the reproductive potential of bolting genotypes, especially in warm regions (Takagi, 1990; Kamenetsky *et al.*, 2004a; Wu *et al.*, 2015, 2016; Kaur and Dhall, 2017). Gibberellin can probably substitute, at least in part, for cold treatments, as it induces axillary bud formation by modifying the plant's soluble sugar and soluble protein content (Ouzounidou *et al.*, 2011; Liu *et al.*, 2019a,b).

Garlic genotypes vary considerably in cold requirements (Shemesh *et al.*, 2008; Dufoo-Hurtado *et al.*, 2015; Kaur and Dhall, 2017; Ben Michael *et al.*, 2018). Varieties adapted to warm regions bolt and form bulbs in response to the cold of the winter without cold storage (Rohkin Shalom *et al.*, 2015), while excessive cold storage of bolting varieties from a temperate climate accelerated plant development and production of smaller bulbs (Bandara *et al.*, 2000; Guevara-Figueroa *et al.*, 2015; Wu *et al.*, 2015).

Morpho-physiological and transcriptome analyses showed that chilling provides a major cue for garlic flowering and bulbing, while photoperiod plays no role in the initial nuclear and metabolic processes, but at the later stages of development. In Israel, propagule vernalization for 12 weeks at 4°C before November planting resulted in flower initiation already in December–January and early blooming and bulbing in March, while plants from non-vernalized cloves did not undergo meristem transition, flowering and bulb formation (Ben Michael *et al.*, 2020a; Fig. 6.6A). Comparative analyses between vernalized and non-vernalized bulb apical meristems revealed ~14,000 differentially expressed genes, coded for major modifications in dormancy release, initiation of active growth, and meristem transition from vegetative to reproductive state (Fig. 6.6B, C). Vernalization caused the reorganization of a large range of cellular components and metabolic processes in garlic apical meristems. Interestingly, many genes associated with light reactions and photoperiodic pathways

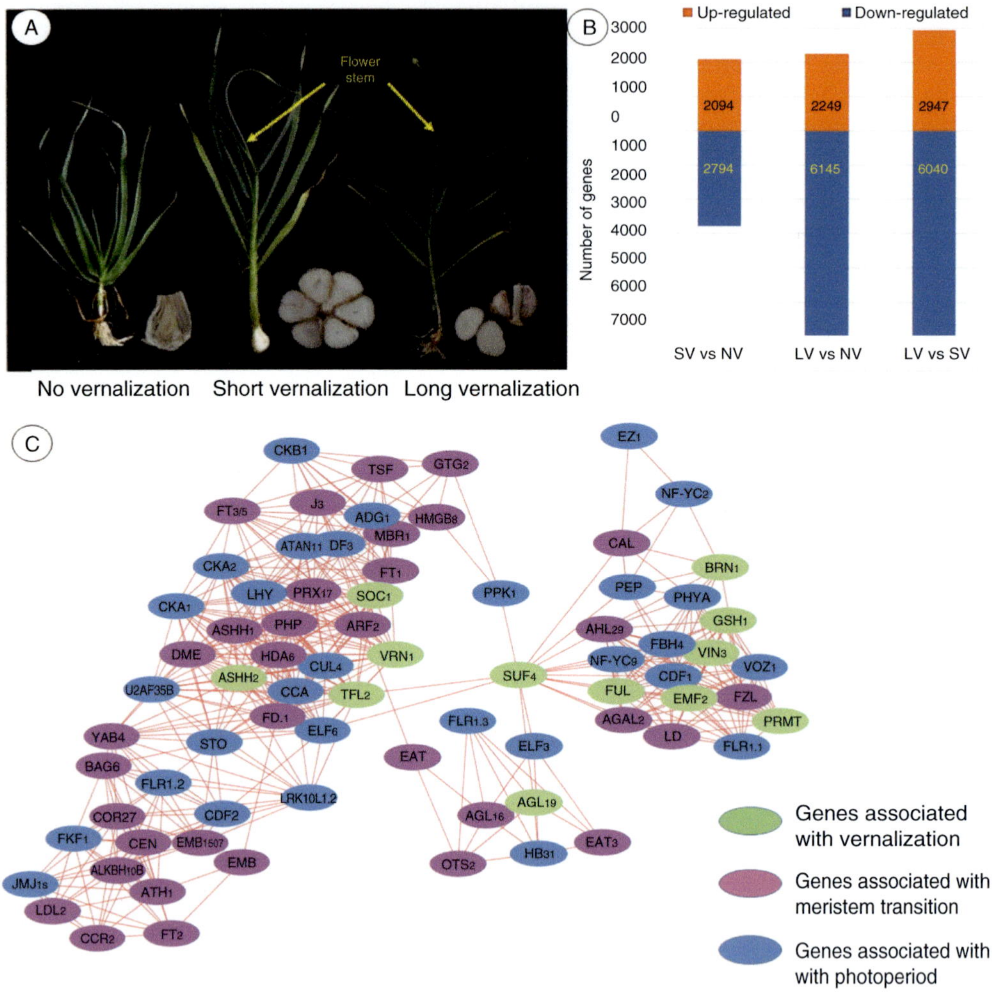

Fig. 6.6. Morpho-physiological and transcriptome analyses of bolting garlic under different vernalization conditions. Plants were subjected to 4°C X 4 or 12 weeks vernalization or ambient conditions. Pictures were taken after 20 weeks of growth at the end of March. (Adapted from Ben Michael *et al.*, 2020a)
A. effect of pre-planting vernalization on leaf development, flowering and bulbing. Apical meristems of non-vernalized plants remained vegetative throughout. Short vernalization sufficed for scape development, and later, long photoperiod promoted bulb formation. Long vernalization promoted flowering and early formation of small bulbs irrespective of photoperiod. B. differential expression of up- and downregulated genes in the apical meristems of short- (SV) and long- (LV) vernalized and non-vernalized (NV) plants.
C. positive co-expression of garlic genes associated with vernalization, photoperiod and meristem transition in response to vernalization.

were expressed in dark-stored vernalized meristems. Thus, in garlic, the genes annotated as involved in 'photoperiodic activity' in model plants, are regulated by low temperature (Ben Michael *et al.*, 2020a; Fig. 6.6C).

Similar to onion (Lee *et al.*, 2013) *FLOWERING LOCUS T* (*FT*) genes are involved in the regulation of bulb formation, foliage development and apical meristem transition to reproductive state, in garlic. Thus, cold storage of Shani led to upregulation of *FLOWERING LOCUS T1* (*AsFT1*) and *AsFT2* (associated with floral transition) while the expression of *AsFT1* antagonist *AsFT4* was two- to threefold lower in the internal bud and storage leaf of cold-treated propagules (Rohkin Shalom *et al.*, 2015).

Storage treatments affected cloves' protein profiles, thus indicating the role of vernalization in protein biosynthesis, energy production and carbohydrate and nucleotide metabolism (Dufoo-Hurtado *et al.*, 2015; Chaturvedi *et al.*, 2018). In addition, vernalization promoted fructan-degradation in the cloves, thus affecting *FT2* expression, the timing of meristem transition and bulb formation (Chaturvedi *et al.*, 2018).

In growing plants, the interaction between long photoperiod (LP) and temperature defines flowering and bulbing quality, soluble proteins, sugars and phenolic compounds (Atif *et al.*, 2019, 2020). A combination of LP with cool or warm temperatures promotes or inhibits bolting and bulbing, respectively (Kamenetsky *et al.*, 2004a; Khokhar, 2014; Wu *et al.*, 2015, 2016). Gradual warming during the growing season resulted in dense inflorescences with viable flowers, while constant moderate or low temperatures caused flower degeneration. Fast temperature upsurge boosted anthesis with a concomitant decline in pollen viability, flowers withering and fast senescence (Shemesh-Mayer *et al.*, 2015a, Fig. 6.5 in this chapter). In addition, LP increased levels of auxin, abscisic acid, gibberellic acid, zeatin and jasmonic acid and resulted in higher bulb diameter and bulbing index (del Pozo and González, 2005; Wu *et al.*, 2016).

Apical meristems' thermosensing is linked to the circadian clock, which, in turn, is associated with the multi-faceted genetic network, involved in the transition to the next developmental stage (Ben Michael *et al.*, 2020b). This network integrates the chilling signals and the expression of the genes associated with vernalization, photoperiod and meristem transition. These findings clearly distinguish onion from garlic in environmental regulation and developmental mechanisms. In onions, chilling promotes flowering while LP induces bulbing. Hence, varieties are categorized as short-, intermediate- and long-day types (Rabinowitch, 1990). In garlic, however, seed stratification and bulb vernalization are required for germination, sprouting and the shift to flowering and bulbing that can occur even under short photoperiod. When garlic vernalization is insufficient, LP acts as a supplementary factor promoting meristem transition and growing cycle. Hence, garlic varieties should better be classified as high- and low-cold demanding types.

5. Propagation

5.1. Vegetative propagation

Despite the advances in fertility restoration and true seed production, the world's garlic industry propagates garlic vegetatively. The compound bulbs with 5–20 cloves are harvested on maturation, and a new production season begins with clove planting.

Clonal propagation suffers many disadvantages including low multiplication rate, thus introduction of a new variety takes five to ten years. Moreover, large volume storage space is required, fields suffer severe sanitary constraints, and perpetuation of diseases and viruses is common (Fereol *et al.*, 2002). Hence, productivity losses are heavy (Manjunathagowda *et al.*, 2017).

In bolting varieties, topsets whose genetic make-up is identical to that of the mother plant can serve for propagation (Popkov, 2012). Topsets' diameter ranges from 0.2 to 2 cm, depending on horticultural groups, growth conditions and cultivar (Hirata *et al.*, 2016; Albuquerque *et al.*, 2017). One mother plant produces 50–200 topsets, thus reducing propagation cost (Kajimura *et al.*, 2000; Zawistowski *et al.*, 2018; Sulastiningsih *et al.*, 2020). Little knowledge is available on topsets biology (development, dormancy and physiological traits) and production. In the first year, plants raised from topsets produce simple round bulbs, which produce marketable bulbs in the second growing season.

It was suggested that topsets are less infected by pathogens, thus potentially produce healthier bulbs compared with cloves (Ebi *et al.*, 2000; Fan *et al.*, 2017). However, Pappu *et al.* (2008) and Parrano *et al.* (2012) reported that topsets are heavily infected with viruses transmitted from the mother plant (Bhusal *et al.*, 2021). Therefore, virus-free topsets can be obtained from virus-free mother plants, solely.

5.2. Biotechnological applications for propagation and virus eradication

The garlic virome contains at least 20 members (Katis *et al.*, 2012; Manjunathagowda *et al.*, 2017) which vary in virulence and the degree of damages to the crop. Breeding for genetic resistance (including transformation, cross-protection) is by far the preferred defence strategy but is currently impractical for garlic. Prophylaxis (quarantine, roguing infected plants, control of natural vectors) is currently the only means of control (Rubio *et al.*, 2020).

In the absence of sexual reproduction, agricultural biotechnology offers different propagation means, including genotypes' replication; mass multiplication; whole plants' regeneration from the genetically modified cells; and elimination of pathogens for the multiplication of 'clean stocks' (Walkey *et al.*, 1987; Wang *et al.*, 2008). Additionally, biotechnological tools serve as an alternative for the preservation and propagation of gene pools (Robledo-Paz and Tovar-Soto, 2012; Fan *et al.* 2017; Bikis, 2018).

Different methods are applied for virus eradication, including meristem-tip culture, thermotherapy, chemotherapy (Salomon, 2002), and cryotherapy (Vieira *et al.*, 2015). Garlic's genetic response to the environment and dormancy should be considered when planning the tissue-culture propagation schedule (Keller *et al.*, 2013).

In vitro propagation requires hardening and thus it takes two to three years for completion (Fig. 6.7, Stages I and II), plus three years to obtain a sufficiently large population for commercial use (Fig. 6.7, Stage III).

5.3. Seed propagation

Large garlic collections represent only a fraction of the species variability and adaptation to local environments. The much larger variability in true seeds facilitates garlic improvement, production, adaptation and global trade. Currently, however, garlic from seed is in its infancy. Non-bolters and semi-bolters do not flower, and most hardneck varieties produce topsets that need to be removed to enable the completion of floral development (Etoh and Simon, 2002). Selection of outstanding seed-producing lines in Israel and the USA (Jenderek and Hannan, 2004; Kamenetsky *et al.*, 2004b) made mass-pollination of several selections possible for breeding and use in commercial farming.

Garlic production from the seed reduces storage and transportation costs, including north–south hemisphere trades, use of pathogen-free propagation material, production flexibility, and more. It should be noted, however, that, currently, the production of quality bulbs from seed takes two growing seasons (Ben Michael *et al.*, 2020b).

6. Seed Production

Standard open-fields seed production practices face common problems, including the need for isolation, and infections of parent stock suppress flowering and seed production (Simon and Jenderek, 2004). Hence, protected farming of healthy mother plants offers a safer environment and consequently higher seed yield with better quality. The individual protandrous flower cannot self-pollinate, but self-pollination is possible within the inflorescence, due to sequential anthesis of flowers in different cymes (Shemesh *et al.*, 2008). Self-pollination leads to poor seed production, low germination and survival rate (Jenderek and Zewdie, 2005), therefore seed producers should promote cross-pollination within the population. In the spring or summer, when flowering occurs, the presence of pollinating insects is essential (Simon and Jenderek, 2004).

Early harvest minimizes loss to shattering, however most seeds do not reach maturity. Therefore, drying umbels should be harvested when the first seed capsules split open, thus seed maturation reaches completion in storage. Garlic seeds are about half the size of onion seeds. Currently, no commercial seed is produced, hence procedures and techniques (cleaning, sorting and quality control) for garlic seed are not yet developed.

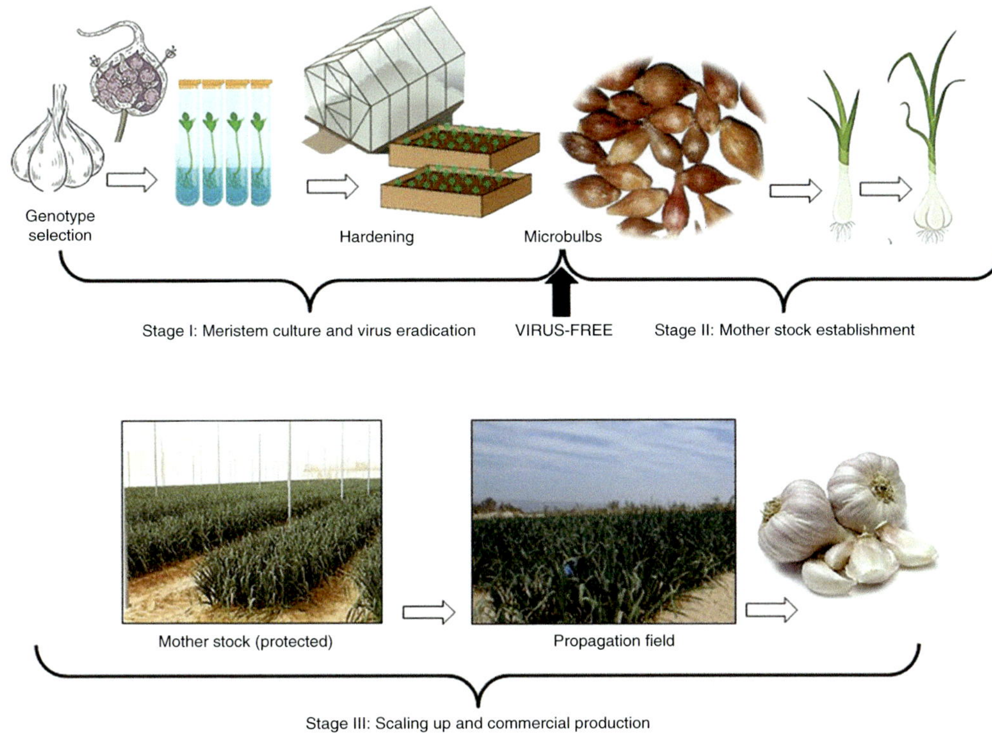

Fig. 6.7. Main stages of pathogen elimination and propagation of disease-free garlic clones using biotechnological tools. Stage I: *in vitro* pathogen eradication (thermo-, chemo- or cryotherapy) and *ex vitro* regenerant hardening. Stage II: establishing protected disease-free mother stock. Stage III: quantitative production of elite disease-free propagules and commercial propagation in isolated fields. Finally, the commercial release of elite propagules.

Seed quality and germination ability decline with storage time and fungal and/or bacterial diseases adversely affect the viability of seed and seedlings (Rozenblat, 2017).

7. Breeding Strategies

Climate change and increased food production in warm regions prioritize breeding crops for tolerance to biotic and abiotic stress (Lamichhane *et al.*, 2015; Khar *et al.*, 2020). Environment strongly affects garlic phenotypic expression, hence distinct varieties can be successfully introduced in diverse areas, provided that diversity extension of traits like bulb size, colour, structure, yield, hardiness and peel ability, shelf life, flavour, pungency and environmental requirements become feasible (Meredith, 2008).

Garlic infertility, however, prohibited genetic improvements, thus living collections and mutations were the only sources of variability. Currently, fertility restoration, seed propagation and modern technologies facilitated the introduction of three main breeding directions, described below.

7.1. Conventional vegetative selection

Random and induced mutations resulted in morphology and chemistry modifications, tolerance to pathogens like *Sclerotium cepivorum* (Perez-Moreno *et al.*, 1991; Al-Safadi *et al.*, 2000), as well as in chromosomal abnormalities (Kebeish *et al.*, 2015; Ezzat *et al.*, 2016; El-Fiki and Adly, 2020; Pangestuti *et al.*, 2020). Polyploidization affected plant and bulb morphology, cell mitotic index, mineral content and secondary metabolites production (Jang *et al.*, 2000; Zhou and Cheng, 2008; Xiao-ling, 2009; Cheng *et al.*, 2012; Dixit and Chaudhary, 2014; Yousef and El-sadek, 2020). Yet none of this research resulted in the release of new and improved garlic varieties.

7.2. Breeding and selection from sexually reproduced populations

Seedling populations from fertile plants exhibit wide variation in environment adaptation, bolting ability and bulb traits. The level of this variability is similar to the one common within vegetative collections (Shemesh *et al.*, 2008). Short-term sexual breeding combines random hybridization within a fertile collection with vegetative propagation of genotypes selected in a wide range of environments (Simon and Jenderek, 2004), as practised in the USA, South Korea and Israel (Simon and Jenderek, 2004; Druker, USA, personal communication 2018; Kwon, South Korea, personal communication, 2019; Shemesh-Mayer and Kamenetsky Goldstein, 2021).

The long-term strategy of sexual breeding is based on generations of fertile breeding lines as common in cross-pollinated crops. Some factors, however, interfere with the task, including germination problems, biannual development, fertility in the F_1 and F_2 generations, and inbreeding depression (Etoh *et al.*, 1988; Etoh and Simon, 2002; Simon and Jenderek, 2004; Jenderek and Zewdie, 2005).

7.3. Variation extension using biotechnological tools

The development of molecular markers associated with economic traits markedly facilitates breeding work. Trait-associated markers were reported for pollen fertility (Zewdie *et al.*, 2005), bolting ability (Ipek *et al.*, 2007; Ovesná *et al.*, 2015; Kaur *et al.*, 2020), resistance to Fusarium basal rot (Rout *et al.*, 2014), and disease resistance and defence responses (Yang and Huang, 2014; Chand *et al.*, 2016, 2017). Currently, however, the large genome size, drawbacks of sexual reproduction, and limited availability of genetic data strongly restrict the application of molecular markers in garlic breeding (Ipek *et al.*, 2020; Khar *et al.*, 2020; see Chapter 17, this volume).

8. Post-harvest

Successful garlic storage depends on a complex of pre- and post-harvest procedures (Madhu *et al.*, 2019). The main pre-harvest factors include genotype, fertigation, chemical agents, bulb size and harvest stage, while post-harvest factors include handling, curing and storage conditions, all of which affect bulb quality.

Traditional ambient storage often encounters 25–40% deterioration; storage at 20–30°C for one to two months resulted in firmness decrease and change in colour due to water loss (Tripathi *et al.*, 2009), while cold storage at 0–1°C and 60–70% RH supports long storage up to nine months (Lawande, 2018).

Much like bulb onion (Miedema, 1994), dormancy is shortest at 5–18°C (Cantwell and Suslow, 2004), while properly cured bulbs maintain high quality at −3°C for a year (Volk *et al.*, 2004b).

Pre-harvest treatment with maleic hydrazide and use of gamma irradiation combined with a controlled atmosphere (less than 0.5% O_2 or 5–10% CO_2) control sprouting in storage and extend shelf life. About 30 minutes' UV exposure reduced up to 8% losses of five months cold-stored bulbs (http://nhrdf.org/en-us).

9. Conclusions and Prospects

Current garlic research aims to increase production efficiency and further improve high-quality market standards. Progress in molecular science and technology significantly advances garlic research, yet only a little impact on the industry is currently visible due to the length of time from laboratory benches to the production field.

Genetic erosion and variability conservation are of utmost importance. Garlic from Central Asian countries and Xingxian of China are probably the most primitive within the species, thus maintaining the highest potential level of intraspecific variation for future improvement programmes. In addition to tolerance to biotic and abiotic stress, these plants bloom and set viable seeds – a trait of the highest scientific and economic importance. Due to the very rapid adoption of modern Sativum group cultivars, this precious gene pool is under severe threat of extinction. Large-scale international rescue actions are essential for *in situ* collections, preservation and evaluation projects.

Various vegetative garlic living collections serve for variability conservation, while molecular tools facilitate the reduction of duplicates, efforts and costs. The exchange and evaluation of germplasm variability could be implemented in adaptation programmes (Viana *et al.*, 2020).

The new era in garlic breeding began in the 1980s with the discovery of fertile genotypes and fertility restoration by environmental manipulations. Current studies focus on understanding the internal and external regulation of flowering, and bulbing processes, and the genetics of several important traits. Future research will deal with hybrid and OP seed production and storage, synchronized and fast seed germination. An additional important breeding goal is biochemical and nutraceutical improvements of garlic as functional food. Transient RNA-based reprogramming of crop plants for improved agronomic performance is fast, tunable and versatile, and can be used throughout the crop cultivation cycle (Torti *et al.*, 2021), especially in plant-pathogen epidemics, but also for speeding up garlic breeding and to facilitate its commercial production from seed.

Modern biological tools improved our understanding of life processes, breeding and production systems. The genes involved in florogenesis, formation of storage organs and dormancy are now revealed. The expected practical outcome is high-yielding, stress-tolerant, superior seed propagated cultivars.

The employment of biotechnological tools can certainly enhance existing garlic cultivars. Today, we are able to cleanse plant material of viruses and diseases and to propagate selected varieties on a large scale using micropropagation techniques. However, garlic regeneration systems are not always reliable. The improvement of regeneration *in vitro* is an important research goal on the way to garlic genome editing. Since the garlic genome is extremely large and repetitive, the genome sequencing remains a challenging task. Therefore, novel methods of genome editing are not available yet and are the key points for future research and applications in garlic genetic engineering.

Acknowledgments

We would like to express our gratitude to Prof. Haim D. Rabinowitch for a long and fruitful partnership in scientific projects. The results of the garlic research of our team ensued from the long-term fruitful collaboration with scientists, growers and students. Special thanks go to Dr C. Kik (The Netherlands), Dr T. Ben Michael (Israel), Dr J. Keller (Germany) and many other colleagues throughout the world.

References

Albuquerque, J.D.R.D.A., Monteiro, H.N.B., Bezerra, A.A.D.C., Humberto Aires Matos Filho, C., Lopes, Â.C.D.A. *et al.* (2017) Agromorphological performance of garlic landraces in Piauí, Brazil. *Ciencia Rural* 47(6). DOI: 10.1590/0103-8478cr20160017.

Al-Safadi, B., Mir, A.N. and Arabi, M.I.E. (2000) Improvement of garlic (*Allium sativum* L.) resistance to white rot and storability using gamma irradiation induced mutations. *Journal of Genetics and Breeding* 54, 175–182.

Anwar, G.M., Mahmoud, M.A., Ata, A.M., Ragab, R.A. and Bakry, H.S. (2020) Studies of molecular variation sources using RAPD and SSR markers in two garlic clones. *Journal of Modern Research* 2(2), 115–122. DOI: 10.21608/jmr.2020.24889.1021.

Atif, M.J., Amin, B., Ghani, M.I., Hayat, S., Ali, M. *et al.* (2019) Influence of different photoperiod and temperature regimes on growth and bulb quality of garlic (*Allium sativum* L.) cultivars. *Agronomy* 9, 879. DOI: 10.3390/agronomy9120879.

Atif, M.J., Amin, B., Ghani, M.I., Ali, M., Liu, X. *et al.* (2020) *Allium sativum* L. (garlic) bulb enlargement as influenced by differential combinations of photoperiod and temperature. *Food Chemistry* 338, 127991. DOI: 10.1016/j.foodchem.2020.127991.

Baitulin, I.O., Agafonova, G., Rabinowitch, H.D. and Kamenetsky, R. (2000) Creation of gene bank of Central Asian species of the genus *Allium* L., their biology and economic potential. In: Granovsky, E.I. and Fain, E.E. (eds) *State and Perspectives of Scientific Collaboration Kazakhstan-Israel* (in Russian). Almaty, Kazakhstan, pp. 87–94.

Bandara, M.S., Krieger, K., Slinkard, A.E. and Tanino, K.K. (2000) Pre-plant chilling requirements for cloving of spring garlic. *Canadian Journal of Plant Science* 80, 379–384. DOI: 10.4141/P99-074.

Ben Michael, T.E., Shemesh-Mayer, E., Kimhi, S., Gershberg, C., Forer, I. *et al.* (2018) Temporal and spatial effect of low pre-planting temperatures on plant architecture and flowering in bolting garlic. *Scientia Horticulturae* 242, 69–75. DOI: 10.1016/j.scienta.2018.07.025.

Ben Michael, T.E., Faigenboim, A., Shemesh-Mayer, E., Forer, I., Gershberg, C. *et al.* (2020a) Crosstalk in the darkness: bulb vernalization activates meristem transition via circadian rhythm and photoperiodic pathway. *BMC Plant Biology* 20, 1–16. DOI: 10.1186/s12870-020-2269-x.

Ben Michael, T.E., Rozenblat, L., Faigenboim, A., Shemesh-Mayer, E., Forer, I. *et al.* (2020b) From embryo to adult: low temperatures affect phase transitions of *Allium sativum* L. from germination to flowering. *Agronomy* 10, 1651. DOI: 10.3390/agronomy10111651.

Bhusal, H. (2021) Topsets in garlic inflorescence: development, virus transmission and use as propagation material. Master's thesis, The Hebrew University of Jerusalem, Jerusalem, Israel.

Bhusal, H., Shemesh-Mayer, E., Forer, I., Kryukov, I., Peters, R. *et al.* (2021) Bulbils in garlic inflorescence: development and virus translocation. *Scientia Horticulturae* 285, 110146. DOI: 10.1016/j.scienta.2021.110146.

Bikis, D. (2018) Review on the application of biotechnology in garlic (*Allium sativum*) improvement. *International Journal of Research Studies in Agricultural Sciences (IJRSAS)* 4(11), 23–33.

Block, E. (2010) *Garlic and Other Alliums: The Lore and the Science*. Royal Society of Chemistry, Cambridge, UK.

Cantwell, M. and Suslow, T. (2004) Garlic. In: *Recommendations for Maintaining Postharvest Quality*. Available at: http:/postharvest. ucdavis. edu/Produce/ProduceFacts (accessed 12 July 2022).

Chand, S.K., Nanda, S. and Joshi, R.K. (2016) Regulation of miR394 in response to *Fusarium oxysporum* f. sp. *cepae* (FOC) infection in garlic (*Allium sativum* L). *Frontiers in Plant Science* 7, 258. DOI: 10.3389/fpls.2016.00258.

Chand, S.K., Nanda, S., Mishra, R. and Joshi, R.K. (2017) Multiple garlic (*Allium sativum* L.) microRNAs regulate the immunity against the basal rot fungus *Fusarium oxysporum* f. sp. *cepae*. *Plant Science* 257, 9–21. DOI: 10.1016/j.plantsci.2017.01.007.

Chaturvedi, A.K., Rohkin Shalom, S., Faigenboim-Doron, A., Teper-Bamnolker, P., Salam, B.B. *et al.* (2018) Differential carbohydrate gene expression during preplanting temperature treatments controls meristem termination and bulbing in garlic. *Environmental and Experimental Botany* 150, 280–291. DOI: 10.1016/j.envexpbot.2018.01.013.

Chen, X., Liu, X., Zhu, S., Tang, S., Mei, S. *et al.* (2018) Transcriptome-referenced association study of clove shape traits in garlic. *DNA Research* 25, 587–596. DOI: 10.1093/dnares/dsy027.

Cheng, Z.H., Zhou, X.J., Khan, M.A., Su, L. and Meng, H.W. (2012) In vitro induction of tetraploid garlic with trifluralin. *Genetics and Molecular Research* 11(3), 2620–2628. DOI: 10.4238/2012.July.10.13.

Cunha, C.P., Hoogerheide, E.S., Zucchi, M.I., Monteiro, M. and Pinheiro, J.B. (2012) New microsatellite markers for garlic, *Allium sativum* (Alliaceae). *American Journal of Botany* 99(1), e17–e19. DOI: 10.3732/ajb.1100278.

del Pozo, A.L. and González, M.I.A. (2005) Developmental responses of garlic to temperature and photoperiod. *Agricultura Técnica* 65, 119–126.

Dixit, V. and Chaudhary, B.R. (2014) Colchicine-induced tetraploidy in garlic (*Allium sativum* L.) and its effect on allicin concentration. *Journal of Horticultural Science and Biotechnology* 89(5), 585–591. DOI: 10.1080/14620316.2014.11513124.

Dufoo-Hurtado, M.D., Huerta-Ocampo, J.Á., Barrera-Pacheco, A., Barba de la Rosa, A.P. and Mercado-Silva, E.M. (2015) Low temperature conditioning of garlic (*Allium sativum* L.) 'seed' cloves induces alterations in sprouts proteome. *Frontiers in Plant Science* 6, 332. DOI: 10.3389/fpls.2015.00332.

Ebi, M., Kasai, N. and Masuda, K. (2000) Small inflorescence bulbils are best for micropropagation and virus elimination in garlic. *HortScience* 35(4), 735–737. DOI: 10.21273/HORTSCI.35.4.735.

Egea, L.A., Mérida-García, R., Kilian, A., Hernandez, P. and Dorado, G. (2017) Assessment of genetic diversity and structure of large garlic (*Allium sativum*) germplasm bank, by diversity arrays technology 'genotyping-by-sequencing' platform (DArTseq). *Frontiers in Genetics* 8, 98. DOI: 10.3389/fgene.2017.00098.

El-Fiki, A. and Adly, M. (2020) Morphological, molecular, and organosulphur compounds characterization in irradiated garlic (*Allium sativum*) by GC–MS and SCoT markers. *Journal of Radiation Research and Applied Sciences* 13, 61–70. DOI: 10.1080/16878507.2019.1697079.

El-Saber Batiha, G., Magdy Beshbishy, A., Wasef, L.G., Elewa, Y.H.A., Al-Sagan, A.A. *et al.* (2020) Chemical constituents and pharmacological activities of garlic (*Allium sativum* L.): a review. *Nutrients* 12(3), 872. DOI: 10.3390/nu12030872.

Engeland, R.I. (1991) *Growing Great Garlic*. Filaree Productions, Okanogan, Washington.

Etoh, T. and Simon, P.W. (2002) Diversity, fertility and seed production of garlic. In: Rabinowitch, H.D. and Currah, L. (eds) *Allium Crop Sciences: Recent Advances*. CAB International, Wallingford, UK, pp. 101–117.

Etoh, T., Noma, Y., Nishitarumizu, Y. and Wakomoto, T. (1988) Seed productivity and germinability of various clones collected in Soviet Central Asia. *Memoirs of the Faculty of Agriculture-Kagoshima University (Japan)* 24, 129–139.

Etoh, T., Johjima, J. and Matsuzoe, N. (1992) Fertile garlic clones collected in Caucasia. In: Hanelt, P., Hammer, K. and Knüpffer, H. (eds) *The Genus Allium – Taxonomic Problems and Genetic Resources*. IPK, Gatersleben, Germany, pp. 49–54.

Ezzat, A.S., Moustafa, Y.M.M., Ahmed, K.Z., Farrag, M.M. and Gadel-Hak, S.-N.H. (2016) Induced variation in some horticultural and cytological characteristics of two garlic (*Allium sativum* L.) genotypes. *Minia Journal of Agricultural Research and Development* 36, 271–292.

Fan, B., He, R., Shang, Y., Xu, L., Wang, N. *et al.* (2017) System construction of virus-free and rapid-propagation technology of Baodi garlic (*Allium sativum* L.). *Scientia Horticulturae* 225, 498–504. DOI: 10.1016/j.scienta.2017.07.042.

FAOSTAT (2021) Crop data. Food and Agriculture Organization of the United Nations. Available at: http://www.fao.org/faostat/en/#home (accessed 12 July 2022).

Fereol, L., Chevelon, V., Causses, S., Michaux-Ferriere, N. and Kahane, R. (2002) Evidence of a somatic embryogenesis process for plant regeneration in garlic (*Allium sativum* L.). *Plant Cell Reports* 21, 197–203. DOI: 10.1007/s00299-002-0498-0.

Fernandez Pappu, M., Ipek, M., Ipek, A. and Simon, P. (2003) Development of a phylogenetic tree in garlic (*Allium sativum* L.) using targeted mtDNA-PCR and RAPD analysis. *Biology* 37, 51–54.

Fritsch, R.M. and Friesen, N. (2002) Evolution, domestication and taxonomy. In: Rabinowitch, H.D. and Currah, L. (eds) *Allium Crop Sciences: Recent Advances*. CAB International, Wallingford, UK, pp. 5–30.

Garcia Lampasona, S., Asprelli, P. and Burba, J.L. (2012) Genetic analysis of a garlic (*Allium sativum* L.) germplasm collection from Argentina. *Scientia Horticulturae* 138, 183–189. DOI: 10.1016/j.scienta.2012.01.014.

Gimenez, M.D. and Lampasona, S.G. (2018) Before-after analysis of genetic and epigenetic markers in garlic: a 13-year experiment. *Scientia Horticulturae* 240, 23–28. DOI: 10.1016/j.scienta.2018.04.044.

Guevara-Figueroa, T., López-Hernández, L., Lopez, M.G., Dufoo Hurtado, M.D., Vázquez-Barrios, M.E. *et al.* (2015) Conditioning garlic 'seed' cloves at low temperature modifies plant growth, sugar, fructan content, and sucrose sucrose fructosyl transferase (1-SST) expression. *Scientia Horticulturae* 189, 150–158. DOI: 10.1016/j.scienta.2015.03.030.

Havey, M.J. and Ahn, Y.K. (2016) Single nucleotide polymorphisms and indel markers from the transcriptome of garlic. *Journal of the American Society for Horticultural Science* 141(1), 62–65. DOI: 10.21273/JASHS.141.1.62.

Hirata, S., Abdelrahman, M., Yamauchi, N. and Shigyo, M. (2016) Diversity evaluation based on morphological, physiological and isozyme variation in genetic resources of garlic (*Allium sativum* L.) collected worldwide. *Genes & Genetic Systems* 91, 161–173. DOI: 10.1266/ggs.15-00004.

Hsiao, J., Yun, K., Moon, K.H. and Kim, S.H. (2019) A process-based model for leaf development and growth in hardneck garlic (*Allium sativum*). *Annals of Botany* 124(7), 1143–1160. DOI: 10.1093/aob/mcz060.

Ipek, M., Ipek, A. and Simon, P.W. (2003) Comparison of AFLPs, RAPD markers, and isozymes for diversity assessment of garlic and detection of putative duplicates in germplasm collections. *Journal of the American Society for Horticultural Science* 128(2), 246–252. DOI: 10.21273/JASHS.128.2.0246.

Ipek, M., Ipek, A., Senalik, D. and Simon, P.W. (2007) Characterization of an unusual cytoplasmic chimera detected in bolting garlic clones. *Journal of the American Society for Horticultural Science* 132, 664–669. DOI: 10.21273/JASHS.132.5.664.

Ipek, M., Sahin, N., Ipek, A., Cansev, A. and Simon, P.W. (2015) Development and validation of new SSR markers from expressed regions in the garlic genome. *Scientia Agricola* 72(1), 41–46. DOI: 10.1590/0103-9016-2014-0138.

Ipek, M., Ipek, A. and Simon, P.W. (2020) Molecular markers development and application in garlic. *Acta Horticulturae* 1297, 653–658. DOI: 10.17660/ActaHortic.2020.1297.86.

Jang, Y., Oh, Y., Choi, I., Song, Y. and Park, J. (2000) Effect of the concentration and treatment period of colchicine on polyploid formation in suspension culture of callus derived from shoot apex of garlic (*Allium sativum* L.). *Journal of Korean Society of Horticultural Science* 41(2), 157–160 (in Korean, with English abstract).

Jenderek, M.M. and Hannan, R.M. (2000) Seed producing ability of garlic (*Allium sativum* L.) clones from two public U.S. collections. In: *Proceedings of the Third International Symposium on Edible Alliaceae*. Athens, Georgia, pp. 73–75.

Jenderek, M.M. and Hannan, R.M. (2004) Variation in reproductive characteristics and seed production in the USDA garlic germplasm collection. *Hortscience* 39, 485–488. DOI: 10.21273/HORTSCI.39.3.485.

Jenderek, M.M. and Zewdie, Y. (2005) Within- and between-family variability for important bulb and plant traits among sexually derived progenies of garlic. *Hortscience* 40, 1234–1236. DOI: 10.21273/HORTSCI.40.5.1234.

Kajimura, Y., Sugiura, T., Suenaga, K., Itakura, Y. and Etoh, T. (2000) A new garlic growing system from bulbils through transplanting. *Journal of Horticultural Science and Biotechnology* 75(2), 176–180. DOI: 10.1080/14620316.2000.11511219.

Kamenetsky, R. (2007) Garlic: botany and horticulture. In: Janick, J. (ed.) *Horticultural Review* 33, 123–172.

Kamenetsky, R., London Shafir, I., Baizerman, M., Khassanov, F., Kik, C. *et al.* (2004a) Garlic (*Allium sativum* L.) and its wild relatives from Central Asia: evaluation for fertility potential. *Acta Horticulturae* 637, 83–91. DOI: 10.17660/ActaHortic.2004.637.9.

Kamenetsky, R., London Shafir, I., Zemah, H., Barzilay, A. and Rabinowitch, H.D. (2004b) Environmental control of garlic growth and florogenesis. *Journal of the American Society for Horticultural Science* 129, 144–151. DOI: 10.21273/JASHS.129.2.0144.

Kamenetsky, R., London Shafir, I., Khassanov, F., Kik, C., Van Heusden, A.W. *et al.* (2005) Diversity in fertility potential and organo-sulphur compounds among garlics from Central Asia. *Biodiversity and Conservation* 14(2), 281–295. DOI: 10.1007/s10531-004-5050-9.

Kamenetsky, R., Faigenboim, A., Mayer-Shemesh, E., Ben-Michael, T., Gershberg, C. *et al.* (2015) Integrated transcriptome catalogue and organ-specific profiling of gene expression in fertile garlic (*Allium sativum* L.). *BMC Genomics* 16(1), 1–16. DOI: 10.1186/s12864-015-1212-2.

Katis, N.I., Maliogka, V.I. and Dovas, C.I. (2012) Viruses of the genus *Allium* in the Mediterranean region. *Advances in Virus Research* 84, 163–208. DOI: 10.1016/B978-0-12-394314-9.00005-1.

Kaur, Y. and Dhall, R.K. (2017) Effect of vernalization on flowering and true seed production behaviour of garlic (*Allium sativum*) under North Indian Plains. *The Indian Journal of Agricultural Sciences* 87, 1554–1558.

Kaur, Y., Dhall, R.K. and Sharma, P. (2020) Characterization of bolting behaviour in garlic (*Allium sativum*) using SNP. *Indian Journal of Agricultural Sciences* 90, 112–117.

Kebeish, R., Deef, H.E. and El-Bialy, N. (2015) Effect of gamma radiation on growth, oxidative stress, antioxidant system, and alliin producing gene transcripts in *Allium sativum*. *International Journal of Research in BioSciences* 3(3), 161–174.

Keller, E.J., Zanke, C.D., Senula, A., Breuing, A., Hardeweg, B. *et al.* (2013) Comparing costs for different conservation strategies of garlic (*Allium sativum* L.) germplasm in genebanks. *Genetic Resources and Crop Evolution* 60(3), 913–926. DOI: 10.1007/s10722-012-9888-5.

Khar, A., Hirata, S., Abdelrahman, M., Shigyo, M. and Singh, H. (2020) Breeding and genomic approaches for climate-resilient garlic. In: Kole, C. (ed.) *Genomic Designing of Climate-Smart Vegetable Crops*. Springer International Publishing, Cham, Switzerland, pp. 359–383.

Khokhar, K.M. (2014) Flowering and seed development in onion: a review. *Library Journal* 1(7), 1–13. DOI: 10.4236/oalib.1101049.

Kim, D.W., Jung, T.S., Nam, S.H., Kwon, H.R., Kim, A. *et al.* (2009) GarlicESTdb: an online database and mining tool for garlic EST sequences. *BMC Plant Biology* 9, 61. DOI: 10.1186/1471-2229-9-61.

Lakshmi, C.S. and Sekhar, C.C. (2018) A review on integrated nutrient management in bulbous vegetable crops. *Journal of Pharmacognosy and Phytochemistry* 7(4), 411–415.

Lamichhane, J.R., Barzman, M., Booij, K., Boonekamp, P., Desneux, N. *et al.* (2015) Robust cropping systems to tackle pests under climate change: a review. *Agronomy for Sustainable Development* 35, 443–459. DOI: 10.1007/s13593-014-0275-9.

Lawande, K.E. (2018) Onion and garlic storage for value addition and supply chain management. *Journal of Allium Research* 1, 1–6.

Lee, R., Baldwin, S., Kenel, F., McCallum, J. and Macknight, R. (2013) FLOWERING LOCUS T genes control onion bulb formation and flowering. *Nature Communications* 4, 2884. DOI: 10.1038/ncomms3884.

Liu, H., Deng, R., Huang, C., Cheng, Z. and Meng, H. (2019a) Exogenous gibberellins alter morphology and nutritional traits of garlic (*Allium sativum* L.) bulb. *Scientia Horticulturae* 246, 298–306. DOI: 10.1016/j.scienta.2018.11.003.

Liu, H., Zhang, Y., Yang, F., Qi, X., Ahmad, H. *et al.* (2019b) Effect of the mode and time of gibberellic acid treatment on plant architecture and bulb structure in garlic (*Allium sativum* L.). *Scientia Horticulturae* 257, 108723. DOI: 10.1016/j.scienta.2019.108723.

Liu, P., Weng, R., Xu, Y., Pan, Y., Wang, B. *et al.* (2020) Distinct quality changes of garlic bulb during growth by metabolomics analysis. *Journal of Agricultural and Food Chemistry* 68(20), 5752–5762. DOI: 10.1021/acs.jafc.0c01120.

Liu, T., Zeng, L., Zhu, S., Chen, X., Tang, Q. *et al.* (2015) Large-scale development of expressed sequence tag-derived simple sequence repeat markers by deep transcriptome sequencing in garlic (*Allium sativum* L.). *Molecular Breeding* 35, 204. DOI: 10.1007/s11032-015-0399-x.

Liu, X.X., Wen, Y.B., Cheng, Z.H. and Mou, S.W. (2017) Establishment of a garlic cryopreservation protocol for shoot apices from adventitious buds in vitro *Scientia Horticulturae* 226, 10–18. DOI: 10.1016/j.scienta.2017.08.016.

Lopez-Bellido, F.J., Lopez-Bellido, R.J., Muñoz-Romero, V., Fernandez-Garcia, P. and Lopez-Bellido, L. (2016) New phenological growth stages of garlic (*Allium sativum*). *Annals of Applied Biology* 169(3), 423–439. DOI: 10.1111/aab.12312.

Ma, K.H., Kwag, J.G., Zhao, W., Dixit, A., Lee, G.A. *et al.* (2009) Isolation and characteristics of eight novel polymorphic microsatellite loci from the genome of garlic (*Allium sativum* L.). *Scientia Horticulturae* 122, 355–361.

Maaß, H.I. and Klaas, M. (1995) Infraspecific differentiation of garlic (*Allium sativum* L.) by isozyme and RAPD markers. *Theoretical and Applied Genetics* 91, 89–97. DOI: 10.1007/BF00220863.

Madhu, B., Mudgal, V.D. and Champawat, P.S. (2019) Storage of garlic bulbs (*Allium sativum* L.): a review. *Journal of Food Process Engineering* 42, e13177. DOI: 10.1111/jfpe.13177.

Manjunathagowda, D.C., Gopal, J., Archana, R. and Asiya, K.R. (2017) Virus-free seed production of garlic (*Allium sativum* L.): status and prospects. *International Journal of Current Microbiology and Applied Sciences* 6, 2446–2456. DOI: 10.20546/ijcmas.2017.604.290.

Martins, N., Petropoulos, S. and Ferreira, I.C. (2016) Chemical composition and bioactive compounds of garlic (*Allium sativum* L.) as affected by pre-and post-harvest conditions: a review. *Food Chemistry* 211, 41–50. DOI: 10.1016/j.foodchem.2016.05.029.

Mathew, B. (1996) *A Review of Allium Sect. Allium*. Royal Botanic Gardens, London.

Meredith, T. (2008) *The Complete Book of Garlic: A Guide for Gardeners, Growers, and Serious Cooks*. Timber Press, Portland, Oregon.

Miedema, P. (1994) Bulb dormancy in onion. I. The effects of temperature and cultivar on sprouting and rooting. *Journal of Horticultural Sciences* 69, 29–39. DOI: 10.1080/14620316.1994.11515245.

Ouzounidou, G., Giannakoula, A., Asfi, M. and Ilias, I. (2011) Differential responses of onion and garlic against plant growth regulators. *Pakistan Journal of Botany* 43, 2051–2057.

Ovesná, J., Mitrová, K. and Kučera, L. (2015) Garlic (*A. sativum* L.) alliinase gene family polymorphism reflects bolting types and cysteine sulphoxides content. *BMC Genetics* 16, 53. DOI: 10.1186/s12863-015-0214-z.

Panda, D., Dehury, B., Sahu, J., Barooah, M., Sen, P. *et al.* (2014) Computational identification and characterization of conserved miRNAs and their target genes in garlic (*Allium sativum* L.) expressed sequence tags. *Gene* 537(2), 333–342. DOI: 10.1016/j.gene.2014.01.010.

Pangestuti, P.W., Sudarsono, S. and Dinarti, D. (2020) Determine the effect of gamma irradiation towards the growth of two local garlic genotypes. *IOP Conference Series: Earth and Environmental Science* 497, 012014.

Pappu, H.R., Hellier, B.C. and Dugan, F.M. (2008) Evaluation of the National Plant Germplasm System's garlic collection for seven viruses. *Plant Health Progress* 9(1), 10. DOI: 10.1094/PHP-2008-0919-01-RS.

Parrano, L., Afunian, M., Pagliaccia, D., Douhan, G. and Vidalakis, G. (2012) Characterization of viruses associated with garlic plants propagated from different reproductive tissues from Italy and other geographic regions. *Phytopathologia Mediterranea* 51, 549–565.

Perez-Moreno, L., Lopez-Munoz, J., Pureco, A. and Hinojosa, J.C. (1991) *Production of Radiation Induced Mutants of Garlic (Allium sativum L.) Resistant to White Rot Caused by the Fungus Sclerotium Cepivorum Berk*. International Atomic Energy Agency (IAEA): IAEA, Vienna.

Pooler, M.R. and Simon, P.W. (1994) True seed production in garlic. *Sexual Plant Reproduction* 7, 282–286. DOI: 10.1007/BF00227710.

Popkov, V.A. (2012) *Garlic: biology, technology, economics*. *Our Idea*. Minsk, Belarus (in Russian).

Portela, J.A. (2001) Genetic and environmental effects influencing branching in garlic (*Allium sativum* L.). *Acta Horticulturae* 555, 175–178. DOI: 10.17660/ActaHortic.2001.555.26.

Putnik, P., Gabrić, D., Roohinejad, S., Barba, F.J., Granato, D. *et al.* (2019) An overview of organosulfur compounds from Allium spp.: from processing and preservation to evaluation of their bioavailability, antimicrobial, and anti-inflammatory properties. *Food Chemistry* 276, 680–691. DOI: 10.1016/j.foodchem.2018.10.068.

Rabinowitch, H.D. (1990) Physiology of flowering. In: Rabinowitch, H.D. and Brewster, J.L. (eds) *Onions and Allied Crops*, vol. 1. CRC Press, Boca Raton, Florida, pp. 113–134.

Ramirez, D.A., Locatelli, D.A., Gonzalez, R.E., Cavagnaro, P.F. and Camargo, A.B. (2017) Analytical methods for bioactive sulfur compounds in Allium: an integrated review and future directions. *Journal of Food Composition and Analysis* 61, 4–19. DOI: 10.1016/j.jfca.2016.09.012.

Robledo-Paz, A. and Tovar-Soto, H.M. (2012) Biotechnological tools for garlic propagation and improvement. *Innovations in Biotechnology* 10, 30636.

Rohkin Shalom, S., Gillett, D., Zemach, H., Kimhi, S., Forer, I. *et al.* (2015) Storage temperature controls the timing of garlic bulb formation via shoot apical meristem termination. *Planta* 242(4), 951–962. DOI: 10.1007/s00425-015-2334-0.

Rubio, L., Galipienso, L. and Ferriol, I. (2020) Detection of plant viruses and disease management: relevance of genetic diversity and evolution. *Frontiers in Plant Science* 11, 1092. DOI: 10.3389/fpls.2020.01092.

Rout, E., Nanda, S., Nayak, S. and Joshi, R.K. (2014) Molecular characterization of NBS encoding resistance genes and induction analysis of a putative candidate gene linked to Fusarium basal rot resistance in *Allium sativum*. *Physiological and Molecular Plant Pathology* 85, 15–24. DOI: 10.1016/j.pmpp.2013.11.003.

Rozenblat, L. (2017) Seed development and germination mechanisms in Allium species. Master's thesis, The Hebrew University of Jerusalem, Jerusalem, Israel.

Salomon, R. (2002) Virus diseases in garlic and the propagation of virus-free plants. In: Rabinowitch, H.D. and Currah, L. (eds) *Allium Crop Science: Recent Advances*. CRC Press, Boca Raton, Florida, pp. 311–329.

Shemesh, E., Scholten, O., Rabinowitch, H.D. and Kamenetsky, R. (2008) Unlocking variability: inherent variation and developmental traits of garlic plants originated from sexual reproduction. *Planta* 227(5), 1013–1024. DOI: 10.1007/s00425-007-0675-z.

Shemesh-Mayer, E. and Kamenetsky Goldstein, R. (2021) Traditional and novel approaches in garlic (*Allium sativum* L.) breeding. In: Al-Khayri, J.M, Jain, S.M. and Johnson, D.V. (eds) *Advances in Plant Breeding Strategies: Vegetable Crops*. Springer, Cham, Switzerland.

Shemesh-Mayer, E., Winiarczyk, K., Błaszczyk, L., Kosmala, A., Rabinowitch, H.D. *et al.* (2013) Male gametogenesis and sterility in garlic (*Allium sativum* L.): barriers on the way to fertilization and seed production. *Planta* 237(1), 103–120. DOI: 10.1007/s00425-012-1748-1.

Shemesh-Mayer, E., Ben-Michael, T., Kimhi, S., Forer, I., Rabinowitch, H.D. *et al.* (2015a) Effects of different temperature regimes on flower development, microsporogenesis and fertility in bolting garlic (*Allium sativum*). *Functional Plant Biology* 42(6), 514–526. DOI: 10.1071/FP14262.

Shemesh-Mayer, E., Ben-Michael, T., Rotem, N., Rabinowitch, H.D., Doron-Faigenboim, A. *et al.* (2015b) Garlic (*Allium sativum* L.) fertility: transcriptome and proteome analyses provide insight into flower and pollen development. *Frontiers in Plant Science* 6, 271. DOI: 10.3389/fpls.2015.00271.

Shen, X., Sun, X., Cao, M., Zhang, Y., Hang, Y. *et al.* (2021) Molecular markers for authentification of *Allium sativum* L. cultivar 'Taicangbaisuan'and genetic relationships among 9 Chinese garlic cultivars. *Genetic Resources and Crop Evolution* 68, 1961–1970. DOI: 10.1007/s10722-021-01109-4.

Simon, P.W. and Jenderek, M.M. (2004) Flowering, seed production, and the genesis of garlic breeding. *Plant Breeding Reviews* 23, 211–244. DOI: 10.1002/9780470650226.ch5.

Sulastiningsih, N.W.H., Ashwani, N. and Hermanto, C. (2020) Agronomic characters evaluation of garlic (*Allium sativum* L.) bulbils. *In: Proceedings International Conference on Green Agro-Industry* 4, 197–203. Available at: http://eprints.upnyk.ac.id/id/eprint/22860 (accessed 12 July 2022).

Sun, X., Zhu, S., Li, N., Cheng, Y., Zhao, J. *et al.* (2020) A chromosome-level genome assembly of garlic (*Allium sativum*) provides insights into genome evolution and allicin biosynthesis. *Molecular Plant* 13, 328–1339. DOI: 10.1016/j.molp.2020.07.019.

Takagi, H. (1990) Garlic *Allium sativum* L. In: Rabinowitch, H.D. and Brewster, J.L. (eds) *Onion and Allied Crops*, vol. III, CRC Press, Boca Raton, Florida, pp. 109–157.

Tchórzewska, D., Deryło, K., Błaszczyk, L. and Winiarczyk, K. (2015) Tubulin cytoskeleton during microsporogenesis in the male-sterile genotype of *Allium sativum* and fertile *Allium ampeloprasum* L. *Plant Reproduction* 28(3–4), 171–182. DOI: 10.1007/s00497-015-0268-0.

Tchórzewska, D., Deryło, K. and Winiarczyk, K. (2017) Cytological and biophysical comparative analysis of cell structures at the microsporogenesis stage in sterile and fertile *Allium* species. *Planta* 245(1), 137–150. DOI: 10.1007%2Fs00425-016-2597-0.

Tooke, F., Ordidge, M., Chiurugwi, T. and Battey, N. (2005) Mechanisms and function of flower and inflorescence reversion. *Journal of Experimental Botany* 56, 2587–2599. DOI: 10.1093/jxb/eri254.

Torti, S., Schlesier, R., Thümmler, A., Bartels, D., Römer, P. *et al.* (2021) Transient reprogramming of crop plants for agronomic performance. *Nature Plants* 7(2), 159–171. DOI: 10.1038/s41477-021-00851-y.

Tripathi, P.C., Sankar, V. and Lawande, K.E. (2009) Effect of storage environment and packing methods on storage losses in garlic. *Indian Journal of Horticulture* 66(4), 511–515.

Tripathi, P.C., Sankar, V. and Lawande, K.E. (2017) Micro-irrigation in onion (*Allium cepa*) and garlic (*A. sativum*): a review. *Current Horticulture* 5(1), 3–14.

Viana, J.P.G., de Jesús Pires, C., Bajay, M.M., dos Santos Valente, S.E., Pinheiro, J.B. *et al.* (2020) Do the importations of crop products affect the genetic diversity from landraces? A study case in garlic (*Allium sativum* L.). *Genetic Resources and Crop Evolution* 68, 1199–1211. DOI: 10.1007/s10722-020-01060-w.

Vieira, R.L., Silva, A.L., Zaffari, G.R., Steinmacher, D.A., Fraga, H.P.F. *et al.* (2015) Efficient elimination of virus complex from garlic (*Allium sativum* L.) by cryotherapy of shoot tips. *Acta Physiologiae Plantarum* 37, 1733–1744. DOI: 10.1007/s11738-014-1733-3.

Volk, G.M., Henk, A.D. and Richards, C.M. (2004a) Genetic diversity among US garlic clones as detected using AFLP methods. *Journal of the American Society for Horticultural Science* 129, 559–569. DOI: 10.21273/JASHS.129.4.0559.

Volk, G.M., Maness, N. and Rotindo, K. (2004b) Crypreservation of garlic (*Allium sativum* L.) using plant vitrification solution 2. *Cryoletters* 25, 219–226.

Walkey, D.G.A., Webb, M.J.W., Bolland, C.J. and Miller, A. (1987) Production of virus-free garlic (*Allium sativum* L.) and shallot (*A. ascalonicim* L.) by meristem-tip culture. *Journal of Horticultural Sciences* 62(2), 211–220. DOI: 10.1080/14620316.1987.11515771.

Wang, Q.C., Cuellar, W.J., Rajamakai, M.L., Hirata, Y. and Valkonen, J.P. (2008) Combined thermotherapy and cryotherapy for efficient virus eradication: relation of virus distribution, subcellular changes, cell survival and viral RNA degradation in shoot tips. *Molecular Plant Pathology* 9(2), 237–250. DOI: 10.1111/j.1364-3703.2007.00456.x.

Winiarczyk, K. and Gębura, J. (2016) Activity of selected hydrolytic enzymes in *Allium sativum* L. anthers. *Plant Physiology and Biochemistry* 102, 37–42. DOI: 10.1016/j.plaphy.2016.02.018.

Winiarczyk, K. and Kosmala, A. (2009) Development of the female gametophyte in the sterile ecotype of the bolting *Allium sativum* L. *Scientia Horticulturae* 121, 353–360. DOI: 10.1016/j.scienta.2009.02.014.

Winiarczyk, K., Jaroszuk-Ściseł, J. and Kupisz, K. (2012) Characterization of callase (β-1, 3-d-glucanase) activity during microsporogenesis in the sterile anthers of *Allium sativum* L. and the fertile anthers of *A. atropurpureum*. *Sexual Plant Reproduction* 25(2), 123–131. DOI: 10.1007/s00497-012-0184-5.

Woodward, P. (2014) *Garlic: An Organic Guide to Knowing, Growing and Using Garlic, from Australian Whites and Tasmanian Purples to Korean Reds and Shandongs*. Hyland House Publishing, Melbourne, Australia.

Wu, C.N., Wang, M.Y., Dong, Y.X., Cheng, Z.H. and Meng, H.W. (2015) Growth, bolting and yield of garlic (*Allium sativum* L.) in response to clove chilling treatment. *Scientia Horticulturae* 194, 43–52. DOI: 10.1016/j.scienta.2015.07.018.

Wu, C., Wang, M., Cheng, Z. and Meng, H. (2016) Response of garlic (*Allium sativum* L.) bolting and bulbing to temperature and photoperiod treatments. *Biology Open* 5(4), 507–518. DOI: 10.1242/bio.016444.

Xiao-ling, X.I.E. (2009) Analysis on the influences of colchicine on the growth of *Allium sativum* and its polyploidinduction effect. *Journal of Anhui Agricultural Sciences* 9, 135 (in Korean, with English abstract).

Yamashita, K., Tsukazaki, H., Kojima, A., Ohara, T. and Wako, T. (2010) Inheritance mode of male sterility in bunching onion (*Allium fistulosum* L.) accessions. *Euphytica* 173(3), 357–367. DOI: 10.1007/s10681-009-0101-7.

Yang, L. and Huang, H. (2014) Roles of small RNAs in plant disease resistance. *Journal of Integrative Plant Biology* 56, 962–970. DOi: 10.1111/jipb.12200.

Yang, X., Su, Y., Wu, J., Wan, W., Chen, H. *et al.* (2021) Parallel analysis of global garlic gene expression and alliin content following leaf wounding. *BMC Plant Biology* 21(1), 1–10. DOI: 10.1186/s12870-021-02948-0.

Yousef, E.A.A. and Elsadek, M.A. (2020) A comparative study of morphological and volatile oil composition characteristics in diploid and tetraploid garlic plants. *Egyptian Journal of Horticulture* 47, 1–14. DOI: 10.21608/ejoh.2019.18311.1118.

Zawistowski, J., Kopec, A., Jedrszczyk, E., Francik, R. and Bystrowska, B. (2018) Garlic grown from air bulbils and its potential health benefits. In: Jayaprakasha, G.K., Patil, B. and Gattuso, G. (eds) *Advances in Plant Phenolics: From Chemistry to Human Health*. American Chemical Society, Washington, DC, pp. 315–328.

Zewdie, Y., Havey, M.J., Prince, J.P. and Jenderek, M.M. (2005) The first genetic linkages among expressed regions of the garlic genome. *Journal of the American Society for Horticultural Science* 130(4), 569–574. DOI: 10.21273/JASHS.130.4.569.

Zhou, X.J. and Cheng, Z.H. (2008) Establishment of induction system of garlic clove-tip by injection with colchicine and effect of mutation. *Acta Agriculturae Boreali-Occidentalis Sinica* 5, 59. (In Korean with English abstract)

7 Bunching Onion

Atsushi Yamasaki[1]* and Hikaru Tsukazaki[2]
[1]*Kyushu Okinawa National Agriculture Research Center/NARO, Kurume, Japan;*
[2]*Tohoku National Agriculture Research Center/NARO, Morioka, Japan*

1. Origin and History

Japanese bunching onion (*Allium fistulosum* L.), also named welsh onion, green onion, spring onion and scallion, is widely distributed from Siberia to tropical Asia, especially in east Asia, with many varieties adapted to local environmental conditions. 'Welsh', ('welsche' in German) does not mean Wales, but strange. Incidentally, the royal badge of Wales illustrates the *A. ampeloprasum* L. leek group (leek), but not bunching onion.

A. altaicum Pall., the assumed ancestral species (Friesen *et al.*, 1999), grows wild in northern China, Mongolia and south Siberia, thus considered to be the centre of origin of bunching onion.

Bunching onion farming in the Chinese plains is mentioned in the old archive *Book of Rites* (Li-ji) and the old regional geography *Classic of Mountains and Seas* (Shan-hai-jing) from the third century BC. These descriptions mention Shandong Province as the centre of cultivation and varietal differentiation – a consequence of adaptation to the wide-ranging climatic conditions in mainland China. Further, new variants have been selected with the introduction of the crop to Japan, probably before the Nara period (710–784 AD) as Akigi (autumn bunching onion) has been described in *Nihonshoki* (*Chronicles of Japan*, 493 AD) and in Korea.

Compared to onion, bunching onion has a stronger flavour and is popular for enhancing the flavour of dishes in Chinese cuisine, kimuchi in Korea and washoku in Japan (Liu *et al.*, 2009).

2. Taxonomy

Great morphological variation is common in bunching onions, with many local varieties adapted to specific climatic conditions in east Asian countries including China, Japan and Korea (Wang *et al.*, 2005) where the temperature is the main limiting distribution factor, due to dormancy and slow growth rate in the winter season (Fig. 7.1).

The classification of bunching onion (Inden and Asahira, 1990) is based on morphological characteristics such as tillering, dormancy and utilization (Table 7.1; Fig. 7.2).

Bunching onion leaves vary in length and width, while varietal differences in leaf surface are visible microscopically (Fig. 7.3). The outermost foliar epidermis is overlaid with cuticular wax whose physiological functions include water state regulation of the whole plant and, consequently, leaf temperature,

*Address for correspondence: yamasaki@affrc.go.jp

 Edible Alliums: Botany, Production and Uses
(eds H.D. Rabinowitch and B. Thomas)
DOI: 10.1079/9781789249996.0007

Fig. 7.1. Typical dormant appearances of bunching onion varieties in winter (8 February, Kyoto, Japan). *Left:* Hanlo (cold region type); *centre:* Asagi Kujo (warm region type); *right:* Bei cong (subtropical region type). (Photo: Prof. H. Inden, Miyazaki University, Japan)

Table 7.1. Simplified classification of bunching onion varieties*

Ecological group	Plant size	Winter dormancy	Morphology	Classified varieties	Utilization
Cold region type	Large to medium	Very strong	Bulbing in cold region types, similar to shallot (Fig. 11.1), and leaf sheath colour. Various tillers	Hanlo, Yeongcheon (Korean), Beijing, Dacong	Whole plant or pseudostem
Cool region type	Large	Strong	Fewer tillers	Zhang qiu, Kaga group: Matsumoto, Shimonita	Blanched pseudostem
Intermediate region type	Large	None	Fewer tillers	Senju group: Ishikura, Kincho, Kuronobori	Blanched pseudostem
Warm region type	Large to medium	None	Many tillers	Kujo group: Asagi, Koshizu, Kannon	Whole plant or leaf blades
Subtropical region type	Medium to small	None	Many tillers	Yunnan, Bei cong	Whole plant or leaf blades

*Classification based on Inden and Asahira (1990).

hence the massive build-up of cuticular wax occurs in summer, in most varieties. This wax undergoes microstructural changes in response to the environment, thus forming three main structures: dendritic, platelets and nodules. Many dendritic varieties are mainly farmed in cold regions, while nodules varieties are cultivated in warm regions (Inden and Yamasaki, 2014). It was thus suggested that the microstructure of the wax is related to temperature adaptation.

The dormant characteristics of bunching onion varieties strongly reflect the winter climate of the place of establishment and the overwintering state of the particularly strong dormant cultivars is similar to that of *A. altaicum* (Fig. 7.4). Probably, as the distribution spread to warm regions, the evolution of varieties progressed while losing dormancy.

3. Genetics and Breeding

3.1. Molecular studies

The annual bunching onion is favoured for genetic studies due to its 28% smaller genome size than that of the bulb onion (Ricroch *et al.*, 2005), while genome synteny is highly conserved between the two *Allium* crops (Tsukazaki *et al.*, 2008).

Fig. 7.2. Local varieties of bunching onion in Japan.

DNA markers are powerful tools for molecular breeding and evaluation of population homogeneity in seeds of established OP and F_1 hybrids. King *et al.* (1998) first produced a bulb onion (*A. cepa* L.) linkage map, based on RAPDs and RFLPs. Hence, QTLs for soluble solids, dry matter, pungency, flavour and antiplatelet aggregation activity were identified (Galmarini *et al.*, 2001), and, later, Ohara *et al.* (2005) constructed bunching onion linkage maps mainly based on those AFLP markers.

In the last two decades, the co-dominant, reliable and abundant SSRs and SNPs are favoured in genetic analyses in *Allium* and other crops. These tandem repeats of short-sequence motifs (SSRs consist of 1–6 bases) are used for constructing linkage maps, marker-assisted selection, phylogenetic studies and genomic synteny analyses (Jones *et al.*, 1997; Bredemeijer *et al.*, 1998) and in *Allium* spp. (Tsukazaki, 2018).

Tsukazaki *et al.* (2008) constructed an SSR-based bunching onion chromosome map using several *Allium* alien addition lines. Consequently, markers present in this map were identified by their chromosomal location in *A. cepa* by alien monosomic addition lines (AMALs) using a complete set of *A. fistulosum* monosomic addition lines obtained by crosses with shallot (2n=16 + 1, FF + 1A to FF + 8A) (Shigyo *et al.*, 1996), and in *A. fistulosum* by alien chromosomes from shallot (*A. cepa* L. Aggregatum group) (2n=23) (Hang *et al.*, 2004). The map was then used as a reference for chromosomal assignment in other populations, by comparing common markers.

Next-generation sequencing (NGS) technologies facilitated genes and QTLs mapping in bulb onion (McCallum *et al.*, 2018). The above AMALs further facilitated the assignment of linkage groups to their relevant chromosomes (van Heusden *et al.*, 2000; Martin *et al.*, 2005). Recently, Jo *et al.* (2017) and Chou *et al.* (2020) independently constructed a high-density genetic linkage maps based on genotyping-by-sequencing (GBS) analyses, which revealed QTLs controlling anthocyanin synthesis and content (Chou *et al.*, 2020). Fujito *et al.* (2021a) constructed high-density genetic linkage map of bulb onion based on transcriptome-based unigene markers, and Finkers *et al.* (2021) published the first genome assembly of the bulb onion.

Employing AFLP-based linkage maps constructed based on reciprocal BC_1 populations revealed 13 QTLs (seven groups for D map and six for J map) for bunching onion seedling growth (Ohara *et al.*, 2009). Tsukazaki *et al.* (2012) located the major pungency QTL on chromosome 2a, and Wako *et al.* (2016) detected two

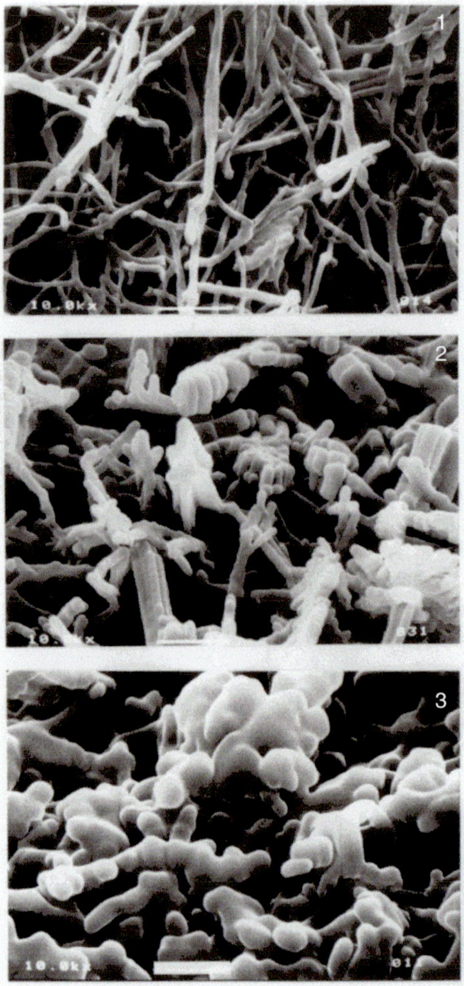

Fig. 7.3. Varietal differences of ultrastructure of epicuticular wax of bunching onion leaf surface (Inden and Yamasaki, 2014).

major QTLs for bolting time in $F_{2:3}$ populations between ever-flowering and late-bolting lines. The QTL located on chromosome 2a was present in hybrid populations between early- and late-bolting lines.

Tsukazaki *et al.* (2017) identified a total of 27 QTLs for six morphological traits: plant height, leaf length, pseudostem length, leaf width, pseudostem width, and the number of leaf sheaths, in 16 chromosomal regions. A major QTL for the number of leaf sheaths is repeatedly detected on chromosome 8, and two QTLs are associated with pseudostem pigmentation.

Fig 7.4. Old perennial *A. fistulosum* forms bulbs resembling bulbs of *A. altaicum*. (Photo: Nikolai Friesen)

A genome-wide association study (GWAS) provides a powerful tool for the identification of genomic regions underlying complex traits (Atwell *et al.*, 2010; Huang *et al.*, 2010). GWAS does not require F_2 populations, but the genome sequence serves as a reference. Chen *et al.* (2018) proposed a transcriptome-referenced association study (TRAS), which does not need a reference genome sequence for characterization of garlic (*A. sativum* L.) clove shape in 102 landraces, and identified 22 candidate transcripts. Later reports reveal that a combined genomic selection scheme with a shortening of generation time improved bulb onion breeding efficiency (Sekine and Yabe, 2020). We thus propose that combining GWAS with a genomic selection scheme is expected to enhance bunching onion breeding although the reference genome sequence was not reported.

To date, no candidate genes of agronomic traits were identified in bunching onion. It is expected that genomes comparisons between a bunching onion and bulb onion will provide a better view of this matter. For example, QTLs for bolting time in bulb onion were detected on chromosomes 1, 3 and 6 (Baldwin *et al.*, 2014), and in bunching onion on chromosomes 1b, 3b and 6a (Wako *et al.*, 2016). Located on chromosome 5 (Baldwin *et al.*, 2014), the *AcFT2* promotes floral transition in response to vernalization of bulb onion (Lee *et al.*, 2013) and may have a homolog in bunching onion; RNA-seq analysis identified four unigenes that code for glossy leaves in *A. fistulosum* due to wax content (Liu *et al.*, 2014), and glossiness in bulb onion leaves is controlled by a single recessive gene on chromosome 8. Moreover, epistasis was detected between chromosomes 1 and 8 (Munaiz and Havey, 2020). The actual genetic control, however, requires a genomes comparison between a bunching onion and bulb onion.

Havey (2018) provided strong evidence for recombinations within interspecific hybrids between a bunching onion and bulb onion but the limited number of common markers hinders a detailed comparison of linkage maps between the two species.

Assignment of common markers indicates a highly conserved macrosynteny between a bunching onion and bulb onion (Tsukazaki *et al.*, 2008), yet the small number of orthologous markers is insufficient (Tsukzaki *et al.*, 2015) to compare the bunching onion linkage map based on transcriptome shotgun assembly (TSA)-derived markers with those of bulb onion (Duangjit *et al.*, 2013; Baldwin *et al.*, 2014).

A minimum of two heteromorphic bivalents was observed in a meiotic analysis of interspecific hybrids between a bunching onion and bulb onion (Peffley, 1986). Hence, chromosomal mutations such as translocation and inversion might be involved in the speciation of the two crops. Microsynteny should facilitate the genome-wide level of such chromosomal mutations, by plotting many more common (orthologous) markers onto the maps of both species. These markers will also further the understanding of *Allium* comparative genomics.

3.2. Breeding

The typical allogamous bunching onion exhibits a high level of outcrossing (Ford-Lloyd and Armstrong, 1993). Consequently, landraces and OP varieties maintain high levels of heterozygosity and heterogeneity (Haishima *et al.*, 1993; Tsukazaki *et al.*, 2006). Hence requiring continuous breeding for improved crop performance and uniformity has been carried out where bunching onion is farmed.

Hybrids' breeding depends on the availability of reliable male sterility. Nishimura and Shibano (1972) identified a spontaneous CMS individual (Moue and Uehara, 1985), which, since the late 1980s, serves for the breeding and seed production of F_1 varieties.

F_1 hybrids between the Senju and Kujo groups show remarkable heterosis (Ohara *et al.*, 2004). Hence, over 80% of bunching onion varieties released in the 21st century in Japan were hybrids (Japan Horticultural Production and Research Institute, 2016; Japan Seed Trade Association, 2021).

Yamashita *et al.* (1999) introgressed by backcrossing CMS plus a dominant male fertility restorer gene from *A. galanthum* Kar. et Kir. into Japanese bunching onion. These may serve as seed parents for the commercial *A. fistulosum* F_1 seed production (Yamashita and Tashiro, 2004). Yamashita *et al.* (2010) identified two CMS lines among 135 accessions from Zhangqiu and Guangzhou populations. These were backcrossed several times for F_1 hybrid breeding (Wako *et al.*, 2017) and used as parent lines of F_1 varieties at NARO.

Important goals for bunching onion breeding are disease and pest resistance, late-bolting, high-quality yields (e.g. low pungency, high sugar content), suitability for mechanized farming, e.g. vigorous growth, short leaves and adaptation to mechanized tillering for improved blanching (Wako *et al.*, 2010, 2017; Fig. 7.5).

Bunching onion has been reported to be resistant to leaf blight (*Botrytis squamosa*) (Currah and Maude, 1984), pink root (*Pyrenochaeta terrestris*) (Netzer *et al.*, 1985), anthracnose (*Colletotrichum gloeosporioides*) (Galván *et al.*, 1997), smut (*Urocystis cepulae*) (Felix, 1933), onion yellow dwarf virus (OYDV) (Costa *et al.*, 1971) and *Stemphylium* leaf blight (Pathak *et al.*, 2001).

Some tolerance to rust (*Puccinia allii* (DC.) F. Rudolphi) was recorded in bunching onion but no immunity was found (Wako *et al.*, 2012). Hence, recurrent selection programmes for improved rust tolerance were initiated (Yamashita *et al.*, 2005; Wako *et al.*, 2012). The Beicong and Kokusen natsuyo accessions are resistant to stone leek leafminer (*Liriomyza chinensis* (Kato)) due to antibiosis that leads to egg-killing (Sueyoshi

Fig. 7.5. Bunching onion variety Fuyu-warabe with short, thick leaves (Wako *et al.*, 2010).

et al., 2006; Takeda *et al.*, 2020). A resistant line was thus selected by repeated self-pollinating Beicong (Fujito *et al.*, 2021b). In addition, Yang *et al.* (2017) reported on an onion thrips-resistant glossy mutant.

4. Growth Physiology

4.1. Temperature

As expected from its centre of origin, bunching onion requires cooler environments than other Alliums. It thrives in cool and dry climates, with 15–25°C for optimum growth. The maximum growth rate occurs at 18–19°C as compared with 25°C for onion and leek (Brewster, 1979, 1994). However, a wide range of cold and heat tolerance and vigorous growth rate under low or high temperatures have been acquired during the long history of bunching onion cultivation and selections.

4.2. Light

Adult bunching onion plants reach maximum photosynthesis rates at 15–20°C while a marked decrease in photosynthesis has been recorded at temperatures >30°C. Assimilation rates increased with CO_2 concentration up to 2000 ppm (v/v) and with light intensity from 0–690 $\mu mol \cdot m^{-2} \cdot s^{-1}$ (Fig. 7.6), but no further effect up to 1500 $\mu mol \cdot m^{-2} \cdot s^{-1}$ was measured caused by stomata closure due to water-stress conditions (Wolff and Coltman, 1990). Most of the assimilated biomass is stored in the leaves and pseudostem.

Compared with 12 and 24 hours at 576 $\mu mol \cdot m^{-2} \cdot s^{-1}$, bunching onion growth rate increased by only 40% with extended light hours (Brewster, 1990), probably due to downregulation by the long days, as reported for bulb onions (van Gestel *et al.*, 2005).

5. Physiology of Flowering

Floral induction of bunching onion requires vernalization (Yamasaki and Tanaka, 2002; Shiraiwa *et al.*, 2007; Dong *et al.*, 2013), while short days can substitute for insufficient cold units (Yamasaki *et al.*, 2000a). These cues are perceived only by post-juvenile plants when seedlings developed a certain number of leaves or a stem of a certain diameter.

It is interesting to note that vernalization cues for floral induction are perceived by seeds of the Taiwanese variety Bei cong (Yamasaki *et al.*, 2013).

The transition of the underground apical meristem to a flower bud and a determinate umbel (Qu *et al.*, 1994) is followed by differentiation into florets, bolting and flowering. Concomitantly, lateral buds sprout and new vegetative growth begins (Yamasaki, 2007).

Following initiation, the early stage of flower development is day-neutral, but when the floret formation stage ends, long-day promotes flower development and elongation of the seed stalk (Yamasaki *et al.*, 2000a).

5.1. Flower-inducing factor: low temperature

Much like bulb onion, bunching onion responds to a wide range of vernalization temperatures (3–15°C) with an optimum ~9°C (Yamasaki *et al.*, 2000b), while high daytime temperatures result in devernalization. Genotypes differ markedly in the duration of the induction period, from minimal for Bei cong (Lin and Chang, 1980) to little or no bolting of Bozu-shirazu. Similarly, genotypes vary in response to

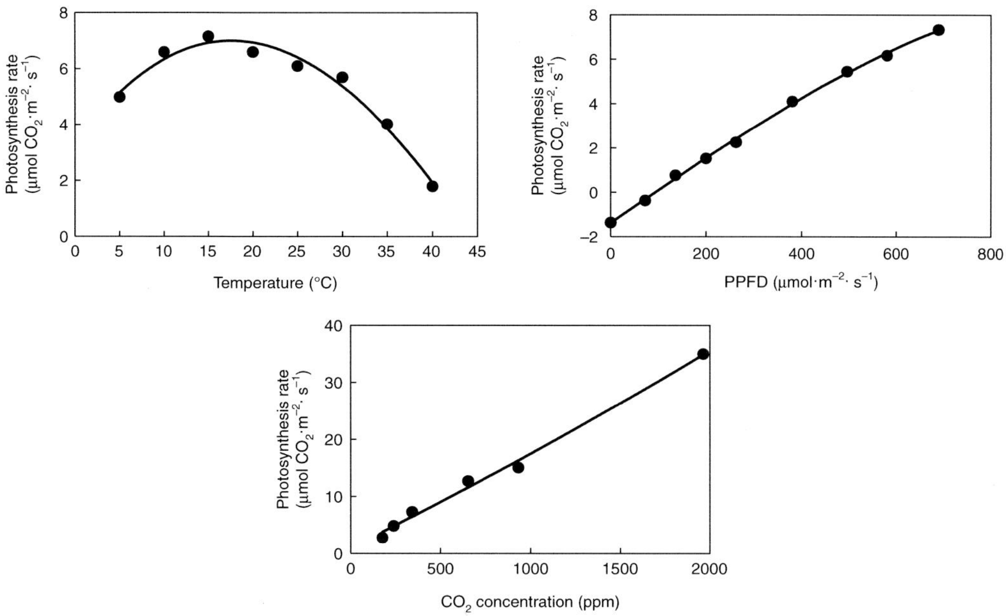

Fig. 7.6. Photosynthetic properties of bunching onion cv. Kincho (Yamasaki *et al.*, unpublished data). Potted plants 140 days after sowing were measured in a closed growth chamber at an initial condition: temperature 25°C, light 690 μmol·m^{-2}·s^{-1}, CO_2 340 ppm.

devernalization. Hence, vernalized Kincho and Asagi-kujo do not bloom only at >26.5°C while Cho-etsu undergoes devernalization at 13.5–20°C.

Spring and early summer harvests were formerly considered out-of-season production due to bolting, yet protected cultivation facilitates a simple environmental control by winter to early spring devernalization (Yamasaki *et al.*, 2000c), thus low tunnel farming permits out-of-season production in warmer regions of Japan (Ando *et al.*, 2002).

5.2. Flower-inducing factor: photoperiod

Bunching onion is a short-day (SD) plant, thus flower development in vernalized plants is suppressed by long-day (LD), particularly under marginal cold treatment. Suppression by LD, however, is genotype-dependent; hence, despite the similar sensitivity to vernalization of Kincho and Asagi-kujo, Asagi-kujo and Cho-etsu respond to a markedly longer LD period than Kincho.

Genotypes vary in inductive requirements for bolting, e.g. Kincho responds better to chilling while Asagi-kujo is more responsive to SD treatment; hence screening genetic resources for photoperiod requirements could benefit the quality production of bunching onion. Vernalization and SD inductions could be applied sequentially, as SD exposure before vernalization promoted flower bud initiation (Yamasaki *et al.*, 2000a), and LD treatment in the nursery could be applied to suppress bolting.

Skilled choice of varieties and management should lower bolting risks and damages. Hence, Asagi-kujo-type varieties that strongly require short-day induction should preferably be farmed under LD conditions rather than Kincho-type varieties. In Japan, the continuous expansion of greenhouse culture and plug-seedling transplanting of bunching onions may facilitate LD artificial application for bolting control.

5.3. Flower-inducing factor: plant size or age

Reliable indicators for crossing the juvenile phase (and becoming receptive to vernalization and SD) were sought, but genotypes vary markedly in post-juvenility age. Yet a general measure of 5 mm pseudostem diameter for vernalization suffices the needs of many varieties, but not all.

The internode at which flower buds differentiate is genotype characteristic, e.g. an eigenvalue of eight leaves represents the post-juvenile physiological age for Kincho and Asagi-kujo, while only plants with 11 leaves cross the juvenile phase for Cho-etsu (Yamaşaki *et al.*, 2012). Hence, physiological age expressed by leaf number is accepted as a good index for individual genotypes (Shiraiwa *et al.*, 2005).

6. Fertilization

The highest rate of bunching onion growth was recorded at ~30 days after transplanting. Thus, a 100 kg nitrogen ha^{-1} basal fertilizer is required to maintain a high crop growth rate after planting and a similar amount is applied as topdressing 1–2 months later for yield maximization (Hayashi *et al.*, 2003).

In addition, it is well known that flower initiation in properly vernalized bunching onion is enhanced by low nitrogen (Shiraiwa *et al.*, 2005; Yamasaki and Tanaka, 2005). Thus, nitrogen deprivation must be avoided at overwintering culture.

7. Production

7.1. Global bunching onion production

Bunching onion is an important vegetable in east Asia, and its popularity worldwide is on the increase because of its soft texture and unique flavours for spice vegetables. It is mainly used either for leaf blades or pseudostems, or both, and market demands dictate the respective cultivation methods.

Statistics on bunching onions are rather poor. China is the largest producer followed by Japan, Korea and Taiwan (Ford-Lloyd and Armstrong, 1993; Wang *et al.*, 2005).

In the Asia Pacific, China is the largest consumer of bunching onion, followed by Japan and Korea (Table 7.2). In China, bunching onion is consumed as seedlings, immature stems, pseudostems and overwintered plants (Wang *et al.*, 2005). North China and the middle and lower reaches of the Yangtze river supply ~65% of the country's total output, of which three northern provinces – Shandong, Henan and Hebei – produce more than 50% (Zhang *et al.*, 2021).

In Japan, it is the third most important vegetable crop (Table 7.2), with an annual output value of ~130 billion yen (MAFF, 2021). Consumers dislike tillered plants, hence varieties with few shoots are preferred. Commonly, these plants produce very large and hard leaf blades that reach ~1 m height at harvest, hence blade-trimming is common (Fig. 7.7).

Bunching onion was first introduced to western Europe during or at the end of the Middle Ages, and from there into Russia, and later to southeast Asia and northeast India (Singh and Ramakrishna, 2017). Currently, bunching onion is used by many communities indistinguishably from green onion (scallions), and as an important seasoning worldwide (Tabassum *et al.*, 2016), yet international trade is very limited, except for exports from China to Japan, which account for 30% of the bunching onion's global trade. Malaysia is another major importer (Leong, 2001).

Table 7.2. Bunching onion production in east Asia (2019). No accurate statistics are available for other east Asian countries.

Country	Year	Area (ha)	Production (1000 tons)	Data resource
China	2019	30,504	1107.3	FAOStat[1]
Japan	2019	22,400	465.3	Government
Korea	2019	17,170	463.7	Government
Taiwan	2020		63.8	Knoema[1,2]

[1]Onion, shallots, green

[2]https://knoema.com/FAOPRDSC2020/production-statistics-crops-crops-processed?tsId=1073810 (accessed September 2022)

Fig. 7.7. Blanched bunching onion field. *Left:* one month after transplanting; *centre:* close view just after earthing-up operation; *right:* distant view of harvesting stage. (Photos: H. Kawashiro, J.A. Zenno, Japan)

8. Blanching

Blanching is applied mainly in open fields. Seedlings are often grown in protected nurseries to the age of three to four leaves. Like leek, bunching onion is transplanted in trenches, followed by mounding soil mixed with fertilizers throughout the growing season up to the branching point of the pseudostem and leaf blade, for blanching. During earthing up, very little rooting develops upward from the true stem (Fig. 7.8). Older leaves wilt while new etiolated leaves emerge through the central core of the buried pseudostem, thus blanching is complete. Optimal temperatures for blanching are around 15°C (Ishiguro, 1967). The standard length of blanched pseudostem ranges from 20–30 cm up to >50 cm (Fig. 7.9). At lower temperatures, no renewal of the outer leaves occurs, chlorophyll is visible and blanching is not completed. At higher temperatures, the blanched pseudostems remain short.

In recent years, the use of light-shielding film blanching became popular in the greenhouse, out-of-season, and/or for high-quality production (Fig. 7.10).

Fig. 7.8. Root system profile of blanched bunching onion. (Photo: S. Takeda and M. Honjo, Akita Agricultural Experiment Station, Akita, Japan)

Fig. 7.9. Package of blanched bunching onion for shipment to wholesale market. (Photo: H. Kawashiro, J.A. Zenno, Japan)

9. Integrated Mechanization System

Bunching onion farming is labour-consuming, especially in the blanching operation, hence an integrated mechanization system (Fig. 7.11) became popular in Japan.

Seedlings are raised in nursery pots, trays or plug trays with 288 or 448 wells (Honjo *et al.*, 2015), up to the three-leaves stage followed by automated planter transplanting. Earthing-up machines for

Fig. 7.10. Blanched bunching onion in greenhouse out-of-season production using artificial types of blanching by a light-shielding film. (Photo: H. Kawashiro, J.A. Zenno, Japan)

Fig. 7.11. Integrated mechanization system of blanched bunching onion cultivation: (a) transplant; (b) weed; (c) chemical spray; (d) earthing up; (e) harvest; (f) trimming of leaves and roots; (g) peeling outer leaves for package. (Photo: the authors and H. Kawashiro, J.A. Zenno, Japan)

blanching are used, and a mechanical harvester digs up and lifts, and transfers, the plants by a conveyor into a container. At harvest, external wilted leaves are removed by compressed air and machines remove roots, leaf blades and external leaves. Plants with hard leaf blades are finely topped before shipment.

10. Non-blanched Farming of Green Bunching Onion

The entire plant, excluding roots, is consumed as green bunching onion (Fig. 7.12). The varieties of choice form soft blades and many tillers. They grow fast at low temperatures and dormancy is short. Produce size varies with consumers' preferences and usage. Some open-field varieties produce many >1 m-high tillers; others are intensively cultivated in a greenhouse for early supply when leaves are 30–50 cm long. In Japan, hydroponic production is a small niche, and young sprouts form a different marketing branch (sprout bunching onion, Sushi toppings etc.; Fig. 7.13).

In recent years, sliced blade tops are used not only in the food industry but also in home cooking. For this market, repeated harvest is based on cutting the tops at the ground level and promoting regrowth (Fig. 7.14).

11. Diseases and Pests

Bunching onion possesses resistance to several diseases that affect common onions (Ford-Lloyd and Armstrong, 1993), e.g. Stemphylium leaf blight (*Stemphylium vesicarium*) (Pathak *et al.*, 2001; Kudryavtseva *et al.*, 2019), Fusarium basal rot (*Fusarium oxysporum* f. sp. *cepae*) (Vivero, 2009), thrips (*Thrips tabaci*), smut (*Urocystis cepulae*) (Jones and Mann, 1963), pink root (*Phoma terrestris*) (Netzer *et al.*, 1985; Martinez *et al.*, 2005), and more (Fig. 7.15).

Clonal propagation is conducive to virus infestation, but many landraces appear to be relatively tolerant. Hence it may be considered a source for improvement of the bulb onion (Jones and Mann, 1963; van der Meer, 1984). Yet it is susceptible to several diseases, many of them common to most *Allium* crops. *Allium fistulosum* is resistant to the onion yellow dwarf virus (OYDV). In Japan and Indonesia, it develops typical mosaic-type symptoms of the welsh onion yellow stripe virus (WoYSV), including chlorotic mottling, streaking and stunting, and distorted flattening of the leaves.

11.1. Alternaria leaf spot (*Alternaria porri* (Ellis.) Cif.)

Purple leaf blotch, Alternaria blight causes characteristic concentric spots on the leaves. Only moderate infection has been recorded in bunching onion (Nanda *et al.*, 2016) but symptoms

Fig. 7.12. Green bunching onion production in greenhouse (*left*) and various sizes of products for wholesale market (*right*).

Fig. 7.13. Special use of bunching onion. *Left and centre:* sprouts; *right:* processed cut for topping on noodles.

Fig. 7.14. Bunching onion top harvester. (Photo: Nishizawa Co., Ltd, Kagawa, Japan)

Fig. 7.15. Diseases and pests. (a) *Alternaria*, (b) rust, and (c) stone leek leafminer.

intensify under high humidity (Inden and Asahira, 1990). Similar symptoms are caused by *Stemphylium vesicarium*, onion leaf blight (Theresa *et al.*, 1993). Typically, purple leaf blotch lesions are often colonized by both pathogens in the field. Both pathogens are seed-borne in bulb onion (Simmons, 1967).

11.2. Fusarium wilt

Fusarium species, and especially *F. oxysporum*, is the most widely spread and economically important soilborne pathogen (Leslie and Summerell, 2006) in many *Allium* crops, including bunching onion (Takakuwa *et al.*, 1977). *F. oxysporum* f. sp. *cepae* causes Fusarium wilt in bunching onion. Screening for pathogenic isolates in wilted bunching onion from across Japan (Dissanayake *et al.*, 2009) revealed a significant variation in virulence, and high levels of resistance to each isolate were recorded in bunching onion and chives (*A. schoenoprasum*). Evaluation of alien monosomic addition lines (AMALs) of bunching onion carrying 2A chromosome (FF+2A) from shallot, showed a shallot-derived specific saponin band and extremely high resistance to Fusarium wilt (Vu *et al.*, 2012). The genes coding for Fusarium wilt resistance are located on chromosome 2, in shallot.

11.3. Rust (*Puccinia allii* (DC.) F. Rudolphi)

Rust seriously injures bunching onion foliage with the consequent marked quality and yield decline. Uredospores are the main inocula for secondary infections throughout the year with frequent epiphytotics in autumn and spring (Takeuchi, 1990). In Japan, *A. cepa* is considered less susceptible to *P. allii* isolates from *A. fistulosum* than bunching onion (Morinaka, 1985). AMALs of bunching onion with shallot 1A chromosome (FF+1A) were highly resistant to rust in controlled-environment experiments at the seedling but not at the adult stage, due to the presence of genes for early rust resistance in shallot chromosome 1 (Wako *et al.*, 2015).

11.4. White rot (*Sclerotium cepivorum* Berk.)

White rot is one of the most damaging pathogens in Alliums, especially in garlic and bulb onion. It is very persistent and survives in the absence of host plants for 20 years in the soil (Coley-Smith *et al.*, 1990). The optimum temperature for disease development in both bunching onion (Ikeda and Sakai, 2020) and bulb onion (Adams and Papavizas, 1971) is ~15°C. Poor, unbalanced nutrition and heavy rains stimulate the development of the pathogen.

11.5. Stone leek leafminer (*Liriomyza chinensis* (Kato) (Diptera: Agromyzidae))

This is one of the most destructive pests of the bunching onion in Japan, Korea, China and other Asian countries (Chen *et al.*, 2003; Choi *et al.*, 2004; Tokumaru, 2006) that causes heavy losses to commercial and home growers alike. The larvae tunnel into the leaves as they move inside tissues. Infested plants exhibit curly, wavy or distorted leaves, followed by secondary infection by fungi or bacteria (Tokumaru, 2006). Recently, a variant of *L. chinensis* (biotype B) was identified in Japan, with a markedly larger number of larvae in the leaves and greater damages than those caused by the conventional strain (biotype A) (Tokumaru and Uesugi, 2019).

11.6. Beet armyworm (*Spodoptera exigua*) and American bollworm (*Heliothis armigera*)

These notorious pests are difficult to control since the larvae find shelter inside the hollow leaves (https://www.prota4u.org/database/protav8.asp?g=pe&p=Allium+fistulosum+L; accessed September 2022).

11.7. Thrips (*Thrips tabaci* (Lindeman) (Thysanoptera: Thripidae))

Thrips is a major pest of Alliums. It injures plants directly by feeding on cell sap and often by vectoring pathogens such as the onion yellow dwarf virus, iris yellow spot virus, and Alternaria leaf spot (Straub and Emmett, 1992; Hsu *et al.*, 2010).

11.8. Viruses

Bunching onion is susceptible to a number of viruses, the two most common being welsh onion yellow stripe virus and thrips-transmitted iris yellow spot virus.

11.8.1. Welsh onion yellow stripe virus (WoYSV)

Common in Japan and Indonesia, welsh onion yellow stripe virus causes mosaic-type symptoms, including chlorotic mottling, streaking and stunting, and distorted flattening of the leaves. Some tolerance was detected in Kujo group (https://www.prota4u.org/database/protav8.asp?g=pe&p=Allium+fistulosum+L; accessed September 2022).

11.8.2. Iris yellow spot virus (IYSV)

This virus was reported in Indonesia and the USA (Pappu and Rauf, 2013; Tabassum *et al.*, 2016). Symptoms consist of elliptical to irregular, straw-coloured or chlorotic foliar lesions (Tabassum *et al.*, 2016).

12. Seed Production

Bunching onion seeds are short-lived, hence seed companies are mindful of seed quality such as viability, energy reserves and seed-borne diseases. In Japan, seed production is located where no or little rainfall occurs during April–June, the time of blooming and seed harvest, or abroad, e.g., Chile and Italy, to reduce seed quality declination by extreme weather.

In a sense, bunching onion hybrid seed production resembles the seed-to-seed method, except for the direct seeding, as selection takes place before transplanting. Plants are vernalized in the field, bolt and produce seed (Fig. 7.16). The selection fields are seeded in late summer, allowing plants to cross the juvenile phase so that vernalization can occur during autumn or winter. Before planting, it is important to ensure that minimum isolation distances are met from bulb onion and bunching onion fields.

CMS is indispensable for practical F_1 hybrid seed production of bunching onion. Field-grown male lines are selected for uniformity. Excelling lines are transferred to seed production plots, in isolated fields, or indoors to prevent unwanted cross-pollination. Strict pest and disease control are applied throughout to avoid seed-borne diseases. To achieve simultaneous flowering of male and female parental lines, different planting dates are frequently required. At the end of blooming time, all-male plants are rogued just before harvest to prevent mechanical mixture.

Bunching onion seeds are mainly imported from Chile, South Africa and Italy (Plant Protection Station, Japan, 2021). Upon arrival, seeds are checked for quality and treated prophylactically with fungicide before planting in F_1 hybrid production fields. A portion of the imported seed is exported to China and Korea for local F_1 hybrid seed production.

13. Future of Bunching Onion Production

Bunching onion is an essential component of Japanese cuisine (washoku). While production of most major vegetables decreases with population decline, the production of bunching onion remains quite steady, especially due to the change of dietary habits in Japan. Hence, the application of repeated harvests cropping system would be increased to fulfill the demand of sliced bunching onion top for spicy topping. Additionally, the popularity of Japanese food is rising worldwide, thus leading to the breeding of locally adapted varieties and cultivation methods all over the world.

Fig. 7.16. F_1 hybrid seed production in Australia. (Photo: S. Ogasawara, Japan)

References

Adams, P.B. and Papavizas, G.C. (1971) Effect of inoculation density of *Sclerotium cepivorum* and some soil environmental factors on disease severity. *Journal of Phytopathology* 61, 1253–1256.

Ando, T., Kota, N. and Okoshi, K. (2002) Characteristics of flower-bud formation and bolting of late-season flowering welsh onion harvested in early summer. *Bulletin of the Chiba Prefectural Agricultural Research Center* 1, 13–23. (In Japanese with English summary)

Atwell, S., Huang, Y.S., Vilhjálmsson, B.J., Willems, G., Horton, M. *et al.* (2010) Genome-wide association study of 107 phenotypes in *Arabidopsis thaliana* inbred lines. *Nature* 465, 627–631.

Baldwin, S., Revanna, R., Pither-Joyce, M., Shaw, M., Wright, K. *et al.* (2014) Genetic analyses of bolting in bulb onion (*Allium cepa* L.). *Theoretical and Applied Genetics* 127, 535–547.

Bredemeijer, G.M.M., Arens, P., Wouters, D., Visser, D. and Vosman, B. (1998) The use of semi-automated fluorescent microsatellite analysis for tomato cultivar identification. *Theoretical and Applied Genetics* 97, 584–590.

Brewster, J.L. (1979) The response of growth rate to temperature in seedlings of several *Allium* crop species. *Annals of Applied Biology* 93, 351–357.

Brewster, J.L. (1990) The effect of the duration of daily irradiation on the growth rates of seedlings of leek (*Allium ampeloprasum* L.) and Japanese bunching onion (*Allium fistulosum* L.). *Scientia Horticulturae* 43, 207–211.

Brewster, J.L. (1994) The physiology of crop growth, development and yield. In: Brewster, J.L. (ed.) *Onions and Other Vegetable Alliums*. CAB International, Wallingford, UK, pp. 63–92.

Chen, X.J., Lang, F., Xu, Z., Jun-Hua, H.E. and Ma, Y. (2003) The occurrence of leafminers and their parasitoids on vegetables and weeds in Hangzhou area, Southeast China. *Biocontrol* 48, 515–527.

Chen, X., Liu, X., Zhu, S., Tang, S., Mei, S. *et al.* (2018) Transcriptome-referenced association study of clove shape traits in garlic. *DNA Research* 25, 587–596.

Choi, I.H., Jan, Y.S., Kim, G.H. and Kim, J.W. (2004) Control effects of some insecticides on different stages of the stone leek leafminer, *Liriomyza chinensis* Kato (Diptera: Agromyzidae). *Korean Journal of Applied Entomology* 43, 169–173. (In Korean with English summary)

Chou, Y., Kim, S. and Lee, J. (2020) Construction of an onion (*Allium cepa* L.) genetic linkage map using genotyping-by-sequencing analysis with a reference gene set and identification of QTLs controlling anthocyanin synthesis and content. *Plants* 9, 616.

Costa, A.S., Costa, C.L., Nagai, H. and Kitajima, E.W. (1971) Welsh onion, a source of the onion yellow dwarf virus. *Biologico.* 37, 157–159.

Coley-Smith, J.R., Mitchell, C.M. and Sansford, C.E. (1990) Long-term survival of sclerotia of *Sclerotium cepivorum* and *Stromatinia gladioli*. *Plant Pathology* 39, 58–69.

Currah, L. and Maude, R.B. (1984) Laboratory tests for leaf resistance to *Botrytis squamosa* in onions. *Annals of Applied Biology* 105, 277–283.

Dissanayake, M.L.M.C., Kashima, R., Tanaka, S. and Ito, S. (2009) Pathogenic variation and molecular characterization of *Fusarium* species isolated from wilted welsh onion in Japan. *Journal of General Plant Pathology* 75, 37–45.

Dong, Y., Cheng, Z., Meng, H., Liu, H., Wu, C. and Khan, A.R. (2013) The effect of cultivar, sowing date and transplant location in field on bolting of welsh onion (*Allium fistulosum* L.). *BMC Plant Biology* 13, 154.

Duangjit, J., Bohanec, B., Chan, A.P., Town, C.D. and Havey, M.J. (2013) Transcriptome sequencing to produce SNP-based genetic maps of onion. *Theoretical and Applied Genetics* 126, 2093–2101.

Felix, E.L. (1933) Disease resistance in *Allium fistulosum* L. *Journal of Phytopathology* 23, 109–110.

Finkers, R., van Kaauwen, M., Ament, K., Burger-Meijer, K., Egging, R. *et al.* (2021) Insights from the first genome assembly of onion *(Allium cepa). G3* 11, jkab243. DOI: https://doi.org/10.1093/g3journal/jkab243.

Ford-Lloyd, B.V. and Armstrong, S.J. (1993) Welsh onion *Allium fistulosum* L. In: Kalloo, G. and Bergh, B.O. (eds) *Genetic Improvement of Vegetable Crops*. Pergamon Press, Oxford, UK, pp. 51–58.

Friesen, N., Pollner, S., Bachmann, K. and Blattner, F.R. (1999) RAPDs and non-coding chloroplast DNA reveal a single origin of the cultured *Allium fistulosum* from *A. altaicum* (Alliaceae). *American Journal of Botany* 86, 554–562.

Fujito, S., Akyol, T.Y., Mukae, T., Wako, T., Yamashita, K. *et al.* (2021a) Construction of an ultra-high-density linkage map and graphical representation of the arrangement of transcriptome-based unigene markers on the chromosomes of *Allium cepa* L. *BMC Genomics* 22, 481.

Fujito, S., Urairi, C., Yamashita, K., Wako, T., Kawai, A. *et al.* (2021b) Japanese bunching onion line with high resistance to the stone leek leafminer, *Liriomyza chinensis* from the 'Beicong' population: evaluation the inheritance of resistance. *Euphytica* 217, 31.

Galmarini, C.R., Goldman, I.L. and Havey, M.J. (2001) Genetic analyses of correlated solids, flavor, and health-enhancing traits in onion (*Allium cepa* L.). *Molecular Genetics and Genomics* 265, 543–551.

Galván, G.A., Wietsma, W.A., Putrasemedja, S., Permadi, A.H. and Kik, C. (1997) Screening for resistance to anthracnose (*Colletotrichum gloeosporioides* Penz.) in *Allium cepa* and its wild relatives. *Euphytica* 95, 173–178.

Haishima, M., Kato, J. and Ikehashi, H. (1993) Isozyme polymorphism in native cultivars of Japanese bunching onion (*Allium fistulosum* L.). *Japanese Journal of Breeding* 43, 537–547.

Hang, T.T.M., Shigyo, M., Yamauchi, N. and Tashiro, Y. (2004) Production and characterization of alien chromosome additions in shallot (*Allium cepa* L. Aggregatum group) carrying extra chromosome(s) of Japanese bunching onion (*A. fistulosum* L.). *Genes & Genetic Systems* 79, 263–269.

Havey, M.J. (2018) Impact of the *Allium* genomes on plant breeding. In: Shigyo, M., Khar, A. and Abdelrahman, M. (eds) *The Allium Genomics*. Springer Nature, pp. 197–214.

Hayashi, T., Hikasa, Y. and Sakamoto, N. (2003) Nitrogen application methods based on growth properties on Japanese bunching onion (*Allium fistulosum* L.) under greenhouse conditions. *Japanese Journal of Soil Science and Plant Nutrition* 74, 407–414. (In Japanese with English summary)

Honjo, M., Takeda, S., Katahira, M., Yashiro, M., Shindo, H. *et al.* (2015) Effects of raising condition using cell tray in an unheated plastic house of cold regions on the development and yield of summer-cropping Japanese bunching onion (*Allium fistulosum* L.). *Horticultural Research (Japan)* 14, 25–35. (In Japanese with English summary)

Hsu, C.L., Hoepting, C.A., Fuchs, M., Shelton, A.M. and Nault, B.A. (2010) Temporal dynamics of iris yellow spot virus and its vector, *Thrips tabaci* (Thysanoptera: Thripidae), in seeded and transplanted onion fields. *Environmental Entomology* 39, 266–277.

Huang, X., Wei, X., Sang, T., Zhao, Q., Feng, Q. *et al.* (2010) Genome-wide association studies of 14 agronomic traits in rice landraces. *Nature Genetics* 42, 961–967.

Ikeda, K. and Sakai, H. (2020) Effect of soil temperature on incidence of welsh onion white rot caused by *Sclerotium cepivorum*. *Japanese Journal of Phytopathology* 86, 148–150.

Inden, H. and Asahira, T. (1990) Japanese bunching onion (*Allium fistulosum* L.). In: Brewster, J.L. and Rabinowitch, H.D. (eds) *Onions and Allied Crops*, vol. 3. CRC Press, Boca Raton, Florida, pp. 159–178.

Inden, H. and Yamasaki, A. (2014) Origin and varietal diversity. In: *Physiology, Ecology and Production of Bunching Onion*, 2nd edn. Seibundo Shinkosha Publishing, Tokyo, pp. 10–17. (In Japanese)

Ishiguro, K. (1967) Studies on stabilizing and labor saving procedure of raising Japanese bunching onions, *Allium fistulosum* L., var. Koshizu. *Bulletin of the Aichi Horticultural Experiment Station* 6, 1–66. (In Japanese with English summary)

Japan Horticultural Production and Research Institute (2016) *Sosai no Shinhinshu*, vol. 19. Seibundo Shinkosha Publishing, Tokyo, pp. 174–175. (In Japanese)

Japan Seed Trade Association (2021) *Yasai Hinshu Meikan*. Japan Seed Trade Association, Tokyo, pp. 160–167. (In Japanese)

Jo, J., Purushotham, P.M., Han, K., Lee, H.R., Gyoungju, N.G. *et al.* (2017) Development of a genetic map for onion (*Allium cepa* L.) using reference-free genotyping-by-sequencing and SNP assays. *Frontiers in Plant Science* 8, 1606.

Jones, C.J., Edwards, K.J., Castaglione, S., Winfield, M.O., Sala, F. *et al.* (1997) Reproducibility testing of RAPD, AFLP and SSR markers in plants by a network of European laboratories. *Molecular Breeding* 3, 381–390.

Jones, H.A. and Mann, L.K. (1963) *Onion and Their Allies*. Leonard Hill Publishers, New York.

King, J.J., Bradeen, J.M., Bark, O., McCallum, J.A. and Havey, M.J. (1998) A low density map of onion reveals a role for tandem duplication in the evolution of an extremely large diploid genome. *Theoretical and Applied Genetics* 96, 52–62.

Kudryavtseva, N., Havey, M.J., Black, L., Hanson, P., Sokolov, P. *et al.* (2019) Cytological evaluations of advanced generations of interspecific hybrids between *Allium cepa* and *Allium fistulosum* showing resistance to *Stemphylium vesicarium*. *Genes* 10, 195.

Lee, R., Baldwin, S., Kenel, F., McCallum, J., and Macknight, R. (2013) FLOWERING LOCUS T genes control onion bulb formation and flowering. *Nature Communications* 4, 2884.

Leong, A.C. (2001) Performance of bunching onion (*Allium fistulosum* l.) cultivars in the lowlands of tropical Malaysia. *ISHS Acta Horticulturae* 555, 99–101.

Leslie, J.F. and Summerell, B.A. (2006) *The Fusarium Laboratory Manual*. Blackwell Publishing, Ames, Iowa.

Lin, M.W. and Chang, W.N. (1980) Interspecific hybridization in genus *Allium*. I. Effect of different temperature on bolting of Japanese bunching onion (*Allium fistulosum* L.). *Chinese Horticulture* 26, 173–176. (In Chinese with English summary)

Liu, Q., Wen, C., Zhao, H., Zhang, L., Wang, J. *et al.* (2014) RNA-seq reveals leaf cuticular wax-related genes in welsh onion. *PLOS ONE* 9, e113290.

Liu, S., He, H., Feng, G. and Chen, Q. (2009) Effect of nitrogen and sulfur interaction on growth and pungency of different pseudostem types of Chinese spring onion (*Allium fistulosum* L.) *Scienta Horticulturae* 121, 12–18.

MAFF (2021) Shuyounousanbutsu no Sanshutsugaku to Kousei Wariai (The ranking of major agricultural products for annual output and their compositions in Japan). Available at: https://www.e-stat.go.jp/stat-search/files?page=1&layout=datalist&toukei=00500206&tstat=000001015617&cycle=7&year=20190&month=0&tclass1=000001019794&tclass2=000001152786&tclass3val=0 (accessed 15 July 2022). (In Japanese)

Martin, W.J., McCallum, J., Shigyo, M., Jakše, J., Kuhl, J.C. *et al.* (2005) Genetic mapping of expressed sequences in onion and in silico comparisons with rice show scant colinearity. *Molecular Genetics and Genomics* 274, 197–204.

Martinez, L.E., Glamarini, C.R. and Masuelli, R.W. (2005) Introgression of *Allium fistulosum* L. into interspecific hybrid backcrosses between A. *fistulosum* L. and A. *cepa* L. *Acta Horticulturae* 688, 109–115.

McCallum, J., Baldwin, S., Macknight, R., Khosa, J. and Shaw, M. (2018) Molecular mapping of genes and QTL: progress to date and development for NGS genetics. In: Shigyo, M., Khar, A. and Abdelrahman, M. (eds) *The Allium Genomics*. Springer Nature, pp. 181–196.

Morinaka, T. (1985) Parasitism of rust fungi of welsh onion and Chinese chive to *Allium* species. *Proceedings of the Kanto Tosan Plant Protection Society* 32, 115–116. (In Japanese)

Moue, T. and Uehara, T. (1985) Inheritance of cytoplasmic male sterility in *Allium fistulosum* L. (welsh onion). *Journal of the Japanese Society for Horticultural Science* 53, 432–437. (In Japanese with English summary)

Munaiz, E.D. and Havey, M.J. (2020) Genetic analysis of epicuticular waxes associated with the glossy foliage of 'White Persian' onion. *Journal of the American Society for Horticultural Science* 145, 67–72.

Nanda, S., Chand, S.K., Mandal, P., Tripathy, P. and Joshi, R.K. (2016) Identification of novel source of resistance and differential response of *Allium* genotypes to purple blotch pathogen, *Alternaria porri* (Ellis) Ciferri. *Plant Pathology Journal* 32, 519–527.

Netzer, D., Rabinowitch, H.D. and Weintal, C. (1985) Greenhouse technique to evaluate pink root disease caused by *Pyrenochaeta terrestris*. *Euphytica* 34, 385–391.

Nishimura, Y. and Shibano, M. (1972) Male sterility in Japanese bunching onion. In: *Abstract of the Japanese Society of Horticultural Science Spring Meeting*, p.180 (In Japanese)

Ohara, T., Wako, T. and Kojima, A. (2004) Diallel analysis of seedling growth in Japanese bunching onion (*Allium fistulosum* L.). *Breeding Research* 6, 11–19.

Ohara, T., Song, Y.S., Tsukazaki, H., Wako, T., Nunome, T. *et al.* (2005) Construction of a genetic linkage map of Japanese bunching onion (*Allium fistulosum*) based on AFLP and SSR markers. *Euphytica* 144, 255–263.

Ohara, T., Tsukazaki, H., Song, Y.S., Wako, T., Yamashita, K. *et al.* (2009) Mapping of quantitative trait loci controlling seedling growth in bunching onion (*Allium fistulosum* L.). *Journal of the Japanese Society for Horticultural Science* 78, 436–442.

Pappu, H.R. and Rauf, A. (2013) First report of Iris yellow spot virus infecting green onion in Indonesia. *Plant Disesase* 97, 1665.

Pathak, C.S., Black, L.L., Cheng, S.J., Wang, T.C. and Ko, S.S. (2001) Breeding onions for *Stemphylium* leaf blight resistance. *Acta Horticulturae* 555, 77–81.

Peffley, E.B. (1986) Evidence for chromosomal differentiation of *Allium fistulosum* and *A. cepa*. *Journal of the American Society for Horticultural Science* 111, 126–129.

Plant Protection Station (2021) *The Annual Statistics of the Plant Quarantine of Japan*. Available at: http://www.pps.go.jp/TokeiWWW/Pages/toukeiList/toukeiInfoList.xhtml (accessed 15 July 2022). (In Japanese)

Qu, Y., Takagi, H., Ogasawara, N. and Etoh, T. (1994) Development and types of inflorescences of garlic and some *Allium* species. *Journal of the Japanese Society for Horticultural Science* 63, 121–130.

Ricroch, A., Yockteng, R., Brown, S.C. and Nadot, S. (2005) Evolution of genome size across some cultivated *Allium* species. *Genome* 48, 511–520.

Sekine, D. and Yabe, S. (2020) Simulation-based optimization of genomic selection scheme for accelerating genetic gain while preventing inbreeding depression in onion breeding. *Breeding Science* 70, 594–604.

Shigyo, M., Tashiro, Y., Issiki, S. and Miyazaki, S. (1996) Establishment of a series of alien monosomic addition lines of Japanese bunching onion (*Allium fistulosum* L.) with extra chromosomes from shallot (*A. cepa* L. Aggregatum group). *Genes & Genetic Systems* 71, 363–371.

Shiraiwa, N., Kashima, Y., Inoue, H., Itai, A. and Tanabe, K. (2005) Effect of liquid fertilizer during flower-bud differentiation on nitrogen concentration, bolting, and yield in welsh onion harvested at early summer. *Horticultural Research (Japan)* 4, 411–415. (In Japanese with English summary)

Shiraiwa, N., Tanabe, K., Kashima, Itai, A., Inoue, H. *et al.* (2007) Effect of soil warming by electrically heated wire on bolting and growth in welsh onion (*Allium fistulosum* L.). *Horticultural Research (Japan)* 6, 459–464. (In Japanese with English summary)

Simmons, E.G. (1967) Typification of *Alternaria*, *Stemphylium*, and *Ulocladium*. *Mycologia* 59, 67–92.

Singh, B.K. and Ramakrishna, Y. (2017) Welsh onion (*Allium fistulosum* L.): a promising spicing-culinary herb of Mizoram. *Indian Journal of Hill Farming* 30, 201–208.

Straub, R.W. and Emmett, B. (1992) Pests of monocotyledon crops. In: McKinlay, R.G. (ed.) *Vegetable Crop Pests*. Palgrave Macmillan, London, pp. 213–262.

Sueyoshi, T., Shimomura, K., Koga, K., Yamamura, Y. and Takemoto, H. (2006) The varietal difference in resistance to stone leek leafminer in welsh onions. *Bulletin of the Fukuoka Agricultural Research Center* 25, 37–41. (In Japanese with English summary)

Tabassum, A., Reitz, S., Rogers, P. and Pappu, H.R. (2016) First report of Iris yellow spot virus infecting green onion (*Allium fistulosum*) in the United States. *Plant Disease* 100, 2539.

Takakuwa, M., Ishizaka, N., Kodama, F. and Saito, I. (1977) Host range of *Fusarium oxysporum* f. sp. *cepae*, causal fungus of fusarium basal rot of onion. *Annals of the Phytopathological Society of Japan* 43, 479–481. (In Japanese with English summary)

Takeda, M., Kawai, A., Mitsunaga, T., Tsukazaki, H., Yamashita, K. *et al.* (2020) A novel method for evaluating the egg killing defenses and varietal resistance of the bunching onion against *Liriomyza chinensis* (Diptera: Agromyzidae) via the artificial inoculation of eggs. *Applied Entomology and Zoology* 55, 93–103.

Takeuchi, T. (1990) Forecasting of occurrence of rust of welsh onion. *Bulletin of the Chiba Agricultural Experimental Station* 31, 73–84. (In Japanese)

Theresa, A.S., Veling, S. and Snyman, H.G. (1993) Infection studies of *Stemphylium vesicarium* on onion leaves. *Mycological Research* 97, 984–988.

Tokumaru, S. (2006) Hymenopterous parasitoids of *Liriomyza chinensis* Kato (Diptera: Agromyzidae) in Kyoto Prefecture. *Japanese Journal of Applied Entomology and Zoology* 50, 63–65. (In Japanese with English summary)

Tokumaru, S. and Uesugi, R. (2019) Occurrence of genotype of *Liriomyza chinensis* Kato in Kyoto prefecture. *Shokubutsu-boueki (Plant Protection)* 73, 581–583. (In Japanese)

Tsukazaki, H. (2018) Simple sequence repeats. In: Shigyo, M., Khar, A. and Abdelrahman, M. (eds) *The Allium Genomics*. Springer Nature, pp. 113–127.

Tsukazaki, H., Fukuoka, H., Song, Y.-S., Yamashita, K.-I., Wako, T. *et al.* (2006) Considerable heterogeneity in commercial F_1 varieties of bunching onion (*Allium fistulosum*) and proposal of breeding scheme for conferring variety traceability using SSR markers. *Breeding Science* 56, 321–326.

Tsukazaki, H., Yamashita, K., Yaguchi, S., Masuzaki, S., Fukuoka, H. *et al.* (2008) Construction of SSR-based chromosome map in bunching onion (*Allium fistulosum*). *Theoretical and Applied Genetics* 117, 1213–1223.

Tsukazaki, H., Yaguchi, S., Yamashita, K., Shigyo, M., Kojima, A. *et al.* (2012) QTL analysis for pseudostem pungency in bunching onion (*Allium fistulosum*). *Molecular Breeding* 30, 1689–1698.

Tsukazaki, H., Yaguchi, S., Sato, S., Hirakawa, H., Katayose, Y. *et al.* (2015) Development of transcriptome shotgun assembly-derived markers in bunching onion (*Allium fistulosum*). *Molecular Breeding* 35, 55.

Tsukazaki, H., Yaguchi, S., Yamashita, K. and Wako, T. (2017) QTL analysis of morphological traits and pseudostem pigmentation in bunching onion (*Allium fistulosum*). *Euphytica* 213, 152.

van der Meer, Q.P. (1984) Exchange of genes between *Allium cepa* L. and *Allium fistulosum* L. *Proceedings of the 3rd Eucarpia Allium Symposium*, 4–6 September. IVT, Wageningen, The Netherlands.

van Gestel, N.C., Nesbit, A.D., Gordon, E.P., Green, C., Paré, P.W. *et al.* (2005) Continuous light may induce photosynthetic downregulation in onion – consequences for growth and biomass partitioning. *Physiologia Plantarum* 125, 235–246.

van Heusden, A.W., Shigyo, M., Tashiro, Y., Vrielink-van Ginkel, R. and Kik, C. (2000) AFLP linkage group assignment to the chromosomes of *Allium cepa* L. via monosomic addition lines. *Theoretical and Applied Genetics* 100, 480–486.

Vivero, G.A.G. (2009) Resistance to Fusarium basal rot and response to arbuscular mycorrhizal fungi in Allium. PhD dissertation, Wageningen University, The Netherlands.

Vu, H.Q., El-Sayed, M.A., Ito, S., Yamauchi, N. and Shigyo, M. (2012) Discovery of a new source of resistance to *Fusarium oxysporum*, cause of Fusarium wilt in *Allium fistulosum*, located on chromosome 2 of *Allium cepa* Aggregatum group. *Genome* 55, 797–807.

Wako, T., Kojima, A., Yamashita, K., Tsukazaki, H., Ohara, T. *et al.* (2010) Breeding and characteristics of a bunching onion (*Allium fistulosum* L.) cultivar 'Fuyuwarabe' with short leaf sheath. *Horticultural Research (Japan)* 9, 273–285. (In Japanese with English summary)

Wako, T., Yamashita, K., Tsukazaki, H., Ohara, T., Kojima, A. *et al.* (2012) Development of 'Negi Chuukanbohon Nou 1', a bunching onion (*Allium fistulosum* L.) parental line with rust resistance. *Bulletin of the National Institute of Vegetable and Tea Science* 11, 55–62. (In Japanese with English summary)

Wako, T., Yamashita, K., Tsukazaki, H., Ohara, T., Kojima, A. *et al.* (2015) Screening and incorporation of rust resistance from *Allium cepa* into bunching onion (*Allium fistulosum*) via alien chromosome addition. *Genome* 58, 135–142.

Wako, T., Tsukazaki, H., Yaguchi, S., Yamashita, K., Ito, S. *et al.* (2016) Mapping of quantitative trait loci for bolting time in bunching onion (*Allium fistulosum* L.). *Euphytica* 209, 537–546.

Wako, T., Yamashita, K., Tsukazaki, H. and Fujito, S. (2017) Development of bunching onion varieties 'Koiwarabe' and 'Suzuwarabe' with short leaf sheaths adaptable to spring and summer production. *Bulletin of the NARO Vegetable and Floriculture Science* 1, 5–21. (In Japanese with English summary)

Wang, D., Gao, J. and Liu, G. (2005) General situation of *Allium* crops in China. *Acta Horticulturae* 688, 327–332.

Wolff, X.Y. and Coltman, R.R. (1990) Productivity of eight leafy vegetable crops grown under shade in Hawaii. *Journal of the American Society for Horticultural Science* 115, 182–188.

Yamasaki, A. (2007) Flower bud initiation and bolting. In: *Nougyo Gijutsu Taikei Vegetables 8*. Noubunkyo, Tokyo, pp. 81–95. (In Japanese)

Yamasaki, A. and Tanaka, K. (2002) Effect of soil temperature on bolting of Japanese bunching onion (*Allium fistulosum* L.). *Horticultural Research (Japan)* 1, 209–212. (In Japanese with English summary)

Yamasaki, A. and Tanaka, K. (2005) Effect of nitrogen on bolting of Japanese bunching onion (*Allium fistulosum* L.). *Horticultural Research (Japan)* 4, 51–54. (In Japanese with English summary)

Yamasaki, A. Miura, H. and Tanaka, K. (2000a) Effect of photoperiods before, during and after vernalization on flower initiation and development and its varietal difference in Japanese bunching onion (*Allium fistulosum* L.). *Journal of Horticultural Science and Biotechnology* 75, 645–650.

Yamasaki, A., Tanaka, K., Yoshida, M. and Miura, H. (2000b) Effects of day and night temperature on flower-bud formation and bolting of Japanese bunching onion (*Allium fistulosum* L.). *Journal of the Japanese Society for Horticultural Science* 69, 40–46.

Yamasaki, A., Tanaka, K., Yoshida, M. and Miura, H. (2000c) Induction of devernalization in mid-season flowering cultivars of Japanese bunching onion (*Allium fistulosum* L.) by high day temperature. *Journal of the Japanese Society for Horticultural Science* 69, 611–613.

Yamasaki, A., Tanaka, K. and Inden, H. (2012) Involvement of growth stage, especially cumulative leaf number, in flower-bud initiation of bunching onion (*Allium fistulosum* L.). *Horticultural Research (Japan)* 11, 257–263. (In Japanese with English summary)

Yamasaki, A., Tanaka, K. and Inden, H. (2013) Seed vernalization-like character in bunching onion (*Allium fistulosum* L.). *Acta Horticulturae* 969, 133–134.

Yamashita, K. and Tashiro, Y. (2004) Seed productivity test of CMS lines of Japanese bunching onion (*Allium fistulosum* L.) possessing the cytoplasm of a wild species, *A. galanthum* Kar. et Kir. *Euphytica* 136, 327–331.

Yamashita, K., Arita, H. and Tashiro, Y. (1999) Cytoplasm of a wild species, *Allium galanthum* Kar. *et* Kir., is useful for developing the male sterile line of *A. fistulosum* L. *Journal of the Japanese Society for Horticultural Science* 68, 788–797.

Yamashita, K., Wako, T., Ohara, T., Tsukazaki, H. and Kojima, A. (2005) Improvement of rust resistance in bunching onion (*Allium fistulosum* L.) by recurrent selection. *Journal of the Japanese Society for Horticultural Science* 74, 444–450.

Yamashita, K., Tsukazaki, H., Kojima, A., Ohara, T. and Wako, T. (2010) Inheritance mode of male sterility in bunching onion (*Allium fistulosum* L.) accessions. *Euphytica* 173, 357–367.
Yang, L., Liu, Q., Wang, Y. and Liu, L. (2017) Identification and characterization of a glossy mutant in welsh onion (*Allium fistulosum* L.). *Scientia Horticulturae* 225, 122–127.
Zhang, J., Wu, J., Kong, F., Shen, C., Zhou, X. *et al.* (2021) Analysis of Chinese vegetable market in 2020 and outlook for 2021. *China Vegetables* 40 (1), 4-10. (in Chinese)

8 Leek and Other Main *Allium ampeloprasum* Crops

Haim D. Rabinowitch*
The RH Smith Institute of Plant Science and Genetics in Agriculture, The Hebrew University of Jerusalem, Rehovot, Israel

1. Introduction

A member of section *Allium* (Amaryllidaceae), *Allium ampeloprasum* L. is a variable morphological and cytological polyploid complex of wild and domesticated species (Keller *et al.*, 2012) composed of four gene pools: wild leek, European leek cultivars, Egyptian kurrat and great-headed garlic (Kik *et al.*, 1997). The complex's ancestry and its ploidy nature (2n=24, 32, 40, 48 or 56), however, are in question, and even controversial (Beminger and Buret, 1967; Koul and Gohil, 1970; Von Bothmer, 1970, 1974; Schweisguth, 1970; Hanelt, 1990; Guern *et al.*, 1991; Khazanehdari *et al.*, 1995; Stack and Roelofs, 1996; Kik *et al.*, 1997; De Clercq and Van Bockstaele, 2002; Fritsch and Friesen, 2002; Bohanec *et al.*, 2005; Jauzein and Tison, 2005; Hirscherger *et al.*, 2010; Khedim *et al.*, 2020; Chapter 1, this volume).

Along with several crops and horticultural varieties (Table 8.1), the biennial tall, hardy, robust herbaceous *A. pyrenaicum* and *A. iranicum* (x=8, 2n=32; Van der Meer and Hanelt, 1990) share origin with the tetraploid *A. truncatum* and Tunisian diploids from the north African genomic pool (Guenaoui *et al.*, 2013; Khedim *et al.*, 2020). Hence, within-group interspecific crosses are successful (Kik *et al.*, 1997).

Other cultivated group members include poireau perpetuél; prei anak (Van der Meer and Hanelt, 1990); Mushuu-ninniku (Ariga *et al.*, 2002) and a Chinese form of the bulbous leek (Bohanec *et al.*, 2005). Vernacular names are listed by Lim (2015); Shelke *et al.* (2020); Prota4U (2021); and PROSEA, (n.d.).

2. Taxonomy, Origin and Distribution

The genus *Allium* comprises ~1000 species (Govaerts *et al.*, 2021) including 25 *Ampeloprasum* spp. (4x, 6x and 8x) group/complex (Von Bothmer, 1970; Guern *et al.*, 1991; Mathew, 1996), formed by three or four gene pools (Chapter 1, this volume). Taxonomically, *A. sativum* is closer to *A. ampeloprasum* than *A. fistulosum* and *A. schoenoprasum* (Guenaoui *et al.*, 2013).

The centre of origin, *sensu lato*, stretches from Iran to Portugal and northern Africa to southern England (Von Bothmer, 1970; Hanelt, 1990; Mathew, 1996), where taxon members have been cultivated by ancient Egyptians, Greeks and Romans. Some of their selected offspring are still farmed worldwide.

The domesticated *A. ampeloprasum*: leek, great-headed garlic, pearl onion, prei-anak, Tarée

*Address for correspondence: haim.rabinowitch@mail.huji.ac.il

 Edible Alliums: Botany, Production and Uses (eds H.D. Rabinowitch and B. Thomas)
DOI: 10.1079/9781789249996.0008

Table 8.1. Major *A. ampeloprasum* crops and main cultivation regions (Fritsch and Friesen, 2002; Dashti *et al.*, 2003, 2005; Mousavi *et al.*, 2006; Guenaoui *et al.*, 2013).

Crop group	Botanical name	Main cultivation region	English name	Ploidy level
Leek	*A. ampeloprasum* var. *porrum* (L.) J. Gay, *A. porrum* L.	Indonesia, Turkey, Korea, Europe, Kazakhstan	Leek	4x[1]
Kurrat	*A. kurrat* Schweinf. ex Krause	Egypt and adjacent areas	Kurrat, salad leek	4x
Pearl onion	*A. ampeloprasum* var. *sectivum* Lued.	Atlantic and temperate Europe	Pearl onion	4x
Great-headed garlic	*A. ampeloprasum* var. *holmense* (Mill.) Aschers. et Graebn.	California, eastern Mediterranean	Great-headed garlic, elephant garlic	6x
Tarée	*A. ampeloprasum* ssp. *persicum*	Iran	Tarée Irani/ Persian leek	4x

[1]Opinions differ on whether leek is segmental allo- (Khazanehdari and Jones, 1997) or auto-tetraploid (Maragheh *et al.*, 2018).

Irani/Persian leek, and mushuu-ninniku (Kik *et al.*, 1997; Hanelt, 2001; Ariga *et al.*, 2002; Hirschegger *et al.*, 2006; Mousavi *et al.*, 2006), were subgrouped as follows:

1. The tetraploids leek (*A. ampeloprasum*); kurrat (*A. kurrat*); Tarée Irani; pearl onion (*A. ampeloprasum* pearl onion group) aka bulbous leek. No evidence for natural hybridization in leek and kurrat was published, while introgression events were identified in pearl onion (Hirschegger *et al.*, 2010).
2. The allopolyploid 6x–8x great-headed garlic (GHG, *A. ampeloprasum* var. *holmense*) (Van der Meer and Hanelt, 1993), aka elephant garlic, giant garlic, English leek, European leek, great round-headed leek, Levant garlic and wild leek (Khazanehdari and Jones, 1997; Fritsch and Friesen, 2002; Brewster, 2008).

Hirschegger *et al.* (2010) argued that GHG is a separate taxon from other *A. ampeloprasum* cultivated forms, whose allopolyploid origin includes *A. iranicum* and *A. polyanthum*.

Based on morphology, Oregon State University (Anon, 2010) experts subclassified leeks as European leek with short, thick pseudostem; Turkish leek with long, thin pseudostem, and Kurrat, pseudostem-less.

Leek, a millennia-old crop (Täckholm and Drar, 1954; Bottéro, 1985; Davies, 1992), was domesticated in the eastern Mediterranean, where many related species and types grow wild (for a historical review see Silvertand, 1996). Linnaeus named leek *Allium porrum* L., while Hanelt (1996) placed it in the *Allium ampeloprasum* complex. Some botanical and ploidy (4x=32) differences from the *A. ampeloprasum* lectotype, however, strongly support taxonomic reconsideration.

3. Economic Importance

Global leek markets and by-country analyses of the group's economically most important crop are available (Tridge Leek Production, 2020; Tridge Leek Export, 2020; Statistica, 2020). Informative reports are available for some countries (e.g. Turkey: Hanci *et al.*, 2018), as well as for continents, e.g. Asia (https://www.globaltrademag.com/leek-market-in-asia-supply-and-demand; accessed September 2022).

Globally, leek hectarage and production increased from 1996 to 2012 (Hanci *et al.*, 2018) and stabilized thereafter at 130,000–140,000 ha and 2.1–2.25 million tons, respectively (Fig. 8.1).

Data from 10 and 40 top leek-producing countries (Figs 8.2, 8.3) reveal marked differences in leek production acreage.

Countries also differ in productivity and efficiency (Fig. 8.4).

Good yield/ha and total production correlations were reported for Indonesia and the Republic of Korea, a negative correlation is evident for Turkey, and no correlation is found for the world's production (Hanci *et al.*, 2018).

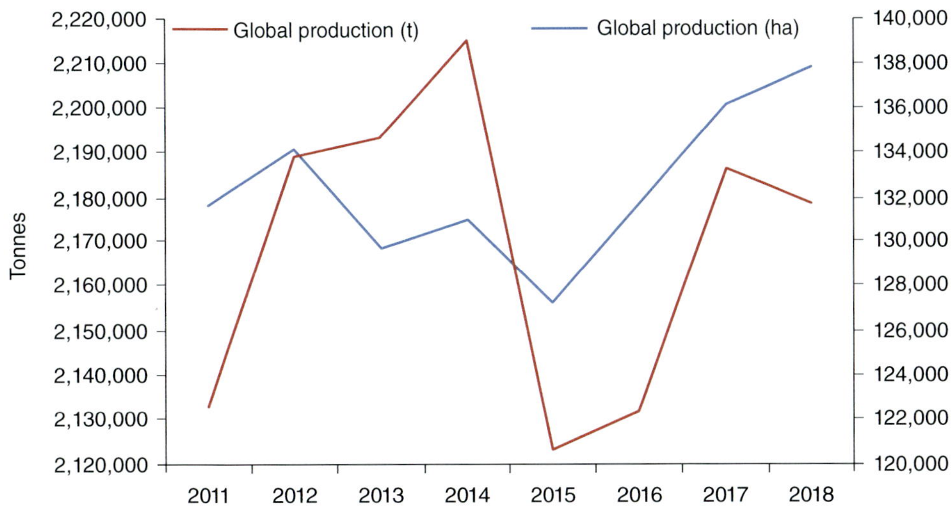

Fig. 8.1. Global leek and related alliaceous vegetables farmed area (ha) and production (t), 2011–2018. (From FAOStat, 2020)

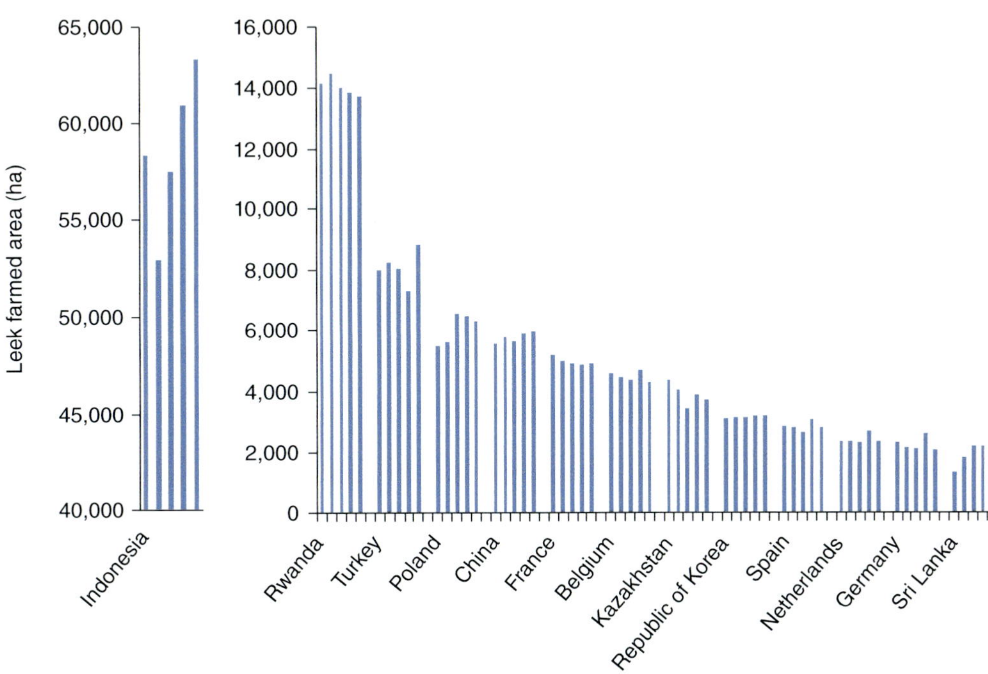

Fig. 8.2. Leek (and related crops) farmed area (ha) in world-leading countries. Country data presented from left to right for 2014–2018, respectively (Processed from FAOStat, 2020)

The Netherlands, China and Belgium lead the international trade with more than 60% of the exported produce (Fig. 8.5).

Economic information on other *A. ampeloprasum* vegetables is scarce, poor or non-existent.

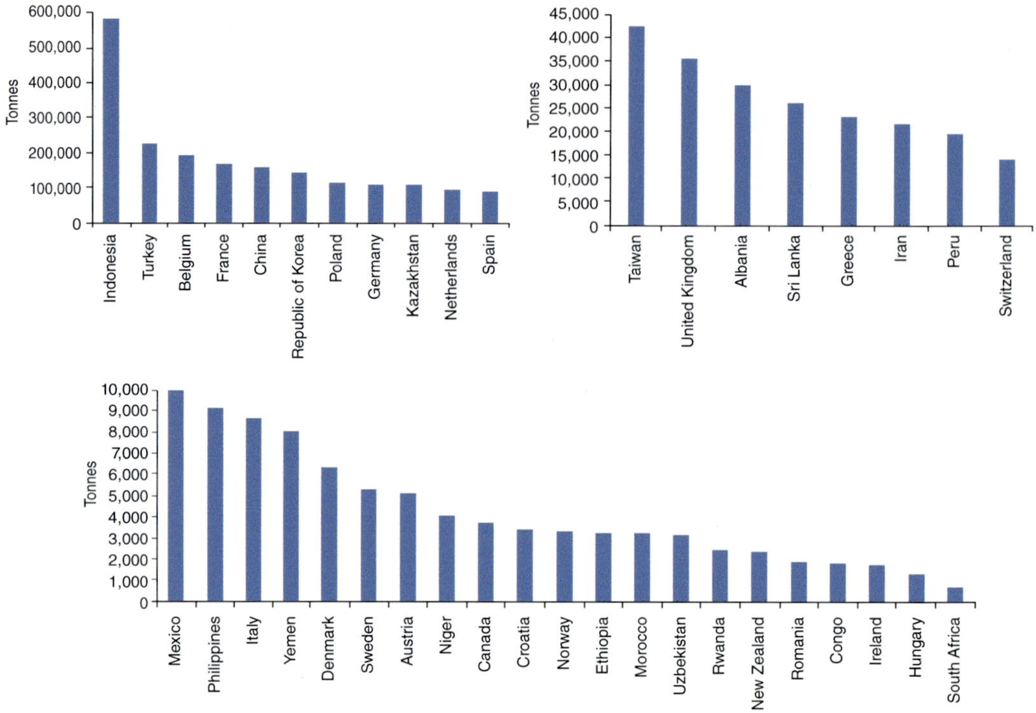

Fig. 8.3. Top 40 leek-producing countries (t) (http://www.fao.org)

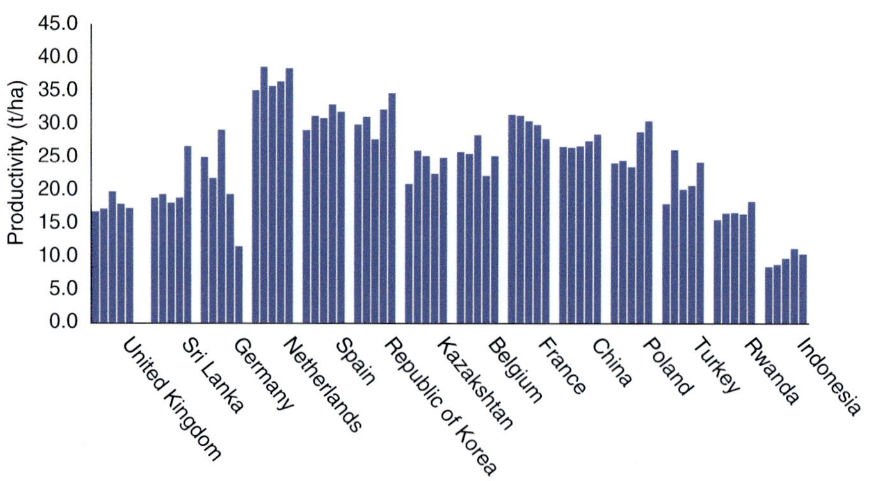

Fig. 8.4. Leek and related Alliums productivity (tonnes/ha); country data presented from left to right for 2014–2018, respectively. (From FAOStat, 2020)

4. Leek and Minor *A. ampeloprasum* Crops – General Description

The leek's leaves are larger and milder than spring onions (scallions), and, like garlic, blades are dorsiventrally flattened and much longer. The false stem (shaft) resembles a scallion (Fig. 8.6).

Under long days (LD) and 12–18°C, most reserves are stored in the dorsal parts of both

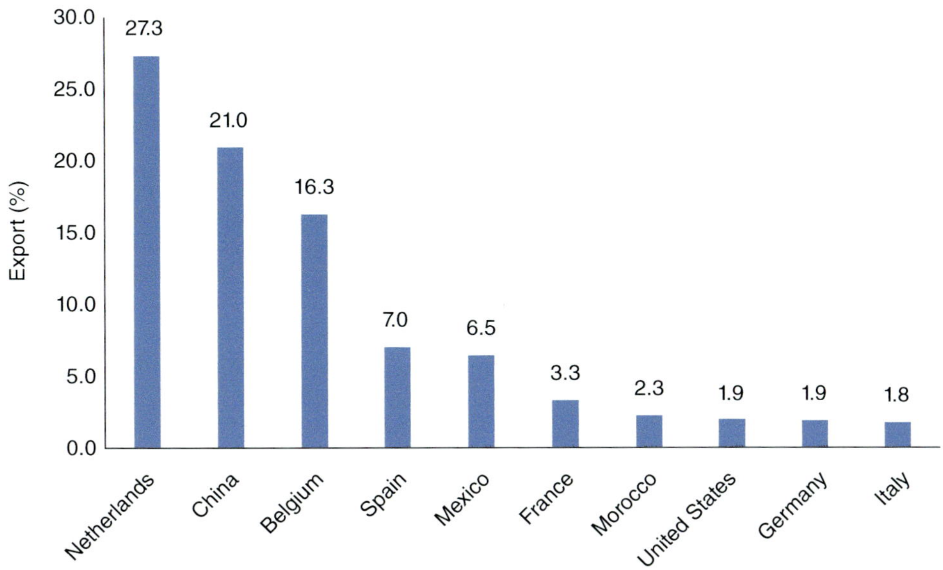

Fig. 8.5. Top 10 leek-exporting countries, 2020. (From Tridge Leek Export, 2020)

Fig. 8.6. Leek vegetative development consists of distichously alternate leaves with tubular sheaths. Together with young blades, they form a white 15– 60 cm long pseudostem. (Original drawing: Asaf Silner)

sheaths and blades, thus forming the pseudostem (shaft), the leek's main storage and edible organ. Some reserves are stored in the bulged leaf-sheath bases (Rubatzky and Yamaguchi, 1997), but neither bulbs nor cloves form in the first growing season.

The longer the white shaft (Fig. 8.7), the higher the leek's economic yield; hence, further

Fig. 8.7. Intact leek with white long pseudostem (*left*) and scallions (*right*) (Buenos Aires market).

to genotypic expression, blanching is forced by light exclusion, e.g. trenching/hilling/enclosing by opaque materials (Gagnebin and Bonnet, 1979; https://www.hobbyfarms.com/4-household-materials-for-blanching-leeks; accessed September 2022).

4.1. Minor crops

The *A. ampeloprasum* centre of origin stretches from Iran to Portugal and northern Africa, where taxon members underwent separate selection pressures with the consequent breeding/development of a variety of crop types farmed for bulbs (great-headed garlic, pearl onion), pseudostem (leek) and leaves (kurrat, Tarée Irani).

Kurrat. The east Mediterranean and southwest Asia tetraploid (2n=4x=32) *Allium ampeloprasum* var. *Kurrat* (Mohamed-Yasseen and Costanza, 1996) or *A. kurrat* Schweinf. ex K. Krause (Jones and Mann, 1963; Guenaoui *et al.*, 2013) is farmed, especially in Egypt (Van der Meer and Hanelt, 1990; Brewster, 1995), for its perpetual supply of green leaves for fresh salads. However, the entire plant, its flowers, roots and bulged-leaf bases are also used fresh or cooked in local dishes. The seed- (or clonal- (Mohamed-Yasseen and Costanza, 1996)) propagated plants are grown as a perennial and reach up to 1 m tall. The hermaphrodite flowers are pollinated by bees and other insects.

Both leaves and seeds are rich in phenolic compounds, thus adding nutraceutical value to the consumer's diet (Abd El-Rehem *et al.*, 2013). Kurrat resembles a bulbous small leek (Van der Meer and Hanelt, 1990). Kadry and Kamel (1955) and Van der Meer (1984a, b) reported that it crosses freely with leek.

Leek yellow stripe virus (LYSV) causes considerable quality losses to leek and garlic yields worldwide (Currah, 1986; Vučurović *et al.*, 2017), thus the resistant kurrat is important for tolerance introgression (Fritsch and Friesen, 2002).

Pearl onion. The small-intermediate, very winter-hardy shaftless leek-like tetraploid (2n=4x=32) plants are raised mainly in Italian, eastern German and Dutch home gardens (Fritch and Friesen, 2002; Temperate Plants Database, 2021). The white flowers set fertile seeds and cross freely with leek. Direct seeding results in great field variability; hence, transplanting of underground daughter bulblets is common (Van der Meer and Hanelt, 1990). Silvery-white skinned bulblets are formed around the main bulb and, like garlic cloves, bulblets consist of a single storage leaf (Shanmugasundaram and Kalb, n.d.). On maturation, when their size approximates to a glass marble (aka steelies) (Fig. 8.8), they undergo a short summer dormancy.

The bulbs are generally used as a condiment (Van der Meer, 1997; Hanelt, 2001) and their mild onion flavour is sweeter than the common onion. Traditionally, bulblets are home-pickled, while some commercial manufacturers pickle densely grown white onion sets (Fritch and Friesen, 2002).

Tarée Irani (*Allium* sp.) is native to the Middle East and Iran where ~5000 tons of green leaves are produced annually, mainly in Iran (Van der Meer and Hanelt, 1990). Cytological (Tahbaz, 1976) and morphology studies determined allopolyploidy (2n=4x=32) differences

Fig. 8.8. The silvery white skins and single scales mature Pearl onion bulblet, the tetraploid A. *ampeloprasum* var. *sectivum*, Pearl-Onion Group.

from a leek in chromosome length and type, and vegetative and floral traits. Roots, stem, leaf and scape anatomies resemble kurrat (Van der Meer and Hanelt, 1990) but the herbaceous perennial Persian leek sheaths form a cylindrical pale-green structure, 0.5–2.0 cm in diameter, which does not store reserves, nor does it form bulbs or cloves (Tahbaz, 1971, 1974). While the leek's 4–12 cm in diameter subglobose umbel bears several hundred flowers, normally without bulbils (*Allium ampeloprasum*, PROSEA, 2016) umbels of Tarée Irani are completely bulbiferous (Van der Meer and Hanelt, 1990) or consist of a mixture of flowers and bulbils (*Allium ampeloprasum*, PROTA, 2018). Hence, Iranian growers raise this vegetable from seeds. The two crops differ in stamen length, thus the Persian leek and its closest wild relative, *A. iranicum* (Mousavi *et al.*, 2006), form different taxonomic groups from the common leek, a distinct subordinate of *A. ampeloprasum*.

Harvest begins when the linear folded green leaves of the relatively short plants (Tahbaz, 1971; Shakouri *et al.*, 2018) are 15–20 cm long and 1–2 cm wide. Thereafter, harvests perpetuate every 3–4 weeks, up to 15 times a year (Mousavi *et al.*, 2006). The mild onion-flavoured leaves are eaten raw, either alone or in food combinations.

In the second growing season, winter cold induces the development of 1–3 round, robust, 50–150 cm long scapes through the pseudostem. A range of bulbiferous inflorescences consists of a mixture of topsets and sterile flowers (Mousavi *et al.*, 2006) and kurrat-like mixed inflorescences of flowers and topsets (*Allium ampeloprasum*, PROTA, 2018). In flowering plants, fertilization of the cross-pollinated 600–1200 flowers results in the setting of black, wrinkled polyhedral 2.3–3 g thousand-weight seeds.

As in onions and leek (Bierhuizen and Wagenvoort, 1974; Brewster, 1979), Tarée Irani seed germination strongly depends on temperature with optimum and maximum at ~20° and ~40°C, respectively (Ramin, 1997). In the Netherlands, these seedlings produce only bulbiferous umbels, indicative of the environmental (LD?) regulation (*Allium ampeloprasum*, PROSEA, 2016).

Great-headed garlic (GHG). The domesticated hexa- and octoalloploid (Hirschegger *et al.*, 2010), high-yielding robust plants (*A. ampeloprasum* L. var. *ampeloprasum* auct.) form large bulbs (up to 1 kg) (https://www.ecpgr.cgiar.org/in-situ-landraces-best-practice-evidence-based-database/crop?cropUid=13453; accessed September 2022), probably due to high heterozygosity rather than polyploidy (Guenaoui *et al.*, 2013). Each bulb (Fig. 8.9) develops 3–6 mild garlic-like, odorous, juicy cloves coated by papery skins, with peripheral small lateral bulblets.

Fig. 8.9. Great-headed garlic (GHG): (a) harvested bulbs; (b) bulb size; (c) *left:* garlic (*A. sativum*); *right:* great-headed garlic (*A. ampeloprasum* var. *holmense*); (d) GHG cloves; (e) withering of harvested GHG; (f) GHG flowering buds; (g) GHG blooming; (h) GHG flowers sold as ornamentals with very long vase life.

Mass spectrometry direct analysis in real time (DART) revealed that GHG synthesizes both allicin, as in garlic and bulb onion, and syn-propanethial-S-oxide (onion lachrymatory factor), produced both by onion and leek (Block *et al.*, 2010). Hence, GHG cloves are consumed as garlic and the leaves serve as a condiment.

GHG produces broad flat leaves and its tall, solid stalk bears a sterile flowerhead that is favoured as an ornamental (Fig. 8.9). The very rarely produced seeds are all sterile. GHG differs markedly from other *A. ampeloprasum* crops in Mediterranean ancestry (Von Bothmer 1970, 1974, 1982; Guern *et al.*, 1991; Jauzein and Tison, 2005), and its chromosome number (48–64) supports the biphyletic evolution theory of the *A. ampeloprasum* crop groups.

5. Botany

The epigeal germination of the bent cotyledon with attached testa begins at the apex. The first true leaf emerges through a pore in the cotyledonary base, followed by the sprouting of distichously alternate leaves with tubular sheaths. Together with young blades, they form a white 15–60 cm long pseudostem (Figs 8.10, 8.11), whose diameter keeps growing with plants' development.

The keeled, V-shaped in cross-section, green linear to linear-lanceolate blades' dimensions are

Fig. 8.10. Leek plant vegetative and reproductive main parts. (Original drawing: Asaf Silner)

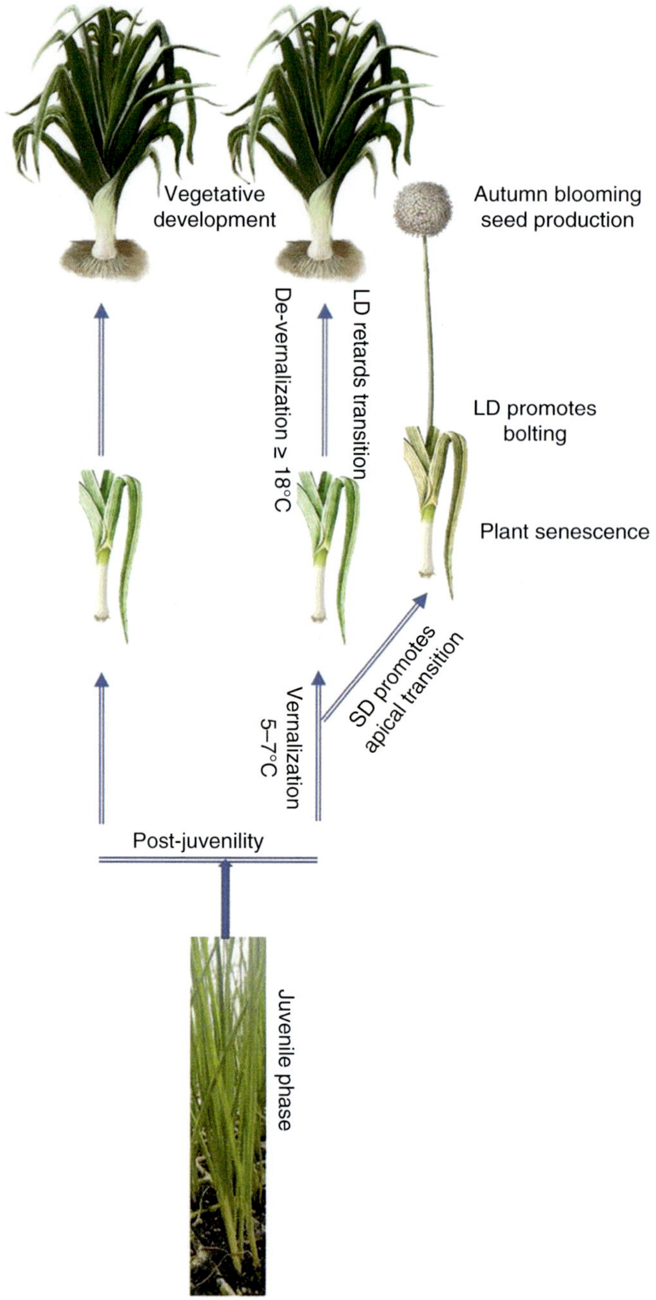

Fig. 8.11. Leek physiology: effect of physiological age, temperature and photoperiod on phase change and the regulation of developmental timing.

up to 50 x 7 cm. Harvest begins when ~15 leaves are visible, or as per consumers' quality demands of pseudostem size (Messiaen and Rouamba, 2004).

Leek has been bred for cloveless cylindrical pseudostems and an OECD common interpretation of leek descriptors guidelines for quality standards is available (OECD, 2019).

Young leeks do not form bulbs, but ethephon-induced swelling of sheath bases was reported (Levy and Kedar, 1970). At the end of the second growing season, two vestigial cloves occasionally form at the scape base, after seed maturation.

The transitioned apical meristem develops into a single internode, solid, terete, up to 1.5 m long stalks, each crowned with an umbel (Figs 8.10, 8.11).

The subtended single inflorescence bears a long-beaked spathe. On anthesis, the prolific globose compact umbel measures 4–12 cm across and develops hundreds of monoecious, campanulate protandrous flowers (Clement *et al.*, 2007), each borne on a 1–5 cm long pedicel. Flowers consist of five whorls: two outer whorls each having three free, 4–6 mm, pink, purple or white tepals; two inner whorls of three stamens each. The central ovoid-globose, superior three-carpels ovary is positioned near the septa's centre (Fig. 8.10). Within an individual flower, the capitate stigma becomes receptive only on completion of anthers' dehiscence, thus no self-pollination occurs, but there is no barrier for pollination between successive fertile flowers or umbels.

Following fertilization, capsules, 2–4 mm across, ranging in shape from depressed globose to ovoid, develop. Each bears up to six small, black, irregular, angular, 2.5–3.5 1000-weight seeds (Paunero *et al.*, 2003, 2004; Anon, 2010). Unlike garlic (Chapter 6, this volume), leek seldomly produces topsets. When produced, they resemble garlic cloves in appearance and texture, but taste like onion/leek and can be used for propagation.

6. Agronomy and Agroecology

Leeks are farmed worldwide, most commonly from seedlings, yet transplanting basal plates' bulbils, bulbs formed after flowering, or topsets stimulated by early flower removal, are also practised (Brewster, 1994). Despite the invigorating osmopriming effect (Parera and Cantliffe, 1992; Coarbineau *et al.*, 1994), nursery-produced leek seedlings, rather than direct sowing, are the preferred choice as field emergence is slow and scattered. Dutch nurseries germinate seeds in June, in an 18–20°C glasshouse, and transfer the deep seedlings trays 1–2 weeks later, to 15°C, for further development and improved performance. Hardened 15–30 cm seedlings are transplanted in loose, deep, organic-matter-rich soil, pH>5.6, in trenches/holes about the height of the seedlings' pseudostems, in the production fields.

For good development and quality, temperate climates (e.g. high altitudes in the tropics/subtropics) are required. In the spring, autumn-planted seedlings are exposed to optimum 20/30°C night/day temperatures, respectively, thus yielding quality produce. Under a warmer environment, leaf senescence can be as rapid as new growth, resulting in no shaft-widening, thus quality production in hot-humid lowlands is rarely successful.

For extra-long white pseudostems, natural blanching and elongation boosting are promoted by light exclusion (hilling/mounding soil throughout production, or wrapping with opaque materials up to the leaf bases). The pseudostem of summer cultivars may reach up to 80 cm while winter cultivars' shafts are half or one-third of this length (Celebi-Toprak and Alan, 2021). Harvest time, however, depends on markets' preferences for size.

Refrigerated packaged leeks maintain their total phenolic content and antioxidant capacity for 13 days (Bernaert *et al.*, 2013). A 2–3-month storage at 0–1°C and 95–100% RH retards shaft elongation and curving, common in warmer conditions, while preservation for 4–5 months at 0°C requires a controlled atmosphere (CA) of 1–3% O_2 and either 2–5% or 5–10% CO_2. CA retards yellowing and decay development, but some quality losses occur (Baugerod, 1980; Gross *et al.*, 2016).

7. Physiology

Leek seeds do not enter dormancy (Van der Meer and Hanelt, 1990) but emergence is low and slow.

Like other *A. ampeloprasum*, they are susceptible to heat and salt stress (Guenaoui *et al.*, 2013), hence the non-uniform emergence, poor stands, great variability (Brocklehurst *et al.*, 1984) and poor transplant production reported under Florida summer conditions.

Apart from moisture, leek germination depends mainly on temperature, with an optimum of 18–22°C (Van der Meer and Hanelt, 1990), and osmotic priming treatments improve emergence and field uniformity (Brocklehurst *et al.*, 1984; Bray *et al.*, 1989; Parera and Cantliffe, 1992; Dearman *et al.*, 2008).

Optimum temperatures for vegetative growth range between 20°C and 25°C (Rubatzky and Yamaguchi, 1997; De Clercq and Van Bockstaele, 2002), while Brewster (1979) reported the highest seedling relative growth rate at 27°C. Populations' growth-rate variability, however, was higher at 27°C than at 23°C.

Foliage withering before blooming (Von Bothmer, 1974) leads to a severe quality drop (Kurki, 1979). Hence, the right choices of cultivars, quality seedlings, timely planting, proper agro-management and harvest are of high priority.

Dragland (1972) and Daniel (1975) proposed that leek is a quantitative LD plant and cold induction is not required for flowering. Others disagreed and reported that flower induction and further inflorescence development depend on 0–18°C post-juvenile vernalization with an optimum at 5°C (Jones and Mann, 1963; Hösslin and Andresen, 1972; Gray and Steckel, 1990, 1993; Brewster, 1994; Wiebe, 1994; De Clercq and Bockstaele, 2002; Fig. 8.11) or near 7°C (Wurr *et al.*, 1999), and effective vernalization length varies with genotype, field environment and physiological age. Leek has an obvious juvenile phase (Van der Meer and Hanelt, 1990), hence nursery temperatures did not affect flowering (Wurr *et al.*, 1999).

The critical size for vernalization is about 2 g fresh weight or five visible leaves, a total of ~10 leaves including initials (Wiebe, 1994). The bolting percentages increase with age/size when vernalization begins (Wurr *et al.*, 1999). During vernalization, LD delayed floral induction compared with SD, and temperatures >18°C caused devernalization (Wiebe, 1994). Autumn and early spring short days (SD) enhance the apical meristem transition from vegetative to reproductive, and the spring–summer LD (Brewster, 1994) together with thermal time (Wurr *et al.*, 1999) stimulate scape elongation. Unlike other plant species, no abortion of already developed flowers occurred (Wiebe, 1989).

Field agro-management, unlike temperature, has little effect on leek seed development and germinability (Gray *et al.*, 1991). Seed development time increased from 55 to 80 days as mean temperatures fell from 30/20°C to 20/10°C, day/night, respectively, and seed/umbel, seed weight and yield/umbel were inversely related to temperature. Mean seed weight increased by 35 µg for every 1°C drop in mean temperature (Gray *et al.*, 1992). Hence, seasonal temperature fluctuations and harvest time contribute to seed weight variation within a single lot (Gray *et al.*, 1992), and thus to plant vigour and field performance. Additionally, like bulb onion (Gray and Steckel, 1984), high temperatures during leek seed development raised the upper temperature limit for germination (Gray *et al.*, 1992), providing additional explanations for the great germination and stand variability in direct-seeded leek fields (Gray *et al.*, 1991).

Seedhead and/or capsule morphologies are widely used to determine harvest dates of *Allium* (Hawthorn and Pollard, 1954), e.g. capsules splitting when moisture content is <40% and change in seed colour. Most leek seeds turned black and were capable of slow germination when seed moisture dropped to 50–60%, ~60 days post-anthesis at 18°C (Gray and Ward, 1987), approximately when the accumulation of reserves in the endosperm begins (Gray and Ward, 1987). Maximum germinability was reached only 30 days later.

Drying leek seeds of all sizes, particularly during the earlier developmental stages, promoted germination. Rapid drying stimulated germination more than slower drying, suggesting that drying acts as a cue for the transition to the germinative mode in leek. For practical purposes, maximum viability and shortest mean germination time occur when seed moisture contents reach ~50% (Gray *et al.*, 1989, 1992) yet harvest delay to the point of 20–30% moisture content has practical advantages (Gray *et al.*, 1992).

Unlike bulb onion, where considerable seed shedding occurs at these moisture contents (Sterner and Akintobi, 1986), leek capsules often remain green and do not fully open, thus only minute seed shedding occurs until very late in the season. Therefore, the highest quality was attained at ambient conditions after 900–950 day-degrees >6°C when seed moisture content reached <30% (Gray *et al.*, 1991, 1992).

8. Edible Organs, Flavour, Nutritional Values and Uses

Nutritional/medicinal properties of the *A. ampeloprasum* complex were reviewed by Swamy and Veere Gowda (2006); Dey and Khaled (2013); Lim (2015); Sakthi and Monika (2018) and Shahrajabian *et al.* (2021).

Table 8.2 lists the main nourishing components (other than water) in leek (Bernaert, 2013; Koca and Tasci, 2016; Chapter 3, this volume).

The leek-flavoured tissues (Yoo and Pike, 1998) are rich in amino acids (glutamine, glutamic acid, aspartic acid, arginine and alanine), antioxidants, vitamins A, B and C (Fattorusso *et al.*, 2001; Ozgur *et al.*, 2011; García-Herrera *et al.*, 2014; Ben Arfa *et al.*, 2015; Koca and Tasci, 2015; Najda *et al.*, 2016), and an array of >20 fatty acids (Tsiaganis *et al.*, 2006) of which linoleic, palmitic, oleic and α-linolenic acids make up 80% of the total lipid contents, sulphur volatiles and other secondary metabolites with potential health benefits. The leek antioxidative properties are comparable with those attributed to kurrat leaves and onion bulbs, while leek blades resemble chives and welsh onion. Leek and pearl onion are rich in methiin and propiin, respectively (Fritch and Keusgen, 2006; Horie and Yamashita, 2006), two important flavour components and precursors of flavours and odours of *Allium* vegetables.

The leek pseudostem, together with the basal parts and the proximal leaf blade sections, is consumed fresh, cooked, fried, pickled or as flavouring condiment. Coarse blades can be sautéed/added to stock or (usually) discarded.

For related crops: GHG can substitute for garlic (Najda *et al.*, 2016); kurrat leaves add flavour to dishes in salads or cooked meals (Musselman, 2002) and pearl onions are used for pickling (Van der Meer and Hanelt, 1990; Pink, 1993).

Overall, *A. ampeloprasum* plants are nutraceutical-rich (Sadeghi *et al.*, 2013a, b; Golubkina *et al.*, 2018; Shelke *et al.*, 2020; Chapter 3, this volume) including health-beneficial compounds, especially sulphur-containing bioactive constituents (GHG: Block *et al.*, 1992; leek: Uchida *et al.*, 2009; Kratchanova *et al.*, 2010). Hence, for over 5000 years, *A. ampeloprasum* plants have been used in folk medicine (Avicenna (Ibn-e Sina), 1998), including anti-inflammatory effects (Adão *et al.*, 2011), protection against some common bacteria, and anti-cancer activity (Zhou *et al.*, 2011; Huang and Ren, 2013).

Table 8.2. Main nutritional components per 100 g leek shaft fresh weight. (Adapted from Swamy and Veere Gowda, 2006; Grzelak-Blaszczyk *et al.*, 2011; Anon, 2020; https://fdc.nal.usda.gov/fdc-app.html#/food-details/1103357/nutrients (accessed September 2022))

Nutrients and energy		Content	Vitamins	Weight unit	Content	Minerals	Weight unit	Content
Energy	kcal	61.00	Vitamin A	mcg	83.00	Ca	mg	48–75
Protein	g	1.5–2.20	β Carotene	mg	1.00	Cu	mg	0.06–0.30
Carbohydrate	g	5–14.15	Lutein + zeaxanthin	mg	1.90	Fe	mg	2.10
Sugars, total	g	3.90	Thiamin	mg	0.06	K	mg	180–347
Dietary fibre, total	g	1–3.20	Riboflavin	mg	0.03	Mg	mg	1–28
Total fat	g	0.1–0.40	Niacin	mg	0.40	Na	mg	5–20
Saturated fatty acids	g	0.04	Vitamin B-6	mg	0.23	P	mg	35.00
Monounsaturated FA	g	0.00	Folate, food	mcg	64.00	Se	mcg	1.00
Polyunsaturated FA	g	0.17	Folate, DFE	mcg	64.00	Zn	mg	0.12
			Choline, total	mg	9.50			
			Vitamin C	mg	12.00			
			Vitamin E, α tocopherol	mg	0.92			
			Vitamin K	mcg	47.00			

FA = fatty acids; DFE = dietary folate equivalent

9. Genetics and Breeding

In tetraploid leek, diploidization (Stack, 1993) and chiasmata terminal localization results in only a few recombinations between the two chromosomal sets (Berninger and Buret, 1967; Schweisguth, 1970; Gohil, 1984), hence, chromosome conservation, genetic stability, limited variability, flawless meiosis and successful sexual reproduction (Kollman, 1972; Gohil and Koul, 1977; Gohil, 1984; Pink, 1993; Jones *et al.*, 1996; Stack and Roelofs, 1996).

Self-pollination at ~20–30% is common in the self-compatible leek (Berninger and Buret, 1967; Prota4U, 2021). Yet open-pollinated cultivars (OPs) show up to 88 proportions of weak to very weak performance (Smith and Crowther, 1995a) due to deleterious gene expression and inbreeding depression (Schweisguth, 1970; Gagnebin and Bonnet, 1979; Currah, 1986; Smith and Crowther, 1995a). Low mean seed weight from self-pollinated progeny results in poor field emergence and survival compared with seedlings from cross-pollinated progeny. Under controlled conditions, however, the germination percentages of both seeds (self- and open-pollination) were similar (Gray and Steckel, 1986).

The strong plant-to-plant variation in OPs (Pink, 1993; De Clercq *et al.*, 2003) results from field environment (Gray *et al.*, 1991), dehydration rate, seed-weight variability, and long germination and emergence (Gray and Steckel, 1986). Therefore, occasionally >50% of the OP population fails to meet market specifications (Smith and Crowther, 1995a, b) with further economic significance, since modern horticulture emphasizes uniformity.

Interestingly, seeds with 62–86% B chromosomes in the cell's nucleus germinated significantly faster than those with 3–43% (Vosa, 1966). Supposedly, selections for early germination favoured the accumulation of the small supernumerary chromosomes, which thus may aid selection for rapid germination (Currah, 1986).

Leek breeding aims at improved yields (Kampe, 1980) and qualitative traits such as winter-hardiness and long shafts (the blanched pseudostem formed by sheaths and emerging leaves; Figs 8.10, 8.11). It is worth noting that the mean heritabilities of shaft length, leaf length and width were 0.45, 0.30 and 0.21, respectively (Schweisguth, 1973).

Autumn and winter types produce narrow-based, long, dark-green pseudostem (well associated with tolerance to biotic and abiotic stress; Silvertand, 1996), erect foliage and frost-hardiness. Cold-hardiness is of special importance for winter harvest and is tightly correlated with a short pseudostem (Anon, 2010). Plants' demand for long vernalization permits a prolonged period of spring harvest (McCollum, 1976), while summer types require earliness and long vernalization to minimize bolting.

Important traits other than yield, size and shape are no bulbing, minimum bolting, tolerance to rust (*Puccinia porri* Wint.) (Pink and Innes, 1984) and high nutraceutical levels (Ament *et al.*, 2018). In Europe, tolerance to LYSV has been known for decades (Van der Meer, 1980; Graichen *et al.*, 1988), thus cvs Siegfried (Clause) and de Verrieres (Vilmorin) are symptom-free (Gagnebin and Bonnet, 1979).

Since the 1940s, Swiss breeding work on leek combined improvements in cold tolerance and long shafts. They argued that maintaining crop heterozygosity for vigorous growth contributed equally to the best cultural conditions (Gagnebin and Bonnet, 1979), as 26–62% vigour loss was recorded after single self-pollination compared with OP populations, while hybrids produced more than twice the OPs' yields.

Inbreds perform poorly, as selfing results in vigour loss by ~35% for S1 and 40% between S1 and S2 (Smith and Crowther, 1995a). Hence, the breeding of partially inbred populations is limited to 2–3 self-pollinated generations (Currah, 1986). Gynogenetic haploidization offers a solution (n=2x=16; Smith *et al.*, 1991; Voorrips, 1991; Schum *et al.*, 1993; Keller and Korzun, 1996; Chen *et al.*, 2011; Kaska *et al.*, 2013; Alan *et al.*, 2016; Akgün, 2018; Celebi-Toprak and Alan, 2021; Chapters 2, 15, 17, this volume). Ovule culture might generate fertile diploids with minimum major defects for OPs that serve as pollinators for F_1 hybrids or as synthetic cultivars. Information on the successful implementation of haploidization in leek commercial breeding programmes, however, is not publicly available.

Uniformity depends on heredity, environment and agro-management. F_1 hybrids and/or excelling inbred selections offer a solution to the genetic component of the equation (Schweisguth, 1970; Smith and Crowther, 1995b). However, climate change and future needs require further and significant progress, achievable through interspecific crosses (Buiteveld *et al.*, 1998; Peterka *et al.*, 2002, 2005; Ament *et al.*, 2018); integration of genomics tools, including saturated linkage maps of desired traits; and the use of marker-assisted advanced breeding (e.g. Khosa *et al.*, 2016). For *A. ampeloprasum*, however, some inherited obstacles exist, including the large genome size, e.g. 12.7–16.7 pg (Ricroch *et al.*, 2005), 24.1–26.8 pg per 2C (Labani and Elkington, 1987; Arumuganathan and Earle, 1991), biennial life cycle; cross-pollinated nature; strong inbreeding depression; as well as small research community (DEFRA, 2004) and low funding.

The leek's vegetative phase is day-neutral. Hence, cultivar classification depends on the morphology, leaf colour, earliness/lateness (harvest period) and winter-hardiness/adaptation to warmer climates (De Clercq *et al.*, 1999). Within temperature limitations, the day-neutrality enables easy adaptation of excelling genotypes in different latitudes (De Clercq and Van Bockstaele, 2002).

The dark side of millennia of selection pressure for the above traits, and globalization, is the deterioration of genetic variation (Van der Meer and Hanelt, 1990; Silvertand, 1996; Chapter 2, this volume).

Currently, no major/advanced breeding work in and for tropical regions has been reported. Seed companies maintain popular traditional OPs under permanent mass-selection pressure (McCollum, 1976) and offer improved F_1 hybrids with heterosis with up to 78% quality marketable yield above OPs, and shaft thickness, but not shaft length (Gagnebin and Bonnet, 1979; Kampe, 1980; Pink, 1993; Smith and Crowther, 1995b), probably by purging inbreeding depression common in OPs.

The improved phenotypic expression (Schweisguth, 1983) and more uniform maturity of F_1 hybrids (Schweisguth, 1970, 1973; Gagnebin and Bonnet, 1979; Kampe, 1980; Smith and Crowther, 1995a, b) suit the needs for a one-off mechanical harvest and sorting. Moreover, hybrids may form combinations of dominantly inherited resistances to diseases (Currah, 1986; Smith and Crowther, 1995b; Havey and Lopes Leite, 1999).

F_1 seed production requires male-sterile parents. Among inbreds from commercial varieties, very low (10^{-4}) frequencies of male sterility coded by one, two, three or more recessive genes were reported (Schweisguth, 1970; Rauber, 1989; Baumunk-Wende, 1985–1990). No reports are available on selections of complete male-sterile lines, and lack of fertility restorers (maintainer) and/or efficient large-scale clonal reproduction halted F_1 breeding programmes. Only advances in asexual propagation (Peterka *et al.*, 2002) enabled, in 1993, the functional use of Brian Smith's male-sterile lines (Horticulture Research International, UK) for the production of commercial hybrids. Currently, ~99% of farmed and consumed leeks in the developed world are hybrids (Nunhems, 2018; Searle, 2018), but not elsewhere (e.g. mostly Italian OPs (Fiume, 2019) are farmed in Indonesia, the largest global leek producer (Fig. 8.2)).

Tissue culture is an expensive, labour-intensive and time-consuming technology, therefore topsets obtained by early flower ablation and 20 hours of photoperiod are occasionally employed. The alternative production of capitulum plantlets may develop disorders and infections, e.g. viruses (Van Cappellen *et al.*, 2009). Hence, the production of hybrid seeds is expensive (Smith and Crowther, 1995b; Peterka *et al.*, 2002).

Cytoplasmic male sterility (CMS) enables efficient male-sterile seed propagation. No CMS has been identified in cultivated *A. ampeloprasum* due to the low cytoplasmic diversity (Kik *et al.*, 1997; Havey and Lopes Leite, 1999; Peterka *et al.*, 2002). Hence, interspecific sexual (Peterka *et al.*, 1997, 2002; Bejo Zaden, 2008) hybridizations with *A. cepa* (Buiteveld *et al.*, 1998), *A. fistulosum* (Peterka *et al.*, 2005) and introgression of garlic mitochondria (Van Cappellen *et al.*, 2009) generated CMS genotypes, but, so far, no commercial applications have been reported.

Phytogametocidals induce male sterility in several Alliums (Currah, 1986; Kaul, 1988), but GA_3 (used in onions: Van der Meer and Van Bennekom, 1973) and maleic hydrazide did not induce male sterility in leek (Jacobsen, 1996).

A list of major institutes engaged in leek research is published in Celebi-Toprak and Alan, 2021).

10. Seed Production

Leeks are propagated by seed or transplants. Higher handling costs of transplants, however, favour the seed-to-seed production method (Chapter 12, this volume). The slow/long germination/emergence and post-juvenility vernalization response, however, should be considered.

Pollination control for hybrid seed production is difficult. The spherical leek umbel, 4–12 cm in diameter, bears 200–600 small, perfect, self-compatible protandrous flowers (Currah and Ockendon, 1978). Flower size, up to six seeds/flower, and the weeks-long span between first and last anthesis make emasculation for commercial hybrid seed production impracticable, thus male sterility is the only workable means for making hybrids (Chapters 12, 14, 15, 17, this volume).

Early sowing and dense plots bear small leek umbels, but seed yield is higher than both wide-spacing and transplanting since population density has little effect on mean seed weight (Gray *et al.*, 1991), and germinability is similar (Adamson, 1952).

Flies and bees (10–20 vigorous hives/ha) are the favoured pollinators of the entomophile *A. ampeloprasum*, hence strongly limiting the choice of chemical fertilization (Waller *et al.*, 1972; Hagler, 1990), chemical control methods and timing before and during flowering.

In leek, anthesis begins later, and seeds' maturation takes longer than onions' (Jones and Mann, 1963), thus leek seed production encounters maturity problems where the growing season is short. The optimum harvest time is arguable, due to conflicting objectives: maximum seed maturation *in situ* and minimum shattering/seed loss (Gray and Steckl, 1986). Harvesting when the liquid transparent endosperm turns opaque and condensed (Rabinowitch, 1990) gives the highest yields (Rusev and Buchvarov, 1975). Therefore, in practice, harvest best begins when ~10–30% of leek heads have a few opened capsules and black, ripened seeds become visible.

In small (e.g. breeding) plots, the weeks-long period from the first to the last anthesis necessitates 2–3 pickings for sufficient quality seeds, while one-off manual or mechanical harvesting is common in commercial production. The harvested heads with ~10–20 cm attached scapes are placed in a drying facility for maturation. Forced or passive aeration removes moisture to minimize heating and microbiologic infection, thus reducing seed deterioration. Rapid drying stimulated germination more than a slower process (Gray *et al.*, 1992), indicating its role as a cue for the germinative process. Dried seed heads are gently threshed to prevent seed injuries and mixed with seeds shattered during drying (Thornton *et al.*, 1993).

11. Prospects

Contrary to popular opinion, the definitions 'minor crop' and 'unimportant/insignificant' are not synonymous. *Ampeloprasum* complex, particularly leek, is a productive (10–40 tons/ha; Fig. 8.4), nutritious crop (Tables 8.1, 8.2), highly valued in the diet of certain populations (e.g. Indonesia, Rwanda, Egypt, Turkey, Iran) and contributes a great culinary value in European cuisine. Yet science and commercial attention are rather limited, resembling 'orphan illnesses'.

The latest, most exciting breakthrough in leek development happened some 20–30 years ago with the development of F_1 hybrids, while the adoption of great advances in biology/genetics stayed far behind. There is certainly scope for seed physiology studies, seed quality and production improvements; CMS is probably not employed in commercial seed production; progress in resistance/tolerance to biotic and abiotic stress (e.g. cold-hardiness, resistance to purple blotch), in quality, stronger demand for cold units for bolting, tolerance for tropical conditions based on the array of subtropical lines, and seed production, are limited to non-existent.

The small research community, shortage of funds, and little interest by most seed companies in minor crops tell the leeks' and other small crops' stories, thus expectations for a significant leap soon are low. It is up to international bodies such as the FAO, the World Bank, the EU and others to pick up the slack by establishing universal multi-disciplinary crop-specific task teams capable of advancing the complex crops' performance by utilizing genetic resources, modern genomics, metabolomics, nutrigenomics, physiology, agronomic technologies, and more. The results should become freely available to the international community for furthering the achievements and the human benefits worldwide. Major international companies, charities and prosperous individuals could contribute/participate in such a border-free 'improve our food' venture, thus profiting mankind and our heritage.

References

Abd El-Rehem, F., Abd El-Rehem, A. and Mohammed Ali, R.F. (2013) Proximate compositions, phytochemical constituents, antioxidant activities and phenolic contents of seed and leaves extracts of Egyptian leek (*Allium ampeloprasum* var. *kurrat*). *European Journal of Chemistry* 4, 185–190. DOI: 10.5155/eurjchem.4.3.185-190.711.

Adamson, R.M. (1952) Effects of various methods in leek seed production. *Scientific Agriculture* 32, 634–637.

Adão, C.R., da Silva, B.P. and Parente, J.P. (2011) A new steroidal saponin with antiinflammatory and antiulcerogenic properties from the bulbs of *Allium ampeloprasum* var. *porrum*. *Fitoterapia* 82, 1175–1180.

Akgün, S. (2018) Studying the effects of different sucrose concentrations on gynogenesis induction in leek (*Allium ampeloprasum* L.). MSc thesis, Pamukkale University, Turkey. Available at: http://acikerisim.pau.edu.tr/xmlui/bitstream/handle/11499/27244/Selda%20%20Akg%C3%BCn.pdf?sequence=1&isAllowed=y (accessed 16 July 2022).

Alan, A.R., Celebi Toprak, F. and Kaska, A. (2016) Production and evaluation of gynogenic leek (*Allium ampeloprasum* L.) plants. *Plant Cell, Tissue and Organ Culture (PCTOC)* 125, 249–259. DOI: 10.1007/s11240-016-0944-2.

Ament, K., Bongers, H., Chretien, M.L., Adriaanse, Schrijver, M. *et al.* (2018) Interspecific hybrid plant resulting from a cross between an *Allium ampeloprasum* plant with an *Allium sativum* plant. EP 3 351 091 A1. Available at: https://data.epo.org/publication-server/document?iDocId=5660607&iFormat=0 (accessed 16 July 2022).

Ariga, T., Kumagai, H., Yoshikawa, M., Kawakami, M., Seki, T. *et al.* (2002) Garlic-like but odorless plant *Allium ampeloprasum* Mushuu-ninniku. *Journal of the Japanese Society of Horticultural Science* 71, 362–369. Available at: https://www.jstage.jst.go.jp/article/jjshs1925/71/3/71_3_362/_article/-char/en (accessed 30 July 2022).

Anon (2010) Leeks. Oregon State University. Available at: https://horticulture.oregonstate.edu/oregon-vegetables/leeks-0 (accessed 30 July 2022).

Anon (2020) Leek, raw. USDA Food Data Central. Available at: https://fdc.nal.usda.gov/fdc-app.html#/food-details/1103357/nutrients (accessed 30 July 2022).

Arumuganathan, K. and Earle, E.D. (1991) Nuclear DNA content of some important plant species. *Plant Molecular Biology Reporter* 9, 208–218. DOI: 10.1007/BF02672069.

Avicenna (Ibn-e Sina)s (1998) *Canon of Medicine* (Qanun). Jamia Hamdard, Hamdard University, New Delhi. Available at: http://data.nur.nu/Kutub/English/Avicenna_Canon-of-Medicine_text.pdf (accessed 16 July 2022).

Baugerod, H. (1980) Atmospheric control in CA storage rooms by means of control diffusion through air-filled channels. *Acta Horticulturae* 116, 179–185. DOI: 10.17660/ActaHortic.1981.116.23.

Baumunk-Wende, E. (1985–1990) *Principles of Leek Breeding (Allium porrum L.)*. Jahresbericht Bundesforschungsanstalt für gartenbauliche Pflanzenzüchtung, Ahrensburg, Germany.

Bejo Zaden, B.V. (2008) Cytoplasmic male sterile leek plants, methods for the preparation and use thereof, and a cytoplasmic male sterile garlic lant. Available at: https://patentimages.storage.googleapis.com/65/a4/49/c73c1a8c0e2b8e/WO2010007059A1.pdf (accessed 16 July 2022).

Beminger, E. and Buret, E. (1967) Etude des déficients chlorophylliens chez deux espèces cultivtées du genre *Allium*: l'oignon A. *cepa* L. et le poireau A. *porrum*. L. *Ann Amélior Plant* 17, 175–194.

Ben Arfa, A., Najjaa, H., Yahia, B., Tlig, A. and Neffati, M. (2015) Antioxidant capacity and phenolic composition as a function of genetic diversity of wild Tunisian leek (*Allium ampeloprasum* L.). *Journal of New Sciences* 3, 15–26. Available at: https://www.jnsciences.org/agri-biotech/29-volume-21/100-antioxidant-capacity-and-phenolic-composition-as-a-function-of-genetic-diversity-of-wild-tunisian-leek-allium-ampeloprasum-l.html (accessed 16 July 2022).

Bernaert, N. (2013) Bioactive compounds in leek (*Allium ampeloprasum var. porrum*): analysis as a function of the genetic diversity, harvest time and processing techniques. PhD thesis, Ghent University, Belgium. Available at: https://biblio.ugent.be/publication/3221552/file/4336560.pdf (accessed 16 July 2022).

Bernaert, N., De Clercq, H., Van Bockstaele, E., De Loose, M. and Van Droogenbroeck, B. (2013) Antioxidant changes during postharvest processing and storage of leek (*Allium ampeloprasum* var. *porrum*). *Postharvest Biology and Technology* 86, 8–16. DOI: 10.1016/j.postharvbio.2013.06.010.

Bierhuizen, J.F. and Wagenvoort, W.A. (1974) Some aspects of seed germination in vegetables. 1. The determination and application of heat sums and minimum temperature for germination. *Scientia Horticulturae* 2, 213–219. Available at: https://reader.elsevier.com/reader/sd/pii/0304423874900296?

token=0284696C170C820FF4C0DC9FAC6464A82DDCD1E8D63BA1AA28A9516649EBF5B9A3636 8101386FC18FA8423F989E2FB24&originRegion=eu-west-1&originCreation=20220717143109 (accessed 30 July 2022).

Block, E., Naganathan, S., Putman, D. and Zhao, S.H. (1992) *Allium* chemistry: HPLC analysis of thiosulfinates from onion, garlic, wild garlic (ransoms), leek, scallion, shallot, elephant (great-headed) garlic, chive, and Chinese chive: uniquely high allyl to methyl ratios in some garlic samples. *Journal of Agricultural and Food Chemistry* 40, 2418–2430. DOI: 10.1021/jf00024a017.

Block, E., Dane, A.J., Thomas, S. and Cody, R.B. (2010) Applications of direct analysis in real time mass spectrometry (DART-MS) in Allium chemistry: 2-propenesulfenic and 2-propenesulfinic acids, diallyl trisulfane s-oxide, and other reactive sulfur compounds from crushed garlic and other Alliums. *Journal of Agricultural and Food Chemistry* 58, 4617–4625. DOI: 10.1021/jf1000106.

Bohanec, B., Jakše, M., Sešek, P. and Havey, M.J. (2005) Genetic characterization of an unknown Chinese bulbous leek-like accession and its relationship to similar *Allium* species. *HortScience* 40, 1690–1694. DOI: 10.21273/HORTSCI.40.6.1690.

Bottéro, J. (1985) The cuisine of ancient Mesopotamia. *Biblical Archaeology* 481, 30–47. DOI: 10.2307/3209946.

Bray, C.M., Davison, P. A., Ashraf, M. and Taylor, R.M. (1989) Biochemical changes during osmopriming of leek seeds. *Annals of Botany* 63, 185–193. Available at: https://www.jstor.org/stable/42758167 (accessed 16 July 2022).

Brewster, J.L. (1979) The response of growth rate to temperature in seedlings of several *Allium* crop species. *Annals of Applied Biology* 93, 351–357. DOI: 10.1111/j.1744-7348.1979.tb06551.x.

Brewster, J.L. (1994) *Onions and Other Vegetable Alliums*. CAB International, Wallingford, UK.

Brewster J.L. (1995) Onions and other vegetable alliums. *Scientia Horticulturae* 62, 145–146.

Brewster, J.L. (2008) *Onions and Other Vegetable Alliums*. CAB International, Wallingford, UK.

Brocklehurst, P.A., Dearman, J. and Drew, R.L.K. (1984) Effects of osmotic priming on seed germination and seedling growth in leek. *Scientia Horticulturae* 24, 201–210. DOI: 10.1016/0304-4238(84)90103-1.

Buiteveld, J., Syo, Y., van Lookeren Campagne, M.M. and Cremers-Molenaar, J. (1998) Production and characterization of somatic hybrid plants between leek (*Allium ampeloprasum* L.) and onion (*Allium cepa* L.). *Theoretical and Applied Genetics* 96, 765–775. Available at: https://link.springer.com/content/pdf/10.1007/s001220050800.pdf (accessed 16 July 2022).

Celebi-Toprak, F. and Alan, A.R. (2021) *In vitro* gynogenesis in leek (*Allium ampeloprasum* L.). In: Segui-Simarro J.M. (ed.) *Doubled Haploid Technology. Methods in Molecular Biology* 2287. Humana, New York. DOI: 10.1007/978-1-0716-1315-3_7.

Chen, J., Cui, L., Malik, A.A. and Mbira, K.G. (2011) *In vitro* haploid and dihaploid production via unfertilized ovule culture. *Plant Cell, Tissue and Organ Culture (PCTOC)* 104, 311–319. DOI:10.1007/s11240-010-9874-6.

Clement, S.L., Hellier, B.C., Elberson, L.R., Staska, R.T. and Evans, M.A. (2007) Flies (Diptera: Muscidae: Calliphoridae) are efficient pollinators of *Allium ampeloprasum* L. (Alliaceae) in field cages. *Journal of Economic Entomology* 100, 131–135.

Coarbineau, F., Picard, M.A. and Come, D. (1994) Germinability of leek seeds and its improvement by osmopriming. *Acta Horticulturae* 371, 45–52. DOI: 10.17660/ActaHortic.1994.371.4.

Currah, L. (1986) Leek breeding: a review. *Journal of Horticultural Sciences* 61, 407–415. DOI: 10.1080/14620316.1986.11515720.

Currah, L. and Ockendon, D.J. (1978) Protandry and the sequence of flower opening in the onion (*Allium cepa* L.). *New Phytology* 81, 419–428. Available at: https://doi.org/10.1111/j.1469-8137.1978.tb02647.x (accessed 30 July 2022).

Daniel, G. (1975) Untersuchungen fiber die Ursachen des Blühens von Lauch (*Allium porrum* L.). Dissertation. Technische Universität München.

Dashti, F., Kashi, A. and Vazvaee, A. (2003) Study on genetic diversity in genotypes of Tarée Irani (*Allium ampeloprasum* ssp. Persicum) using morphological characters. *Journal of Seedling and Seed* 19(1), 87–100. (In Farsi)

Dashti, F., Kashi, A., Vazvaee, A., Mousavi, A. and Ershadi, A. (2005) Study on genetic diversity in genotypes of Tarée Irani (*Allium ampeloprasum* ssp. Persicum) using RADP marker. *Journal of the Institute of Water, Soil and Plant in Agriculture* 5(3), 1–11. (In Farsi)

Davies, D. (1992) *Alliums, the Ornamental Onions*. Timber Press, Portland, Oregon.

De Clercq, H. and Van Bockstaele, E. (2002a) Leek: advances in agronomy and breeding. In: Rabinowitch, H.D. and Currah, L. (eds) *Allium Crop Science: Recent Advances*. CAB International, Wallingford, UK, pp. 431–468.

De Clercq, H. and Van Bockstaele, E. (2002b) Leek: advances in agronomy and breeding. In: Rabinowitch, H.D. and Currah, L. (eds) *Allium Crop Science: Recent Advances*. CAB International, Wallingford, UK, pp. 431–458.

De Clercq, H., Baert, J. and Van Bockstaele, E. (1999) Breeding potential of Belgian landraces of leek (*Allium ampeloprasum* L. var. *porrum*). *Euphytica* 106, 101–109. DOI: 10.1023/A:1003544231948.

De Clercq, H., Peusens, D., Roldán-Ruiz, I. and Van Bockstaele, E. (2003) Causal relationships between inbreeding, seed characteristics and plant performance in leek (*Allium porrum* L.). *Euphytica* 134, 103–115. DOI: 10.1023/A:1026198910662.

Dearman, J., Brocklehurst, P.A. and Drew, R.L.K. (2008) Effect of osmotic priming and aging on the germination and emergence of carrot and leek seed. *Annals of Applied Biology* 111, 717–722. DOI:10.1111/j.1744-7348.1987.tb02029.x

DEFRA (2004) DNA marker in ledk improvement: final report. Available at: http://randd.defra.gov.uk/Document.aspx?Document=HH0914TFV_3851_FRP.pdf (accessed 16 July 2022).

Dey, P. and Khaled, K.L. (2013) An extensive review on *Allium ampeloprasum*, a magical herb. *International Journal of Scientific Research* 6, 14. Available at: https://www.ijsr.net/archive/v4i7/SUB156374.pdf (accessed 16 July 2022).

Dragland, S. (1972) Effect of temperature and day-length on growth, bulb formation and bolting in leeks (*Allium porrum* L.). *Norges Landbrukshogskole* 51(21), 1–23.

Fattorusso, E., Lanzotti, V., Taglialatela-Scafati, O. and Cicala, C. (2001) The flavonoids of leek, *Allium porrum*. *Phytochemistry* 57, 565–569. DOI: 10.1016/S0031-9422(01)00039-5.

Fiume, F. (2019) Leek. Available at: http://francescofiume.altervista.org/Leek.html (accessed 16 July 2022).

Fritsch, R.M. and Friesen, N. (2002) Evolution, domestication, and taxonomy. In: Rabinowitch, H.D. and Currah, L. (eds) *Allium Crop Science: Recent Advances*. CAB International, Wallingford, UK, pp. 5–30.

Fritsch, R.M. and Keusgen, M. (2006) Occurrence and taxonomic significance of cysteine sulphoxides in the genus Allium L. (Alliaceae). *Phytochemistry* 67, 1127–1135. DOI: 10.1016/j.phytochem.2006.03.006.

Gagnebin, E. and Bonnet, J.C. (1979) Quelque considérations sur la culture et l'amélioration du poireau. *Revue Horticole Suisse* 52, 112–116.

García-Herrera, P., Morales, P., Fernández-Ruiz, V., Sánchez-Mata, M.C., Cámara, M. *et al.* (2014) Nutrients phytochemicals and antioxidant activity in wild populations of *Allium ampeloprasum* L., a valuable underutilized vegetable. *Food Research International* 62, 272–279. DOI: 10.1016/j.foodres.2014.03.004.

Gohil, R.N. (1984) Extent of recombination possible in the cultivated leek. In: *Eucarpia 3rd Allium Symposium*, 4–6 September, pp. 99–105.

Gohil, R.N. and Koul, A.K. (1977) The cause of multivalent suppression in *Allium ampeloprasum* L. *Beiträge zur Biologie der Pflanzen* 53, 473–478.

Golubkina, N.A., Seredin, T.M., Antoshkina, M.S., Kosheleva, O.V., Teliban, G.C. *et al.* (2018) Yield, quality, antioxidants and elemental composition of new leek cultivars under organic or conventional systems in a greenhouse. *Horticulturae* 4, 39. DOI: 10.3390/horticulturae4040039.

Govaerts, R., Kington, S., Friesen, N., Fritsch, R.M., Snijman, D.A. *et al.* (2005–2021) World checklist of Amaryllidaceae (2021). Available at: http://apps.kew.org/wcsp (accessed 16 July 2022).

Graichen, K., Kegler, H., Schmidt, H.E., Richter, J., Meyer, U. *et al.* (1988) Quantitative resistance of leek to leek yellow stripe virus. *Archiv für Gartenbau* 36(2), 77–81.

Gray, D. and Steckel, J.R.A. (1984) Viability of onion seed as influenced by temperature during seed growth. *Annals of Applied Biology* 104, 375–382. DOI: 10.1111/j.1744-7348.1984.tb05623.x.

Gray, D. and Steckel, J.R.A. (1986) Self- and open-pollination as factors influencing seed quality in leek (*Allium porrum*). *Annals of Applied Biology* 108, 167–170. DOI: 10.1111/j.1744-7348.1986.tb01977.x.

Gray, D. and Steckel, J.R.A. (1990) Effects of temperature of plant raising and nitrogen nutrition on the growth and flowering of leek. *Acta Horticulturae* 267, 235–240. DOI: 10.17660/ActaHortic.1990.267.29.

Gray, D. and Steckel, J.R.A. (1993) Effect of the plant raising environment, transplant weight and date of transplanting on the subsequent growth and development of leeks (*Allium porrum* L.) grown for early production. *Journal of Horticultural Sciences* 68, 955–965. DOI: 10.1080/00221589.1993.11516436.

Gray, D. and Ward, J.A. (1987) A comparison of leek (*Allium porrum*) and onion (*Allium cepa*) seed development. *Annals of Botany* 60, 181–187. DOI: 10.1093/oxfordjournals.aob.a087435.

Gray, D., Steckel, J.R.A., Chrimes, J.R. and Davies, A.C.W. (1991) Density and harvest date effects on leek seed quality. *Seed Science and Technology* 19, 331–340. Available at: http://pascal-francis.inist.fr/vibad/index.php?action=getRecordDetail&idt=5209151 (accessed 16 July 2022).

Gray, D., Steckel, J.R.A. and Hands, L.J. (1992) Leek (*Allium porrum* L.) seed development and germination. *Seed Science Research* 2, 89–95. DOI: 10.1017/S0960258500001185.

Gross, K.C., Wang, C.Y. and Saltveit, M. (2016) The commercial storage of fruits, vegetables and florist and nursery stocks. In: *USDA Agricultural Handbook*, vol. 66. Washington, DC. Available at: https://www.ars.usda.gov/arsuserfiles/oc/np/commercialstorage/commercialstorage.pdf (accessed 16 July 2022).

Grzelak-Błaszczyk, K., Kołodziejczyk, K., Badełek, E. and Adamicki, F. (2011) Changes in the contents of mono-, di- and oligosaccharides in leek plants stored in cold room. *European Food Research and Technology* 232, 1027–1033. DOI: 10.1007/s00217-011-1476-y.

Guenaoui, C., Mang, S., Figliuolo, G. and Neffati, M. (2013) Diversity in *Allium ampeloprasum*: from small and wild to large and cultivated. *Genetic Resources and Crop Evolution* 60, 97–114. DOI: 10.1007/s10722-012-9819-5.

Guern, M., Lecorff, J. and Boscher, J. (1991) Comparative karyology of the *Allium ampeloprasum* complex in France. *Bulletin de la Société botanique de France* 138, 303–313.

Hagler, J.R. (1990) Honey bee (*Apis mellifera* L.) response to simulated onion nectars containing variable sugar and potassium concentrations. *Apidologie* 21, 115–121.

Hanci, Ü.F., Pinar, Ü.H. and Uzun, A. (2018) The leek: an analysis of production and trade market in Turkey and world. In: *Conference: EJONS V – Internat. Congr, Math. – Engineer. – Nat. & Med. Sci. At: Gaziantep*. pp. 808–817. Available at: https://www.researchgate.net/publication/329416830_THE_LEEK_AN_ANALYSIS_of_PRODUCTION_and_TRADE_MARKET_WORLDWIDE_and_in_TURKEY (accessed 16 July 2022).

Hanelt, P. (1990) Taxonomy, evolution, and history. In: Rabinowitch, H.D. and Brewster, J.L. (eds) *Onions and Allied Crops: I. Botany, Physiology, and Genetics*. CRC Press, Boca Raton, Florida, pp. 1–26.

Hanelt, P. (1996) Proposal to conserve the name *Allium ampeloprasum* against *A. porrum* (Liliaceae). *TAXON* 45, 691–692. Available at: https://onlinelibrary.wiley.com/doi/pdf/10.2307/1224259 (accessed 30 July 2022).

Hanelt, P. (2001) Alliaceae. In: Hanelt, P. (ed.) *Mansfeld's Encyclopedia of Agricultural and Horticultural Crops*, vol. 4. Springer, Vienna, pp. 2250–2269.

Havey, M.J. and Lopes Leite, D. (1999) Toward the identification of cytoplasmic male sterility in leek: evaluation of organellar DNA diversity among cultivated accessions of *Allium ampeloprasum*. *Journal of the American Society for Horticultural Science* 124, 163–165. Available at: https://journals.ashs.org/jashs/view/journals/jashs/124/2/article-p163.xml (accessed 16 July 2022),

Hawthorn, L.R. and Pollard, L.H. (1954) *Vegetable and Flower Seed Production*. The Blakiston Company Inc., New York.

Hirschegger, P., Galmarini, C. and Bohanec, B. (2006) Characterization of a novel form of fertile great headed garlic (*Allium* sp.). Available at: https://doi.org/10.1111/j.1439-0523.2006.01279.x (accessed 30 July 2022).

Hirschegger, P., Jakše. J., Trontelj, P. and Bohanec, B. (2010) Origins of *Allium ampeloprasum* horticultural groups and molecular phylogeny of the section *Allium* (*Allium*: Alliaceae). *Molecular Phylogenetics and Evolution* 54, 488–497. DOI: 10.1016/j.ympev.2009.08.030.

Horie, H. and Yamashita, K. (2006) Non-derivatized analysis of methiin and alliin in vegetables by capillary electrophoresis. *Journal of Chromatography A* 1132, 337–339. DOI: 10.1016/j.chroma.2006.09.018.

Hösslin, R.V. and Andresen, F. (1972) Überwinterungsversuche mit Porree. *Gemüse* 8, 265–267.

Huang, Z. and Ren, J. (2013) Antibacterial activity of elephant garlic and its effect against U2OS human osteosarcoma cells. *Iranian Journal of Basic Medical Sciences* 16, 1088–1094.

Jauzein, P. and Tison, J. (2005) Le complexe d'*Allium ampeloprasum* L. en France. *Lejeunia* 178, 457–4184. Available at: https://popups.uliege.be/0457-4184/index.php?id=886&file=1&pid=883 (accessed 16 July 2022).

Jones, H.A. and Mann, L.K. (1963) *Onions and Their Allies: Botany Cultivation and Utilization*. Interscience Publishers, New York.

Jones, G.H., Khazanehdari, K.A. and Ford-Lloyd, B.V. (1996) Meiosis in the leek (*Allium porrum* L.) revisited. II. Metaphase I observations. *Heredity* 76, 186–191. DOI: 10.1038/hdy.1996.26.

Kadry, A.R. and Kamel, S.A. (1955) Cytological studies in the two tetraploid species *Allium kurrat* Schweinf and A. *porrum* L. and their hybrid. *Svensk Botanisk Tidskrift* 49, 314–324.

Kampe, R. (1980) Untersuchungen zum Ausmass von Hybrideffekten bei Porree. *Archiv. Züchtungsforsch* 10, 123–131.

Kaska, A., Celebi Toprak, F. and Alan, A.R. (2013) Gynogenesis induction in leek (*Allium ampeloprasum* L.) breeding materials. *Current Opinion in Biotechnology* 24(suppl 1), S42. DOI: 10.1016/j.copbio.2013.05.090.

Kaul, M.L.H. (1988) *Male Sterility in Higher Plants*. Springer Verlag, Berlin. Available at: https://books.google.co.il/books?hl=en&lr=&id=aCX7CAAAQBAJ&oi=fnd&pg=PA3&dq=Kaul,+M.L.H.+(1988)+Male+sterility+

in+Higher+Plants.+Springer+Verlag,+Berlin&ots=PxcxH1LjYi&sig=hvaYNwcSHRiTW1jf1NQkxl1yOIs&redir_esc=y#v=onepage&q&f=false (accessed 16 July 2022).

Keller, J. and Korzun, L. (1996) Haploidy in onion (*Allium cepa* L.) and other *Allium* species. In: Jain, M.S., Sopory, S.K. and Veilleux, R.E. (eds) *In Vitro Haploid Production in Higher Plants*, vol. 3. Kluwer Academic Publishers, Dordrecht, The Netherlands, pp. 51–75. Available at: https://link.springer.com/chapter/10.1007/978-94-017-1858-5_4 (accessed 16 July 2022).

Keller, E.R.J., Blattner, F.R., Fritsch, R., Pistrick, K., Senula, A. and Zanke, C.D. (2012) The genus *Allium* in the Gatersleben plant collections – progress in germplasm preservation, characterization and phylogenetic analysis. *Acta Horticulturae* 969, 273–287. Available at: http://dx.doi.org/10.17660/ActaHortic.2012.969.36 (accessed 30 July 2022).

Khazanehdari, K.A. and Jones, G.H. (1997) The causes and consequences of meiotic irregularity in the leek (*Allium ampeloprasum* spp. *porrum*): implications for fertility, quality and uniformity. *Euphytica* 93, 313–319. DOI: 10.1023/A:1002914808150.

Khazanehdari, K.A., Jones, G.H. and Ford-Lloyd, B.V. (1995) Meiosis in *Allium porrum* L. (the leek) revisited. I. Prophase I pairing. *Chromosome Research* 3, 433–439. DOI: 10.1007/BF00713894.

Khedim, T., Aïnouche, A., Amirouche, N., Ourari, M., Keller, J. *et al.* (2020) Tracking the evolutionary history of the *Allium ampeloprasum* L. complex (section *Allium*) provides evidence of the contribution of North African diploids to the formation of allopolyploid horticultural groups. *Genetic Resources and Crop Evolution* 67, 1885–1904. DOI: 10.1007/s10722-020-00948-x.

Khosa, J.S., McCallum, J., Dhatt, A.S. and Macknight R.C. (2016) Enhancing onion breeding using molecular tools. *Plant Breeding* 135, 9–20. DOI: 10.1111/pbr.12330.

Kik, C., Samoylov, A.M., Verbeek, W.H.J. and Van Raamsdonk L.W.D. (1997) Mitochondrial DNA variation and crossability of leek (*Allium porrum*) and its wild relatives from the *Allium ampeloprasum* complex. *Theoretical and Applied Genetics* 94, 465–471. DOI: 10.1007/s001220050438.

Koca, I. and Tasci, B. (2016) Mineral composition of leek. *Acta Horticulturae* 1143, 147–152. Available at: https://www.actahort.org/members/showpdf?session=22735 (accessed 16 July 2022).

Kollmann, F. (1972) *Allium ampeloprasum*: a polyploid complex II. Meiosis and relationships between the ploidy types. *Caryologia* 25, 295–312. DOI: 10.1080/00087114.1972.10796484.

Koul, A.K. and Gohil, R.N. (1970) Cytology of the tetraploid *Allium ampeloprasum* with chiasma localization. *Chromosoma* 29, 12–19. DOI: 10.1007/BF01183658.

Kratchanova, M., Nikolova, M., Pavlova, E., Yanakieva, I. and Kussovski, V. (2010) Composition and properties of biologically active pectic polysaccharides from leek (*Allium porrum*). *Journal of the Science of Food and Agriculture* 90, 2046–2051. DOI: 10.1002/jsfa.4050.

Kurki, L. (1979) Leek quality change in CA-storage. *Acta Horticulturae* 93, 85–90. Available at: https://www.actahort.org/members/showpdf?session=19147 (accessed 16 July 2022).

Labani, R.M. and Elkington, T.T. (1987) Nuclear DNA variation in the genus *Allium* L. (Liliaceae). *Heredity* 59, 119–128. Available at: https://www.nature.com/articles/hdy1987103.pdf (accessed 16 July 2022).

Levy, D. and Kedar, N. (1970) Effect of ethrel on growth and bulb initiation in onion. *Horticultural Science* 5, 80–82.

Lim, T.K. (2015) *Allium ampeloprasum*. In: Lim, T.K. (ed.) *Edible Medicinal and Non-Medicinal Plants. Vol. 9: Modified Stems, Roots, Bulbs*. Springer. pp. 103–123. DOI: 10.1007/978-94-017-9511-1.

Maragheh, F.P., Janus, D., Senderowicz, M., Haliloglu, K. and Kolano, B. (2018) Karyotype analysis of eight cultivated *Allium* species. *Journal of Applied Genetics* 60, 1–11. DOI: 10.1007/s13353-018-0474-1.

Mathew, B. (1996) *Review of Allium Section Allium*. Royal Botanic Gardens, London.

McCollum, G.D. (1976) Onions and allies. In: Simmonds, N.W. (ed.) *Evolution of Crop Plants*. Longman, London and New York, pp. 186–190.

Messiaen, C.M. and Rouamba, A. (2004) *Allium ampeloprasum* L. In: Grubben, G.J.H. and Denton, O.A. (eds) *PROTA*, Wageningen, The Netherlands.

Mohamed-Yasseen, Y. and Costanza, S.H. (1996) Clonal propagation of kurrat (*Allium ampeloprasum* var. *Kurrat*). *In Vitro Cellular & Developmental Biology of Plants* 32, 100–102. Available at: https://www.jstor.org/stable/20064885 (accessed 16 July 2022).

Mousavi, A., Kashi, A., Davoodi, D. and Shariatpanahi, M. (2006) Characterization of an Allium cultivated in Iran: the Persian leek. *Belgian Journal of Botany* 139, 115–123. Available at: http://www.jstor.org/stable/20794599 (accessed 16 July 2022).

Musselman, L.J. (2002) Is *Allium kurrat* the leek of the Bible? *Economic Botany* 56, 399–400. DOI: 10.1663/0013-0001(2002)056[0399:IAKTLO]2.0.CO;2.

Najda, A., Błaszczyk, L., Winiarczyk, K., Dyduch, J. and Tchórzewska, D. (2016) Comparative studies of nutritional and health-enhancing properties in the 'garlic-like' plant *Allium ampeloprasum* var. *ampeloprasum* (GHG-L) and *A. sativum*. *Scientia Horticulturae* 201, 247–255. DOI: 10.1016/j.scienta.2016.01.044.

Nunhems (2018) A giant leek for mankind. *Hort News*. Available at: https://hortnews.com/giant-leek-mankind/ (accessed 16 July 2022).

OECD (2019) *Leeks: International Standards for Fruit and Vegetables*. OECD Publishing, Paris. Available at: https://doi.org/10.1787/dbf45fde-en-fr (accessed 16 July 2022).

Ozgur, M., Akpinar-Bayaziy, A., Ozcan, T. and Afolayan, A.J. (2011) Effect of dehydration on several physico-chemical properties and the antioxidant activity of leeks (*Allium porrum* L.). *Notulae Botanicae Horti Agrobotanici Cluj-Napoca* 39,144–151. DOI: 10.15835/nbha3915861.

Parera, C. and Cantliffe, D.J. (1992) Priming leek seed for improved germination and emergence at high temperature. *HortScience* 27, 1077–1079. DOI: 10.21273/HORTSCI.27.10.1077.

Paunero, I.E., Corbino, G., Bazzigalupi, O. and Uviedo, R.H. (2003) Organic seed production of leek (*Allium porrum* L.) in the Northeastern of Buenos Aires. 1. Yield components and quality. *Spanish Journal of Agricultural Research* 1, 49–54. Available at: https://doi.org/10.5424/sjar/2003012-20 (accessed 31 July 2022).

Paunero, I.E., Bazzigalupi, O., Gonzalez, J. and Uviedo, R.H. (2004) Organic production of leek (*Allium porrum* L.) seeds in the Northeast of the Province of Buenos Aires. 2: Phenology, yield and quality. *Spanish Journal of Agricultural Research* 2, 564–569. Available at: https://doi.org/10.5424/sjar/2004024-113 (accessed 31 July 2022).

Peterka, H., Budahn, H. and Schrader, O. (1997) Interspecific hybrids between onion (*Allium cepa* L.) with S-cytoplasm and leek (*Allium ampeloprasum* L.). *Theoretical and Applied Genetics* 94, 383–389. DOI: 10.1007/s001220050427.

Peterka, H., Budahn, H., Schrader, O. and Havey, M. (2002) Transfer of a male-sterility-inducing cytoplasm from onion to leek (*Allium ampeloprasum*). *Theoretical and Applied Genetics* 105, 173–181. DOI: 10.1007/s00122-002-0935-z.

Peterka, H., Budahn, H. and Schrader, O. (2005) Interspecific hybridization for leek (*Allium ampeloprasum*) improvement. *Acta Horticulturae* 688, 101–108. DOI: 10.17660/ActaHortic.2005.688.10.

Pink, D.A.C. (1993) Leek. *Allium ampeloprasum* L. In: Kalloo, G. and Bergh, B.O. (eds) *Genetic Improvement of Vegetable Crops*. Pergamon Press, Oxford, UK, pp. 29–34. Available at: https://doi.org/10.1016/B978-0-08-040826-2.50007-2 (accessed 16 July 2022).

Pink, D.A.C. and Innes, N.L. (1984) Recent trends in the breeding of minor field vegetable crops for the UK. *Plant Breeding Abstracts* 54, 197–212.

PROSEA. (n.d.) *Allium ampeloprasum* (PROSEA). Available at: https://uses.plantnet-project.org/en/Allium_ampeloprasum_(PROSEA) (accessed 16 July 2022).

Prota4U (2021) *Allium ampeloprasum* L. Available at: https://www.prota4u.org/database/protav8.asp?g=pe&p=Allium+ampeloprasum+L. (accessed 16 July 2022).

Rabinowitch, H.D. (1990) Seed development. In: Rabinowitch, H.D and Brewster, J.L. (eds) *Onions and Allied Crops. I. Botany, Physiology, and Genetics*, pp. 151–159. Available at: https://www.taylorfrancis.com/chapters/edit/10.1201/9781351075169-7/seed-development-haim-rabinowitch (accessed 16 July 2022).

Ramin, A.A. (1997) The influence of temperature on germination of Tarée Irani (*Allium ampeloprasum* L. *spp iranicum W.*). *Seed Science and Technology* 25, 419–426.

Rauber, M. (1989) Untersuchungen zur männlichen Sterilität und in vitro *Massenvermehrung bei Allium schoenoprasum L. und Allium ampeloprasum L. var. porrum*. Dissertation. Universität Hannover, Germany.

Ricroch, A., Yockteng, R., Brown, S.C. and Nadot, S. (2005) Evolution of genome size across some cultivated *Allium* species. *Genome* 48, 511–520. DOI: 10.1139/G05-017.

Rubatzky, V.E. and Yamaguchi, M. (1997) *World Vegetables: Principles, Production, and Nutritive Values*, 2nd edn. Chapman & Hall, London.

Rusev, D. and Buchvarov, S. (1975) Studies on leek seed production. III. Effect of the date and method of seed plant harvesting on seed yield and quality. *Horticultural Science Abstracts* 45, 8321.

Sadeghi, M., Zolfaghari, B., Senatore, M. and Lanzotti, V. (2013a) Antifungal cinnamic acid derivatives from Persian leek (*Allium ampeloprasum* subsp. *persicum*). *Phytochemistry Letters* 6, 360–363. DOI: 10.1016/j.phytol.2013.04.007.

Sadeghi, M., Zolfaghari, B., Senatore, M. and Lanzotti, V. (2013b) Spirostane, furostane and cholestane saponins from Persian leek with antifungal activity. *Food Chemistry* 141, 1512–1521. DOI: 10.1016/j.phytol.2013.04.007.

Sakthi, A.M. and Monika, N. (2018) *Allium porrum*: a review. *World Journal of Pharmaceutical and Life Sciences* 4(3), 28–40. Available at: https://www.researchgate.net/publication/348778474 (accessed 16 July 2022).

Schum, A., Mattiesch, L., Timmann, E.-M. and Hofmann, K. (1993) Regeneration of dihaploids via gynogenesis in *Allium porrum* L. *Gartenbauwissenschaft* 58 (5), 227–232. Available at: http://pascal-francis.inist.fr/vibad/index.php?action=getRecordDetail&idt=6372779 (accessed 30 July 2022).

Schweisguth, B. (1970) Études préliminaires à l'amélioration du poireau A. *porrum* L. Proposition d'une methode d'amélioration. *Ann Amélior Plant* 22, 215–231.

Schweisguth, B. (1973) Étude de l'hérédité de trois caractères quantitatifs chez le poireau (*Allium porrum* L.). *Ann Amélior Plant* 23, 45–57.

Schweisguth, B. (1983) Which varietal structure for leek? In: *Proceedings of the 1st Allium Konferenz*, Freising, Germany, 19–23 July, pp. 264–276.

Searle, F. (2018) Breeders celebrate 25th anniversary of first hybrid leek. *Fresh Produce Journal.* Available at: http://www.fruitnet.com/fpj/article/176813/breeders-celebrate-25th-anniversary-of-first-hybrid-leek (accessed 10 May 2017).

Shahrajabian, M.H., Sun, W. and Cheng, Q. (2021) A review of leek (A. *ampeloprasum* L.): an important vegetable and food ingredient with remarkable pharmaceutical activities. *Pharmacognosy Communications* 11, 9–12. DOI: 10.5530/pc.2021.1.3.

Shakouri, M.J., Hassandokht, M.R., Jari, S.K., Larijani, K. and Ghanati, F. (2018) Studying the morphological properties, nutrients, and allicin content of Iranian Tarée accessions (*Allium ampeloprasum* Tarée group). *Eurasian Journal of Biosciences* 12, 13–18.

Shanmugasundaram, S. and Kalb, T. (n.d.) *Suggested Cultural Practices for Onion*. AVRDC Training Guide. Available at: https://web.archive.org/web/20111110013545/http://www.avrdc.org/LC/onion/practices.pdf (accessed 16 July 2022).

Shelke, P.A., Rafiq, S.M., Bhavesh, C., Rafiq, S.I., Swapnil, P. *et al.* (2020) Leek (*Allium ampeloprasum* L.). In: Nayik G.A. and Gull A. (eds) *Antioxidants in Vegetables and Nuts: Properties and Health Benefits*. Springer, Singapore, pp 309–331. DOI: 10.1007/978-981-15-7470-2_16.

Silvertand, B. (1996) Induction, maintenance and utilization of male sterility in leek (*Allium ampeloprasum* L.). Thesis. Landbouwuniversiteit, Wageningen, The Netherlands. Available at: https://core.ac.uk/download/pdf/29333338.pdf (accessed 16 July 2022).

Smith, B.M. and Crowther, T.C. (1995a) Inbreeding depression and single cross hybrids in leeks (*Allium ampeloprasum* ssp. *porrum*). *Euphytica* 86, 87–94. DOI: 10.1007/BF00022013.

Smith, B.M. and Crowther, T.C. (1995b) Breeding F_1 hybrids to improve uniformity in leeks (*Allium ampeloprasum* ssp. *porrum*). *Acta Horticulturae* 433, 317–321. Available at: https://www.actahort.org/members/showpdf?session=26596 (accessed 16 July 2022).

Smith, B.M., Godwin, R.M., Harvey, E. and Werner, C.P. (1991) Gynogenesis from whole flower buds in bulb onions *(Allium cepa* L.) and leeks *(Allium porrum* L.). *Journal of Genetics and Breeding* 45, 353–358. Available at: https://espace.library.uq.edu.au/view/UQ:354509 (accessed 16 July 2022).

Stack, S.M. (1993) Diploidization in autotetraploid *Allium porrum* by restricted crossing over. *American Journal of Botany* 80(6), abstract 231. Available at: https://www.jstor.org/stable/2445684?refreqid=excelsior%3A9746a88e97b7d32cd0ce13d7ae9570d8&seq=78#metadata_info_tab_contents (accessed 16 July 2022).

Stack, S.M. and Roelofs, D. (1996) Localized chiasmata and meiotic nodules in the tetraploid onion *Allium porrum*. *Genome* 39, 770–783. DOI: 10.1139/g96-097.

Statistica (2020) Available at: https://www.statista.com/statistics/625243/leek-production-in-belgium (accessed 16 July 2022).

Sterner, J.J. and Akintobi, D.C. (1986) Effect of harvest maturity on viability of onion seed. *Horticultural Science* 5, 1220–1221. Available at: https://agris.fao.org/agris-search/search.do?recordID=US8708133 (accessed 16 July 2022).

Swamy, K.R.M. and Veere Gowda, R. (2006) Leek and shallot. In: *Handbook of Herbs and Spices*, vol. 3. Woodhead Publishing Series in Food Science Technology and Nutrition, pp. 365–389. Available at: https://doi.org/10.1533/9781845691717.3.365 (accessed 16 July 2022).

Täckholm, V. and Drar, M. (1954) Flora of Egypt, *Allium* in ancient Egypt, Cairo. *Bulletin of Faculty of Pharmacy, Cairo University* 3(12), 93–136.

Tahbaz, F. (1971) L'*Allium* 'Tarée irani' du groupe *ampeloprasum* L. cultivé en Iran, région de Téhéran. *Bulletin de la Société botanique de France* 118, 753–761. DOI: 10.1080/00378941.1971.10838942.

Tahbaz, F. (1974) Comparative morphology and anatomy of different organs in Tareh Irani (Persian leek) and leek. *Periodical of the College of Agriculture, Tehran University* 1, 25–37. (In Persian with English abstract)

Tahbaz, F. (1976) Étude comparée caryotypes des *Allium* du group '*ampeloprasum*' cultivés en Iran. *Comptes Rendus de l'Académie des Sciences Series* 253, 1185–1188.

Temperate Plants Database (2021) Ken Fern. Available at: temperate.theferns.info/plant/Allium+ampeloprasum+Pearl+Onion+Group (accessed 16 July 2022).

Thornton, M.K., Mohan, S.K., Wilson, D.O., Beaver, R.G. and Colt, W.M. (1993) Onion and leek seed production. *Pacific Northwest Extension Bulletin.* Available at: https://research.libraries.wsu.edu/xmlui/bitstream/handle/2376/4849/pnw_433_1993_onion_and_leek_seed_production.pdf?sequence=1 (accessed 16 July 2022).

Tridge Leek Export (2020) Available at: https://www.tridge.com/intelligences/leek/export (accessed 16 July 2022).

Tridge Leek Production (2020) Available at: https://www.tridge.com/intelligences/leek/production (accessed 16 July 2022).

Tsiaganis, M.C., Laskari, K. and Melissari, E. (2006) Fatty acid composition of *Allium* species lipids. *Journal of Food Composition and Analysis* 19, 620–627. DOI: 10.1016/j.jfca.2005.06.003.

Uchida, A., Ogihara, J., Kumagai, H., Akao, M., Matsufuji, H. *et al.* (2009) Mechanisms of sulfide components expression and structural determination of substrate precursor in jumbo leek (*Allium ampeloprasum* L.). *The Japanese Society for Food Science and Technology* 56, 280–285. Available at: https://www.cabdirect.org/cabdirect/abstract/20093220572 (accessed 16 July 2022).

Van Cappellen, W., Adriaanse, M., Langedijk, A.E., Chretien, H., Bongers, M.L. *et al.* (2009) Cytoplasmic male sterile leek plants, methods for the preparation and use thereof, and a cytoplasmic male sterile garlic plant. Available at: https://patents.google.com/patent/WO2010007059A1 (accessed 16 July 2022).

Van der Meer, Q.P. (1984a) Resistance to leek yellow stripe virus in wild leek (*Allium ampeloprasum*). *Zaadbelangen* 34(6), 173.

Van der Meer, Q.P. (1984b) Breeding for resistance to yellow stripe virus in leeks (*Allium porrum* L.): a progress report. In: *3rd Eucarpia Allium Symposium*, 4–6 September, Wageningen, The Netherlands.

Van der Meer, Q.P. (1997) Old and new crops within edible *Allium*. *Acta Horticulturae* 433, 17–31. DOI: 10.17660/ActaHortic.1997.433.1.

Van der Meer, Q.P. and Hanelt, P. (1990) Leek (*Allium ampeloprasum*). In: Brewster, J.L. and Rabinowitch, H.D. (eds) *Onions and Allied Crops. III. Biochemistry, Food Science, and Minor Crops*. CRC Press, Boca Raton, Florida, pp. 179–196.

Van der Meer, Q.P. and Van Bennekom, J.L. (1973) Gibberellic acid as a gametocide for the common onion (*Allium cepa* L.). *Euphytica* 22, 239–243.

Von Bothmer, R. (1970) Cytological studies in *Allium*. I. Chromosome numbers and morphology in *Allium* Sect. *Allium* from Greece. *Botaniska Notiser* 123, 518–550. Available at: https://www.cabdirect.org/cabdirect/abstract/19711604042 (accessed 16 July 2022).

Von Bothmer, R. (1974) Biosystematic studies in the *Allium ampeloprasum* complex. *Opera Botanica a Societate Botanica Lundensi* 34, 1–104.

Von Bothmer, R. (1982) Karyotype variation in *Allium commutatum* (*Liliaceae* s. lato). *Plant Systematics and Evolution* 140, 179–189. DOI: 10.1007/BF02407296.

Voorrips, R.E. (1991) Unpollinated ovule and ovary culture in leek (*Allium porrum* L.). *Allium Improvement Newsletter* 1, 75–76.

Vosa, C.G. (1966) Seed germination and B-chromosome in the leek (*Allium porrum*). In: Darlington, C.D. and Lewis, K.R. (eds) *Chromosome Manipulation and Plant Genetics*. Oliver and Boyd, Edinburgh and London, pp. 24–27.

Vučurović, I., Nikolić, D., Radović, N., Vučurović, A., Ristić, D. *et al.* (2017) Incidence and distribution of leek yellow stripe virus in *Allium* crops in Serbia. *Pesticides and Phytomedicine* 32(3–4), 145–155. DOI: 10.2298/PIF1704145V.

Waller, G.D., Carpenter, E.W. and Ziehl, O.A. (1972) Potassium in onion nectar and its probable effect on attractiveness of onion flowers to honey bees. *Journal of the American Society of Horticultural Science* 97, 535–539.

Wiebe, H.-J. (1989) Vernalisation von wichtigen Gemfisearten – Ein Uberblick. *Gartenbauwissenschaft* 54(3), 97–104.

Wiebe, H.-J. (1994) Effects of temperature and daylength on bolting of leek (*Allium porrum* L.). *Scientia Horticulturae* 59, 177–185. Available at: https://reader.elsevier.com/reader/sd/pii/0304423894900116?token=C3353108D28C16DDAD0ACA0416B890EF7A8199E96D0B2AF6086AD3A16B90B18029021 6476D33A4F38E097068E642CDE7&originRegion=eu-west-1&originCreation=20210511071825 (accessed 16 July 2022).

Wurr, D.C.E., Fellows, J.R., Hambidge, A.J. and Fuller, M.P. (1999) Growth, development and bolting of early leeks in the UK. *The Journal of Horticultural Science and Biotechnology* 74, 140–146, DOI: 10.1080/14620316.1999.11511087.

Yamaguchi, M. (1983) *World Vegetables: Principles, Production and Nutritive Values*. Van Nostrand Reinhold Co., New York.

Yoo, K.S. and Pike, L.M. (1998) Determination of flavor precursor compound S-alk(en)yl-l-cysteine sulfoxides by an HPLC method and their distribution in *Allium* species. *Scientia Horticulturae* 75(1–2), 1–10. DOI: 10.1016/s0304-4238(98)00107-1.

Zhou, Y., Zhuang, W., Hu, W., Liu, G.J., Wu, T.X. *et al.* (2011) Consumption of large amounts of *Allium* vegetables reduces risk for gastric cancer in a meta-analysis. *Gastroenterology* 141, 80–89. DOI: 10.1053/j.gastro.2011.03.057.

9 Chives (*Allium schoenoprasum*)

Haim D. Rabinowitch[1]* and Brian Thomas[2]
[1]*The RH Smith Institute of Plant Science and Genetics in Agriculture, The Hebrew University of Jerusalem, Rehovot, Israel;* [2]*School of Life Sciences, University of Warwick, Coventry, UK*

'He who bears chives on his breath is safe from being kissed to death.'
– Marcus Valerius Matialis, 80 AD

1. Introduction

1.1. Origin, distribution and morphology

Poulson (1983, 1990), Chen (2006), and Singh *et al.* (2018) composed comprehensive reviews on chives, the smallest species of Alliaceae. Having these monographs in the background, we limited our discussion to brief coverage of common knowledge and focused on current developments.

The herbaceous perennial chive is the only *Allium* species native to both the New and the Old World (Europe and Asia) and the sub-Arctic belt (McCollum, 1976; Friesen and Blattner, 2000; Xu and Kamelin, 2000). It most probably originated from Siberia, and therefrom it spread to Asia, Europe, and, in the latter Ice Age, via the Beringian Bridge to North America (Štajner *et al.*, 2011; Chapter 1, this volume). Hence, Asian plants do not differ genetically from North American populations (Nikolai Friesen, personal communication).

Two subspecies are native to Europe: *A. schoenoprasum* ssp. *alpinum* (DC) Čelak and *A. schoenoprasum* L. ssp. *schoenoprasum*; and three subspecies from Iberia – *A. schoenoprasum* subsp. *orosiae*, *A. schoenoprasum* subsp. *latiorifolium* and *A. schoenoprasum* subsp. *schoenoprasum* (Maxted and Rhodes, 2016) – are primary wild relatives of chives.

Breakthroughs in molecular genetics have advanced scientists' understanding of the *Allium* taxonomy and the classification of three evolutionary lineages, which led to the greatest revision of *Allium* classification (Friesen *et al.*, 2006; Chapter 1, this volume). Henceforth, chives were included in the subgenus Cepa, section Schoenoprasum; *Allium schoenoprasum*.

1.2. Origins and usage

Chives have been cultivated in Europe since the Middle Ages, although evidence of their usage in Asia dates back 4000 (Mahr, 2021) or 5000 years (Trowbridge Filippone, 2006). Now the crop is farmed in Austria, Canada, China, Fenno-Scandinavia, France, Germany, the UK, Israel, Italy, The Netherlands, the USA (Singh *et al.*, 2018) and elsewhere (Maxted and Rhodes,

*Address for correspondence: haim.rabinowitch@mail.huji.ac.il

 Edible Alliums: Botany, Production and Uses (eds H.D. Rabinowitch and B. Thomas)

DOI: 10.1079/9781789249996.0009

2016). In India, it grows at an altitude of 250–350 m in the western Himalayas from Kashmir to Kumaon (Chopra *et al.*, 1992). China is the largest world producer of chives almost year-round, and Germany is the largest in Europe, but commercial production is limited to the summer months (Fig. 9.1). The figure shows shortages in chives supply in May and December.

Chives' hollow and tubular leaves, immature inflorescence, and open flowers are consumed as a common herb, while its ball-like, lavender-coloured umbels serve both as food and as ornamentals (Chen, 2006; Zdravković-Korać *et al.*, 2010; Fig. 9.2). Hence, chives are popular plants in many gardens, and yet, globally, a great share of the produce is farmed commercially (Štajner *et al.*, 2011). For instance, ~12 million potted chives were bought in 2010 in Germany, and by extrapolation from data published by the Ministry of Agriculture, ~1000 ha were farmed (Pude *et al.*, 2013).

The name 'schoenoprasum' originated from the Greek *skhoínos* (grass) and *prason* (leek), while the name 'chive' was adopted from the French: *cive* (Charles, 2013), which, in turn, originated from the Latin word *cepa* (onion). It is also known as snow mountain garlic or Kashmiri garlic (Koul *et al.*, 2016).

Other alliaceous crops were also named chives, e.g. Korean chives (*Allium thunbergii* G. Don, Amaryllidaceae; Choi and Oh, 2011; Wikipedia, 2021). Unlike chives (*A. schoenoprasum*), however, with round, 1–3 mm-in-diameter hollow leaves (Yamaguchi, 1983; Poulson, 1990), Korean chives produce flat, light, fan-shaped green leaves. It is used in folk medicine for a variety of ailments (Yin and Tsao, 1999; Teuscher *et al.*, 2006).

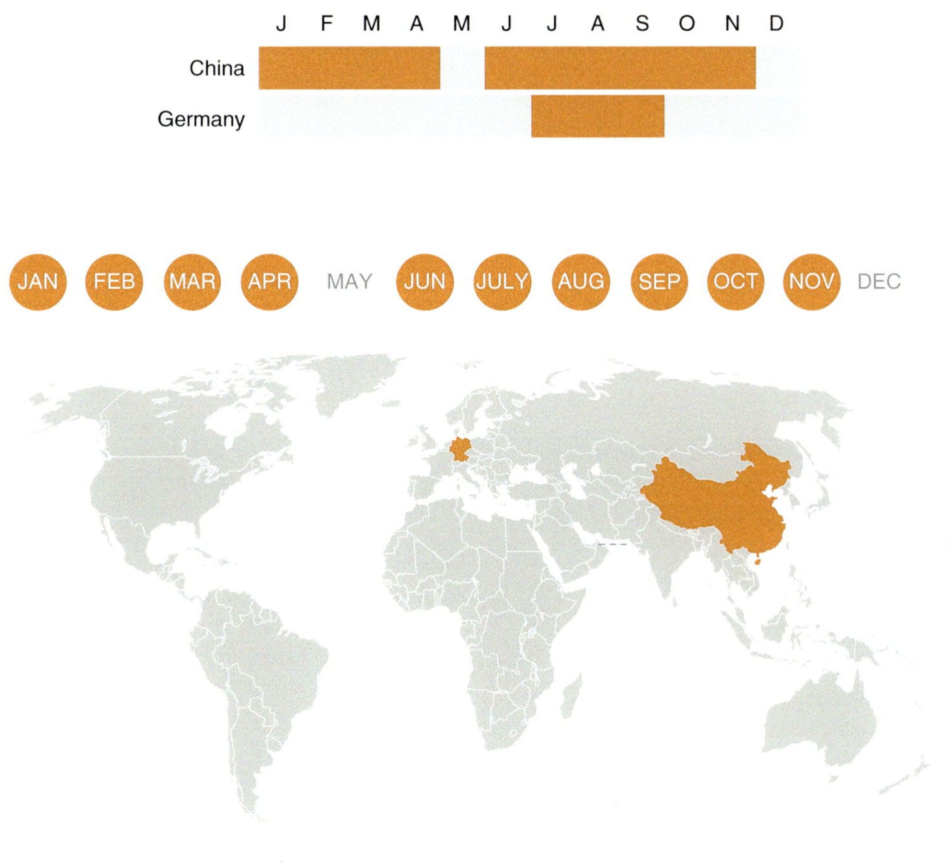

Fig. 9.1. The two leading chive-producing countries and time of supply. (From NedSpice Group, Rotterdam, The Netherlands)

Fig. 9.2. (a) Chives production plot in greenhouse; (b) intact plant; (c) chives in bloom (iStock); (d) freshly produced chive bunches (iStock); and (e) scapes (iStock).

Upon disruption, chives' leaves emit a distinct onion aroma, while the scent of the flowers resembles violets. Hence, it is used for both food and ornamentals. Chives are grown extensively in China as food and thus nicknamed Chinese chives (although this very name is commonly used for *A. tuberosum* (Chapter 10, this volume)).

2. Physiology and Agronomy

2.1. Growth cycle

Chives initially develop from seeds, usually planted in the spring, and develop into plantlets with thin tubular leaves whose bases are slightly swollen to form small bulbs, sitting on a rhizome. Thereafter, they can be considered perennial, with plants overwintering and growing in the second year or planted as individual bulbs. The latter can be stored and planted throughout the year, under protection, for year-round production of leaves, for culinary uses, domestically or in restaurants, and for prepared meals.

In the centre of origin, seeds germinate in the spring, grow and develop throughout summer, and enter winter dormancy in late autumn (Poulsen, 1990; Kamenetsky and Rabinowitch, 2006), thus completing a production cycle already in the first growing season. Winter vernalization induces the perennial chives to flower in the second growing season and annually thereafter, provided that plants receive sufficient cold units and temperatures do not exceed 18°C (Rubatzky and Yamaguchi, 1997), and light markedly affects the process (Chen, 2006; Fig. 9.3).

About two or three months after sowing, bulbs begin to form, and a pair of axillary buds develop on the main axis from which two new plants are formed. Following further growth, the unit perpetually divides to produce two new growth centres and the process repeats several times to produce more than 80 leaves on the cluster in the first year (Poulsen, 1990). Around mid-summer, new leaves that will emerge and grow the following year are initiated.

2.2. Dormancy

In late summer, the shortening days trigger the onset of dormancy. The optimum temperature

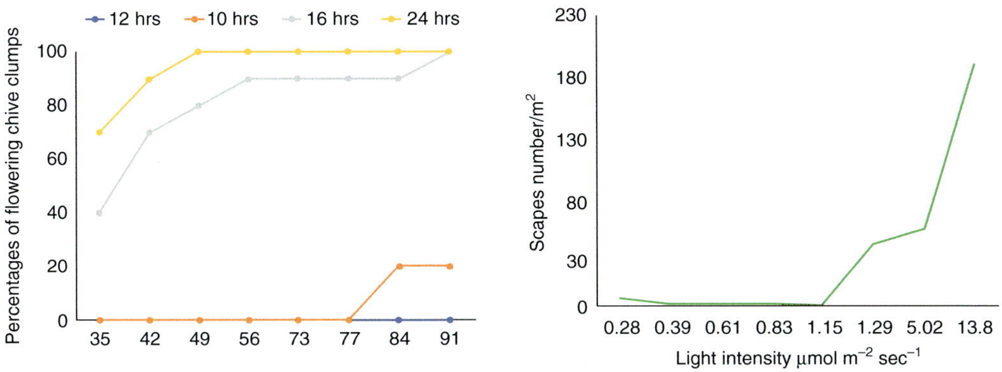

Fig. 9.3. Effect of daylength and light intensity on flowering of chives. For light intensity, plants were grown in the ground in a high tunnel and with a gradual supplement of incandescent light. For daylength experiments, chive clumps were grown in pots in the RH Smith Faculty of Agriculture, Food and Environment, Rehovot (FAHU). Each group was exposed to the assigned hours of light. The figures were adapted from Itay Rabinovich's thesis at FAHU (2009), under the supervision of Y. Dudai, R. Kamaentsky and H.D. Rabinowitch.

for dormancy induction is in the mid-teens, and both higher (>20°C) and lower <10°C) temperatures slow or prevent the onset of dormancy. In controlled environments, daylengths of 12–14 hours are most efficient in causing dormancy, which is inhibited by longer daylengths. Extending daylengths with incandescent light is effective in preventing dormancy. This mirrors the classic long-day response to light rich in far-red wavelengths, e.g. in onion bulb initiation (Chapter 4, this volume).

Production out of season in the winter can be sustained through forcing which is most effective after a rest period, thus enabling assimilates to be moved from old leaves to bulbs and roots, making them available for subsequent growth. However, the natural rest period needs to be shortened to enable new growth, which can be achieved through heat treatment given as hot water or warm-air treatments (Fölster and Krug, 1977).

In the subtropics, autumn germination (usually in a commercial nursery) is followed by transplanting in the production plots, and growth and development continue throughout the year (no dormancy). Branching commences when lateral initiation occurs after developing every two or three leaves (Poulson, 1990), and thus plants develop shoot clusters.

The terate fistulose leaves, 3–6 per plant, are 2–6 mm wide; fistulose, smooth leaves are slightly shorter than the 10–50 cm and up to 60 cm long inflorescences. The cross-pollinated flowers produce only a small amount of pollen and, commonly, seed yield is poor.

2.3. Forcing

Successful forcing of chives is supported by a preceding rest period. This promotes the translocation of assimilates from the foliage to the bulbs and especially to the roots. Furthermore, the old leaves die at the beginning of the rest period, followed by the developing new leaves. Forcing temperatures of 22–24°C gave the highest growth rates.

For early forcing, the rest period has to be broken by a cold- or, more successfully, by a hot-temperature treatment, e.g. immersing in 40°C water for 8–16 hours, followed by gradual cooling; or by hot air: 33–36°C for 2–3 days, RH ~80%. Non- or partially rested plants (mostly end of September/beginning of October) are more susceptible to heat damage than deeply rested plants (mid-October–November).

Chives can be harvested about once a month with two to four harvests in cold climates, increasing to five to seven harvests a year in warmer regions (Chen, 2006).

2.4. Flowering

As with many *Allium* species, chive does not usually flower in the first year, unless planted early in January for glasshouse production. Chives normally flower in the second and subsequent years, depending on cold exposure. Flowering takes place in May but can be prevented to allow for forcing of leaf production, by maintaining the temperature above 18°C (Rubatzky and Yamaguchi, 1997). Pollen production and the formation of viable seeds is limited. Chives are out-crossers and are normally insect-pollinated although some self-pollination occurs.

In the subtropics, winter cold induction plus spring increase in daylength and light intensity promote summer flowering in the first year of cultivation. The bases of the smooth terete scapes are 10–40 cm and up to 60 cm long, and they are wrapped with sheaths one-third to half their length.

Ten to fifty flower buds are densely packed in the subglobose ornate umbel. Pedicels vary in size and are shorter than the perianths. Each flower is complete, having six stamens, three carpels united with one pistil, and six perianth segments. United at the base, filament heights range between one-third and two-thirds of the perianth length, and adnate-to-perianth segments of 1–1.5 mm. The inner three anthers dehisce first and the outer whorl follows suit. Each of the three compartments of the pistil contains two ovules. The inner flowers of an umbel develop faster than the outer ones. Nectaries are located at the base of the subglobose ovary. The style remains within the perianth and flowering (in temperate zones) occurs in late summer between July and September. The self-compatible, hermaphrodite plant easily experiences geitonogamous selfing (adapted from *Flora of China*, n.d., and other sources).

The black hard chive seeds are very similar to onion seeds. Fresh seeds germinate evenly, but under ambient conditions, the older they get, vigour is lost rapidly and germination, and thus field uniformity, gets poorer and slower. Indeed, when stored at 20.3 ± 2.38°C and 50.5 ± 6.3%, chives had the least storable seed of the 18 tested crop species. Chive germination in the first year was 62.4 ± 22.9% and reached zero by year 6. In in-field experiments, seed germination during the first storage year was 24.4 ± 10.1% (Nagel and Börner, 2010). Similarly, of the hundreds of tested genotypes, the lowest seed half-life (P_{50}) value of 1.9 years was recorded for chive seed (Solberg *et al.*, 2020). When stored in a cool and dry location, chive seeds remain viable for one to two years.

With advanced development, elongated and false bulbs are formed, usually in clusters. Each shoot forms slender ovoid-cylindric bulbs, 0.5–1 cm in diameter, wrapped in tunics. The greyish-brown or tinged-with-yellow-shade tunics are papery, laciniate, sometimes fibrous at the apex, and with a cardboard-like texture. Like the original shoots, bulbs are clustered into clumps developing close to the soil surface, hence mulching to conserve moisture and weeding are essential (Fig. 9.4).

3. Response to the Environment

Chives have a wide range of environmental tolerance, being resistant to extreme heat and cold. The optimum temperature for leaf growth in controlled environments is between 19°C and 25°C, while for root growth it is 21–25°C. This resilience enables chive production to be spread across a wide geographical range. Even in historically low-production countries such as Finland, the potential for viable production has been confirmed, subject to variety, fertilization regimes, and appropriate agronomic techniques, such as using plastic mulch (Suojala, 2003). Chives can be grown in a wide range of soils, with the most suitable being sandy loams and loams containing organic matter. But they demand high levels of nitrogen and phosphorus for practical production (Chen, 2006).

Chives are resilient to short-term drought treatments (Egert and Tevini, 2002) and have effective antioxidant systems to protect them from oxidative damage during drought stress. This resilience has led to them being used in the provision of living walls (Martensson *et al.*, 2016) and green roof coverings. The concept of using chives for green roofing has been further developed by showing that it can be used successfully as a roof vegetable garden species (Whittinghill *et al.*, 2013). They showed that yields of chive remained consistent after being established in the first year. Eksi *et al.* (2020) demonstrated 100% chives survival when established on a range of potential roofing substrates.

As water becomes an increasingly valuable global resource, there is increasing use of saline

Fig. 9.4. Separated and clumped chive bulbs. (Original drawings: Asaf Sillner)

water for irrigation. Thus, there is increasing environmental stress on chives as a salt-sensitive crop with an optimum conductivity of 0.38 dSm^{-1} and a maximum threshold of 1.13 dSm^{-1} when irrigated in soil (Arslan *et al.*, 2018). In hydroponic circulating solutions, increased solute concentrations led to an increase in root-to-shoot dry-matter accumulation, with overall detriment to the shoots, confirming that chives are a salt-sensitive species (da Silva *et al.*, 2019).

Chives need to be germinated at a temperature of 15–20°C and kept moist. In temperate zones, seeds can be started indoors a few weeks before the average date of the last frost, before being planted outside, or can be directly seeded once the soil is warm.

Bulb clumps should be split once every two to three years to prevent overcrowding. Each division should consist of more than five bullets. These divisions can be used for propagation and transplanted 30–50 cm apart.

Once established, chives are drought-tolerant but respond favourably to regular moisture. Chives are fairly shallow-rooted, so be careful when cultivating around the plants or applying mulch for weed control. They have low nutrient requirements, so do not need regular fertilization. Agro-management of chives varies with country and climate. References for selected sites are provided in Table 9.1.

4. Breeding and Seed Production

4.1. Genetics

Information on European genetic resources of chives, classification and characterization is publicly available (Thomas *et al.*, 2005). Different varieties of chives can have different levels of ploidy. The most common is 2n=16, but both triploids and tetraploids with chromosome numbers near 30 were also found. The sect. *Schoenoprasum* has the smallest genome size in the subgenus Cepa as well as in the whole *Allium* genus (mean 2C DNA per genome -17.96 pg; Ohri *et al.*,1998). Chives (2n=16; 24, 32; Chinnappa and Basappa, 1986; Chen 2006) carry up to 20 B chromosomes (Houben, 2017) and displays a high level of structural polymorphism for their B chromosome (Bougourd and Parker, 1979b), which, unlike the negative effects in rye, are considered as adaptive and beneficial with selective attributes (Bougourd and Jones, 1997). B chromosomes are common in many chive populations (Bougourd and Jones, 1997) and are quite stable during meiosis (Bougourd and Parker, 1979a; Holmes and Bougourd, 1989). They may exist in 55–65% of the population and more, where they contribute to nucleotide polymorphism and are involved deeply in germination

Table 9.1. Effect of varieties and hort-management on chive production worldwide.

Country/continent	Source of information	Country	Source of information
Africa	Greenlife (n.d.)	California, USA	Johanstone (2021)
China		Missouri, USA	Whittinghill *et al.* (2016)
Egypt	Algharib *et al.* (2021)	New Jersey USA	Ramu (2012)
Finland	Suojala (2003)	Oregon, USA	Scannell (2003)
Israel	Wininger *et al.* (2003)	Virgin Islands	Palada *et al.* (2004)
Turkey	Arslan *et al.* (2018)		

rates and, consequently, in conferring a selective advantage (Bougourd *et al.*, 1995).

Recent genetic studies are limited. FISH mapping showed that the alliinase gene is located on the long arm of chromosome 4 in *A. schoenoprasum*, the same location as in *A. cepa* (Khrustaleva *et al.*, 2019). Bulb formation in chives was studied by crossing *A. schoenoprasum*, which produces poorly defined bulbs, with the bulb-forming variety foliosum. Progeny were either non-bulbing, intermediate or bulb-forming, indicating the involvement of at least two recessive genes in bulb formation (Xiao *et al.*, 2010). Stevens and Bougourd (1988a) studied the control of flower pigmentation, and reported on three non-linked diallelic genes: W/w – responsible for the presence/absence of pigment, D/d – codes for the distribution of pigment in the petals, and P/p – controls hue expression. There is no evidence of linkage between these three loci.

More information on the heritabilities of various traits and the genetics of some specific enzymes was detailed by Tatlioglu (1993).

4.2. Male sterility

Most breeding has been targeted at the forcing crop, aiming for crops with uniform erect leaves, high yields and exhibiting late flowering. Chives are protandrous but still have a high level of self-pollination and suffer from inbreeding depression. This can be overcome via male-sterile plants to enable F_1 hybrid development (Poulson, 1990).

Five types of cytoplasm are known in chives, three normal, N_1, N_2 and N_3, and two sterility-inducing, S_1 and S_2 (Engelke and Tatlioglu, 2004). In chives, male sterility is seen in two forms. Genetic male sterility (GMS) is controlled by a single nuclear gene with recessive inheritance (Engelke and Tatlioglu, 2000). The fertility of some male-sterile plants is temperature-sensitive, controlled by a dominant allele (t). Production of viable pollen can be restored by exposure to a constant temperature of 24°C. The major male-sterility system in chives is, however, cytoplasmic (CMS) (Tatlioglu and Wricke, 1988; Engelke *et al.*, 2004), which depends on the interaction between the cytoplasm (S) and a single nuclear fertility restoration locus X, which blocks an 18 kd mitochondrial unique polypeptide involved in male sterility in chives. Hence, only Sxx genotypes are sterile (Tatlioglu, 1993). A detailed cytomorphological description of male-sterile flowers was reported by Singh and Kobabe (1969).

4.3. Seed production

The protandrous, nectar-secreting flowers attract a large number of insects, in particular bees, hoverflies and butterflies. Protandry enhances cross-pollination, but a certain rate of geitonogamy (self-pollination) results in severe inbreeding depression (0.718) (Stevens and Bougourd, 1988b). The rates of self-pollination vary with population, time and environment. Stevens and Bougourd (1988b) reported on <10% selfing, while Poulsen (1990) stated that high rates of self-pollination are common in chives, with the consequent inbreeding depression. Hence the appeal of breeding hybrid varieties. Indeed, Tatlioglu and Wricke (1980) reported that hybrids have a superior yield and leaf dimension compared with standard OPs.

Seed production commonly begins from seeds rather than transplants, and yields are harvested for 3–4 years before deterioration due to weed competition and pathogen infestation. The soil must be well drained and pH neutral.

Due to the sizable variation in a cross-pollinated crop and the concomitant high level of inbreeding depression (Stevens and Bougourd, 1988b; Poulsen, 1990; Tatlioglu, 1993), there is an increased tendency to grow commercial chives from hybrid seeds.

Chives are insect-pollinated. The honeybee is especially important for pollination. An isolation distance of 800 m or more should be strictly maintained to avoid cross-contamination of pollen.

Seeds are sown 5–10 mm deep. After emergence, weeds must be controlled using herbicides and hoeing. Seeds are harvested at maturation when the capsules start to open, and black seeds can be seen. This is normally at the beginning of July in Denmark. Umbels are cut and then dried on racks for at least three weeks. After threshing, the seed is further dried. The yield is typically 300 kg/ha in Denmark. In New Zealand, a seed yield of 60 kg/ha is reported. Seeds can become dormant.

Since seeds are not used for food, pathogen control is usually efficient, yet, with time, damages accumulate and production decreases.

5. Biochemistry and Nutraceutics

Chive flavour is milder than that of onion or garlic, yet its therapeutic properties are similar. Thus, chives are used as an alternative in folk medicine (Chopra *et al.*, 1992; Werner, 2006; Zdravković-Korać *et al.*, 2010; Charles, 2013). Negative effects are rarely reported, although digestive problems may occur following overconsumption (Štajner *et al.*, 2011). Chives possess insect-repelling properties, which make them useful in gardens to control pests (Štajner *et al.*, 2011) and they have been reported to have larvicidal properties against the mosquito malarial vector (Sanei-Dehkordi *et al.*, 2019). They also have anti-microbial properties against bacteria that cause food poisoning, while *Lactobacillus acidophilus*, often used as a probiotic, is unaffected (Rattanachaikunsopon and Phumkhachorn, 2008; Restuana Sihombing *et al.*, 2018).

5.1. Composition

Apart from water, carbohydrates and proteins constitute the major components of chive leaves. Additionally, the leaves are rich in many vitamins and minerals (Table 9.2). Volatile sulphur [S-alk(en)yl cysteine sulphoxides (ACSOs)] and glucosides compounds generate the unique flavour due to complex enzymatic and non-enzymatic interactions among them (Chapter 3, this volume). Chives synthesize high levels of beta-carotene and ascorbic acid, flavonoids and antioxidants, and are rich in minerals (Rubatzky and Yamaguchi, 1997; Justasen, 2000; Table 9.2).

Four ACSOs: 2-methyl-2-butenal, 2-methyl-2-pentenal, methyl-propyl disulphide, and dipropyl disulphide were identified in chives' green leaves plus propenyl-propyl disulphide (Chen, 2006) but not allyl disulphide (Wahlroos and Virtanen, 1965).

In chives, of the total 0.19 μmol/g wet (fresh) thiosulfate, the major components are n-propyl groups (~75%), methyl (~10%), and 1-propenyl groups (~12 %) (Block *et al.*, 1992), about the middle level of the edible species of *Allium* (Chen, 2006). The values-band phenotypic expression, however, depends on genetic, ecological and agro-management; hence the season-to-season variation in flavour intensity.

The major precursors for chives' and onion's pungency, and chives' flavour, are mainly the propyl group, and the species' unique odour results from quantitative and qualitative differences in the thiosulfates. The flavour and lachrymatory properties depend on the concentration of n-propyl and 1-propyl they contain, respectively (Chen, 2006).

Chive leaves contain 16.7 mg/10 g−1 f.w. antioxidative flavonoid glycosides, mainly kaempferol glucosides (di- and tri-glycosides), dominant as glucose and galactose. The 3-beta-D-glucosides of kaempferol, quercetin and isorthamnetin were isolated (Starke and Herrmann, 1976), and both PAR and UV-N lights affect their synthesis (Nitz *et al.*, 2004; Chen, 2006). Chive seeds contain glutamyl peptides [N,N'-bis-(γ-glutamyl)-cysteine, N,N'-bis- (γ-glutamyl)-3,3'-(2-methylethylene-1,2-dithio)-dialanine, γ-glutamyl-S-propylcystiene and S-(I-propenyl) cysteinesulfoxide], a substituted amino analog of alliin (Matikkala and Virtanen, 1962, 1963, 1965; Hiromu and Seiji, 1983). Further characterization and structure were published by Fossen *et al.* (2000) and Singh *et al.* (2018).

Table 9.2. Nutritional composition of 100g raw chives. (From USDA Chives, 2019)

Nutrient	Unit	Raw chives/100g	Nutrient	Unit	Raw chives/100g
Water	g	90.6	Vitamin C[3]	mg	58.1
Energy	Kcal/kJ	30/126	Thiamin	mg	0.078
Protein	g	3.27	Riboflavin	mg	0.115
Total lipid (fat)	g	0.73	Niacin	mg	0.647
Carbohydrates[1]	g	4.35	Vitamin B6	mg	0.138
Fibre[2]	g	2.5	Folate, DFE[4]	µg	DFE 105
Sugars, total	g	1.85	Vitamin B12	µg	0.00
Calcium, Ca	mg	92	Vitamin A, RAE[5]	µg	RAE 218
Copper, Cu	mg	0.157	Vitamin A	IU	4353
Iron, Fe	mg	1.6	Vitamin E[6]	mg	0.21
Magnesium, Mg	mg	42	Vitamin D (D_2+D_3)	IU	0.0
Manganese, Mn	mg	0.373	Vitamin D	IU	0
Phosphorus, P	mg	58	Vitamin K[7]	µg	212.7
Potassium, K	mg	296	Saturated FA[8]	g	0.146
Selenium, Se	µg	0.9	Unsaturated FA[9]	g	0.095
Sodium, Na	mg	3	Polyunsaturated[10]	g	0.267
Zinc, Zn	mg	0.56			

[1]by difference; [2]total dietary; [3]ascorbic acid, total; [4]DFE = dietary folate equivalents (DFE); [5]RAE = retinol activity equivalents; [6]α-tocopherol; [7]phylloquinone; [8]total saturated fatty acids; [9]total monounsaturated fatty acids; [10]total polyunsaturated fatty acids.

6. Future Prospects

Chives are a relatively minor crop with a small research community. Growing systems and market usage are well established. Future developments are likely to be in novel uses for the crop in urban and green growing systems, exploiting its environmental resilience. Exploiting the health-related properties of chives, including anti-microbial and anti-inflammatory properties (Parvui *et al.*, 2014), are also areas for future research. As with all vegetable crops, maintaining effective crop protection in chemical-free growing systems and in a warming environment will throw up new challenges for the growing community.

References

Algharib, A.M., El-Gohary, A.E. and Hendawy, S.F. (2021) Response of chive (*Allium schoenoprasum* L.) plant to natural fertilizers. *Journal of Ecological Engineering* 22, 200–208. DOI: 10.12911/22998993/140262.

Arslan, H., Kiremit, M.S. and Güngör, A. (2018) Impacts of different water salinity levels on salt tolerance, water use, yield, and growth of chives (*Allium schoenoprasum*). *Communications in Soil Science and Plant Analysis* 49, 20, 2614–2625. DOI: 10.1080/00103624.2018.1526949.

Block, E., Naganathan, S., Putman, D. and Zhao, S.H. (1992) Allium chemistry: HPLC analysis of thiosulfinates from onion, garlic, wild garlic (Ramsoms), leek, scallion, elephant (Great headed garlic), chive, and Chinese chive. Uniquely high allyl to methyl ratios in some garlic samples. *Journal of Agricultural and Food Chemistry* 40, 2418–2430.

Bougourd, S.M. and Jones, R.N. (1997) B chromosomes: a physiological enigma. *New Phytologist* 137, 43–54. Available at: https://nph.onlinelibrary.wiley.com/doi/pdf/10.1046/j.1469-8137.1997.00823.x (accessed 31 July 2022).

Bougourd, S.M. and Parker, J.S. (1979a) The B-chromosome system of *Allium schoenoprasum*. II. Stability, inheritance and phenotypic effects. *Chromosoma* 75, 369–383.

Bougourd, S.M. and Parker, J.S. (1979b) The B-chromosome system of *Allium schoenoprasum*. III. *An abrupt change in B-frequency. Chromosoma* 75, 385–392. DOI: 10.1007/BF00293479.

Bougourd, S.M., Plowman, A.B., Ponsford, N.R., Elias, M.L., Holmes, D.S. *et al.* (1995) The case for unselfish B chromosomes: evidence from *Allium schoenoprasum*. In: Brandham, P.E. and Bennett, M.D. (eds) *Kew Chromosome Conference IV*. Kew, London.

Charles, D.J. (2013) *Antioxidant Properties of Spices, Herbs and Other Sources*. Springer Verlag, New York, pp. 225–230. DOI: 10.1007/978-1-4614-4310-0_18.

Chen, H. (2006) Chives. In: Peter, K.V. (ed.) *Handbook of Herbs and Spices*. Woodhead Publishing Limited, Sawston, UK, pp. 337–346. DOI: 10.1533/9781845691717.3.337.

Chinnappa, C.C. and Basappa, G.P. (1986) Cytology of some Western Canadian Allium species. *American Journal of Botany* 73, 529–534. Available at: https://www.jstor.org/stable/pdf/2444258.pdf?refreqid=-excelsior%3Ad4967ca0e96f761f033c58a04e2c6289 (accessed 31 July 2022).

Choi, H.J. and Oh, B.U. (2011) A partial revision of *Allium* (Amaryllidaceae) in Korea and north-eastern China. *Botanical Journal of the Linnean Society* 167, 153–211. DOI: 10.1111/j.1095-8339.2011.01166.x.

Chopra, R.N., Nayar, S.L. and Chopra, I.C. (1992) *Glossary of Indian Medicinal Plants*. Council of Scientific and Industrial Research, New Delhi, pp. 11–12.

Da Silva, F.J., Santos, J.A., Silva, M.M.D., Silva, Ê.F.D.F. and Souza, E.R.D. (2019) Water relations of chives in function of salinity and circulation frequency of nutrient solutions. *Revista Brasileira de Engenharia Agrícola e Ambiental.* 23, 359–365. Available at: https://www.scielo.br/j/rbeaa/a/k7Pg4gv7xttmyfCG-P8mshSw/?lang=en (accessed 31 July 2022).

Egert, M. and Tevini, M. (2002) Influence of drought on some physiological parameters symptomatic for oxidative stress in leaves of chives (*Allium schoenoprasum*). *Environmental and Experimental Botany* 48(1), 43–49. DOI: DOI:10.1016/S0098-8472(02)00008-4.

Eksi, M., Sevgi, O., Akburak, S., Yurtseven, H. and Esin, I. (2020) Assessment of recycled or locally available materials as green roof substrates. *Ecological Engineering* 156, 105966. DOI: 10.1016/j.ecoleng.2020.105966.

Engelke, T. and Tatlioglu, T. (2000) Genetic analyses supported by molecular methods provide evidence of a new genic (st1) and a new cytoplasmic (st2) male sterility in *Allium schoenoprasum* L. *Theoretical and Applied Genetics* 101, 478–486. Available at: https://link.springer.com/content/pdf/10.1007/s001220051506.pdf (accessed 31 July 2022).

Engelke, T. and Tatlioglu, T. (2004) The fertility restorer genes X and T alter the transcripts of a novel mitochondrial gene implicated in CMS1 in chives (*Allium schoenoprasum* L.). *Molecular Genetics and Genomics* 271, 150–160. DOI: 10.1007/s00438-003-0966-4.

Engelke, T., Gera, D. and Tatlioglu, T. (2004) Determination of the frequencies of restorer- and maintainer-alleles involved in CMS1 and CMS2 in German chive varieties. *Plant Breeding* 123, 51–59. DOI: 10.1046/j.1439-0523.2003.00948.x.

Flora of China (n.d.) 24, 195. Science Press, Beijing and Missouri Botanical Garden, St. Louis, Missouri. Available at: http://www.efloras.org/florataxon.aspx?flora_id=2&taxon_id=200027528 (accessed 31 July 2022).

Fölster, E. and Krug, H. (1977) Influence of the environment on growth and development of chives (*Allium schoenoprasum* L.). ii. Breaking of the rest period and forcing. *Scientia Horticulturae* 7, 213–224. DOI: 10.1016/0304-4238(77)90018-8.

Fossen, T., Slimestad, R., Ovstedal, D.O. and Andersen, O.M. (2000) Covalent anthocyanin-flavonol complexes from flowers of chive, *Allium schoenoprasum* L. *Phytochemistry* 54, 317–323. DOI: 10.1016/s0031-9422(00)00102-3.

Fredotovic, Z. and Puizina, J. (2019) Edible allium species: chemical composition, biological activity and health effects. *Italian Journal of Food Science* 31, 19–39. DOI: 10.14674/IJFS-1221.

Friesen, N. and Blattner, F.R. (2000) Geographical isolation predominates over ecological differentiation in the phylogeny of *Allium* section Schoenoprasum (Alliaceae). *Plant Biology* 2, 297–305.

Friesen, N., Fritsch, R.M. and Blattner, F.R. (2006) Phylogeny and new intrageneric classification of *Allium* (Alliaceae) based on nuclear ribosomal DNA ITS sequences. *Aliso* 22, 372–395. DOI: 10.5642/aliso.20062201.31.

Greenlife (n.d.) Chive production. Available at: https://www.greenlife.co.ke/chives-production (accessed 31 July 2022).

Hiromu, K. and Seiji, H. (1983) Two sulfur constituents from *Allium schoenoprasum*. *Phytochemistry* 22, 294–295. DOI: 10.1016/S0031-9422(00)80109-0.

Holmes, D.S. and Bougourd, S.M. (1989) B-chromosome selection in *Allium schoenoprasum*. I. Natural populations. *Heredity* 63, 83–87. DOI: 10.1038/hdy.1989.78.

Houben, A. (2017) B chromosomes – a matter of chromosome drive. *Frontiers in Plant Science.* 8: article 210. DOI: 10.3389/fpls.2017.00210.

Johanstone, G. (2021) How to grow and care for chives. Available at: https://www.thespruce.com/chives-plant-care-growing-guide-5078626 (accessed 31 July 2022).

Justasen, U. (2000) Negative atmospheric pressure chemical ionisation low-energy collision activation mass spectrometry for the characterisation of flavonoids in extracts of fresh herbs. *Journal of Chromatography A* 902, 369–397. DOI: 10.1016/s0021-9673(00)00861-x.

Kamenetsky, R. and Rabinowitch, H.D. (2006) The genus *Allium*: a developmental and horticultural analysis. *Horticultural Review* 32, 329–378.

Khrustaleva, L., Kudryavtseva, N., Romanov, D., Ermolaev, A. and Kirov, I. (2019) Comparative Tyramide-FISH mapping of the genes controlling flavor and bulb color in *Allium* species revealed an altered gene order. *Scientific Reports* 9, 12007. DOI: 10.1038/s41598-019-48564-9.

Koul, M., Meena, S., Kumar, A., Sharma, P.R. Singamaneni, V. *et al.* (2016) Secondary metabolites from endophytic fungus *Penicillium pinophilum* induce ROS-mediated apoptosis through mitochondrial pathway in pancreatic cancer cells. *Planta Medica* 82, 344–355. DOI: 10.1055/s-0035-1558308.

Mahr, S. (2021) Chive, *Allium schoenoprasum*. The University of Wisconsin-Madison, Madison, Wisconsin.

Martensson, L.M., Fransson, A.M. and Emilsson, T. (2016) Exploring the use of edible and evergreen perennials in living wall systems in the Scandinavian climate. *Urban Forestry & Urban Greening* 15, 84–88. DOI: 10.1016/j.ufug.2015.12.001.

Matikkala, E.J. and Virtanen, A.I. (1962) A new gamma-glutamylpeptide, gamma-l-glutamyl-S-(prop-l-enyl)-lcysteine, in the seeds of chives (*Allium schoenoprasum*). *Acta Chemica Scandinavica* 16, 2461–2462. Available at: http://actachemscand.org/pdf/acta_vol_16_p2461-2462.pdf (accessed 31 July 2022).

Matikkala, E.J. and Virtanen, A.I. (1963) New gamma-glutamylpeptides isolated from seeds of chives (*Allium schoenoprasum*) n, n-bis-(y-glutamyl)-cystine, n, n-bis-(gamma-glutamyl)-3, 3-(2-methylethylene-1, 2-dithio)-dialanine, gamma-glutamyl-s-propylcysteine. *Acta Chemica Scandinavica* 17, 1799–1801. DOI: 10.3891/acta.chem.scand.17-1799.

Matikkala, E.J. and Virtanen, A.I. (1965) y-Glutamylpeptidase (Glutaminase) in germinating seeds of chive (*Allium schoenoprasum*). *Acta Chemica Scandinavica* 19, 1258–1261. DOI: 10.3891/acta.chem.scand.19-1258.

Maxted, N. and Rhodes, L. (2016) Chives. *Allium schoenoprasum var. Scarberrimum Regel. The IUCN Red List of Threatened Species 2016*. Available at: https://dx.doi.org/10.2305/IUCN.UK.2016-3.RLTS.T172256A19391728.en (accessed 31 July 2022).

McCollum, G. (1976) *Allium Liliaceace*. In: Simmonds, N.W. (ed.) *Evolution of Crop Plants*. Longmans, London pp. 186–196.

Nagel, M. and Börner, A. (2010) The longevity of crop seeds stored under ambient conditions. *Seed Science Research* 20, 1–12. DOI: 10.1017/S0960258509990213.

Nitz, G.M., Grubmüller, E. and Schnitzle, W.H. (2004) Differential flavonoid response to PAR and UV-B light in chive (*Allium schoenoprasum* L.). *Acta Horticulturae* 659, 825–830. DOI: 10.17660/ActaHortic.2004.659.107.

Ohri, D., Fritsch, R.M. and Hanelt, P. (1998) Evolution of genome size in *Allium* (Alliaceae). *Plant Systematics and Evolution* 210, 57–86. DOI: 10.1007/BF00984728.

Palada, S.M.A., Davis, M.C., Crossman, A.M., Robles, C. and Chichester, E.A. (2004) Sustainable crop management practices for improving production of culinary herbs in the Virgin Islands. *Acta Horticulturae* 629, 289–298. Available at: https://www.actahort.org/members/showpdf?session=49738 (accessed 31 July 2022).

Parvui, A.E., Parvu, M., Vlase, L., Miclea, P., Mot, A.C. and Silaghi-Dumitrescu, R. (2014) Anti-inflammatory effects of *Allium schoenoprasum* l. Leaves. *Indian Journal of Physiology and Pharmacology* 65, 309–315. PMID: 24781739.

Poulsen, N. (1983) Chives (*Allium schoenoprasum* L.): a literature review. Et litteraturstudium, *Tidsskrift for Planteavls Specialserie* 1657. Available at: https://dcapub.au.dk/pub/S_beretning_1656_1983.pdf (accessed 31 July 2022).

Poulsen, N. (1990) Chives (*Allium schoenoprasum*). In: Rabinowitch, H.D. and Brewster, J.L. (eds) *Onions and Allied Crops. III. Biochemistry, Food Science, and Minor Crops*. CRC Press, Boca Raton, Florida, pp. 231–250.

Pude, R., Hehnen, D. and Blum, H. (2013) Rheinische Friedrich-Wilhelms-Universität Bonn Landwirtschaftliche Fakultät. Forschungsbericht Nr. 173. Available at: https://repository.publisso.de/resource/frl:5497274-1/data (accessed 31 July 2022).

Rabinovich, I. (2009) Vegetative and reproductive development of chives (*Allium schoenoprasum* L.). Thesis, The Hebrew University of Jerusalem.

Ramu, S. (2012) Chives and flowers (*Allium schoenoprasum*). In: *Ethnic Herbs and Greens Crop Production Manual*. Rutgers University, New Brunswick, New Jersey, pp. 52–57. Available at: https://njaes.rutgers.edu/ultra-niche-crops/roselle/Ethnic-Herbs-and-Greens-Crop-Production-Manual-v3.1.pdf (accessed 31 July 2022).

Rattanachaikunsopon, P. and Phumkhachorn, P. (2008) Diallyl sulfide content and antimicrobial activity against food-borne pathogenic bacteria of chives (*Allium schoenoprasum*). *Bioscience, Biotechnology and Biochemistry* 72, 2987–2891. DOI: 10.1271/bbb.80482.

Restuana Sihombing, D., Herla, R., Suryanto, D., Rosa Tampubolon, S. and Yanti, S. (2018) Antimicrobial effects of chive extracts against bacteria pathogen and *Lactobacillus acidophilus*. In: *International Conference on Agribusiness, Food and Agro-Technology*. DOI:10.1088/1755-1315/205/1/012049.

Rubatzky, V.E. and Yamaguchi, M. (1997) *World Vegetables*, 2nd edn. Springer, New York, pp. 325–326.

Sanei-Dehkordi, A., Soleimani-Ahmadi, M., Abadi, Y.S. and Paksa, A. (2019) Wild chive oil is an extremely effective larvicide against malaria mosquito vector *Anopheles stephensi*. *Asian Pacific Journal of Tropical Medicine* 12, 170–174. DOI: 10.4103/1995-7645.257117.

Scannell, E. (2003) *Chives (Allium schoenoprasum)*. Oregon State University, Jackson County Master Gardener Association.

Singh, V.P. and Kobabe, G. (1969) Cyto-morphological investigation on male sterility in *A. schoenoprasum* L. *Indian Journal of Genetics and Plant Breeding* 29, 241–247. Available at: https://www.isgpb.org/documents/archive/12-1.pdf (accessed 31 July 2022).

Singh, V., Chauhan, G., Krishan, P. and Shri, R. (2018) *Allium schoenoprasum* L.: a review of phytochemistry, pharmacology and future directions. *Natural Product Research* 32, 2202–2216. Available at: https://doi.org/10.1080/14786419.2017.1367783 (accessed 31 July 2022).

Solberg, S.Ø., Yndgaard, F., Andreasen, C., von Bothmer, R., Loskutov, I.G. *et al.* (2020) Long-term storage and longevity of orthodox seeds: a systematic review. *Frontiers in Plant Science* DOI: 10.3389/fpls.2020.01007.

Štajner, D., Popović, B.M., Ćalić-Dragosavac, D., Malenčić, Đ., Zdravkovi, Ć. and Korać, S. (2011) Comparative study on *Allium schoenoprasum* cultivated plant and *Allium schoenoprasum* tissue culture organs antioxidant status. *Phytotherapy Research* 25, 1618–1622. DOI: 10.1002/ptr.3394.

Starke, H. and Herrmann, K. (1976.) Flavonole und Flavone der Gemüsearten. VII. Flavonole des Porrees, Schnittlauchs, und Knoblauchs. *Z Lebensm Unters Forch.* 161, 25–30. DOI: 10.1007/BF01145416.

Stevens, J.P. and Bougourd, S.M. (1988a) Genetic analysis of flower colour variation in *Allium schoenoprasum* L. (wild chives). *Heredity* 60, 253–256. DOI: 10.1038/hdy.1988.39.

Stevens, J.P. and Bougourd, S.M. (1988b) Inbreeding depression and outcrossing rate in natural populations of *Allium schoenoprasum* L. (wild chives). *Heredity* 60, 257–261. Available at: https://www.nature.com/articles/hdy198840.pdf (accessed 31 July 2022).

Suojala, T. (2003) Yield potential of chives: effect of cultivar, plastic mulch and fertilization. *Agricultural and Food Science* 12, 95–105. Available at: https://orgprints.org/id/eprint/17091/1/suojala.pdf (accessed 31 July 2022).

Tatlioglu, T. (1993) Chive *Allium schoenoprasum*. In: Kalloo, G. and Bergh, B.O. (eds) *Genetic Improvement of Vegetable Crops*. Pergamon Press, pp. 3–13.

Tatlioglu, T. and Wricke, G. (1980) Genetisch züchterische Untersuchungen am Schnittlauch (*Allium schoenoprasum* L.). *Gartenbauwissenschaft* 45, 278–292. Available at: http://www.jstor.org/stable/43387840 (accessed 31 July 2022).

Tatlioglu, T. and Wricke, G. (1988) Genetic control of tetracycline sensitivity of cytoplasmic male sterility (cms) in chives (*Allium schoenoprasum* L.). *Plant Breeding* 100, 34–40. DOI: 10.1111/j.1439-0523.1988.tb00214.x.

Teuscher, E., Bauermann, U. and Werner, M. (2006) *Medicinal Spices: A Handbook of Culinary Herbs, Spices, Spice Mixtures, and Their Essential Oils*. Wissenschaftliche.

Thomas, G., Astley, D., Boukema, I., Daunay, M.C., Del Greco, A. *et al.* (2005) Report of a Vegetables Network. Joint Meeting with an ad hoc group on Leafy Vegetables, 22–24 May 2003, Skierniewice, Poland. International Plant Genetic Resources Institute, Rome. Available at: http://archive-ecpgr.cgiar.org/fileadmin/bioversity/publications/pdfs/1057_Report_of_a_vegetables_network-cache=1415188977.pdf (accessed 31 July 2022).

Trowbridge Filippone, P. (2007) Chives. In: *The About.com. Guide to Home Cooking*. Adams Media Corporation.

USDA (2019) Chives, raw. Available at: https://fdc.nal.usda.gov/fdc-app.html#/food-details/169994/nutrients (accessed 31 July 2022).

Wahlroos, O. and Virtanen, A.I. (1965) Volatiles from chives (*Allium schoenoprasum*). *Acta Chemica Scandinavica* 6, 1327–1332. Available at: http://actachemscand.org/pdf/acta_vol_19_p1327-1332.pdf.

Werner, M. (2006) *Medicinal Spices: A Handbook of Culinary Herbs, Spices, Spice Mixture and Their Essential Oils*. Medpharm. Scientific Publications, Stuttgart, Germany.

Whittinghill, L.J., Rowe, D.B. and Cregg, B.M. (2013) Evaluation of vegetable production on extensive green roofs. *Agroecology and Sustainable Food Systems* 37(4), 465–484. DOI: 10.3390/agriculture8110168.

Whittinghill, L.J., Bradley Rowe, D., Ngouajio, M. and Cregg, B.M. (2016) Evaluation of nutrient management and mulching strategies for vegetable production on an extensive green roof. *Agroecology and Sustainable Food Systems* 40, 297–318. DOI: 10.1080/21683565.2015.1129011.

Wikipedia (2021) *Allium Thunbergii*. Available at: https://en.wikipedia.org/w/index.php?title=Allium_thunbergii&oldid=1016736857 (accessed 31 July 2022).

Xiao, J., Ureshino, K., Hosoya, M., Okubo, H. and Suzuki, A. (2010) Inheritance of bulb formation in *Allium schoenoprasum* L. *Journal of the Japanese Society for Horticultural Science* 79, 282–286. DOI: 10.5109/26156.

Xu, J. and Kamelin, R.V. (2000) *Allium Linnaeus*, Sp. Pl: 294. 1753. In: Wu, Z.D. and Raven, P.H. (eds) *Flora of China*, vol. 24. Missouri Botanical Garden Press, St. Louis, Missouri, pp. 165–202.

Yamaguchi, M. (1983) *World Vegetables: Principles, Production and Nutritive Values*. Avi Publishing, Westport, Connecticut.

Yin, M. and Tsao, S. (1999) Inhibitory effect of seven *Allium* plants upon three *Aspergillus* species. *International Journal of Food Microbiology* 49, 49–56. DOI: 10.1016/s0168-1605(99)00061-6.

Zdravković-Korać, S., Milojević, J., Tubić, L., Ćalić-Dragosavac, D., Mitić, N. *et al.* (2010) Somatic embryogenesis and plant regeneration from root sections of *Allium schoenoprasum* L. *Plant Cell, Tissue and Organ Culture* 101, 237–244. DOI: 10.1007/s11240-010-9682-z.

10 Crop Development – Chinese Chive

Yoshihiro Imahori*
Osaka Prefecture University, Osaka, Japan

1. Introduction

Chinese chive is the second most important *Allium* leafy vegetable crop in east Asia (Oyuntsetseg *et al.*, 2012). In recent years, it is becoming more economically valuable due to beneficial nutritional and functional compounds for human health (Wang *et al.*, 2020). Chinese chive's popularity has increased throughout the world, over a vast geographic range from Asia through the Middle East to Europe and in North America (Imahori *et al.*, 2004, 2007), especially in many other Asia countries, Caribbean islands, the USA and Europe (Li *et al.*, 2010).

Presumably, Chinese chive originated in China where it has a long history of cultivation. Both leaves and flowering umbels are consumed as vegetables, spices, ornamentals and traditional medicines. It is also grown and commonly used in Korea, India, Thailand and Japan as a spice for soups, salads and other traditional dishes in Asian cuisines and for treating common colds, headaches and cardiovascular diseases (Zhou *et al.*, 2015). Hence, among the alliaceous crops, it is of high economic importance (Fritsch and Friesen, 2002).

Local names vary from Nira, to oriental garlic, Chinese leek, garlic chives and flowering leek.

Chinese chive has a unique flavour and aroma due to thioester compounds (Gao *et al.*, 2018). The vitamins, fibre and mineral-rich leaves have a distinctive mild garlic flavour, and flower buds and young scapes are nutritious. Together they are consumed as foods.

This chapter provides a comprehensive and up-to-date review of Chinese chive as well as current information on its physiology, genetics, breeding, production, processing and health benefits.

2. Taxonomy

Chinese chive is a member of one of the largest monocotyledonous genera: *Allium* – a species-rich, taxonomically complicated genus. It consists of about 60 taxonomic groups at subgeneric, sectional and sub-sectional ranks (Fritsch *et al.*, 2010) and about 1000 accepted species (Govaerts *et al.*, 2021). *Allium* is a member of the family Amaryllidaceae J.St.-Hil, subfamily Alliodeae Herb, and the only in tribe Allieae (Friesen *et al.*, 2006; Li *et al.*, 2010). Chinese chive belongs to the subgenus *Rhizirideum* and comprises two main morphospecies, *Allium tuberosum* Rottler ex Spreng and *Allium ramosum* L., which have

*Address for correspondence: imahori@omu.ac.jp

 Edible Alliums: Botany, Production and Uses (eds H.D. Rabinowitch and B. Thomas)
DOI: 10.1079/9781789249996.0010

been variously considered either as one or two genetic entities (Fritch and Friesen, 2002), as well as various transitional forms (Fritsch and Friesen, 2002; Yamashita *et al.*, 2005, Chapter 1, this volume). The late-flowering *Allium tuberosum* (aka Nira in east Asia), develops small tepals, while the early-flowering *Allium ramosum* (aka Djusai in central Asia) develops large tepals. Molecular analyses, however, indicated that the two species are a sister group to wild species (Fritsch and Friesen, 2002; Blattner and Friesen, 2006; Oyuntsetseg *et al.*, 2012).

Most *A. tuberosum* cultivated strains are tetraploids or triploids, and often develop apomictic seeds.

3. Distribution

Chinese chive origin is not clear (Jones, 1990; Oyuntsetseg *et al.*, 2012) but central-northern Asia and northern and eastern India indigenous plants are assumed to be ancestors of the modern crop (van der Meer, 1997). Domestication occurred in northern China more than three millennia ago (Hanelt, 2001) where populations of *Allium* plants similar to this modern vegetable crop grow wild in steppes and dry meadows in southern Siberia, Mongolia, northern China and North Korea. These wild plants were distinguished as a separate species, *Allium ramosum*, while the vegetable crop is referred to as *Allium tuberosum* (Blattner and Friesen, 2006; Oyuntsetseg *et al.* 2012; Chapter 1, this volume).

4. Morphology

The garlic-like flat-leaf Chinese chive has a distinctive growth habit and the mild flavour reminds one of onions. The cultivated plants differ slightly from the wild types in morphology and life history (Blattner and Friesen, 2006; Pandey *et al.*, 2014). In summer, the former develop umbels with tens of white flowers whose perianth segments have green or yellowish-green midvein, narrow tepals and short filaments. The latter flowers in the autumn and has broader tepals and long filaments. Moreover, there is a marked difference at the molecular level between the two. The former are mostly tetraploids or triploids that bear mostly apomictic seeds while the latter are diploids (Fritsch and Friesen, 2002).

5. Development

Chinese chive propagates by producing horizontally spreading perennial rhizomes and also from seeds (Pandey *et al.*, 2014). Perennial rhizome branches bear many thick fibrous roots and produce clusters of new leaves and shoots each growing season (Takagi, 1994). The smooth, slim, ~5 mm wide, flattish, and slightly keeled green leaves grow as clumps (Saito, 1990). Each leaf clump reaches up to 40 cm in length while bending down at the tip under its weight (Fig. 10.1).

The shoots tillers vigorously, and when reaching the physiological age of six foliage leaves, each forms a new lateral every two or three nodes. In warmer areas, Chinese chive may remain green year-round while in cold climates, aerial parts die back, while only underground roots/rhizomes overwinter and re-sprout in the spring (Pandey *et al.*, 2014).

In late summer, the plants form lateral vegetative bulbs, which vary in size with genotype (Saito, 1990), and produce many white fragrant flowers (Mahr, 2010).

Unlike many other *Allium* species, the rhizomatous, herbaceous, clump-forming perennial plants (Imahori *et al.*, 2004, 2007) form poorly developed firm and fibrous small bulbs attached to a stout rhizome (DeMason, 1990) covered by a pale, brown, fibrous reticulum made of the remains of older leaf bases (Saito, 1990).

The branched rhizomes produce inflorescences every second to fourth leaf (Saito, 1990). Following the apical transition from vegetative to reproductive, sympodial growth continues rapidly from the axillary shoot along the primary shoot axis from the lateral bud just below the inflorescence.

Initiated from the apical bud, the rounded umbel develops, at the top of a sturdy terete, 30–60 cm long angular cross-sectional stalk covered with leaf sheaths at the base. The flat-topped rounded umbel, 5 cm across, consists of small, creamy-white, star-shaped, fragrant, perfect (hermaphrodite) flowers with brown striped tepals that appear in loose clusters (Mahr, 2010).

Fig. 10.1. (a) Chinese chives 'Nira'; (b) Poor bulbs and rhizomes in Chinese chives (Mahr, 2010).

Stigma receptivity and pollen shed occur simultaneously in both *Allium tuberosum* and *Allium ramosum*. The flowers' scent is attractive to many insect pollinators, and following fertilization they set dark, papery capsules filled with up to six hard, triangular-shaped black seeds.

6. Cytogenetics

Alliums were domesticated millennia ago and exhibit a wide phenotypic and genetic diversity due to numerous generations of selections under a variety of environments and cultural habitats.

The cultivated Chinese chive species is a tetraploid (2n=4x=32), but wild plants and double haploid populations are also reported (Sharma and Gohil, 2013). The crop is characterized by mitotic and metacentric complement of the four sets of chromosomes.

In *A. tuberosum* most meiosis cells contain 32 bivalents while ~5% of cells remain univalent. The bivalents separate into two groups of 32 chromosomes at anaphase I and produce egg cells without further division, thus developing into apomictic seeds (Jones, 1990). Based on esterase analyses, Kojima *et al.* (1991) concluded that Chinese chive is a facultative apomict, with a calculated degree of apomixis higher than 90%. In comparison, male meiosis is irregular. It forms a variety of multivalents, i.e. quadrivalents, trivalent bivalents, as well as univalent, which remain unpaired during the meiotic division (Jones, 1990).

In Chinese chive, 2125 simple-sequence repeats (SSRs) were identified, ranging in lengths from 12 to 136 nucleotides. Lengths of many tri-nucleotide repeats ranged from 15 to 30 bp, and a total of 1937 primer pairs were designed from the 2125 SSRs (Zhou *et al.*, 2015). Tang *et al.* (2017) reported on a total of 2553 SSRs from the species' transcriptome, with 626, 643 and 536 of the markers located in coding sequences, 5` and 3` untranslated regions. Among the 2553 SSRs, di-AC/GT and tri-nucleotide

GAA/TTC motifs were the most abundant (52.3% and 45.6%, respectively). PCR amplification indicated that when using 100 SSR primer pairs, 94% of the markers had good quality and 83–88 of markers could be amplified in six other *Allium* species.

7. Physiology

Like many other *Allium* species (Kamenetsky and Rabinowitch, 2002), Chinese chive is a long-day plant. High temperature and long photoperiod are required for flower initiation and development, whereas bolting is inhibited by a photoperiod shorter than 12 h, and seed formation fails at 10 h or less (Kawagishi *et al.*, 2009). A brief interruption with short-day treatment led to the development of only a small number of short flower stalks and flower bud degeneration before emergence, and exposure of floral-initiated plants to an 8-h photoperiod resulted in complete inhibition of florogenesis.

In the Japanese variety Tairyo of Chinese chive, flower initiation and differentiation are induced by days longer than 8.5 h, and flower stalk elongation and inflorescence formation require the same qualitative long photoperiod (Kawagishi *et al.*, 2009). The photoperiodic requirements and response, however, vary with genotype. Hence, south Asian varieties sprout and flower under photoperiods that do not suffice the needs of Japanese cultivars (Saito, 1990).

Short photoperiods induce dormancy in Japanese varieties (Nira) from late autumn to winter, while low winter temperatures promote dormancy break and sprouting (Saito, 1990); whereas in the subtropics, Chinese chive is evergreen (Pandey *et al.*, 2014).

8. Cultivation

Chinese chive is commonly farmed in open fields. The plants are multiplied by seeds or by old plants' rhizomes which produce numerous pseudostems with 4–5 leaves each. The optimum germination temperature is 15–20°C, and 20°C for foliage development. In the early spring, seedlings are raised in nurseries for three or more months before transplanting in the production fields (Fig. 10.2) as annuals with four or five summer harvests of green leaves, as quality deteriorates with time. Yet, perennial three years' farming is also common (Takagi, 1994).

Farming of climate-adapted varieties in rich, well-fertilized, black mulched fields resulted in a marked yield and quality increase (Adamczewska-Sowińska and Turczuk, 2016).

Chinese chive grows well in hot conditions and plants bloom in the late summer, while growth is slow in cold climates. Hence, winter production is practised in plastic tunnels or greenhouses for foliage development and autumn/winter harvest of weak-dormant cultivars (e.g. Japan; Fig. 10.2; Takagi, 1994). Winter crops supply the autumn to early spring demand yet chilling conditions and low light intensities lead to low yields of inferior quality and chemical composition compared with the open fields' spring and autumn crops (Saito, 1990).

Fig. 10.2. Chinese chives 'Nira' growing (a) in plastic greenhouse; (b) in open fields.

In China, Chinese chive plantations are farmed for 7–20/30 years as a perennial crop. Initially, seeds are sown 10 cm deep, 25 cm apart, in 1 m wide beds with a restricted watering regime to promote deep rooting. No harvest is practised in the first year to maximize root reserves. Since its roots tend to grow upward, the bed surface should be hilled up with 1–5 cm soil (van der Meer, 1997; Saito, 1990). Early in the second season, older and decayed leaves are removed.

In the cold seasons, weak-dormant cultivars are raised indoors (polythene tunnels/glasshouses) to supply the demand from late autumn to early spring (Takagi, 1994).

In China, blanched leaves are favoured. To that end, the pseudostem is hilled up with 4–6 cm sandy soil to blanch the leaf sheaths, while etiolated blades are produced by dark growth (light exclusion) under straw mulching, roofing tiles or dark paper cover. In the winter, in colder areas, blanching is practised in heated greenhouses (Saito, 1990).

In recent years, the Chinese production of Chinese chive increased by adopting a multi-layer plastic greenhouse system for the winter supply (Wang *et al.*, 2020) where beds are protected by glass or plastic structures. In open fields, no autumn harvest is practiced, thus allowing the build-up of root reserves, and winter raking of old and dead roots is common before growth renewal in the spring. Thereafter, 20 cm long blades are harvested, at 20–30 day intervals, and fertilizers are added in-between harvests.

Allium vegetables, such as onion, synthesize organosulphur compounds beneficial to human health. Their levels depend, however, on genotype and environment and are markedly affected by sulphur availability (Perner *et al.*, 2011). Soil nitrogen as NH_4 and NO_3 affects the synthesis of the *S*-compounds precursors, particularly the level of *S*-alk(en)yl-L-cysteine-sulfoxides, provided that sulphur availability is not limited (Coolong and Randle, 2003). Both elements contribute to the richness of organosulphur compounds in Chinese chive (Perner *et al.*, 2011).

9. Breeding and Propagation

Most Chinese chive cultivars are autotetraploid and exhibit a high degree of apomixis with pseudogamous seed development (Yamashita *et al.*, 2012), and cross-breeding is rare (Yamashita *et al.* 2005). Crosses made between tetraploid varieties revealed that parthenogenesis is regulated by a single dominant gene (Nakazawa *et al.* 2006), whose expression depends on the presence of a diplospory gene (Yamashita *et al.* 2012).

Some Japanese (e.g. Nakazawa *et al.*, 2005) and Chinese (e.g. Chengfu *et al.*, 1995) researchers published short accounts on classical breeding of Chinese chive by crossing male-sterile plants with fertile genotypes and selection of excelling offspring, with or without molecular markers.

Using pollen parent-specific Random Amplified Polymorphic DNA (RAPD) markers, Oshima *et al.* (2017) bred a new cultivar, 'Yume Midori', from a segregating cross-bred population. This new cultivar is characterized by a very low crossing rate, high-yielding, broad leaves, long leaf sheath, weak dormancy and fast growth under low temperatures.

US patent 5,367,111 describes means for production of Chinese chive hybrids by culturing fertilized ovule-embryo (Ohsumi and Hyashi, 1994).

Many wild and cultivated forms of *Allium* species, including accessions of *Allium ramosum*, are amphimictic, thus contributing to the increase of Chinese chive genetic variability through cross-breeding, and *Allium scabriscapum*, a member of the secondary gene pool, may also contribute to Chinese chive breeding and improvement (Yamashita *et al.*, 2005).

In China, vigorous, mid-season-flowering, healthy-looking plants are selected for seed production, with >90% apomictic seed (van der Meer, 1997). Ripening of a single umbel and populations spreads over a long time whereas capsules split open and shed seeds on ripening. Hence, mostly unripe seed heads are cut with 30 cm of stem attached, air- or sun-dried, thus allowing the completion of seed maturation, while the early harvest lowers seed loss. Thereafter, mature seeds are threshed and cleaned. The short-lived seeds serve usually in the following spring (Saito, 1990). In Japan, the local cultivar, Nira, is usually propagated from seed in commercial nurseries, yet planting of clump divisions is also practised (Takagi, 1994).

Tissue culture techniques are being employed for clonal multiplication and *in vitro* conservation of many *Allium* species including Chinese chive (Zee *et al.*, 1977; Shuto *et al.* 1993;

Matsuda and Adachi, 1996; Xue *et al.*, 1997; Zhang *et al.*, 2002; Alizadeh *et al.*, 2013; Chapter 2, this volume). *In vitro* propagation serves as an important means in apomictic plants for the development of (somaclonal) variation and transfer of desirable genes in mutated plants for genotypic and phenotypic improvements.

10. Chemical Composition

Dry matter contents of Chinese chive leaves amounts to ~4%, mainly sugars (fructose, sucrose and glucose) and vitamins B_1 and B_2, carotene and vitamin C (Saito, 1990), fibres, minerals and organosulphur compounds with important antimicrobial properties (Imahori *et al.*, 2004; Table 10.1). Cultural conditions, however, have marked quantitative and qualitative effects.

Chinese chive's biological functions and health benefits mainly depend on richness in polyphenols, sulphur-containing compounds, saponins, flavonoids and vitamins A, C and E (Putnik *et al.*, 2019; Riaz *et al.*, 2020). It's essential oils consist of 24 sulphur-containing compounds, four aldehydes, three alcohols, three ketones and hydrocarbons (Yang *et al.* 2011; Zeng *et al.* 2017). Hence, the vegetable serves as an important source of dietary phytochemicals with proven anti-oxidative activity. The main phenolic compounds are gallic acid, catechin, p-coumaric acid, ferulic acid, isoquercitrin and rutin (Riaz *et al.*, 2020). Of the 44 vegetables analysed for concentrations of free-soluble and bound-insoluble phenolic acids, Chinese chive was listed third in hydroxycinnamic acid and hydroxybenzoic acid contents (94.42 μg g^{-1} fresh weight) after purple cabbage and broccoli (Zhang *et al.*, 2019). Hence, together with thiosulfinates and tuberoside M, Chinese chive is functional as an anti-tumor and anti-inflammation agent. The thiosulfinates inhibit the proliferation and activation of human colon cells by the caspase-independent apoptotic pathways, and seed-extracted tuberoside M inhibits the growth of human leukemia cells (Zeng *et al.*, 2017).

Additionally, water-soluble amino-styrene-acrylic acids and their glycosides from leaves have a strong anti-coagulant effect, whereas the butanol extract from the leaves promotes blood coagulation and activates blood circulation and hemostasis (Zeng *et al.*, 2017). Chinese chive roots are rich in two phenylpropane glycosides and four flavonoids (Han *et al.*, 2015), and six spirostanol saponins from the roots have anti-bacterial activity against *Bacillus subtilis* and *Escherichia coli* (Fang *et al.*, 2015). These functional components and molecular anti-cancer mechanisms are exceptionally high among *Allium* species.

Table 10.1. Nutritional composition and organic sulphur compounds of Chinese chive 'Nira' for 100 g^{-1} fresh weight (Yabuki *et al.*, 2010; Ministry of Education, Culture, Sports, Science and Technology, Japan, 2020).

Compound	Content	
Energy	21.0	kcal
Water	92.6	g
Proteins	1.7	g
Total lipids	0.3	g
Carbohydrate	4.0	g
Total sugars	1.7	g
Total dietary fiber	2.7	g
Minerals		
Sodium, Na	1.0	mg
Potassium, K	510.0	mg
Calcium, Ca	48.0	mg
Magnesium, Mg	18.0	mg
Phosphorus, P	31.0	mg
Iron, Fe	0.7	mg
Zinc, Zn	0.3	mg
Cupper, Cu	0.07	mg
Manganese, Mn	0.39	mg
Vitamins		
Vitamin A (retinol activity equivalents)	290	μg
Vitamin E (α-tocopherol)	2.5	mg
Vitamin K	180	μg
Vitamin B_1	0.06	mg
Vitamin B_2	0.13	mg
Niacin	0.6	mg
Vitamin B_6	0.16	mg
Folate	100	μg
Vitamin C	19	mg
Organic sulphur compounds		
Methiin	307	mg
Alliin	43.7	mg

11. Volatile Compounds

Chinese chives have a unique flavour and aroma, due to sulfides, disulfides, trisulfides, tetrasulfides, pentasulfides, oxygenated compounds and terpene volatiles (Pino *et al.*, 2001). Sulphur compounds accounted for 88% of the total volatiles in the leaf extract, of which the most abundant compound is allyl methyl disulfide (39.3%) followed by dimethyl disulfide (15%) and dimethyl trisulfide (12.6%).

Steam distillate of Chinese chive revealed 29 sulphur-containing volatiles, the major ones being dimethyl disulfide (32%), dimethyl trisulfide (16%), allyl methyl trisulfide (10%) and allyl methyl disulfide (8%) (Iida *et al.*, 1983). The differences in reports on volatile composition probably result from a large variation in plant organs (Pino *et al.*, 2001).

When Chinese chives are crushed, a garlic-like distinctive odour immediately develops. A study of leaf volatiles from 12 Chinese varieties, 4 Japanese, and 6 commercial cultivars revealed 11 sulphur-containing volatiles, five disulfides, two trisulfides and two vinyldithins (Yabuki *et al.*, 2010). Sulfides with a methyl group predominated over those with an allyl group among the sulphur compounds detected (Yabuki *et al.*, 2010).

Volatile organosulphur compounds in *Allium* vegetables are reported to play preventive roles in the development of human pathologies, such as cancer, cardiovascular and inflammatory diseases (Yabuki *et al.*, 2010; Chapter 3, this volume). The volatile compounds are produced by enzymatic reactions which also release pungency and non-volatile precursor compounds, *S*-alk(en)yl-L-cysteine sulfoxides, such as methiin (*S*-methyl-L-cysteine sulfoxide), alliin (*S*-(2-propenyl)-L-cysteine sulfoxide), and isoalliin (*S*-(*E*-1-propenyl)-L-cysteine sulfoxide). *Allium* vegetables biosynthesize large amounts of *S*-alk(en)yl-L-cysteine derivatives (Yamazaki *et al.*, 2011) via enzymatic hydrolysis by alliinase (cysteine sulfoxide lyase) which release alk(en)yl sulfenic acids. These sulfenic acids are transformed via continuous reactions to thiosulfinates and other organosulphur volatile compounds to produce the characteristic flavours (Yoo and Pike, 1998; Yamazaki *et al.*, 2011).

The alk(en)yl groups are mainly a combination of propyl, 1-propenyl, allyl, and methyl groups, depending on species (Pino *et al.*, 2001). *S*-methylcysteine sulfoxides are the main pungency precursors in Chinese chive and both *S*-2-propenyl(allyl)-L-cysteine sulfoxide and *S*-1-propenyl-L-cysteine sulfoxide were also recorded (Yoo and Pike, 1998). Data from 12 Chinese varieties, 4 Japanese varieties and 6 commercially available cultivars showed greater amounts of methiin than alliin (Yabuki *et al.*, 2010). *S*-methylcysteine sulfoxide, *S*-2-propenyl(allyl)-L-cysteine sulfoxide, and *S*-1-propenyl-L-cysteine sulfoxide in Chinese chive amounted to 80%, 13% and 7% in molar ratios, respectively, hence the Chinese chive specific flavours compared with other *Allium* vegetables (Yamazaki *et al.*, 2011).

In addition to the above, the isolated extracts contained a group of 27 novel volatile sulphur, including sulfide, disulfides, trisulfides and tetrasulfides with ethyl, butyl and pentyl groups (Pino *et al.*, 2001). The main odorous component identified in Chinese chive is thioether compounds with a typical pungent odour that is not characteristic of Chinese chive (Gao *et al.*, 2018).

12. Post-harvest Physiology and Storage

Chinese chives are a very delicate highly perishable vegetable. They lose freshness quickly after harvest, hence have a very limited shelf life, especially in summer.

Initial senescence deterioration symptoms include chlorophyll loss and leaf yellowing, accelerated protein breakdown with the consequent liberation and accumulation of amino acids, and degradation of many compounds essential for sensory and nutritional qualities, such as ascorbate and lipids. The senescence rate depends on several factors including, but not limited to, cultivar, leaf age at harvest, time in storage, temperature and atmosphere composition (Imahori *et al.*, 2004).

The deterioration of scapes is characterized by flower opening, dehydration, yellowing and becoming harder. Respiration rate increases, lignification of senescence tissues is enhanced, the fresh-cut becomes woody and the process progresses upward during transportation and storage (Wu *et al.*, 2009).

Cold storage is most widely used for effective shelf life extension of horticultural crops by deceleration of many metabolic processes (Endo *et al.*, 2019), such as flower opening. Hence, only slightly more than 50% of the stored chives had opened flowers after 18 days of storage at 10°C, and leaf quality remained high for upto 14 days at 0–5°C (Ishii and Okubo, 1984). Others report that Chinese chive stored well for

26 days at about 3°C (Wu, 1998). It is generally recommended, however, to store Chinese chive leaves at 0°C with 95–100% relative humidity for one to two weeks (Thompson, 2014).

Low O_2 storage atmosphere slows down post-harvest senescence by effective reduction of chlorophyll degradation and of free amino acids accumulation with no acetaldehyde and ethanol production, due to an increase in pyruvate decarboxylase and alcohol dehydrogenase.

However, anoxia treatment enhanced CO_2 production – a typical result of anaerobic metabolism, which injured the leaf tissue and generated off-flavours (Imahori *et al.*, 2004).

A CO_2-rich atmosphere retarded post-harvest senescence, delayed leaf yellowing and accumulation of free amino acids. Succinate dehydrogenase (SDH) catalyzes the conversion of succinate into fumarate in the tricarboxylic acid cycle. This enzyme is most significantly inhibited by high CO_2 levels. Yet, 30% CO_2 induced anaerobic metabolism and reduced SDH activity with the consequent injuries to the leaf tissue (Imahori *et al.*, 2007). Thus, the accumulation of succinate results in toxic effects in plant tissues (Mathooko, 1996).

A combination of low O_2 (5%) and high CO_2 (10%) by controlled-atmosphere (CA) storage significantly suppressed respiration rate, yellowing, and decay of Chinese chive leaves (Imahori, 2012), and was more effective in extending shelf life compared to the separate effect of a single gas treatment (Suzuki *et al.*, 2004).

Plastic packing of cold-stored Chinese chive leaves was effective in maintaining freshness. Quality loss, decay and flower opening were reduced to less than 1% days at 3°C in polypropylene bag (film thickness of 0.05 mm) after 26 days in storage (Wu, 1998).

Similar results were obtained by modified-atmosphere packaging (MAP) thus maintaining quality and extending the shelf life of fruits and vegetables by slowing down senescence and preventing pathological disorders (Imahori, 2012). Here, partially sealed package bags, a new MAP type, were more effective in maintaining the freshness of Chinese chive leaves than commercial perforated polypropylene bags (Fig. 10.3).

Fig. 10.3. Partial seal packaging.

This packaging design modifies the non-perforated polypropylene bag's internal atmosphere by partial sealing with alternating 0.4 × 4 mm of fused followed by 0.6 × 4 mm (length × width, respectively) of non-fused part width. The gas permeability of the partially sealed bag was 4110 and 9300 mL m^{-2} $atmosphere^{-1}$ day^{-1} for oxygen and CO_2, respectively (Fig. 10. 4; Suzuki *et al.*, 2003).

After five days of storage below 20°C, O_2 and CO_2 levels in the partially sealed bag were about 5% and 11%, respectively. The contents of sugar and ascorbate remained high and yellowing leaves and decay were markedly weaker than the control (Table 10.2; Suzuki *et al.*, 2003).

A 0.5 μlL^{-1} 1-methylcyclopropene (1-MCP) treatment of Chinese chive scapes significantly delayed weight loss and opening rate of flowers,

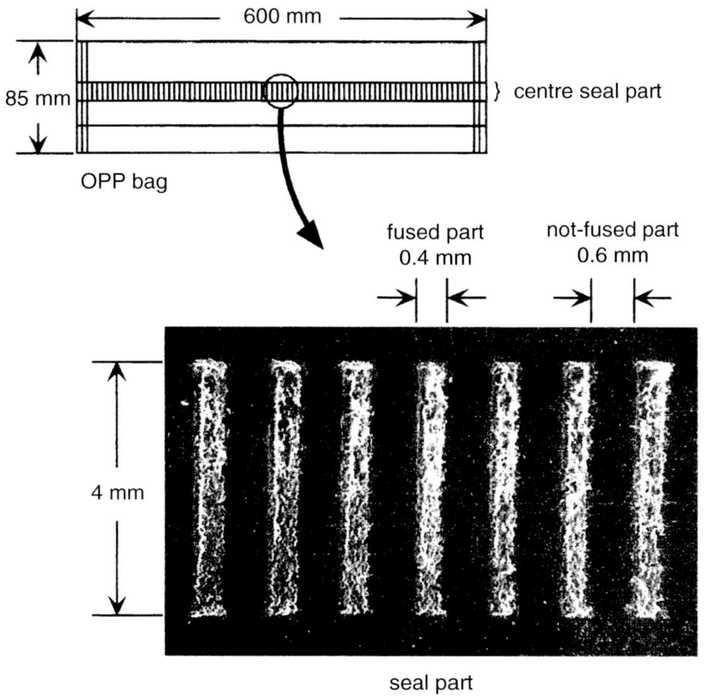

Fig. 10.4. Structure of partial seal packaging (Suzuki *et al.*, 2003).

Table 10.2. Effects of partial seal packaging on in-package gas concentration and freshness of Chinese chive after 5 days storage at 20°C (Suzuki *et al.* 2003).

	Gas concentration (%)*		Occurrence (%)*		Contents (%)[z,**]			
Packaging	O_2	CO_2	Yellow leaves	Decay leaves	Chlorophyll	Sugar	Ascorbate	Off-flavour score[y,*]
Partial seal	5.3 ± 0.3	11.4 ± 0.2	0.6 ± 0.3	0.3 ± 0.2	94.6 ± 1.1	79.5 ± 5.9	83.7 ± 1.4	0.0 ± 0.0
Complete seal	0.8 ± 0.0	8.0 ± 0.2	0.0 ± 0.0	0.3 ± 0.3	93.9 ± 2.2	92.0 ± 6.2	87.7 ± 3.1	2.1 ± 0.2

[z]Percentage of initial (78.9 mg $100g^{-1}$ fresh weight for chlorophyll, 0.921 g $100g^{-1}$ fresh weight for sugar, and 25.2 mg $100g^{-1}$ fresh weight for ascorbate).
[y]Off-flavour was scored from 0 to 4 with 0 = no off-flavour, 1 = slight, 2 = moderate, 3 = severe, 4 = extreme.
[x]Mean ± SE (*: n = 6, **: n = 10).

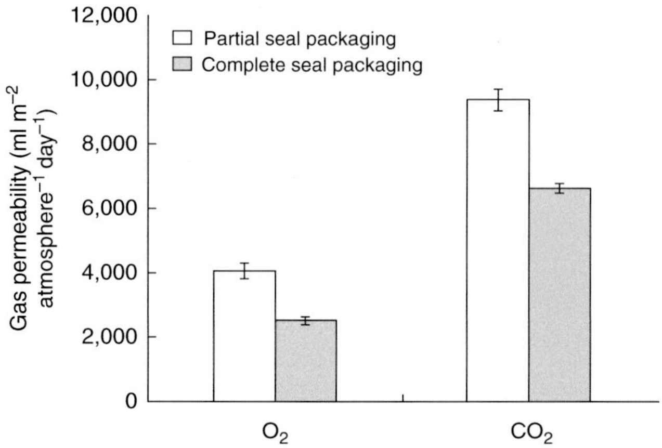

Fig. 10.5. Gas permeability of partial seal packaging and complete seal packaging at 20°C. The vertical bars represent standard error (n=10) (Suzuki *et al.*, 2003).

maintained higher chlorophyll and ascorbate contents, reduced respiration rate, and had lower activities of phenylalanine ammonia-lyase, cinnamyl alcohol dehydrogenase, and peroxidase. 1-MCP treatment also slowed down lignin and cellulose accumulation, thus maintaining the post-harvest quality of these scapes (Wu *et al.*, 2009).

References

Adamczewska-Sowińska, K. and Turczuk, J. (2016) Yielding and biological value of garlic chives (*Allium tuberosum* Rottl. ex Spreng.) depending on the type of mulch. *Journal of Elementology* 21, 7–19. DOI:10.5601/jelem.2015.20.2.910mentol.

Alizadeh, B., Royandazagh, S.D., Khawar, K.M. and Özcan, S. (2013) Micropropagation of garlic chives (*Allium tuberosum* Rottl. ex Sprang) using mesocotyl axis. *Journal of Animal and Plant Sciences* 23, 543–549. Available at: https://www.researchgate.net/publication/288634097 (accessed 20 July 2022).

Blattner, F.R. and Friesen, N. (2006) Relationship between Chinese chive (*Allium tuberosum*) and its putative progenitor *A. ramosum* as assessed by random amplified polymorphic DNA (RAPD). In: Zeder, M.A., Bradley, D., Emshwiller, E. and Smith, B.D. (eds) *Documenting Domestication: New Genetic and Archaeological Paradigms*. University of California Press, Berkeley, California, pp. 133–142. Available at: https://www.degruyter.com/document/doi/10.1525/9780520932425-013/html (accessed 20 July 2022).

Chengfu, T., Peifen, G., Fengyue, M. and Qun, W. (1995) Breeding and utilization of male sterile lines in Chinese chive. *Acta Horticulturae* 402, 423–430. DOI: 10.17660/ActaHortic.1995.402.70.

Coolong, T.W. and Randle, W.M. (2003) Sulfur and nitrogen availability interact to affect the flavor biosynthetic pathway in onion. *Journal of the American Society for Horticultural Science* 128, 776–783. DOI: 10.21273/JASHS.128.5.0776.

DeMason, D.A. (1990) Morphology and anatomy of *Allium*. In: Rabinowitch, H.D. and Brewster, J.L. (eds) *Onions and Allied Crops*, vol. I. CRC Press, Boca Raton, Florida, pp. 27–52. Available at: https://www.taylorfrancis.com/chapters/edit/10.1201/9781351075169-2/morphology-anatomy-allium-darleen-demason (accessed 20 July 2022).

Endo, H., Ose, K., Bai, J. and Imahori, Y. (2019) Effect of hot water treatment on chilling injury incidence and antioxidative responses of mature green mume (*Prunus mume*) fruit during low-temperature storage. *Scientia Horticulturae* 246, 550–556. DOI: 10.1016/j.scienta.2018.11.015.

Fang, J.P., Cai, Y.S., Li, L., Wang, Y., Xiao, H. *et al.* (2015) Spirostanol steroids from the roots of *Allium tuberosum*. *Steroids* 100, 1–4. DOI: 10.1016/j.steroids.2015.03.015.

Friesen, N., Fritsch, R.M. and Blattner, F.R. (2006) Phylogeny and new intrageneric classification of *Allium* (Alliaceae) based on nuclear ribosomal DNA its sequences. *Aliso* 22, 372–395. DOI:10.5642/aliso.20062201.31.

Fritsch, R.M. and Friesen, N. (2002) Evolution, domestication and taxonomy. In: Rabinowitch, H.D. and Currah, L. (eds) *Allium Crop Science: Recent Advances*. CAB International, Wallingford, UK, pp. 5–30.

Fritsch, R.M., Blattner, F.R. and Gurushidze, M. (2010) New classification of *Allium* L. subg. *Melanocrommyum* (Webb & Berthel) Rouy (*Alliaceae*) based on molecular and morphological characters. *Phyton* 49, 145–220.

Gao, Q., Li, X.B., Sun, J., Xia, E.D., Tang, F. *et al.* (2018) Isolation and identification of new chemical constituents from Chinese chive (*Allium tuberosum*) and toxicological evaluation of raw and cooked Chinese chive. *Food and Chemical Toxicology* 112, 400–411. DOI: 10.1016/j.fct.2017.02.011.

Govaerts, R., Kington, S., Friesen, N., Fritsch, R.M., Snijman, D.A. *et al.* (2005–2021) World checklist of Amaryllidaceae (2021). Available at: http://apps.kew.org/wcsp (accessed 20 July 2022).

Han, S.H., Suh, W.S., Park, K.J., Kim, K.H. and Lee, K.R. (2015) Two new phenylpropane glycosides from *Allium tuberosum* Rottler. *Archives of Pharmacological Research* 38, 1312–1316. DOI: 10.1007/s12272-015-0603-x.

Hanelt, P. (2001) Alliaceae. In: Hanelt, P. (ed.) *Mansfeld's Encyclopedia of Agricultural and Horticultural Crops*, vol. 4, 3rd edn. Springer-Verlag, Vienna, pp. 2250–2269.

Iida, H., Hashimoto, S., Miyazawa, M. and Kameoka, H. (1983) Volatile flavor components of Nira (*Allium tuberosum* Rottl.). *Journal of Food Science* 48, 660–661. DOI: 10.1111/j.1365-2621.1983.tb10818.x.

Imahori, Y. (2012) Postharvest stress treatments in fruits and vegetables. In: Ahmad, P. and Prasad, M.N.V. (eds) *Abiotic Stress Responses in Plants*. Springer, New York, pp. 347–358. DOI: 10.1007/978-1-4614-0634-1.

Imahori, Y., Suzuki, Y., Kawagishi, M., Ishimaru, M., Ueda, Y. and Chachin, K. (2007) Physiological responses and quality attributes of Chinese chive leaves exposed to CO_2-enriched atmosphere. *Postharvest Biology and Technology* 46, 160–166. DOI: 10.1016/j.postharvbio.2007.04.008.

Imahori, Y., Suzuki, Y., Uemura, K., Kishioka, I., Fujiwara, H. *et al.* (2004) Physiological and quality responses of Chinese chive leaves to low oxygen atmosphere. *Postharvest Biology and Technology* 31, 295–303. DOI: 10.1016/j.postharvbio.2003.09.004.

Ishii, K. and Okubo, M. (1984) The keeping quality of Chinese chive (*Allium tuberosum* Rottler) by low temperature and seal packaging with polyethylene bag. *Journal of the Japanese Society for Horticultural Science* 53, 87–95. DOI: 10.2503/jjshs.53.87.

Jones, R.N. (1990) Cytogenetics. In: Rabinowitch, H.D. and Brewster, J.L. (eds) *Onions and Allied Crops*, vol. I. CRC Press, Boca Raton, Florida, pp. 199–214. Available at: https://www.taylorfrancis.com/chapters/edit/10.1201/9781351075169-10/cytogenics-jones?context=ubx&refId=0b0fb4ec-61f2-4fb1-bb7a-4bf3b1c2b42b (accessed 20 July 2022).

Kamenetsky, R. and Rabinowitch, H.D. (2002) Florogenesis. In: Rabinowitch, H.D. and Currah, L. (eds) *Allium Crop Science: Recent Advances*. CAB International, Wallingford, UK, pp. 31–57.

Kawagishi, K., Abe, T., Ubukata, M. and Kato, S. (2009) Inhibition of flower stalk elongation and abnormal flower development by short-day treatment in Japanese variety of Chinese chive (*Allium tuberosum* Rottler ex Sprengel). *Scientia Horticulturae* 119, 197–202. DOI: 10.1016/j.scienta.2008.07.018.

Kojima, A., Nagatom, Y. and Hinata, K. (1991) Degree of apomixis in Chinese chives (*Allium tuberosum*) estimated by esterase isozyme analysis. *Japanese Journal of Breeding* 41, 73–83. DOI: 10.1270/jsbbs1951.41.7.

Li, Q.Q., Zhou, S.D., He, X.J., Yu, X., Zhang, Y.C. *et al.* (2010) Phylogeny and biogeography of *Allium* (Amaryllidaceae: Allieae) based on nuclear ribosomal internal transcribed spacer and chloroplast *rps*16 sequences, focusing on the inclusion of species endemic to China. *Annals of Botany* 106, 709–733. Available at: https://academic.oup.com/aob/article/106/5/709/142613 (accessed 20 July 2022).

Mahr, S. (2010) Garlic chives, *Allium tuberosum*. Available at: https://hort.extension.wisc.edu/articles/garlic-chives-allium-tuberosum (accessed 20 July 2022).

Mathooko, F.M. (1996) Regulation of respiratory metabolism in fruits and vegetables by carbon dioxide. *Postharvest Biology and Technology* 9, 247–264. DOI: 10.1016/S0925-5214(96)00019-1.

Matsuda, Y. and Adachi, T. (1996) Plant regeneration via embryogenesis in commercial cultivars of Chinese chive (*Allium tuberosum* Rottl.). *Plant Science* 119, 149–156. DOI; 10.1016/0168-9452(96)04452-4.

Ministry of Education, Culture, Sports, Science and Technology, Japan (2020) Standard Tables of Food Composition in Japan. Available at: www.mext.go.jp/a_menu/syokuhinseibun/index.htm (accessed 20 July 2022).

Nakazawa, Y., Namai, K., Sakai, M., Tasaki, K. *et al.* (2005) Selection of hybrid plants from cross populations of Chinese chives (*Allium tuberosum* Spr.) by RAPD marker. *Bulletin of the Tochigi-Ken Agricultural Experimental Station* 55, 27–32. Available at: https://agris.fao.org/agris-search/search.do?recordID=JP2006006273 (accessed 20 July 2022).

Nakazawa, Y., Namai, K., Kojima, A., Kobayashi, S., Tasaki, K. and Amagai, K. (2006) Estimating the inheritance for parthenogenesis in tetraploid Chinese leek (*Allium ramosum,* syn. *A. tuberosum*). *Breeding Research* 8, 89–98. DOI: 10.1270/jsbbr.8.89.

Oshima, K., Aoki, M., Sato, R., Saito, Y., Nihei, Y. *et al.* (2017) Development of a novel cultivar, 'Yume Midori', via the use of pollen parent-specific DNA markers in Chinese chive. *Horticulture Research* 16, 401–408. DOI: 10.2503/hrj.16.401.

Ohsumi, C. and Hyashi, T. (1994) Hybrid plants of onion and garlic or Chinese chive and method for breeding and propagating the same. *United States Patent* 5,367,111. Available at: https://patentimages.storage.googleapis.com/2f/86/21/25a7b566e430e8/US5367111.pdf (accessed 20 July 2022).

Oyuntsetseg, B., Blattner, F.R. and Friesen, N. (2012) Diploid *Allium ramosum* from East Mongolia: a missing link for the origin of the crop species A. *tuberosum*? *Erforschung biologischer Ressourcen der Mongolei* 12, 415–424. Available at: https://digitalcommons.unl.edu/biolmongol/38/ (accessed 20 July 2022).

Pandey, A., Pradheep, K. and Gupta, R. (2014) Chinese chives (*Allium tuberosum* Rottler ex Sprengel): a home garden species or a commercial crop in India. *Genetic Resources and Crop Evolution* 61, 1433–1440. DOI: 10.1007/s10722-014-0144-z.

Perner, H., Schwarz, D., Krumbein, A. and George, E. (2011) Influence of sulfur supply, ammonium nitrate ratio, and arbuscular mycorrhizal colonization on growth and composition of Chinese chive. *Scientia Horticulturae* 130, 485–490. DOI: 10.1016/j.scienta.2011.07.028.

Pino, J.A., Fuentes, V. and Correa, M.T. (2001) Volatile constituents of Chinese chive *(Allium tuberosum* Rottl. ex Sprengel) and Rakkyo (*Allium chinense* G. Don). *Journal of Agricultural and Food Chemistry* 49, 1328–1330. DOI: 10.1021/jf9907034.

Putnik, P., Gabrić, D., Roohinejad, S., Barba, F.J., Granato, D. *et al.* (2019) An overview of organosulfur compounds from *Allium* spp.: from processing and preservation to evaluation of their bioavailability, antimicrobial, and anti-inflammatory properties. *Food Chemistry* 276, 680–691. DOI: 10.1016/j,foodchem.2018.10.068.

Riaz, A., Lagnika, C., Luo, H., Nie, M., Dai, Z. *et al.* (2020) Effect of Chinese chives (*Allium tuberosum*) addition to carboxymethyl cellulose based food packaging films. *Carbohydrate Polymers* 235, 115944. DOI: 10.1016/j.carbpol.2020.115944.

Saito, S. (1990) Chinese chives *Allium tuberosum* Rottl. In: Brewster, J.L. and Rabinowitch, H.D. (eds) *Onions and Allied Crops*, Volume III, CRC Press, Boca Raton, Florida, pp. 219–230. Available at: https://www.taylorfrancis.com/chapters/edit/10.1201/9780429355752-12/chives-allium-schoenoprasum-poulsen?context=ubx&refId=0ad87537-5e3b-493c-9032-8f620fe9d930 (accessed 20 July 2022).

Sharma, G. and Gohil, R.N. (2013) Double hypoploid of *Allium tuberosum* Rottl. ex Spreng. (2n = 4x = 30): its origin and cytology. *Genetic Resources and Crop Evolution* 60, 2283–2292. DOI: 10.1007/s10722-013-9995-y.

Shuto, H., Abe, T. and Sasahara, T. (1993) *In vitro* propagation of plants from root apex-derived calli in Chinese chive (*Allium tuberosum* Rottler) and garlic (*Allium sativum* L.). *Japanese Journal of Breeding* 43, 349–354. DOI: 10.1270/jsbbs1951.43.349.

Suzuki, Y., Okabayashi, H., Ishikawa, Y., Imahori, Y. and Ueda, Y. (2003) Partial seal packaging as a new type of modified atmosphere packaging for keeping the freshness of Chinese chive (*Allium tuberosum* Rottler) (Studies on the Development of Partial Seal Packaging, part I). *Food Preservation Science* 29, 141–146. DOI: 10.5891/jafps.29.141.

Suzuki, Y., Nagata, M., Imahori, Y. and Ueda, Y. (2004) Effects of low oxygen, high carbon dioxide and ethylene on respiration and quality of Chinese chives (*Allium tuberosum* Rottler) (Studies on the Development of Partial Seal Packaging, part II). *Food Preservation Science* 30, 173–177. DOI: 10.5891/jafps.30.173.

Takagi, H. (1994) Vegetables peculiar to Japan. In: *XXIVth International Horticultural Congress Publication: Horticulture in Japan*. Asakura, Tokyo, pp. 111–112.

Tang, Q., Yi, L., Yuan, X. and Li, F. (2017) Large-scale development, characterization, and cross-amplification of EST-SSR markers in Chinese chive. *Genetics and Molecular Research* 17, gmr16039861. DOI:10.4238/gmr16039861.

Thompson, A.K. (2014) *Fruit and Vegetables: Harvesting, Handling and Storage*, Volume 2, 3rd edn. Wiley Blackwell, Chichester, UK.

van der Meer, Q.P. (1997) Old and new crops within edible Allium. *Acta Horticulturae* 433, 17–31. DOI: 10.17660/ActaHortic.1997.433.1.

Wang, C., Lv, J., Coulter, J.A., Xie, J., Yu, J. *et al.* (2020)Slow-release fertilizer improves the growth, quality, and nutrient utilization of wintering Chinese chives (*Allium tuberosum* Rottler ex Spreng.). *Agronomy* 10, 381. DOI: 10.3390/agronomy10030381.

Wu, C., Du, X., Wang, L., Wang, W., Zhou, Q. *et al.* (2009) Effects of 1-methylcyclopropene on postharvest quality of Chinese chive scapes. *Postharvest Biology and Technology* 51, 431–433. DOI: 10.1016/j.postharvbio.2008.08.005.

Wu, P. (1998) Studies on postharvest physiology of Chinese chive scapes. *Acta Horticulturae* 467, 379–386. DOI: 10.17660/ActaHortic.1998.467.43.

Xue, H.M., Araki, H., Kanazawa, T., Harada, T. and Yakuwa, T. (1997) Callus formation and plantlet regeneration through *in vitro* culture of immature embryo and seedling in Chinese chive (*Allium tuberosum* Rottler). *Journal of the Japanese Society for Horticultural Science* 66, 353–358. DOI: 10.2503/jjshs.66.353.

Yabuki, Y., Mukaida, Y., Saito, Y., Oshima, K., Takahashi, T. *et al.* (2010) Characterization of volatile sulphur-containing compounds generated in crushed leaves of Chinese chive (*Allium tuberosum* Rottler). *Food Chemistry* 120, 343–348. DOI: 10.1016/j.foodchem.2009.11.028.

Yamashita, K., Tsukazaki, H. and Kojima, A. (2005) Interspecific hybrids between amphimictic diploid Chinese chive (*Allium ramosum* L.) and *A. scabriscapum* Bois. et Ky. *Journal of the Japanese Society for Horticultural Science* 74, 127–133. DOI: 10.2503/jjshs.74.127.

Yamashita, K., Nakazawa, Y., Namai, K., Amagai, M., Tsukazaki, H. *et al.* (2012) Modes of inheritance of two apomixis components, diplospory and parthenogenesis, in Chinese chive (*Allium ramosum*) revealed by analysis of the segregating population generated by back-crossing between amphimictic and apomictic diploids. *Breeding Science* 62, 160–169. DOI: 10.1270/jsbbs.62.160.

Yamazaki, Y., Iwasaki, K., Mikami, M. and Yagihashi, A. (2011) Distribution of eleven flavor precursors, *S*-alk(en)yl-L-cysteine derivatives, in seven *Allium* vegetables. *Food Science and Technology Research* 17, 55–62. DOI: 10.3136/fstr.17.55.

Yang, M.Y., Zheng, F.P., Duan, Y. and Xie, J.C. (2011) Analysis of volatiles in wild Chinese chive flowers by solvent extraction/solvent-assisted flavor evaporation coupled with gas chromatography-mass spectrometry. *Food Science* 32, 211–216.

Yoo, K.S. and Pike, L.M. (1998) Determination of flavor precursor compound *S*-alk(en)yl-L-cysteine sulfoxides by an HPLC method and their distribution in *allium* species. *Scientia Horticulturae* 75, 1–10. Available at: https://www.sciencedirect.com/science/article/pii/S0304423898001071?via%3Dihub (accessed 20 July 2022).

Zee, S.Y., Fung, A. and Yue, S.B. (1977) Tissue culture and differentiation of Chinese chive. *HortScience* 12, 264. Available at: https://agris.fao.org/agris-search/search.do?recordID=US7855262 (accessed 20 July 2022).

Zeng, Y., Li, Y., Yang, J., Pu, X., Du, J. *et al.* (2017) Therapeutic role of functional components in Alliums for preventive chronic disease in human being. *Evidence-Based Complementary and Alternative Medicine. Article ID* 9402849. DOI: 10.1155/2017/9402849.

Zhang, L., Li, Y., Liang, Y., Liang, K., Zhang, F. *et al.* (2019) Determination of phenolic acid profiles by HPLC-MS in vegetables commonly consumed in China. *Food Chemistry* 276, 538–546. DOI: 10.1016/j.foodchem.2018.10.074.

Zhang, S., Da, K. and Chengxing, C.L. (2002) Efficient plant regeneration via root tip culture of Allium tuberosum Rottl.ex Spreng. *Acta Horticulturae Sinica* 29, 141–144.

Zhou, S.M., Chen, L.M., Liu, S.Q., Wang, X.F. and Sun, X.D. (2015) *De novo* assembly and annotation of the Chinese chive (*Allium tuberosum* Rottler ex Spr.) transcriptome using the Illumina platform. *PLOS ONE* 10, Article ID e0133312. DOI: 10.1371/journal.pone.0133312.

11 Agronomy and Production of Edible Alliums

Irwin L. Goldman*
Department of Horticulture, University of Wisconsin-Madison, Wisconsin, USA

1. Domestication, Distribution and Founding Populations

1.1. Origins

Bulb onion and shallot originated in rocky sites with shallow soils in the Irano-Turanian floristic region (Fig. 11.1) with the Mediterranean as a secondary centre of origin (Hanelt, 1990; Fritsch and Friesen, 2002). Domestication of *Allium cepa* started with some of the earliest human civilizations (Chapter 1, this volume). Like all Alliums, they form characteristic organosulphur compounds derived from cysteine sulfoxides that provide unique flavours (Brewster, 2008; Chapter 3, this volume).

Evolution within the subgenus *Cepa* section *Cepa* resulted in the formation of a vertical rhizome, a disc-like stem, from which leaves and roots grow. Crops from sections *Cepa* and *Schoenoprasum* have thickened basal leaf sheaths along with thickened bladeless leaves (prophylls, e.g. onion), or clusters of thickened bladeless storage buds (cloves) surrounded by dry scales (garlic), or clusters of thickened basal leaf sheaths and thickened bladeless leaves surrounded by dry scales (shallot). They may possess basal leaves and very reduced pseudostems arising from the true stem (chive, Japanese bunching onion, Chinese chive), and in leek, concentric sheaths from substantial intercalary growth, resulting in a storage pseudostem.

Onions were popular in ancient Egypt, Greece and Rome, and spread from there into northern Europe (Brewster, 2008, Fig. 11.1), Russia by the 13th century, and the New World in the 16th century. Introductions by early English colonists to North America (Goldman *et al.*, 2000) became the basis of long day (LD) onion cultivars in temperate regions (Magruder *et al.*, 1941a, b). Later, some of these onions spread west and eventually into Asia from Europe and North America (Havey, 2018). Possibly due to daylength sensitivity for bulbing and limited channels for crop introduction, only a small group of founding populations comprise the majority of global diversity (Magruder *et al.*, 1941a, b; Goldman *et al.*, 2000; Havey and Ghavami, 2018). Key founding populations: bulb onion 'Early Grano' and 'Grano' were selected from cvs 'Valencia' and 'Babosa', respectively (Goldman *et al.*, 2000; Havey and Ghavami, 2018, Fig. 11.1). Long-storing New Zealander cvs 'Pukekohe Long Keeper' and Australia's 'Creamgold' were derived from Grano types. The Argentinian cv. 'Valcatorce' and similar other intermediate day (ID) mild populations worldwide were selected from 'Sweet Spanish' populations, e.g. 'Cebolla Valenciana'.

*Address for correspondence: ilgoldma@wisc.edu

DOI: 10.1079/9781789249996.0011

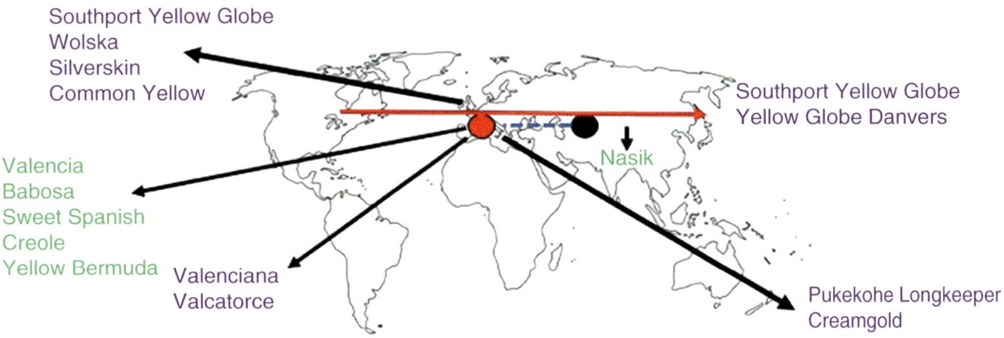

Fig. 11.1. Movement of founding onion populations from primary and secondary centres of origin.

The sources of dehydrator onions and part of the Granex hybrids were the Italian 'Creole' and 'Yellow Bermuda' introductions into the southern USA, respectively. The dehydrated onions include both LD and short day (SD) cultivars, e.g. 'Southport White Globe' and 'White Creole', respectively.

Modern SD germplasm started in the 1940s with 'Texas Early Grano' and an extended range of maturity dates were selected from crosses of 'Texas Grano' with the late-maturing Israeli 'Ben Shemen' (Pike, 1986). In the 1950s, Jones released the F_1 Granex series, a cross between the cytoplasmic-genic sterile 'Bermuda' and 'Grano' for autumn sowing and spring harvest. They are typically mild and store poorly.

All modern US LD germplasm has European origins, e.g. 'Southport Yellow Globe', 'Silverskin', 'Common Yellow' and 'Wolska' and was later distributed to Japan and East Asia. They are sown in the spring, produce long-storing, pungent onions, and harvested in autumn. In northern Europe, the Dutch 'Rijnsburger' is the progenitor of many onion cultivars.

The only key founding population that was not derived from European founders was the Indian SD 'Nasik Red', which arrived in India via a separate path from Central Asia (Havey and Ghavami, 2018).

1.2. Key founding populations: other Alliums

Garlic has been consumed for nearly six millennia (Zohary *et al.*, 2012) while being maintained clonally, due to sterility. True garlic seed has been reported for collections from Central Asia (Etoh, 1986) and through horticultural manipulation (Kamenetsky *et al.*, 2004, Chapter 6, this volume).

Engeland (1991) developed the hardneck and softneck classification system. Typically grown in cooler climates, hardneck (ssp. *ophioscorodon*) garlic bolts and forms a flowering scape before bulbing and later forms a single whorl of large pungent cloves. Commonly produced worldwide, softnecks (ssp. *sativum*) do not form a scape and produce many mild-tasting cloves. Today, the distinction between the groups exists, but the subspecies designations are not always followed (Table 11.1).

Leek, shallot and welsh onion, chives, and Chinese chives are discussed separately in dedicated chapters.

2. Adaptation and Photoperiodic Responses

2.1. Onion bulb formation

Onion production and its unique features have been reviewed extensively (Brewster, 1990a, b, 1997, 2008; Bosch Serra and Currah, 2002; Rabinowitch and Kamentsky Goldstein, 2020; Chapter 4, this volume). Although the bulbing signal is determined by night-length, onion cultivars are categorized by their response to daylength. Hence, LD cultivars require ~15–16 h, while the minimum daylength for ID and SD cultivars is 12–13 h and 11–12 h, respectively.

Table 11.1. Primary types of garlic and their growth habits and qualities (Volk *et al.*, 2004; Drucker, 2020).

Type	Habit	Skin, scape, and bulb characteristics
Artichoke	Softneck	Thick skin, leaves with horizontal orientation
Silverskin	Softneck	Tight skins, vertical leaf orientation, long storability
Rocambole	Hardneck	Several coils in scape, topsets produced, up to 11 cloves per bulb
Purple Stripe	Hardneck	Full loop in coiled scape, easily peeled, up to 12 cloves per bulb
Porcelain	Hardneck	Random curling patters of scape, up to six cloves per bulb
Asiatic/Turban	Hardneck	Reduced scapes, up to 9 cloves per bulb
Creole	Hardneck	Short scapes with many topsets

Bulbing induction is photoperiod-dependent, but interactions with temperature (Garner and Allard, 1920) and other environmental factors strongly affect the process, and the rate of bulbing increases with plant age (Sobeih and Wright, 1986; Chapter 4, this volume). The ratio of the bulb's/pseudostem's maximum and minimum diameters is commonly used to quantify the bulb formation process. Bulbing is considered to have started when the bulbing ratio is 2 and considered complete when it is equal to 8.

Light quality perceived by phytochrome pigments, and specifically a lower ratio of red to far-red (R:FR) light, promotes bulbing (Austin, 1972; Mondal *et al.*, 1986). Shading of leaves by other onion plants or weeds in the canopy promotes bulbing because leaves absorb more red than far-red wavelengths (Brewster, 2008).

Rabinowitch and Kamenetsky Goldstein (2020) have reviewed bulb formation in onion. In brief, under appropriate photoperiods, leaf bud initiation and elongation cease as bulbing is initiated, and at this point the buds become true scales (prophylls) while photosynthates translocate from the oldest leaves to the younger ones and from the blades to the sheaths and through the basal plate to the prophylls (Davis and Jones, 1944). As older leaf sheaths and blades export their stored assimilates, turgor is lost and the neck becomes weak, eventually collapsing. This is known as 'tops down', and the outer empty sheaths become dry tunics.

Bulbing is enhanced by warm temperatures and inhibited by cold and very high temperatures and can be predicted by models that make use of time, temperature and photoperiod (Searle and Reid, 2016). Background genetic effects, however, together with myriad environmental factors, e.g. plant density, fertilizers, irrigation, salinity, soilborne fauna, competition from weeds, and other stresses, affect the rates of bulbing (Rabinowitch and Kamenetsky Goldstein, 2020; Chapter 6, this volume), emphasizing that predictive models have limitations.

Around the equator, bulb formation begins soon after emergence, but low temperatures at high elevation slow down bulbing. At more extreme latitudes, the production season of SD and ID cultivars begins in autumn at higher elevations, and mature bulbs are harvested in late spring to early summer while LD cultivars are typically spring-planted at higher latitudes and harvested in late summer to early autumn (Fig. 11.2).

2.2. Other Alliums

Garlic bulbs consist of several cloves, each made of two leaves and a vegetative bud. Planting densities range from about 15 to 40 plants m^{-2}. Mature bulb weight correlates with clove weight at planting, but the size is smaller at densities >40 plants m^{-2}. Density, however, may be increased in fields for processing where bulb size is not important. Garlic production is similar to onion production in that the crop is very shallowly rooted and has similar fertility needs (see Chapter 4).

Bulb formation is similar to onion bulbs produced from sets (Brewster, 2008; Chapters 4 and 6), and is induced by LD, but long storage and/or field vernalization of less than 15°C (and preferably around 2–5°C) is required. These temperature effects, however, are reversible. In temperate climates, photoperiods and low ratios of red:far-red light are conducive to garlic bulbing when temperatures are warm enough, such as in the late spring. Bulbing is a consequence of

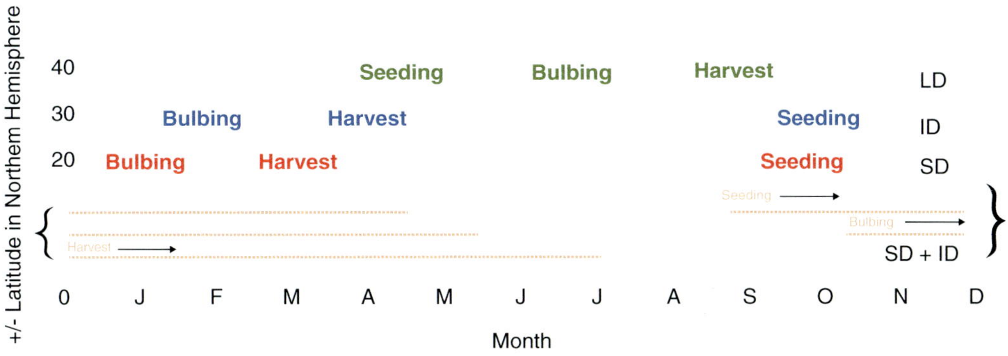

Fig. 11.2. Approximate seeding, bulbing and harvest months for long (LD), intermediate (ID) and short day (SD) onion cultivation at approximately 40° latitude (green), 30° latitude (blue), 20° latitude (red) and below 20° latitude (orange) in the northern hemisphere. Both SD and ID types may be grown below 20° latitude. Seeding of all types may involve the use of seed, transplants or sets, and the latter two approaches may shorten the time to harvest. Seeding of SD and ID types below 20° latitude may take place from September through May.

axial bud development in inner foliage leaves, and the bulbing ratio is ~5 at maturation.

Hardneck clones typically flower before bulbing, as inflorescence initiation occurs before leaf bases begin to swell, hence the apical dominance breaks with the consequent formation of axillary bulbs and cloves (Brewster, 2008). For non-bolting plants, the apical dominance breaks before leaf bases swell. Side-shoots may form under cool temperatures and SDs, or excessive nitrogen. These shoots form stiff necks or cloves that appear rough on their exterior, which is undesirable. Long cold storage before planting may be among the causes of rough bulb formation.

LD promotes bolting and flowering, and, together with warm temperatures, the development of flowers and topsets is promoted, as is the case with onion.

Leek is produced for its pseudostem, which is made from the intercalary growth of a series of non-swollen leaf bases. Leaf initiation at the apex correlates with temperature, yet the growth rate is slow compared to other crops (Brewster and Sutherland, 1993), hence its long growing season. Each successive leaf contributing to pseudostem diameter grows a bit longer than its predecessor though their rate of elongation is essentially the same. The process has been modelled by Hay and Kemp (1992). Flowering in leek is similar to onion, with similar vernalization patterns and bolting influenced by both temperature and daylength (Chapter 8).

Japanese bunching onion is primarily a leaf crop, though pseudostems are also often consumed (Chapter 7). Bolting is undesirable, and induced by vernalizing temperatures similar to an onion, with an optimum of approximately 7°C (Yamasaki *et al.*, 2000). In temperate climates, northern-adapted types have winter dormancy. In warmer regions, growth continues through winter but not during the heat of summer. A third type grows at most temperatures and is not winter-dormant (Inden and Asahira, 1990).

Chives and Chinese chives do not form bulbs (Chapters 9 and 10). In temperate zones, the crops enter winter dormancy as the photoperiod shortens and temperatures decline, but not in the subtropics. Long cold exposure or very high temperatures (e.g. 40°C for 3 hours) can break dormancy. Chinese chives' dormancy can be broken by extended exposure to low temperatures.

2.3. Environmental effects

Similar to the photoperiodic nature of flowering (Taylor *et al.*, 2010), daylength perception in onion, and possibly in all edible Alliums, occurs in leaf tissue with a corresponding response in the apical meristem (Brewster, 2008). Taylor *et al.* (2010) suggested a signal that moves from the leaf to the meristem. Lee *et al.* (2013) demonstrated that the two *FLOWERING LOCUS T* (*FT*) genes produce signalling proteins that regulate flowering and other aspects of plant development in several plant species and also control bulbing and flowering in onion. The *AcFT1* gene is responsible for bulb formation, while *AcFT4* prevents the upregulation of *AcFT1* and prevents bulbing. Onion plants in LD have *AcFT4* downregulated and *AcFT1* upregulated, which results in bulb formation (Fig. 11.3).

Ar Rashid *et al.* (2019) compared the daylength sensitivity of onion plants for bulbing and for flowering. Both processes depend on a leaf-perceived LD signal, followed by a meristematic response that is promoted by far-red light activation of phytochrome; both are reversible. Researchers have therefore suggested that similar genes are involved in the regulation of flowering and bulbing. Various *AcFT* genes were found associated with the perception of daylength in the leaf while the *LFY* genes were associated with leaf basal tissues (Fig. 11.3). Dalvi *et al.* (2016) reported that in an SD cultivar 'JISL-5', *AcFT6* might be involved in the initiation of bulbing with others like *AcFT1*, *AcFT2* and *AcFT3* active in later stages of bulbing.

3. Bulb Characteristics: Colour, Shape, Flavour, Storability and Dormancy

3.1. Colour

Bulb onion colour is determined by alleles at the *R*, *C*, *L*, *G* and *I* loci (Havey, 2018), through the regulation of several enzymes including chalcone synthase, dihydroflavonol 4-reductase and anthocyanidin synthase (El-Shafie and Davis, 1967; Kim *et al.*, 2004, 2005). Yellow and brown pigments are flavonoids, while red and purple pigments are anthocyanins and anthocyanidins, all from the same phenylpropanoid pathway. White and chartreuse colours are produced from the interaction of the above genes.

	Increasing daylength →		Decreasing daylength →	
Year 1				
Season	**Spring**	**Summer**	**Autumn**	**Winter**
Crop stage	Seeding	Bulbing	Bulb Harvest	Vernalization, Dormancy
Key genes	*AcFT4* expressed in leaves, inhibits bulbing *AcLFY* expressed in leaf basal tissues *Go3ox1* expressed in leaves	*AcFT4* suppressed *AcFT1* upregulated to promote bulbing *AcFT5*, *AcFT6*, *Go3ox1* expressed in leaves		*AcFT2* upregulated to promote flowering
Year 2				
Season	**Spring**	**Summer**	**Autumn**	
Crop stage	Bolting	Flowering	Seed Harvest	
Key genes	*AcFT2* expression	*AcFT2* expression		

Fig. 11.3. FLOWERING LOCUS T and GIBBERELLIN OXIDASE 3 genes involved at critical crop stages during vegetative (Year 1) and reproductive (Year 2) phases of the onion life cycle for LD onion production (Lee *et al.*, 2013; Ar Rashid *et al.*, 2019). Similar patterns exist for onion production under SDs, though gene expression differs at various time points. Importantly, for both LD and SD schemes, *AcFT4*, which is expressed in leaves during early crop growth, is inversely correlated with *AcFT1* expression during later crop-growth stages, which serves to promote bulb formation, while *AcFT2* is expressed during bolting and flowering.

Fig. 11.4. Common onion bulb shapes from 19th-century Europe, reprinted from Ernst Benary Album, Tables 16 and 22, respectively, Erfurt, Germany, 1876 (Courtesy TU Berlin Universitatsbibliothek). A (*left column top to bottom*): thick flat, flat, Granex, thick flat; (*middle column*): high globe, spindle; (*right column*): flat, thick flat, flattened globe, thick flat. B (*left column top to bottom*): the flat, thick flat, flattened globe; (*middle column*): globe, thick flat; (*right column*): flat, flat, spindle.

3.2. Shape

Bulb shapes vary from round to thick flat. The shapes and market classes popular in Europe in the 1870s (Fig. 11.4) suggest that similar types have been preferred for the last 150 years. Other than for size grades, no attempt has yet been made to objectively standardize market class shape categories, and they are largely viewed as a function of the genetic background of the variety. Bulb height and shape are highly heritable while diameter and weight have low heritability (McCollum, 1971; Dowker and Fennell, 1974), suggesting that shape is more easily manipulable through breeding than is yield.

3.3. Carbohydrates

Onions accumulate a range of carbohydrates including fructose, glucose, sucrose and polymers of fructose known as inulin (Jaime *et al.*, 2002; Shiomi *et al.*, 2005). The reducing sugars fructose and glucose are inversely correlated with concentration of the storage carbohydrate fructan in the bulb (McCallum *et al.*, 2006). Fructans offer stress protection to the plant (Vijn and Smeekens, 1999; Shiomi *et al.*, 2005) as well as provide some human health benefits since they act like fibre in the intestinal tract and may promote the growth of beneficial microbes in the gut.

A major gene (*Frc*) on chromosome 8 is responsible for high bulb fructan concentrations (McCallum *et al.*, 2006), fructans' degradation by acid invertases and fructan exohydrolases increase during post-harvest storage, and both are associated with the breaking of dormancy (Havey, 2018).

3.4. Flavour

Soil sulphur in the form of sulphate is taken up by onion plants and incorporated into the cytoplasmic S-alkyl cysteine sulfoxides

(ACSOs), which are also known as flavour precursors (Block, 1992; Chapter 3). Upon cells' rupture, the vacuole-stored alliinase lyses the substrate and a cascade is initiated that results in the formation of thiosulfinates. ACSOs' enzymatic lysis forms pyruvic acid (EPY) as a by-product. During EPY formation, propanethial S-oxide is formed from the action of two copies of the lachrymatory factor synthase enzyme, otherwise known as LFS (Imai *et al.*, 2002; Masamura *et al.*, 2012). Propanethial S-oxide causes tearing when cells are broken, and downregulate LFS, resulting in a tearless onion (Eady *et al.*, 2008) with no effect on the amounts of sulphur-based flavour compounds in the bulb.

The common measurement of bulb pungency is pyruvic acid (EPY) (Schwimmer and Weston, 1961), and onion cultivars and production regions are often characterized by EPY levels for marketing purposes and to demonstrate to the consumer how pungent the onion bulb might be.

Different ACSOs accumulate in different types of onions, resulting in different flavour profiles. Pungent onions accumulate more S-1-propenyl – l-cysteine sulfoxide than those with lower pungency (Kopsell *et al.*, 1999). However, it is their by-product, the thiosulfinates, that imparts flavour. Pungency strongly interacts with environmental conditions such as sulphur fertilizer and temperature and is to some extent determined by genetic background (Chapter 4). SD onion cultivars in Georgia (e.g. cv. 'Vidalia'), Texas, and Washington State (e.g. cv. 'Walla Walla') are often considered low-pungency types, in large part because of the careful management of sulphur fertility and water stress.

3.5. Dormancy and storability

Onion bulbs exhibit endodormancy following harvest, which in temperate zones overlaps with the period of cold induction required for flowering. In the tropics, subtropics and centre of origin, Alliums are typically dormant in summer and green in winter. Dormancy facilitates the storability of onion bulbs, a very valuable trait for the marketing of the crop. Certain LD onion cultivars store well for >6 months, while many short-day cultivars have very short dormancy, even at ideal storage conditions.

Storability is influenced by the genetics of the crop, harvest maturity, bulb-curing conditions, the presence of pathogens, storage atmosphere and temperature (Brewster, 2008), and field application of sprout inhibitors, while improved storability was attained with certain biostimulants (Shehata *et al.*, 2017). Although harvest maturity is often associated with storability, Grevsen and Sorenson (2004) showed that, in certain circumstances, early harvest at 25–50% top-down may reduce post-harvest sprouting, while Wright *et al.* (2001) showed that the best bulb quality and storability for New Zealand onions were obtained when harvest took place at 60–80% top-down and bulbs were field-cured followed by foliage removal. Petropolous *et al.* (2017) showed that preventing moisture deficit can result in improved turgor during bulb storage, and excessive nitrogen can cause large necks and reduced storability.

The chemical composition of the bulb during storage changes, e.g. quercetin concentration increased three-fold after one month of storage at 20–25°C (Sharma *et al.*, 2014), and similar increases for pungency and medicinal traits were reported (Debaene *et al.*, 1999).

3.6. Correlated traits

Significant correlations among soluble solids, bulb pungency, bulb dormancy, total dry matter and some medicinally oriented traits have been described (Foskett and Peterson, 1950; Debaene *et al.*, 1999; Galmarini *et al.*, 2001; Sance *et al.*, 2008). LD cultivars are typically long-keeping and high in pungency and soluble solids, possibly due to pleiotropic effects of fructans' polymerization and the consequent carbohydrate concentration (Darbyshire and Henry, 1979; Galmarini *et al.*, 2001). It has also been hypothesized that separate, but closely linked, genes are responsible for these highly correlated traits. A few exceptions exist, hence the mild long-keeping LD onion cultivar 'Evermild' (Scott Hendricks, Seminis Vegetable Seeds); the mild, tearless onion 'Sunion' (Rick Watson of Nunhem's); and the mild-flavoured sweet onion 'Goldies', aka 'Smile Ball' (House Foods Group, Japan) (Keller, 2020). Similarly, the pungency of the USDA release 'Sweet Sandwich' decreases during storage (Peterson *et al.*, 1986). Perhaps the exceptions described above are examples of those linkages being broken, and more research is needed to understand how these traits are genetically governed.

3.7. Vernalization and seed production

Endodormancy was broken when bulbs were soaked in 20% H_2O_2 for 4 h soon after harvest, with consequent rooting (D'Angelo and Goldman, 2018). Provided that these bulbs reached a post-juvenile phase of 10–14 leaves (Heath and Mathur, 1944), they then respond to 14 weeks at 10°C vernalization to promote flowering without selection for bolting susceptibility (Pike, 1986; D'Angelo and Goldman, 2019a) thus shortening the breeding and seed production cycles. This approach would be effective if indoor or controlled-environment seed production was possible. Interestingly, non-vernalized bulbs were able to flower in this study, suggesting that the relationship between vernalization and flowering in onion is facultative rather than obligate.

Breeding and seed production typically require two growing seasons: one from seed to bulb and one from bulb to seed, which onion researchers have sought to compress into a single year (Scully *et al.*, 1945; Chapter 12). Annualization of onion breeding by vernalizing seedlings and inducing flowering before bulbing has been suggested and practised (Scully *et al.*, 1945; Van Kampen, 1970). D'Angelo and Goldman (2019b) managed to achieve annualization and complete a full generation in less than 12 months for all eight varieties examined.

4. Environments, Nutrition and Planting Material

4.1. Soils

Edible Alliums may be grown in a range of soil types including organic soils with pH levels ~4.0, though a pH in the 5–5.5 range is preferred. Onion bulbs are very sensitive to salinity (Wannamaker and Pike, 1987; Shannon and Grieve, 2000) and waterlogging, yet soil environments where moisture can be retained but also provide good drainage are excellent for production. Liming acidic soils is important for improving bulb quality and storability.

Green manure crops and cover crops are often used to improve soil structure and reduce weed pressures. Organic matter promotes soil structure thus benefitting onion production, while bulb production in clay-rich soils is difficult. Peat or muck-type soils with high levels of organic matter and excellent soil structure allow for strong raised-bed formation and provide an excellent growing environment for *Allium* crops (Fig. 11.5).

4.2. Fertility

Compared to many other vegetable crops, onions require fairly high levels of nutrients to perform well (Brewster, 2008), probably due to the relatively small root mass and lack of root hairs, which necessitates higher levels of nutrients for adequate diffusion to the root surface. Many *Allium* crops are similarly shallow-rooted and thus have similar nutrient needs.

Phosphorus and potassium may be applied before seeding. Phosphorus availability is temperature-dependent, and autumn-seeded crops may require less P than winter-seeded ones. The majority of nitrogen and phosphorus uptake, as well as growth in dry weight, takes place during the final half of the growth period; therefore N needs must be carefully managed during the bulbing phase. Yet, nitrogen should not be applied within six weeks of bulb harvest, as yields increase with N availability, but N recovery by the crop is only ~37% of the application rates needed for maximum yield. This inefficiency may be due to the onions' shallow root system and the accelerated root death with ripening. Hence, inevitably, the quantity of residual N in the soil remains high (Brewster, 2008).

In Georgia, nitrogen fertilization at 150 or 205 kg ha^{-1} did not affect yield, regardless of variety or population density (Boyhan *et al.*,

Fig. 11.5. Onion cultivation systems: (a) raised-bed shaping, seeding and fertilization in LD onion production on muck soil in Wisconsin (Photo: I. Goldman); (b) onion production on sandy soils with six rows planted across each raised bed (Photo: D. Brotslaw); (c) beginning the bulbing phase in a four-row production system on sandy soils (Photo: D. Brotslaw); (d) four-row production system on sandy soils in New Mexico (Photo: C.S. Cramer); (e) onion production on mineral soil with six rows planted across each raised bed using a Stanhay precision seeder (Photo: D. Brotslaw); (f) sprinkler irrigation system (Photo: D. Brotslaw).

2009), whereas with cv. 'Bombay Red' in the Tahtai Koraro district of Ethiopia, both N (0, 50, 100 and 150 kg N ha^{-1}) and intra-row spacing (4, 6, 8, 10 cm) significantly affected bolting, leaf length, bulb diameter and marketable yield (Gebretsadik and Dechassa, 2018). The highest marketable yield of 26.72 t ha^{-1} was achieved at 100 kg N and 833,300 plants ha^{-1}.

Knowledge of soil nitrate levels before planting is important, as nitrate applications increase salinity, thus resulting in a smaller size. Hence, many growers split N applications between pre-sowing and post-emergence. Manure applications can supply both nitrogen and phosphorus, but they may also result in high levels of salinity.

Where *Allium* crops are planted in P-rich soils, mycorrhizal fungi may be of limited value (Winkler *et al.*, 2019); however, in organic systems or P-poor soils, vesicular-arbuscular mycorrhizal fungi hyphae enhance phosphorus uptake by the hairless roots, thus providing great benefit for the growth of leek (Sorenson *et al.*, 2005; Wertheim *et al.*, 2014) and garlic (Al-Karaki, 2006).

Micronutrient problems may arise in certain soils. Manganese deficiency results in gradual, interveinal chlorosis, stunting and twisting of the top growth in onions and other alliums. Zinc deficiency is commonly encountered because of its naturally low levels in many production environments, and symptoms include overall plant stunting, a twisting and outward bending of the leaves, and a faint yellowing of the tissue between the veins on the younger leaves (Delahaut and Newenhouse, 1998). Copper and molybdenum deficiencies are also common. The micronutrient deficiencies can be easily corrected with seed coatings.

4.3. Sulphur

Sulphur may have the largest impact of any nutrient on the flavour of *Allium* crops (Freeman and Mossadeghi, 1970, 1971). Increasing sulphur raised ACSO levels in onion and garlic while increasing nitrogen did not (Bloem *et al.*, 2004). Pungency doesn't necessarily respond to increasing sulphur applications because most onion-growing environments contain adequate sulphate. However, mild onions may be produced by reducing sulphate in both soil and water to less than 60 ppm (Randle and Lancaster, 2002). Mild-flavoured onion cultivars can shunt a larger portion of sulphur to sulphate, keeping it separate from the ACSO-flavour-pathway (Chapter 3). Flavour profiles change by the increase of methyl CSO as sulphur fertility decreases, while propyl CSO increases under these conditions (Randle *et al.*, 1995).

Leaf sulphur concentration increased during early plant development but dropped during bulb maturation, and correlations were poor between leaf and bulb sulphur levels and between leaf sulphur levels and pungency (Randle *et al.*, 1993). The longer sulphate was available, the greater bulb pungency remained, and the concentrations of 1-PRENSCO (associated with greater mouth burn when eating onions) increased, especially when sulphate was provided close to harvest (Randle *et al.*, 2002). Hence, reducing sulphate seven weeks before the harvest was sufficient to reduce pungency and prevent the formation of harsher flavours. McManus *et al.* (2012) found that levels of transcripts of the ferredoxin-sulfite reductase (*SiR*) gene were significantly higher in the roots of S-sufficient when compared with S-deficient plants of the pungent inbred line W202A but not the milder cultivar Texas Grano. In both hydroponic and field studies, and when sulphur was sufficient, they found greater enzyme activity in pungent, as compared to less pungent, cultivars. Hence, both genetic and environmental factors underlie the relationship between sulphur fertility and flavour development in onion.

Selenium, an element closely related to sulphur, competes with sulphur for plant uptake. Adding sodium selenate to roots was similar in impact to low levels of sulphur fertility (Kopsell and Randle, 1999). Selenium replaces sulphur in organosulphur compounds, and it is therefore possible to produce mild onions in seleniferous soils, but they may contain metallic flavours.

Organic cultivation of Alliums may benefit from plant-based nitrogen including a variety of legumes, with nitrogen levels of ~3–7% or with animal-based sources with nitrogen levels of 1–22%. Other fertility products may include gypsum as a source of calcium, calcitic or dolomitic lime to raise pH, sulphur to lower pH, potassium sulphate or granite dust as a source of potassium, and rock phosphate as a source of phosphorus (Thompson, 2018).

5. Production Systems and Planting Materials

5.1. Production systems

Comparing organic and conventional production systems for six cropping cycles showed significantly lower yields in the former, up to >40% lower than control. The organic crop, however, was richer in phenols, flavonoids, ascorbic acid and quercetin glucoside. Organic plots had higher levels of microbial populations and soil organic carbon, suggesting longer-term sustainability (Arunachalam *et al.*, 2018).

Allium crops are weak competitors with most weeds, hence weed pressure is the most significant factor in organic production. Soil solarization improves pre-planting weed control. Plastic or compost mulch can also be used to smother weeds before emergence Most weed control in organic systems during early seedling growth is carried out manually, but it may be possible to mechanically cultivate during mid-season.

Organic *Allium* crop production relies on suitable rotations to limit pest and disease pressure, and overhead irrigation systems should be avoided. Certain certified copper-based products may be used as bactericides and certain bacterial products, such as *Bacillus subtilis*, can exhibit fungicidal properties as well. Spinosad (derived from the bacterium *Saccharopolyspora spinosa*) provides some control against thrips and other insects.

Edible Alliums benefit from a homogenous, uncompacted soil structure at the surface. Well-drained soils prevent erosion, facilitate irrigation (Fig. 11.5f), and keep *Allium* crops from becoming waterlogged as high soil moisture can foster pathogen growth. Some well-drained soils, such as sandy soils, may leach nitrogen and require more frequent irrigation, but the benefits of good drainage in coarse-textured soils are often an important advantage over fine-textured soils.

Soil crusting may interfere with onion emergence. Low-intensity irrigation or the use of various products such as plastic, vermiculite, phosphogypsum or polyacrylamide may reduce crusting (Bosch Serra and Currah, 2002). Soil crusting is often avoided on highly organic soils (e.g. Canada and the upper mid-western USA) because of the fine soil structure and relatively moist conditions.

Soil solarization has been used to improve onion and garlic crops by mulching fields with plastic and allowing high temperatures to kill pathogens, and both weeds and weed seeds. Solarization offers a sustainable alternative to pesticides for difficult-to-control pathogens such as white rot *(Sclerotium cepivorum)*. In this system, solarization is accomplished during the season before the *Allium* crop would be planted (Basallote-Ureba and Melero-Vara, 1993).

Rotation is among the most valuable options for sustainable *Allium* crop production, as it reduces the initial amount of soilborne pathogens and may provide benefits from nitrogen credits and weed management. Many farmers attempt a five-year rotation for the most effective disease control.

5.2. Irrigation

The shallow-rooted edible Alliums often require irrigation, depending on rainfall. Furrow irrigation requires more water and is less efficient than sprinkler irrigation. The latter is particularly useful in production fields with a significant slope. It is also generally a favourable method for improving seed germination and seedling growth but less so for later-season growth because leaf wetness may promote foliar diseases (Walker *et al.*, 2009). Drip-irrigation is the most efficient method because of the minimization of water supply and less is lost to evaporation. Irrigation may be scheduled based on calculations of evapotranspiration and estimates (Walker *et al.*, 2009).

Growers in the USA must comply with the Produce Safety Rule, which requires that growers train employees on health and hygiene practices, food safety practices, and safe handling of produce. Water that contacts the harvestable portion of the crop, and soil amendments that touch the harvestable part of the crop, must meet certain safety standards, as do implements, tools, equipment and buildings used for onion production. The Rule also covers packing material used to store the crop and equipment used to handle the crop during storage.

5.3. Planting materials

The three methods of onion planting are direct seeding, transplanting of seedlings and planting of sets (Brewster, 2008). Other than daylength and temperature, perhaps the most significant factor for determining the onion harvest date is the sowing method. For bulb onions, the lowest cost and most common method of direct seeding result in the longest production period, thus providing enough time for full bulb development. Drawbacks of direct seeding are the difficulty in establishing exact population densities due to uneven germination and emergence, the challenges associated with weed management during the early growth of the crop, and the potential need for thinning. Onion seedlings can tolerate very cold temperatures, down to –8°C (Brewster, 2008). Both early spring and late autumn seeding may encounter cooler temperatures, and late autumn seeding is common in the subtropics.

Increasingly, seed conditioning, priming and coating with pesticides, fertilizers and germination enhancers (Pedrini *et al.*, 2020) have become commonplace for direct-seeded onions. Additionally, farmers prefer pellet coating to make the seed more spherical and uniform for flow-through planting machinery. It should be mentioned that *Allium* crops themselves produce valuable seed-priming materials due to fungicidal properties.

Leek germinates poorly in warm conditions and priming using polyethylene glycol and mannitol significantly improve germination (Parera and Cantliffe, 1992).

A shorter growing season or a need for a short period in which the crop is in the field requires the use of transplants that have been produced indoors on-farm or obtained from specialized suppliers. Substantial labour is required to produce, harvest and transplant seedlings; however, they represent a significant shortening of the crop production cycle. Further, transplants may allow the production of larger bulbs than with direct seeding because they constitute a 'head start' on the crop production schedule. The use of transplants guarantees the desired population density, eliminates the need for thinning and improves the opportunities for weed control since the more developed crop is more competitive with weeds from the time it is planted.

Bulb quality is also impacted by the transplants' size. Several studies have demonstrated that multiple seedlings per cell, especially when plants are allowed to grow for up to ten weeks, may result in improved performance (Leskovar and Vavrina, 1999). Hence, groups of onion seedlings are transplanted as a clump (Chung, 1989). Larger cell sizes, however, appear to be correlated with a higher percentage of bolting in some studies.

Onion and shallot sets, typically less than 20 mm in diameter, offer an alternative method that results in the shortest time from planting to harvesting. Sets are produced during a previous growing season and planted in-furrow at the desired stand for the mature crop. They grow rapidly and form large bulbs early in the growing season.

Shallot sets produce multiple shallots whereas onion sets produce a single or doubled bulb (Rabinowitch, 1979). In general, sets retain the same advantages as transplants but produce a crop faster. On the other hand, they may carry their pests and diseases to new fields, compromising crop production.

In two onion cultivars, bolting correlated with set size and storage temperature, and seed yield was positively impacted by cold storage and was highest at 5°C. Optimum storage duration for large sets for 120 days at 5°C provided the highest seed yields (Khokhar, 2009). By contrast, bulb yields were lowest when sets were stored at 5°C and highest when bulbs were stored at 25°C. A series of very detailed studies (Heath and Mathur, 1944) revealed that, regardless of size, the number of swollen scales in an onion set is nearly constant at three. They concluded that large plants resulting from large sets are due primarily to the large size of leaves rather than the number of leaf initials. They suggested that the bolting of plants grown from sets is likely based on the number of leaves that emerged during set production.

Garlic is exclusively produced from cloves (Fig. 11.6) but over the past 30 years, new opportunities have opened for production from true seed. Shallot may be produced from sets/cloves or seeds. Leek may be direct-seeded or produced from transplants. Chinese chives and chives are grown from seed but are perennial once established. Japanese bunching onion is grown from seed as an annual plant. Many of the same soil types and production conditions used for bulb onion may be used for all of these *Allium* crops.

Fig. 11.6. (a) garlic clove arranged for planting; (b) with an emerging shoot; (c) breaking a basal plate and root emergence; (d) being planted in a trench; (e) garlic shoot emergence from a planted clove; (f) root mass of young garlic plant; (g) garlic plants maturing in mulched plots; (h, i) garlic flowering scapes showing characteristic coiling; and (j) bulbil formation. (Photo credit: S. Mahr.)

6. Population Density and Spacing

Population density is yet another major factor that determines leaf number, length, bulb size and yield, and bulb maturity is hastened by increasing density (Rumpel and Felczynski, 1998). Up to a certain point, marketable yield increases asymptotically with plant density, and bulb size decreases in a corresponding fashion (Brewster, 2008; Anon, 2010), but this varies with genotype (Anon, 2020). Using densities from 52 to 185 plants m^{-2} of LD Rijnsburger onions,

Brewster developed equations for predicting bulb weight of each size grade from a field of onions. These equations have held up for different cultivars and growing environments at densities of 22 and 130 plants m^{-2}.

In general, bulb production in many parts of the world is accomplished in this range of density, in the 50–100 plants m^{-2} range, except for sets production where densities of 2000–4000 plants m^{-2} are common. Densities below 100 plants m^{-2} result in lower yields of medium- to larger-sized bulbs (Brewster, 2008), and densities of 25–50 plants m^{-2} yield a higher fraction of jumbo- or colossal-size bulbs.

Hand weeding and mechanical cultivation require precise between-row spacing, but planting arrangements are more flexible with chemical herbicide. Equidistant cultivation within and between rows seems to be optimal as the exploitation of light, water and nutrients should be maximal with this planting arrangement (Brewster, 2008). Greater rectangularity was associated with yield reductions (Brewster and Salter, 1980). Others suggested that between-row widths should be reduced to optimize yield (Rickard and Wickens, 1979); however, when compared to random within-row spacing, a uniform between-plant distance within a row did not increase yield or yield of bulbs >40 mm in diameter at densities of 86–108 plants m^{-2}. Uniformity of bulb size is not increased by equidistant cultivation, following the planting of seed or sets (Brewster and Salter, 1980).

Size-grade categories for onions vary with the country (Table 11.2). In general, ~40–75 mm-in-diameter onion bulbs are sought for marketing, though it is possible to market smaller and larger bulbs for speciality markets.

Production of colossal- and super-colossal-sized bulbs for the fresh market or speciality applications in the restaurant trade occurs in certain countries such as the USA, planting density being the most prominent factor for the larger-size grades, such as colossal (Shock *et al.*, 2004). In the USA, this grade is reserved for bulbs >95 mm in diameter, which would typically be used for speciality processing applications such as onion rings, onion blossoms, or for the fresh market. Boyhan *et al.* (2009) reported large transplants (13–14 g $plant^{-1}$) produced the best yields of such bulbs, followed by medium-sized transplants. Smaller transplants (2–3 g $plant^{-1}$) produced low yields and a few bolters and doubles. Increased plant populations from 78,280 to 274,000 plants ha^{-1} increased yield. Yields also increased when plant populations were increased from 156,500 plants ha^{-1} to 274,000 plants ha^{-1}, but only when environmental conditions favoured lower bolting percentages (Boyhan *et al.*, 2009). Approximately 200,000 plants ha^{-1} are needed for ID onions transplanted on two rows on 1 m raised bed, with plants spaced 10 cm apart in the row (Walker *et al.*, 2009). Large bulbs are typically produced at densities less than 500,000 plants ha^{-1}, and maximum densities do not generally exceed 920,000 plants ha^{-1}.

Commonly, LD production in the US uses direct sowing in multiple rows per bed, with in-row spacings of ~2.5–5 and 7.5–15 cm for medium-size and jumbo- and colossal-size onions, respectively. Hence, 4–6 rows are planted on 80–122 cm wide beds. SD production environments in the US employ similar numbers of rows per bed but shorter distances between beds.

The traditional seeding system with raw seeds tends to yield a wider range of bulb sizes than that from a precision sowing (Wohleb and Waters, 2016). When a variety of grade categories is required, multiple fields planted at different spacings supply the demand for desired bulb sizes (Table 11.2).

Rainfall may be more common in spring-planted onions compared to an autumn-planted crop, which is often planted in warm weather. Stand establishment depends on surface moisture; hence, irrigation may be needed. Watering is also required when a hard crust forms following heavy rain before seedling emergence.

Generally speaking, the strategy for ensuring continual harvest is to combine different cultivar maturities with different seeding dates to allow for a long harvest window throughout the spring and summer months. The key to the autumn sowing dates is to skilfully choose the right genotypes and minimize conditions that would favour bolting.

For many ID and SD production systems, seeding is carried out in September and October for harvesting from March to June, and in January to February for June to August harvest. Cold weather damage to overwintered crops is not common in the subtropics, yet planting after October can result in conditions where seedlings are killed by cold temperatures.

Table 11.2. Example size-grade categories for onions in several countries.

Category	Minimum bulb diameter[1] (mm)	Maximum bulb diameter (mm)
United States[2]		
Small	25.4	57.2
Prepack	44.5	76.2
Medium	50.8	82.6
Large/jumbo	76.2	No requirement
Colossal	95.3	No requirement
Canada[3]		
No. 1 Pickling	12.7	38.1
No. 1	44.5	76.3[4]
No. 1 Small	31.8	44.5
No. 1 Jumbo	>76.3	
South Africa[5]		
Pickler	10	35
Small	35	50
Medium	40	70
Large	70	90
Extra large	>90	
India[6]		
Category C	35	50
Category B	50	60
Category A	>60	

[1]Onion bulb diameter is measured at right angles to the longitudinal axis of the bulb.
[2]Anonymous (2014)
[3]The Canadian Food Safety for Industry. Available at: https://inspection.canada.ca/eng/1297964599443/1297965645317 (accessed September 2022).
[4]This category may have onions of any size between 44.5 and 76.3 mm as long as the package does not indicate the size.
[5]The Small Scale Market Farmer Marketing Series
[6]The Indian Institute of Horticultural Research

Table 11.3. Size-grade distribution and significance of main effects and interaction from a three-year onion trial of 54 onion varieties in Washington, USA (Wohleb and Waters, 2016[1]).

Year or source of variation	Yield per hectare or p value	Diameter (mm)				
		<57.2	57.2–76.2	76.2–101.6	>101.6	Culls[2]
	kg/ha	%				
2012	127,104	2	18.3	65.6	9.4	4.7
2013	123,518	0.5	11.1	77.6	10.6	0.9
2014	106,705	2.2	24.4	66.9	2.4	4.2
Year	0.0001	0.0001	0.0001	0.0001	0.0001	0.0001
Cultivar	0.0001	NS	0.0001	0.0007	0.0001	NS
Year x cultivar	NS[3]	0.0001	NS	0.0001	0.0001	NS

[1]Wohleb, C.H. and Waters, T.D. (2016) Yield, Quality, and Storage Characteristics of Onion Cultivars in the Columbia Basin of Washington in 2012–2014. *HortTechnology* 26, 230–243.
[2]Double bulbs, split bulbs, or bolted plants
[3]NS = not significant at $p<0.05$

An interesting observation from New Mexico shows that yield and bulb size from transplanted ID cultivars are higher than for the same varieties direct-seeded in spring despite the earlier maturation of the former. Also, bulb shapes of transplanted onions were different from the same varieties produced from autumn seeding (Walker *et al.*, 2009). Overwinter, autumn-sown plants develop a large root system that pushes the bulb into a more top-shaped form, while transplants tend to produce a more rounded bulb shape.

Selection against bolting is important for overwintered onions. Recommendations suggest that seedlings of bolting-resistant cultivars with a diameter smaller than that of a pencil (an indication for a physiological juvenile phase) be planted from mid-January onwards, without a bolting problem, while bolting-susceptible cultivars should not be transplanted until after mid-February, and plant diameter should be smaller than pencil size.

Garlic cloves are typically planted at least 10–20 cm apart in a row, and rows are typically at least 20–30 cm apart. In temperate climates, the production goal in autumn is to get some root growth before the cold winter begins and straw mulching, or with other materials, is common to prevent freezing damage. Snow cover protects the cloves from freezing as long as there is sufficient snowfall for continuous coverage during very cold weather. As the plant matures in the summer, leaves will turn yellowish-brown, and harvest should take place when 40–60% of the leaves have yellowed. In temperate climates, harvest typically takes place mid-summer. Plants are often cured with tops on by hanging in dry, warm places for up to several weeks (Figs 11.7–11.9). In temperate zones, bulbs stored for propagation are generally kept at slightly warmer temperatures than those aimed for marketing.

When grown from individual sets, shallots may be produced in a manner and at a density similar to garlic, though spacing may be slightly thinner for optimal results. Leeks are started from seed and transplanted once they reach the appropriate size at 20 and 30 plants m^{-2} depending on desired marketable size (Swamy and Veere Gowda, 2006). Bunching onion or scallion production may be accomplished by seeding at very high densities. It is possible to produce transplants with multiple seedlings per cell in a plug tray and transplant them into the field in clumps spaced up to 35 cm apart.

7. Seed Production

Onion seed production is typically accomplished in dry, warm environments to reduce the potential for disease development and support the activity of pollinators such as bees and flies. Most commonly, the seed is produced from bulbs grown in a previous season, known as the bulb-to-seed method (Chapter 12). Following standard production, bulbs are vernalized at 5–10°C and planted in the second year at least 30 cm apart. Elsewhere, in autumn-planted bulbs, winter-cold vernalization suffices for bolting and floral development. Many of the diseases common in bulb production, such as downy mildew, purple blotch, *Stemphylium* and *Botrytis*, are potential problems in seed production.

The seed-to-seed method, which allows post-juvenile plants from early-autumn seeding to become field-vernalized over winter and spring bolting in a single production season, is also practised (Chapter 12).

The optimal temperature for induction of meristem transition and bolting is 9°C (Shishido and Saito, 1975), and photoperiodic involvement is debatable (Van Kampen, 1970; Shishido and Saito, 1975a, b). Nitrogen levels may be inversely correlated with the initiation of bolting (Shishido

Fig. 11.7. Garlic bulbs bunched and hanging for curing. (Photo: S. Mahr)

Fig. 11.8. (a) an early maturing LD genotype showing an early phase of foliage collapse in July following seeding in April (Photo: I. Goldman); (b) top-down (foliage collapse) as SD onions mature (Photo: D. Brotslaw); (c) near-complete top-down (foliage collapse) as SD onions mature (Photo: D. Brotslaw); (d) windrowed SD onions ready for harvest (Photo: D. Brotslaw).

Fig. 11.9. (a) harvest and packing of SD onion; mechanical onion harvest (Photo: D. Brotslaw); (b) manual harvest and packing onion bulbs in crates in the field (Photo: C. Cramer).

and Saito, 1976) as low nitrogen and LD at 9°C accelerated bolting (Brewster, 1983). Thereafter, inflorescence development was enhanced by LD and increases in temperature to a certain level but inhibited by low nitrogen (Brewster, 1983).

Onion seed size correlates with vigour (Gamiely *et al.*, 1990; Rodo and Marcos-Filho, 2003), and planting time affected thousand seed weight (El-Helaly and Karam, 2012; Teshome *et al.*, 2014; Gebeyehu, 2016), and Brewster

(2008) reported on the increase in seed yield with the size of the mother bulb at a constant stand because the number of umbels per plant increases.

Hybrid seed production requires interplanting of male-sterile rows, and fertile rows, at a 4:1 ratio, respectively. Isolation of up to 5 km between seed production fields is important because pollen is moved by insects. Beehive density of approximately 15–30 hives ha^{-1} is common (Fig. 11.10).

In general, seed maturation is correlated with increasing temperature, yet high temperatures around the umbels can cause ovule abortion. Umbels are clipped with part of the scape when the capsules begin to dehisce and when most capsules ripen, they are threshed manually or by machine. Onion seed is particularly susceptible to reduced germination with high temperatures and high humidity. Even under ideal moisture conditions of approximately 6%, the seed may only remain viable for several years.

Fig. 11.10. Onion flowering and seed production: (a) onion seed production field (Photo: W.H. Gabelman); (b) flowers beginning to open in the onion umbel (Photo: M. Havey); (c) seed maturation on onion umbel (Photo: C. D'Angelo).

8. Weeds, Pests and Diseases

8.1. Viral diseases

Onion yellow dwarf potyvirus (OYDV), transmitted by aphids, causes yellow striping and yellowing of leaves, as well as reducing bulb yields in onion, garlic and shallot, worldwide. The virus survives in bulb tissues and is easily transmitted during vegetative reproduction. The thrips (*Thrips tabaci*)-transmitted Iris yellow spot virus (IYSV) (Fig. 11.11) is one of the most

Fig. 11.11. Onion plants infected with Iris yellow spot: (a) infected seed production field; (b) examples of onion leaf waxiness; and (c) infection. (Photos: M. Havey)

damaging diseases of onion crops worldwide. The disease manifests as leaf and scape lesions that can girdle leaves and cause lodging of scapes.

No genetic resistance has been identified for IYSV, yet onion germplasm does differ in field-tolerance to the thrips vector due to variation in the kinds and amounts of epicuticular leaf wax. Glossy leaves have lesser amounts of epicuticular wax and exhibit lower levels of thrips feeding damage than waxy leaves (Fig. 11.11). Lower wax genotypes possess the recessive *gl* allele. A second allele at this locus, gl^{w}, causes a semi-glossy phenotype (Munaiz *et al.*, 2020). Hentriacontanone-16 (H16), octaconasol-1 and triacontanol-1 were the most prevalent foliage waxes of these genotypes. Semi-glossy types displayed similar amounts of total epicuticular wax as waxy types, due primarily to lower amounts of H16 and higher amounts of other waxes, but they were less affected by thrips feeding damage. Hsu *et al.* (2011) and Malik *et al.* (2009) suggested that lower nitrogen rates result in lower densities of thrips populations.

Other aphid-transmitted viral diseases include leek yellow stripe virus (LYSV) that causes leaf yellowing and yield loss, as well as reduced storability, and the shallot latent virus (SLV), a ubiquitous virus capable of infecting all edible *Allium* crops. Infective aphids generally feed on overwintered plants such as those in home gardens (Brewster, 2008), thus bridging between cropping cycles. Breaking such bridges is a good way to prevent their build-up.

Viruses infect many clones of garlic and shallot, and these are difficult to remove. With the advent of true seed production in these crops, it may be possible to disrupt the virus life cycle and produce virus-free planting material. It is also possible to eliminate viruses by producing virus-free plantlets in tissue culture.

8.2. Insects

Onion maggot (*Delia antiqua*) and the Allium leafminer (*Phytomyza gymnostoma*) are of critical importance for onion production. Onion maggots lay eggs in the soil or the onion leaves. Hatched maggots migrate to roots, bulbs and stems to feed, thus causing stunted plant growth. They typically complete three generations per year and stay in or near the same fields and soil, year after year, making their control difficult particularly where crop rotations are short or non-existent. Non-chemical control of onion maggot is difficult but is routinely practised, e.g. removing cull onions and crop rotation. Organophosphate insecticides, e.g. chlorpyrifos, sometimes applied by aerial spray (Fig. 11.12), targets eggs to suppress the development of maggots before they enter plant tissue. In some areas of the USA, however, onion maggots became resistant to chlorpyrifos in the 1990s. In 2020, chlorpyrifos was banned for use in California and Hawaii, and due to its proven toxicity, it is being phased out in other states also.

The Allium leafminer originally became a pest in Europe and parts of the Middle East. It is capable of feeding on many *Allium* crops including leek, garlic, chive, shallot, bunching onion and bulb onion. The pest has recently been introduced to the USA where it has been very damaging on onion and leek crops. Feeding by this insect results in weakened and softened plant tissues, which increase the risk of microbial infection and reduce product quality. The leafminer overwinters as pupae in plant material or soil and emerges in the spring. Some growers have found success with the use of row covers to prevent contact with the pest. Timing of plantings to avoid flight activity of the leafminer can be advantageous. Some growers have found success with systemic or contact insecticides.

Leek moth *Acrolepiopsis assectella* caterpillars mine in the crown of leek, onion, garlic, chive and shallot, thus enabling secondary pathogens to enter. Also, the moth can damage seed crops. The moth undergoes as many as five generations per growing season and overwinters as either pupa or adult. Pheromone traps are used to determine the timing of insecticidal sprays.

The shallot aphid *Myzus ascalonicus* can cause tissue damage in stored onion and shallot bulbs as well as transmit viruses, and treatments must target post-harvest storages of bulbs to be effective.

The garlic mite *Aceria tulipae* can cause damage in stored bulbs of garlic, shallot and onion by destroying tissue and transmitting viruses.

Fig. 11.12. Aerial application of pesticides in onion production. (Photo: D. Brotslaw)

8.3. Bacterial diseases

Bacterial diseases of onion include Xanthomonas leaf blight, soft rot, centre rot, slippery skin and sour skin. Xanthomonas/bacterial leaf blight in onions is caused by *Xanthomonas axonopodis*. Lesions begin as white or pale spots with water-soaked margins that become tan to brown and may expand, fuse and cause damage to leaves.

The seedborne pathogen is favoured by warm temperatures and high moisture and may enter the plant through stomata, abraded leaves or wounded tissue. The pathogen also survives on plant debris. Control methods include rotation, sanitation of equipment and irrigation water, the use of clean seed and plant materials, and careful handling to avoid wounding. Overhead irrigation systems favour the development of the disease, especially in humid and warm conditions. Copper products, sometimes along with carbamate fungicides, are used for disease control, and in semi-arid environments, acibenzolar-*S*-methyl is used, which produces systemic acquired resistance. In organic production systems, the use of bacteriophage treatments may be effective.

Bacterial soft rot *Erwinia carotovora* ssp. *carotovora* infects the pseudostem or enters through a wound during the late stages of the growing season. Onion maggots facilitate the pathogen's spread by creating wounds where bacteria may enter. Similarly, wounds caused by hail or topping can create an entry point for the pathogen. Thereafter, it moves down the leaf, decaying tissue on one or perhaps two leaves. When squeezed, the diseased bulbs may collapse under pressure, often releasing fluid with a rotten smell. Should such damage occur, copper-based fungicides may be used to protect bulbs from damage. Drying and curing onion plants until the necks have tight closure is important for preserving bulb quality.

Four *Pantoea* species: *ananatis*, *agglomerans*, *allii* and *stewartii* ssp. *indologenes* cause centre rot of onion (Stumpf *et al.*, 2018). They can be seedborne and/or transmitted by thrips: *Frankliniella fusca* (Hinds) and *T. tabaci*. The disease starts with foliage water-soaked lesions that can wilt and damage leaves. The pathogen then spreads to the bulb, where damage occurs. Centre rot can be managed with copper-based products mixed with carbamates. It is also important to control weeds in onion production fields since weeds are a common host for the pathogen.

Slippery skin and sour skin are caused by several species of *Pseudomonas* and *Erwinia*, which cause bulbs to rot and also produce a sulphurous smell. Often bulbs appear fine on the outside but have rotted inner scales. One or two leaves may wilt in the centre of the plant, and eventually may turn yellow and die back from the tip. When internal scales rot, the bulb may shrink. Compressing the plant's base leads to sliding of the rotted inner portions out through the neck, hence the name slippery skin. These pathogens are more active in warm, wet conditions and are more severe when onion tissues are damaged, such as by onion maggot. Storing dried and cured bulbs usually prevents the development of these diseases. Overhead irrigation, flooding and splashing of bacteria during rains can contribute to pathogen spread.

Bactericides have become more widely studied as protective agents against bacterial diseases in onion. They cannot cure an infection but rather act as preventives (Du Toit *et al.*, 2008). They are also non-systemic, so that coverage of the product on the plant is critical for efficacy, but protection is not guaranteed (Du Toit *et al.*, 2020). Some bactericides, such as Kocide 3000-O (a copper hydroxide product) and Lifeguard WG (*Bacillus mycoides*), resulted in improved marketable bulb yields.

8.4. Fungal diseases

Several fungal diseases severely affect the quality and yield of onion. Strains of Fusarium basal rot (*Fusarium oxysporum* f.sp. *cepae*) are common worldwide, infecting onion, garlic and Japanese bunching onion. Diseased plants suffer damping-off of seedlings and leaf necrosis, and discolouration and necrosis of tissues in the bulbs' basal plate or leaves, leading to root and plant death. Crop rotation, resistant cultivars, soil fumigation and fungicides have been used to control the disease.

Pink root (*Phoma terrestris*) is another damaging pathogen of onion, garlic and shallot roots. Losses caused by the two pathogens are especially serious at higher soil temperatures, and as global temperatures increase, their impact increases in *Allium* crop production. Both fungi survive for long periods in the soil, making rotations to other crops less effective. In past decades, soil fumigants such as methyl bromide were used to control these two diseases (Hartz *et al.*, 1989); however, this option is no longer available. Resistance alleles have been identified in onion for both diseases and are routinely used as part of an integrated control strategy.

Stemphylium vesicarium causes both purple blotch and Stemphylium leaf blight. Purple blotch was common in onion, Japanese bunching onion and leek production, but it has largely been supplanted by Stemphylium leaf blight, which appears as brown lesions on the leaf, similar to lesions caused by *Psuedomonas syringae*. Farmers controlled this pathogen with strobilurin fungicides, but newer products such as the broad-spectrum fungicide Miravis® Prime effectively reduce this disease in the field, and fungicide programmes combining broad-spectrum Bravo® with Miravis® Prime provided control to LD onions grown in muck soils in Michigan (Higgins and Hausbeck, 2020).

Botrytis leaf blight (*Botrytis squamosa*) is a bulb onion-specific pest that produces white lesions with water-soaked appearance and greenish-white haloes and may display slits along with the lesions. When lesions are numerous, leaf tips die back and the entire leaf may be at risk. The fungus thrives in warm, wet conditions, and sclerotia can persist in soil for years. Control is best achieved by removing all onion debris from fields and eliminating cull piles. Fungicides such as azoxystrobin and chlorothalonil may be used to control the disease. These are the same fungicides that control downy mildew (*Peronospora destructor*), another common foliar disease in onion, shallot and Japanese bunching onion.

The soilborne black mould (*Aspergillus niger*) can infect during the latter part of bulb development but primarily affects bulbs and leeks in storage. It invades at the neck or the roots. Bruised or damaged bulbs provide openings for the fungus to enter the bulb and manifest as small black spots on bulb scales. Sanitation is the best method of black mould control, including rotations and care during harvest.

White rot (*Sclerotium cepivorum*) is a persistent pest on *Allium* crops especially onion and garlic. Leaves wilt and die back, decaying and collapsing, beginning with the older leaves. Mycelial growth causes a white mass around the true stem or leaves, and eventually, sclerotia form on the mycelial mat. The severity of white rot is associated with cool, moist conditions. Sclerotia may remain dormant in the soil without hosts for over 20 years. *Allium* root exudates stimulate the sclerotia growth that spreads easily through irrigation water and plant debris. Rotations, fungicides and careful sanitation have only minimal effect, hence control is quite difficult. An innovative approach is the use of microbial antagonists such as *Sporidesmium sclerotivorum* and *Trichoderma harzianum*, which can attack white rot sclerotia. Application of stimulants (e.g. diallyl disulfide) to promote sclerotia germination in the absence of an *Allium* host may reduce the risk when an *Allium* crop is planted. Breeding for resistance to white rot in edible Alliums, however, is a critical objective for the future success of these crops.

8.5. Weeds

Weed management is among the most critical aspects of *Allium* production, as *Allium* crops are not generally competitive with weeds and standard spacings allow for substantial weed emergence. Also, herbicide options are limited and hand weeding is very costly. In general, growers try to limit weed pressure before seeding by solarizing soil, using stale seedbeds, planting in locations that are as weed-free as possible, and using crop rotations with cover crops. Cover crops must be tilled and allowed to break down before crop seeding. Flaming can also be used to suppress germinated weed growth after seeding and before crop emergence (Chehade *et al.*, 2018). The first few weeks of onion growth are the most critical for weed competition (Zandstra, 2019) and leek should be kept weed-free from 7–13 to 60–85 days after transplanting to prevent >5% yield loss (Tursun *et al.*, 2007). The stale seedbed method is practised by allowing weeds to emerge and then cultivating shallowly to remove these weeds but not bring new weed seeds to the soil surface.

Herbicide options are limited beyond bulb onion. Pre-emergent herbicides (e.g. Pendimethalin and Bromoxynil) are often applied after onion seeding but before emergence, and Pendimethalin and Sethoxydim are used for post-emergent weeds, at their one- to two-leaf stage. It is also common for growers to seed onions with a nurse crop such as barley, which protects the young onion seedlings from wind-blown soil particles and sand and is killed with herbicide after the crop is established. After planting, cultivation can be used when planting arrangements allow and as long as the rows on a bed are not disturbed.

References

Al-Karaki, G.N. (2006) Field response of garlic inoculated with arbuscular mycorrhizal fungi to phosphorus fertilization. *Journal of Plant Nutrition* 25, 747–756. DOI: 10.1081/PLN-120002956.

Anon (2010) *Vegetable Production Guides.* Oregon State University Department of Horticulture. Available at: https://horticulture.oregonstate.edu/vegetable-production-guides (accessed 22 July 2022).

Anon (2014) United States Standards for Grades of Onions (other than Bermuda-Granex-Grano and Creole Type). Available at: https://www.ams.usda.gov/grades-standards/onions-other-bermuda-granex-grano-and-creole-type-grades-and-standards (accessed 24 November 2014).

Anon (2020) *Onion Planting Density*. Seminis Co. Available at: https://www.vegetables.bayer.com/au/en-au/resources/growing-tips0/agronomic-spotlights/onion-planting-density.html (accessed 22 July 2022).

Ar Rashid, M.H., Chend, W. and Thomas, B. (2019) Temporal and spatial expression of Arabidopsis gene homologs control daylength adaptation and bulb formation in onion (*Allium cepa* L.). *Scientific Reports* 9, 14629.

Arunachalam, A., Kalyani, T., Shabeer, G., Savalekar, R.K., Banerjee, K. *et al.* (2018) Comparison of organic and conventional farming for onion yield, biochemical quality, soil organic carbon, and microbial population. *Archives of Agronomy and Soil Science* 64, 219–230. DOI: 10.1080/03650340.2017.1341045.

Austin, R.B. (1972) Bulb formation in onions as affected by photoperiod and spectral quality of light. *Journal of Horticultural Sciences* 47, 493–504.

Basallote-Ureba, M. and Melero-Vara, J.M. (1993) Control of garlic white rot by soil solarization. *Crop Protection* 12, 219–223.

Block, E. (1992) The organosulfur chemistry of the genus Allium – implications for organic sulfur chemistry. *Angewandte Chemie International Edition in English* 31, 1135–1178.

Bloem, E., Silvia, H. and Ewald, S. (2004) Influence of nitrogen and sulfur fertilization on the alliin content of onions and garlic. *Journal of Plant Nutrition* 27, 1827–1839. DOI: 10.1081/LPLA-200026433.

Bosch-Serra, A.D. and Currah, L. (2002) Agronomy of onions. In: Rabinowitch, H.D. and Currah, L. (eds) *Allium Crop Science: Recent Advances*. CAB International, Wallingford, UK, pp. 187–232.

Boyhan, G., Torrance, R., Cook, J., Riner, C. and Hill, C. (2009) Plant population, transplant size, and variety effect on transplanted short-day onion production. *HortTechnology* 19, 145–151.

Brewster, J.L. (1983) Effects of photoperiod, nitrogen nutrition and temperature on inflorescence initiation and development in onion (*Allium cepa* L.). *Annals of Botany* 51, 429–440.

Brewster, J.L. (1990a) Physiology of crop growth and bulbing. In: Rabinowitch, H.D. and Brewster, J.L. (eds) *Onions and Allied Crops*, vol. 1. CRC Press, Boca Raton, Florida, pp. 53–88.

Brewster, J.L. (1990b) Cultural systems and agronomic practices in temperate climates. In: Rabinowitch, H.D. and Brewster, J.L. (eds) *Onions and Allied Crops*, vol. 2. CRC Press, Boca Raton, Florida, pp. 1–30.

Brewster, J.L. (1997) Onions and garlic. In: Wien, H.C. (ed.) *The Physiology of Vegetable Crops*. CAB International, Wallingford, UK, pp. 581–619.

Brewster, J.L. (2008) *Onions and Other Vegetable Alliums*, 2nd edn. CAB International, Wallingford, UK.

Brewster, J.L. and Salter, P.J. (1980) A comparison of the effects of regular versus random within row spacing on the yield and uniformity of size of spring-sown bulb onions. *Journal of Horticultural Sciences* 55, 235–238.

Brewster, J.L. and Sutherland, R.A. (1993) The rapid determination in controlled environments of parameters for predicting seedling growth rates in natural conditions. *Annals of Applied Biology* 122, 123–133.

Chehade, L.A., Fontanelli, M., Martelloni, L., Frasconi, C., Raffaelli, M. *et al.* (2018) Effects of flame weeding on organic garlic production. *HortScience* 28, 502–508. DOI: 10.21273/HORTTECH04081-18.

Chung, B. (1989) Multi-plant module transplants of bulb onions. *Acta Horticulturae* 247, 187–191.

Dalvi, V.S., Patil, Y.A., Krishna, B., Sane, P.V. and Sane, A.P. (2016) Identification of bulbing related genes in SDSD, non-vernalization requiring onion. *Acta Horticulturae* 269–276. DOI: 10.17660/Acta Hortic.2016.1143.38.

D'Angelo, C.J. and Goldman, I.L. (2018) Temporal aspects of vernalization and flowering in long-day storage onion. *Journal of the American Society for Horticultural Science* 143, 446–453.

D'Angelo, C.J. and Goldman, I.L. (2019a) Breaking onion bulb endodormancy with hydrogen peroxide. *HortScience* 54, 1694–1702. DOI: 10.21273/HORTSCI14118-19.

D'Angelo, C.J. and Goldman, I.L. (2019b) Annualization of the LDLD onion breeding cycle through threshold vernalization and dormancy disruption. *Crop Breeding, Genetics and Genomics* 1, e190009. DOI: 10.20900/cbgg20190009.

Darbyshire, B. and Henry, R.J. (1979) The association of fructans with high percentage dry weight in onion cultivars suitable for dehydrating. *Journal of the Science of Food and Agriculture* 30, 1035–1038.

Davis, G.N. and Jones, H.A. (1944) *Experiments with Transplant Onion Crop in California*. Bulletin of California Agriculture Experimental Station, 682.

Debaene, J.P., Goldman, I.L. and Yandell, B. (1999) Postharvest flux and genotype x environment effects for onion induced antiplatelet activity, pungency, and soluble solids in long-day onion during postharvest cold storage. *Journal of the American Society for Horticultural Science* 124, 366–372.

Delahaut, K.A. and Newenhouse, A.C. (1998) *Growing Carrots, Beets, Radishes, and Other Root Crops in Wisconsin: A Guide for Fresh-market Growers*. *University of Wisconsin Cooperative Extension.* Available at: http://learningstore.uwex.edu/assets/pdfs/A3686.PDF (accessed 22 July 2022).

Dowker, B. and Fennell, J. (1974) Heritability of bulb shape in some north European onion varieties. *Annals of Applied Biology* 77, 61–65.

Drucker, A. (2020) Garlicana. Available at: https://www.garlicana.com/ (accessed 22 July 2022).

du Toit, L., Holms, B., Twomey, M. and Gatch, E. (2008) Evaluation of seed treatments for soilborne damping-off pathogens and seedborne fungi of onion. *Plant Disease Management Reports* 3, ST021.

du Toit, L., Waters, T., Derie, M. and Darner, J. (2020) Battling onion bacterial diseases with bactericides. *Onion World, pp.* 6–11. Available at: https://issuu.com/columbiamediagroup/docs/onion_world_december_2020 (accessed 31 July 2022).

Eady, C.C., Kamoi, T., Kato, M., Porter, N.G., Davis, S. *et al.* (2008) Silencing onion lachrymatory factor synthase causes a significant change in the sulfur secondary metabolite profile. *Plant Physiology* 147, 2096–2106.

El-Helaly, M.A. and Karam, S.S. (2012) Influence of planting date on the production and quality of onion seeds. *Journal of Horticultural Science & Ornamental Plants* 4, 275–279.

El-Shafie, M. and Davis, G. (1967) Inheritance of bulb color in *Allium cepa*. *Hilgardia* 9, 607–622.

Engeland, R.L. (1991) *Growing Great Garlic*. Filaree Productions, Okanogan, Washington, DC.

Etoh, T. (1986) Fertility of the garlic clones collected in Soviet Central Asia. *Journal of the Japanese Society for Horticultural Science* 44, 312–319.

Foskett, R.L. and Peterson, C.E. (1950) Relation of dry matter content to storage quality in some onion varieties and hybrids. *Proceedings of the American Society for Horticultural Science* 55, 314–318.

Freeman, G.G. and Mossadeghi, N. (1970) Effect of sulphate nutrition on flavour components of onion (*Allium cepa*). *Journal of the Science of Food and Agriculture* 21, 610–615.

Freeman, G.G. and Mossadeghi, N. (1971) Influence of sulphate nutrition on the flavour components of garlic (*Allium sativum*) and wild onion (*A. vineale*). *Journal of the Science of Food and Agriculture* 22, 330–334.

Fritsch, R.M. and Friesen, N. (2002) Evolution, domestication and taxonomy. In: Rabinowitch, H.D. and Currah, L. (eds) *Allium Crop Science: Recent Advances*. CAB International, Wallingford, UK, pp. 5–30.

Galmarini, C.R., Goldman, I.L. and Havey, M.J. (2001) Genetic analyses of correlated solids, flavor, and health-enhancing traits in onion (*Allium cepa* L.). *Molecular Genetics and Genomics* 265, 543–551.

Gamiely, S., Smittle, D.A. and Mills, H.A. (1990) Onion seed size, weight, and elemental content affect germination and bulb yield. *Horticultural Science* 25, 522–523.

Garner, W.W. and Allard, H.A. (1920) Effect of the relative length of day and night and other factors of the environment on growth and reproduction in plants. *Journal of Agricultural Research* 18, 553–606.

Gebeyehu, M.T. (2016) Effect of planting time on growth, yield components, seed yield and quality of onion (*Allium cepa* L.) at Tehuledere district, northeastern Ethiopia. MSc thesis, Jimma University, Ethiopia. Available at: https://cgspace.cgiar.org/bitstream/handle/10568/90445/Maria.pdf?sequence=1&isAllowed=y (accessed 22 July 2022).

Gebretsadik, K. and Dechassa, N. (2018) Response of onion (*Allium cepa* L.) to nitrogen fertilizer rates and spacing under rain fed condition at Tahtay Koraro, Ethiopia. *Scientific Reports* 8, 9495 DOI: 10.1038/s41598-018-27762.

Goldman, I.L., Schroeck, G. and Havey, M.J. (2000) History of public onion breeding programs and pedigree of public onion germplasm releases in the United States. *Plant Breeding Reviews* 20, 67–103.

Grevsen, K. and Sorensen, N.J. (2004) Sprouting and yield in bulb onions (*Allium cepa* L.) as influenced by cultivar, plant establishment methods, maturity at harvest and storage conditions. *The Journal of Horticultural Science and Biotechnology* 79, 877–884. DOI: 10.1080/14620316.2004.11511860.

Hanelt, P. (1990) Taxonomy, evolution and history. In: Rabinowitch, H.D. and Brewster, J.L. (eds) *Onions and Allied Crops*, vol. 1. CRC Press, Boca Raton, Florida, pp. 1–26.

Hartz, T.K., Bogle, C.R., Bender, D.A. and Avila, F.A. (1989) Control of pink root disease in onion using solarization and fumigation. *Journal of the American Society for Horticultural Science* 114, 587–590.

Havey, M.J. (2018) Onion breeding. *Plant Breeding Reviews* 42, 39–85.

Havey, M.J. and Ghavami, F. (2018) Informativeness of single nucleotide polymorphisms and relationships among onion populations from important world production regions. *Journal of the American Society for Horticultural Science* 143, 33–44. DOI: 10.21273/JASHS04277-17.

Hay, R.K.M. and Kemp, D.R. (1992) The prediction of leaf canopy expansion in the leek from a simple model dependent on primordial development. *Annals of Applied Biology* 120, 537–545.

Heath, O.V.S. and Mathur, P.B. (1944) Studies in the physiology of the onion plant. II. Inflorescence initiation and development, and other changes in the internal morphology of onion sets, as influenced by temperature and day length. *Annals of Applied Biology* 31, 173–187.

Higgins, D. and Hausbeck, M. (2020) *Fungicides Help Onions Against Stemphylium Leaf Blight*. *Vegetable Growers News*, p. 30.

Hsu, C.L., Reiners, S., Hoepting, C.A. and Nault, B.A. (2011) Relationship between nitrogen rate, thrips and yield in New York. In: *Proceedings of the 2011 Empire State Fruit and Vegetable Expo*, 25–27 January. Cornell Cooperative Extension and New York State Vegetable Growers Association, Syracuse, New York.

Imai, S., Tsuge, N., Tomotake, M., Nagatome, Y., Sawada, H. *et al.* (2002) An onion enzyme that makes the eyes water. *Nature* 419, 685.

Inden, H. and Asahira, T. (1990) Japanese bunching onion (*Allium fistulosum* L). In: Brewster, J.L. and Rabinowitch, H.D. (eds) *Onions and Allied Crops*, vol. III. *Biochemistry, Food Science, and Minor Crops*. CRC Press, Boca Raton, Florida, pp. 159–178.

Jaime, L., Martinez, F., Martín-Cabrejas, M.A., Mollá, E., López-Andréu, F.J. *et al.* (2002) Study of total fructan and fructooligosaccharide content in different onion tissues. *Journal of the Science of Food and Agriculture* 81, 177–182.

Kamenetsky, R., London Shafir, I., Zemah, H., Barzilay, A. and Rabinowitch, H.D. (2004) Environmental control of garlic growth and florogenesis. *Journal of the American Society for Horticultural Science* 129, 144–151.

Keller, D. (2020) Tearless onions: a golden opportunity? *Onion World*, 1 December. Available at: https://onionworld.net/2020/12/01/tearless-onions-a-golden-opportunity/ (accessed 28 July 2022).

Khokhar, K.M. (2009) Effect of set-size and storage temperature on bolting, bulbing and seed yield in two onion cultivars. *Scientia Horticulturae* 122, 187–194. DOI: 10.1016/j.scienta.2009.05.008.

Kim, S., Binzel, M.L., Park, S., Yoo, K.-S. and Pike, L.M. (2004) Inactivation of DFR (Dihydroflavonol 4-reductase) gene transcription results in blockage of anthocyanin production in yellow onions (*Allium cepa*). *Molecular Breeding* 14, 253–263.

Kim, S., Jones, R., Yoo, K.-S. and Pike, L.M. (2005) The L locus, one of complementary genes required for anthocyanin production in onions (*Allium cepa*), encodes anthocyanidin synthase. *Theoretical and Applied Genetics* 111, 120–127.

Kopsell, D.A. and Randle, W.M. (1999) Selenium affects the S-alk(en)yl cysteine sulfoxides among SDSD onion cultivars. *Journal of the American Society for Horticultural Science* 124, 307–311.

Kopsell, D.E., Randle, W.M. and Eiteman, M.A. (1999) Changes in the S-alk(en)yl cysteine sulfoxides and their biosynthetic intermediates during onion storage. *Journal of the American Society for Horticultural Science* 124, 177–183.

Lee, R., Baldwin, S., Kenel, F., McCallum, J. and Macknight, R. (2013) Flowering Locus T genes control onion bulb formation and flowering. *Nature Communications* 4, 2884.

Leskovar, D.I. and Vavrina, C.S. (1999) Onion growth and yield are influenced by transplant tray cell size and age. *Scientia Horticulturae* 80, 133–143.

Magruder, R., Webster, R.E., Jones, H.A., Randall, T.E., Hawthorn, L.R. *et al.* (1941a) *Descriptions of Types of Principal American Varieties of Onions.* USDA Miscellaneous Publication no. 435. USDA, Washington, DC.

Magruder, R., Hawthorn, L.R., Webster, R., Brown, H.D., Snyder, G.B. *et al.* (1941b) *Storage Quality of the Principal American Varieties of Onions.* USDA Circular no. 618. USDA, Washington, DC.

Malik, M.F., Nawaz, M., Ellington, J., Sanderson, R. and El-Heneidy, A.H. (2009) Effect of different nitrogen regimes on onion thrips, *Thrips tabaci* Lindemann, on onions, *Allium cepa* L. *Southwestern Entomologist* 34, 219–225.

Masamura, N., McCallum, J., Khrustaleva, L., Kenel, F., Pither-Joyce, M. *et al.* (2012) Chromosomal organization and sequence diversity of genes encoding lachrymatory factor synthase in *Allium cepa* L. *G3: Genes, Genomes, Genetics* 2, 643–651.

McCallum, J., Clarke, A., Pither-Joyce, M., Shaw, M., Butler, R. *et al.* (2006) Genetic mapping of a major gene affecting onion bulb fructan content. *Theoretical and Applied Genetics* 112, 958–967.

McCollum, G. (1971) Heritability of onion bulb shape size: estimates from half-sib families. *Journal of Heredity* 62, 101–104.

McManus, M.T., Joshi, S., Searle, B., Pither-Joyce, M., Shaw, M. *et al.* (2012) Genotypic variation in sulfur assimilation and metabolism of onion (*Allium cepa* L.) III. Characterization of sulfite reductase. *Phytochemistry* 83, 34–42. DOI: 10113/57709; 10.1016/j.phytochem.2012.07.028.

Mondal, M.F., Brewster, J.L., Morris, G.E.L. and Butler, H.A. (1986) Bulb development in onion II: the influence of red:far-red spectral ratio and photon flux density. *Annals of Botany* 58, 197–206.

Munaiz, E.D., Groves, R.L. and Havey, M.J. (2020) Epicuticular leaf waxes among onion accessions selected for reduced damage by onion thrips. *Journal of the American Society for Horticultural Science* 145, 30–35.

Parera, C.A. and Cantliffe, D.J. (1992) Priming leek seed for improved germination and emergence at high temperature. *HortScience* 27, 1077–1079.

Pedrini, S., Balestrazzi, A., Madsen, M.D., Bhalsing, K., Hardegree, S.P. *et al.* (2020) Seed enhancement: getting seeds restoration-ready. In: Pedrini, S., Dixon, K.W. and Cross, A.T. (eds) *Standards for Native Seeds in Ecological Restoration*. Special Issue 28, S3, pp, S213–S303. DOI: 10.1111/rec.13184.

Peterson, C.P., Simon, P.W. and Ellerbrock, L. (1986) 'Sweet Sandwich' onion. *HortScience* 21, 1466.

Petropoulos, S.A., Ntatsi, G. and Ferreira, I.C.F.R. (2017) Long-term storage of onion and the factors that affect its quality: a critical review. *Food Reviews International* 33, 62–83. DOI: 10.1080/87559129.2015.1137312.

Pike, L.M. (1986) Onion breeding. In: Bassett, M.J. (ed.) *Breeding Vegetable Crops*. AVI Publishing, Westport, Connecticut, pp. 357–394.

Rabinowitch, H.D. (1979) Doubling of onion bulbs as affected by size and planting date of sets. *Annals of Applied Biology* 93, 63–66. DOI: 10.1111/j.1744-7348.1979.tb04727.x.

Rabinowitch, H. and Kamenetsky Golstein, R. (2020) Allium crops. In: Wien, H.C. and Stutzel, H. (eds) *Physiology of Vegetable Crops*, 2nd edn. CAB International, Wallingford, UK, pp. 423–454.

Randle, W.M. and Lancaster, J.E. (2002) Sulphur compounds in Alliums in relation to flavour quality. In: Rabinowitch, H.D. and Currah, L. (eds) *Allium Crop Science: Recent Advances*. CAB International, Wallingford, UK, pp. 329–356.

Randle, W.M., Bussard, M.L. and Warnock, D.F. (1993) Ontogeny and sulfur fertility affect leaf sulfur in short-day onions. *Journal of the American Society for Horticultural Science* 118, 762–765.

Randle, W.M., Lancaster, J.E., Shaw, M.L., Sutton, K.H., Hay, R.L. *et al.* (1995) Quantifying onion flavor compounds responding to sulfur fertility, sulfur increases levels of alk(en)yl cysteine sulfoxides and biosynthetic intermediates. *Journal of the American Society for Horticultural Science* 120, 1075–1081.

Randle, W.M., Kopsell, D.E. and Kopsell, D.A. (2002) Sequentially reducing sulfate fertility during onion growth and development affects bulb flavor at harvest. *HortScience* 37, 118–121.

Rickard, P.C. and Wickens, R. (1979) Effect of row arrangement and plant population on the yield of ware-sized bulb onions. *Experimental Horticulture (UK)* 31, 1–9.

Rodo, A.B. and Marcos-Filho, J. (2003) Onion seed vigor in relation to plant growth and yield. *Horticultura Brasileira* 21, 220–226. DOI: 10.1590/S0102-05362003000200020.

Rumpel, J. and Felczynski, K. (1998) Effect of plant density on yield and bulb size of direct-sown onions. *Acta Horticulturae* 533, 179–185.

Sance, M.M., González, R.E., Soto, V.C. and Galmarini, C.R. (2008) Relationships between antiplatelet activity, dry matter content, and flavor in onion cultivars. *Journal of Agriculture and Environment* 6, 41–46.

Schwimmer, S. and Weston, W.J. (1961) Enzymatic development of pyruvic acid in onion as a measure of pungency. *Journal of Agricultural and Food Chemistry* 9, 301–304.

Scully, N.J., Parker, M.W. and Borthwick, H.A. (1945) Interaction of nitrogen nutrition and photoperiod as expressed in bulbing and flower-stalk development of onion. *Botanical Gazette* 107, 52–61.

Searle, B.P. and Reid, J.B. (2016) Photothermal models to predict onion (*Allium cepa* L.) bulbing. *Acta Horticulturae* 1118, 65–72. DOI: 10.17660/ActaHortic.2016.1118.10.

Shannon, M.C. and Grieve, C.M. (2000) Options for using low-quality water for vegetable crops. *HortScience* 35, 1058–1062.

Sharma, K., Assefa, A.D., Kim, S., Ko, E.Y. and Park, S.W. (2014) Change in chemical composition of onion (*Allium cepa* L. cv. Sunpower) during post-storage under ambient conditions. *New Zealand Journal of Crop and Horticultural Science* 42, 87–98. DOI: 10.1080/01140671.2013.860039.

Shehata, S.A., Karima, F.A. and El-Mogy, M.M. (2017) Quality and shelf-life of onion bulbs influenced by biostimulants. *International Journal of Vegetable Science* 23, 4, 362–371. DOI: 10.1080/19315260.2017.1298170.

Shiomi, N., Benkeblia, N. and Onodera, S. (2005) The metabolism of the fructooligosaccharides in onion bulbs: a comprehensive review. *Journal of Applied Glycoscience* 52, 121–127.

Shishido, Y. and Saito, T. (1975a) Studies on the flower bud formation in onion plants. I. Effects of temperature, photoperiod and light intensity on the low-temperature induction of flower buds. *Journal of the Japanese Society for Horticultural Science* 44, 122–130.

Shishido, Y. and Saito, T. (1975b) Studies on the flower bud formation in onion plants. II. Effects of the physiological condition on the low-temperature induction of flower buds on green plants. *Journal of the Japanese Society for Horticultural Science* 44, 160–167.

Shishido, Y. and Saito, T. (1976) Studies on the flower bud formation in onion plants. III. Effects of physiological conditions on the low-temperature induction of flower buds on green plants. *Journal of the Japanese Society for Horticultural Science* 45, 160–167.

Shock, C., Feibert, E. and Saunders, L. (2004) Plant population and nitrogen fertilization for subsurface drip-irrigated onion. *HortScience* 39, 1722–1727.

Sobeih, W.Y. and Wright, C.J. (1986) The photoperiodic regulation of bulbing in onions (*Allium cepa* L.). II. Effects of plant age and size. *Journal of Horticultural Sciences* 61, 3, 337–341, DOI: 10.1080/14620316.1986.11515710.

Sorensen, J.N., Larsen, J. and Jakobsen, I. (2005) Mycorrhiza formation and nutrient concentration in leeks (*Allium porrum*) in relation to previous crop and cover crop management on high P soils. *Plant and Soil* 273, 101–114. DOI: 10.1007/s11104-004-6960-8.

Stumpf, S., Kvitko, B., Gitaitis, R. and Dutta, B. (2018) Isolation and characterization of novel *Pantoea stewartia* subsp. *indologenes* strains exhibiting center rot in onion. *Plant Disease* 102, 727–733. DOI: 10.1094/PDIS-08-17-1321-RE.

Swamy, K.R.M. and Veere Gowda, R. (2006) Leek and shallot. In: Peter, K.V. (ed.) *Handbook of Herbs and Spices*, vol. 3. Woodhead Publishing in Food Science, Technology and Nutrition, pp. 365–389.

Taylor, A., Massiah, A.J. and Thomas, B. (2010) Conservation of *Arabidopsis thaliana* photoperiodic flowering time genes in onion (*Allium cepa L.*). *Plant and Cell Physiology* 51, 1638–1647.

Teshome, A., Derbew, B., Sentayehu, A. and Yehenew, G. (2014) Effects of planting time and bulb size on onion (*Allium cepa* L.) seed yield and quality at Kobo Woreda, Northern Ethiopia. *International Journal of Agricultural Research* 9, 231–241.

Thompson, C. (2018) *E3371 Extension Bulletin Organic Garlic Production*. Michigan State University.

Tursun, N., Bekir, B., Sinan, K., Mathieu, N. and Hüsrev, M. (2007) Critical period for weed control in leek (*Allium porrum* L.). *HortScience* 42, 106–109. DOI: 10.21273/HORTSCI.42.1.106.

Van Kampen, J. (1970) Shortening the breeding cycle in onions. *Medelingen Proefstation voor de Groenteteelt in de Vollegrond* (47).

Vijn, I. and Smeekens, S. (1999) Fructan: more than a reserve carbohydrate? *Plant Physiology* 120, 351–359.

Volk, G.M., Henk, A.D. and Richards, C.M. (2004) Genetic diversity among U.S. garlic clones as detected using AFLP methods. *Journal of the American Society for Horticultural Science* 129(4), 559–569.

Walker, S., Ashigh, J., Cramer, C., Sammis, T. and Lewis, B. (2009) *Bulb Onion Culture and Management for Southern New Mexico*. Extension publication CR 563. College of Agricultural, Consumer, and Environmental Sciences, New Mexico State University, Las Cruces, New Mexico.

Wannamaker, M.J. and Pike, L.M. (1987) Onion responses to various salinity levels. *Journal of the American Society for Horticultural Science* 112, 49–52.

Wertheim, F., Douds, D., Handley, D. and Hutton, M. (2014) Evaluating the potential of arbuscular mycorrhizal fungi to boost yields in field-grown leeks. *Journal of the National Association County Agricultural Agents* 7. Available at: https://www.nacaa.com/journal/index.php?jid=281 (accessed 22 July 2022).

Winkler, L., Du Toit, L. and Waters, T. (2019) Mycorrhizal inoculants: Yea or nay? *Onion World*, 6 March. *Available at*: https://onionworld.net/2019/03/06/mycorrhizal-inoculants-yea-or-nay/#:~:text=The%20results%20suggest%20that%20AMF,when%20plants%20are%20under%20stress (accessed 22 July 2022).

Wohleb, C.H. and Waters, T.D. (2016) Yield, quality, and storage characteristics of onion cultivars in the Columbia Basin of Washington in 2012–14. *HortTechnology* 26, 230–243. DOI: 10.21273/HORTTECH.26.2.230.

Wright, P.J., Grant, D.F. and Triggs, C.M. (2001) Effects of onion (*Allium cepa*) plant maturity at harvest and method of topping on bulb quality and incidence of rots in storage. *New Zealand Journal of Crop and Horticultural Science* 292, 85–91. DOI: 10.1080/01140671.2001.9514166.

Yamasaki, A., Miura, H. and Tanaka, K. (2000) Effect of photoperiods before, during and after vernalization on flower initiation and development and its varietal difference in Japanese bunching onion (*Allium fistulosum* L.). *The Journal of Horticultural Science and Biotechnology* 75, 645–650. DOI: 10.1080/14620316.2000.11511301.

Zandstra, B. (2019) Extension Bulletin E-433, *Onion Weed Control Guide for Vegetable Crops*. Available at: https://www.canr.msu.edu/news/onion-weed-control-for-2019 (accessed 22 July 2022).

Zohary, D., Hopf, M. and Weiss, E. (2012) *Domestication of Plants in the Old World*. Oxford University Press, Oxford, UK.

12 Onion Seed Production

Ross Peters*
Agricultural Research Organization, The Volcani Institute, Israel

1. Introduction

This chapter is best placed in the context of related aspects of allium production. Within this book, explanations of the underlying physiology of bulbing and flowering of onions can be found in Chapters 4–10. Information on producing bulbs for cropping is found in Chapter 11. Chapter 13 covers crop protection for bulb production, with the caveat that onion seed production is more susceptible than the bulb crop to just about everything.

Books covering various aspects of onions and their seed production have been written several times since Jones and Mann published the first one in 1963. They include Rabinowitch and Brewster (1999) and Brewster (2008). Rabinowitch and Currah's (2002) book was quite comprehensive but did not include a chapter on seed production specifically. More recently, Khokhar (2019) was published and covers onion seed production in great depth. A number of publications describing or making recommendations for onion seed production, especially locally, have been published: Voss *et al.* (2013), University of Arizona (2006), Johan van der Westhuizen and Sons (PTY) Ltd (https://www.jwseeds.co.za/downloads_docs/onionseed.pdf; accessed September 2022), Pathak (2000), Lal and Kanwar (2017), Nikus and Mulugeta (2010), Lemma (1998), and Limeneh (2021) and Augagneur *et al.* (2020).

2. Seed Production Systems

Onion seed is generally produced in a biennial cycle, but in subtropical and Mediterranean climates, and in the phytotron, it can be produced in an annual cycle. High-latitude climates have spring-planted seed crops versus the autumn-planted seed crops found in Mediterranean, subtropical and tropical environments. Like most biennial crops, the plant can be tricked into producing seed in one year by planting it early enough in the autumn that it will enter the cold-weather period large enough to be vegetatively mature and sensitive to vernalization. This results in seed production, but prevents bulbing and dormancy from occurring. The physiology of this subject is covered in Chapter 4 on commercial bulb production, as well as Chapter 5 on onion and shallot physiology.

*Address for correspondence: rossthyra@gmail.com

 Edible Alliums: Botany, Production and Uses (eds H.D. Rabinowitch and B. Thomas)
DOI: 10.1079/9781789249996.0012

2.1. Bulb to seed

The biennial bulb-to-seed crop is the more traditional method of seed production. For many varieties and locations, it is the only method possible. It has several advantages over the seed-to-seed method. These advantages include much lower quantities of stock seed being needed in many cases, the ability to select the mother bulbs for maturity, colour and shape that are true to type for the variety, as well as for shelf life. It also allows for natural selection to improve the bolting resistance of the mother bulb population (Tarpaga *et al.*, 2011). The disadvantages include the longer planning time needed for production, the higher cost of planting material and the risk of losing the bulb crop to theft and rot, as well as losing the crop to the whims of nature for two years instead of one (Voss *et al.*, 2013).

2.2. Seed to seed

In some places, like California, seed-to-seed production is common practice (Voss *et al.*, 2013). Where the crop can be grown from seed to seed, the advantages include shorter planning time and potentially higher seed yields. The disadvantages include the necessity for a very high level of agriculture, higher stock-seed requirements (compensation for multiple flower stalks from one mother bulb where this occurs), and a longer period in the field for the seed production growing season. This results in higher water and fertilizer inputs as well as exposure to environmental risks for periods of the year that may be more difficult climatically. These come from the earlier planting and the later harvest time.

2.3. Set to seed

Efforts have been made to reap some of the benefits of both worlds by producing seed from small bulbs or sets (Peluffo *et al.*, 2016; Khokhar *et al.*, 2007a, b). This reduces the risks of losing the mother bulbs in storage since small bulbs store much better than large bulbs. It also allows for more flexibility in planning production for autumn-planted types, since planting for set production is usually in the early spring rather than in the autumn for mother bulbs.

2.4. Hybrid

Both the annual and biennial cycles can be used for open-pollinated (OP) and hybrid seed production. Hybrid seed production brings its own complications to the seed production process including getting a good 'nick'/match in flowering time between the male and female, the loss of seed from the portion of the field used for producing pollen, male fertile off-types in the female and increased danger from poor isolation from other varieties. These will be covered in the relevant sections of this chapter.

2.5. Bulbs produced in seed crop

In many varieties and environments, the seed-crop plants will produce bulbs as well as seed. These bulbs are generally not used for another seed crop, but they can be *in situ* or after transplanting to a new field. Leaving them *in situ* can cause issues with bulb preservation and with the final stand. Lazko and Yakimova (2021) discuss when the use of these bulbs might be worth the added risk and the effect of bulb orientation at planting, both on the current seed crop and on *in situ* production the following year.

3. Operations

3.1. Suitability of varieties to various locations and climates

No location in the world is suitable for producing seed of every onion variety. This is due to the huge cultivar variation in vernalization requirements and daylength influence. The seed production location needs to match the physiological requirements of the variety. Just like onion varieties need to be screened for bulb production purposes, they must also be screened for seed production purposes (Nikus and Mulugeta, 2010; Fufa *et al.*, 2021). This is true for bulb-to-seed production and even more so for seed-to-seed production (Branca and Ruggeri, 1994; Lemma, 1998). Generally, the location where a variety was bred is good for its seed production, but other locations are often used to spread risks, increase isolation distances and reduce costs.

Shallots and some other onions were usually produced by vegetative multiplication. True seed production of shallots was introduced by the Bejo Seed Co. in the 1980s. This practice reduces the accumulation of viral, bacterial and fungal diseases in the multiplication material. For this reason, the process of developing varieties suitable for seed production continues in Indonesia. When this is done, one of the main considerations is vernalization requirements (Wahyuni and Padang, 2020; Rosliani *et al.*, 2021). Khokhar (2009) looked at the effects of various temperature and daylength combinations on the flowering of two long-day set varieties, Stuttgarter and Sturon. They found that the longer the daylength (8–17 h) and the lower the temperature (10–25°C), the higher the number of florets produced. They did not produce seed. Whereas the longest daylength and highest temperature gave the earliest date for spathe opening. Khokhar (2008) did a similar experiment with eight varieties of various daylength reactions and received results with similarly shaped curves but with various degrees of slope and various maximum and minimum values for each variety. This speaks to the subject of what seed production locations are suitable for what varieties as well as to the possibility of shortening breeding time by using a phytotron.

Onion seed production is often limited by disease even more than bulb production (Edgerton, 1921; Msika *et al.*, 1994), since it is in the field for a longer period of time than bulb production and may carry disease inoculum with it on mother bulbs. In addition to these problems, at the beginning of the growing season the foliage is much more developed than in a bulb crop due to the earlier planting date and/or the larger planting material. This blocks wind flow, which prolongs the presence of dew on the leaves and thus presents a greater danger of foliar disease. Therefore, seed production fields need to be chosen with care, both climatically – drier regions, rain shadows etc. – and locations that can provide good ventilation and vernalization. If mother bulbs are vernalized before planting at another location or artificially, the seed field and planting time should be selected so that high temperatures at planting do not neutralize vernalization.

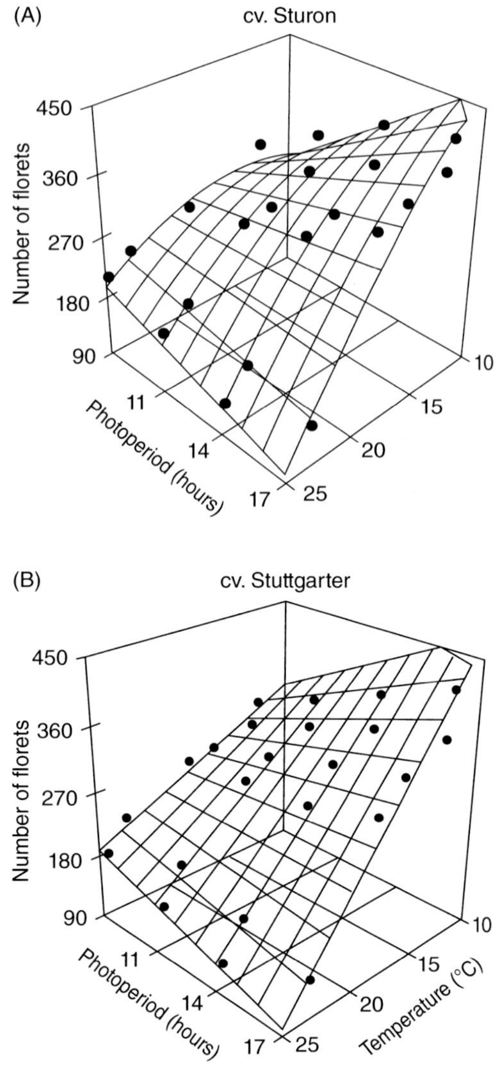

Fig. 12.1. Effect of temperature and photoperiod on number of florets per umbel. The plane was fitted by multiple regression analysis: (A) (NFU= -254.61 (+/-77.96) +40.7 (+/-7.68)T +45.73 (+/-3.5) P -0.78 (+/-0.20)T^2 -2.22 (+/-0.19)TP, r^2+0.93, 67 d.f.); (B) (NFU=149.66 (+/-37.45) +4.52 (+/-2.03)T +34.3 (+/-2.9)P -1.76 (+/-0.16)TP, r^2=0.95, 68 d.f.); where NFU, T and P represent number of florets per umbel, temperature and photoperiod, respectively (Khokhar, 2008).

Fig. 12.2. A good onion seed field location.

3.2. Preparation and storage of mother bulbs

The subject of mother bulbs is not relevant to annual-cycle seed production. That said, mother bulb production is very similar to production of commercial bulbs, but with several important differences. The objective of mother bulb production is good storage and high yields of the bulb size desired for seed production. This is generally smaller than the market prefers. For commercial bulbs, the objective is high yields, high quality (uniformity, good skins etc.) and hitting the market when prices are best, which may involve early harvest or long shelf life. Bulb production practices can be read about in chapters 4 and 11 on onion physiology and bulb crop production.

Spring-planted seed production generally results in a single umbel per bulb. Autumn-planted varieties often split more before umbels emerge. Production of highly single-centred varieties may compensate for the reduced number of umbels per bulb with a higher number of smaller bulbs. This is not possible for varieties with high resistance to bolting; they need larger mother bulbs that are more sensitive to vernalization, i.e. Japanese overwintering varieties. Larger bulbs have a lower vernalization requirement than smaller ones. Interestingly, larger onion plants need less long-day and high-temperature exposure to mature. In general, the larger the bulb, the more seed is produced (Ali *et al.*, 1998, 2015; Badawi *et al.*, 2010; Asaduzzaman *et al.*, 2012a, b; Haile *et al.*, 2017; Amare et al., 2020; Das *et al.*, 2020). This increase does not usually compensate for the additional cost. In general, higher populations of lower-weight bulbs give a better return, if price per ton of bulbs is the same. With sets, smaller sizes may not flower at all (Ashrafuzzaman *et al.*, 2009). The ideal size for sets for two varieties was investigated by Khokhar (2009).

In tropical countries, where winter temperatures may not provide sufficient conditions for vernalization, artificial vernalization can be applied before planting (Kimani *et al.*, 1993; Msika *et al.*, 1994; Khokhar *et al.*, 2007a, b; Badawi *et al.*, 2010; Ami *et al.*, 2013; Khokhar, 2014; Khatun *et al.*, 2020; Lokesh *et al.*, 2020). The consensus is that storage at 10–12°C for two months or more is usually optimal, but this can vary greatly with the requirements of individual varieties and the number of cold units the environment provides after planting. Even in England, cold-temperature treatment of sets was found to increase flowering rates (Khokhar *et al.*, 2007a). In some countries, mother bulbs for planting can be a major part of the expense for growing onion seed. Where labour is cheap, autumn-planted mother bulbs may be divided to obtain a higher seed yield from the same planting material (Ghosh *et al.*, 2018). Cutting onions horizontally is sometimes done to (i) inspect centres for rot and excessive splitting or to evaluate a portion of the bulb for soluble solids, pyruvic acid and other molecules; (ii) provide edible material; and (iii) encourage sprouting. This practice may improve yield, especially when breaking dormancy extends the growing season (Green, 1972). Generally, a quarter to one third of the onion is sacrificed. Drying the cut surface with ash can reduce subsequent rotting (Abrha *et al.*, 2020).

3.3. Time of planting

In general, the earlier the planting date, the longer the growing season and the higher the potential yield (Tesfaye *et al.*, 2018; Tekle *et al.*, 2020; Ashagrie *et al.*, 2021). In autumn-planted crops, very early planting of early varieties may not produce the best results (Anisuzzaman *et al.*, 2009; Das *et al.*, 2020). They run the risk of bulbing, topping and dormancy before winter vernalization comes. This leads to very late flowering and low yields. Late planting, conversely, reduces the length of the growing season and potential yield. In spring-planted varieties, planting too early can lead to freeze damage and rot. Therefore, the selection of the best planting date needs to find a happy medium between too early and too late, on average, over several years.

When there is flowering/bolting at the mother bulb stage, the bulbs that do not bolt are naturally selected for higher bolting resistance. In contrast, if flowering during the seed production stage is not 100%, natural selection will be for easy bolting. Therefore, optimal planting dates for preserving bolting resistance should be used at both stages (Tarpaga *et al.*, 2011).

For hybrid production, synchronized flowering of the male and female parents is essential. It is generally preferred that the male starts flowering slightly before the female so that any fertiles in the female, and pollen coming from a lack of isolation, must compete with pollen from the male rather than pollinating all of the seed until the male starts flowering. A good supply of pollen is essential for the entire flowering period of the female for the same reasons. Split plantings of the male can be used to ensure a continuous and adequate supply of pollen. Flowering date is much less influenced by daylength than the maturation of bulbs. Therefore, flowering date can be adjusted to a certain extent by the planting date. The influence of planting date on flowering date is greater at the beginning of the planting window than at the end.

The planting window for annual or seed-to-seed production is quite narrow due to the problem of sowing too early and getting bulbing, as well as the problem of sowing too late and the resulting plants still being juvenile during the cold period. Flowering in the annual cycle is generally later than flowering from bulbs. This reduces the window during which flowering occurs. Even so, some manipulation of flowering date with sowing date is possible.

3.4. Choosing a field

Selecting a field within a region/location is similar to that of any crop, i.e. rotation, good soil, freedom from weeds. See Chapters 4 and 11 on the development of onions. As far as onion seed production is concerned, drier, good drainage locations with more sun exposure are desired. The field should not have airflow blocked by surrounding trees and/or windbreaks. All of these will contribute to the health of the field as far as foliar diseases are concerned. This is very important since they are generally the most limiting factor for onion seed production. When the field is used for biennial production, freedom from stones and pebbles is less important since the bulbs can better manage to root and establish themselves in coarse soil than can the small seeds of onions.

Because onions are a cross-breeding and insect-pollinated crop, isolation from other onion fields is very important. Various regions have different mechanisms for alerting seed producers to isolation problems so that they can co-operate in obtaining good isolations from each other. Even bulb-onion crops can be problematic if they have bolting. Most organizations recommend isolations of 1.5–5 km between fields depending upon the difference between the varieties being produced; similar varieties may be closer, less similar should be further apart. Where isolation is problematic due to the small size of the production and the large number of varieties to be produced, insect-proof cage isolation is used (Voss *et al.*, 2013).

3.5. Depth of planting and bulb orientation

In very cold locations onions may be planted in the autumn despite freezing temperatures in the winter. They are usually planted deep enough in the soil to insulate them from freezing. Lazko and Yakimova (2020) looked at planting depths in Russia and concluded that while 10 cm depth protected from freezing, it did not prevent lodging of the flower stalks. They found that a 15 cm depth was ideal for both.

3.6. Spacing and male-to-female ratios

Generally speaking, the higher the number of umbels per unit of land area, the higher the yield. This generalization hits an upper limit when the density is so high that foliar diseases become difficult to control. In ideal locations, around 700,000 umbels per ha is recommended. The plant stand for annual cycle production is also guided by that. Similarly, the number of bulbs planted takes into account the number of umbels per bulb expected (Asaduzzaman *et al.*, 2012a, b; El Abas *et al.*, 2015; Haile *et al.*, 2017).

Some growers leave empty beds in their fields to allow spray rigs to apply chemicals with less damage to the growing plants.

Hybrid seed producers are concerned with the amount of land 'wasted' on the male. Generally, a 1:3 or 1:4 ratio is used. The male plants may be more crowded than the females, especially if no seed is to be harvested from them. The distance between rows devoted to the males usually ranges between 6 m and 8 m (Rabinowitch and Brewster, 1990).

3.7. Fertigation – dry land versus irrigated

Of all of the possible irrigation methods, drip-irrigation is increasing the most, both because it saves water and fertilizer and because of the lower humidity created, thus reducing foliar disease risk (Dingre *et al.*, 2012). Rainfed production is still used in some locations. Where it is, stands are generally much lower so that each plant has a larger potential root volume from which to extract water if rains are not frequent. In Israel, dry-land onion crops did not generally get any rain of more than a few millimetres from April, when significant rains cease, to June, when seed was harvested. Investigations into the best irrigation regime for a particular location and soil should be conducted (Bhonde *et al.*, 1996).

Mulching is generally not practised in large-scale farming, but small-scale farmers use it frequently to reduce water use, control weeds and maintain soil structure. Both artificial and natural mulches are used. Anisuzzaman *et al.* (2009) investigated the effect on onion seed yield by representatives of both.

Fertilization for seed production is usually rather minimal compared to bulb production and primarily directed at the period before flowering. High levels of P and K can accumulate in the nectar and repel honeybees. As in bulb production, fertilization levels need to be appropriate to the yield levels expected or desired. 150 kg/ha of N and 120 kg/ha of K

Fig. 12.3. Rice-straw mulched onions.

returned over 500 kg of seed/ha when investigated by Ali *et al.* (2007). Various types of fertilizer were investigated by Adagale *et al.* (2009), nitrogen levels by El Abas *et al.* (2015) and cow dung fertilization by Ali *et al.* (2008). A multi-factorial experiment was conducted by Khang *et al.* (2011) looking at soybean rotation, *Azotobacter* augmentation, and organic and chemical fertilizers. Zinc and boron are essential minor elements necessary for best seed production yields. Haque *et al.* (2014) found that 7 kg and 6 kg/ha, respectively, were sufficient for good seed yields (approx. 1000 kg/ha) under their conditions. Howlader *et al.* (2012) confirmed the important contribution of boron to high seed yields. Amare *et al.* (2020) found that 115 P_2O_5 & 114 N kg ha^{-1} fertilizers could be recommended for high yield and quality onion seed production.

Sometimes fertilization by land application or through the irrigation system can be difficult. When that happens, foliar application can be effective (Mansoor *et al.*, 2019).

Nanoparticles of minor fertilizer elements are being experimented with to increase their availability. Nanoparticles are expanding in use greatly. Many tons of both metal and organic nanoparticles are produced every year. While having many advantages, they can be poisonous to many living creatures. The dangers of soil and other types of contamination are possibly a greater threat to our environment than GMOs (Hayat *et al.*, 2020). Nano zinc was found to increase seed yields and the vigour of seedlings produced from treated plants (Laware and Raskar, 2014). Unfortunately, chelated zinc was not a control in this experiment.

3.8. Roguing

Roguing, or the removal of undesirable plants, is an added expense. Obviously, pure stocks of basic seed and mother bulbs can save a lot of money if they eliminate the need for roguing. Roguing to remove individuals not true to type can be conducted at many stages in the production process; in the mother bulb production field, during planting of the bulbs, in the seed field during the sprouting and growing of the crop; and at flowering. Roguing is generally for colour off-types, but it can also be for foliage type, plant height, flower height and fertiles in the female. Each type of roguing is most efficiently conducted at a specific stage in the process. Electronic eye sorting of mother bulbs is probably the least expensive, since it is done by machine.

3.9. Diseases, pests and weeds

Pest control for onion seed production is similar to that for bulb production, but has the big limitation that it can seriously interfere with insect pollination. Insecticides and their residues, even when not lethal to bees and other pollinators, can repel them. Additionally, they are suspected of having deleterious effects on onion flowers (Gent *et al.*, 2006; Long and Morandin, 2011; Gillespie *et al.*, 2014).

Weed control in the seed crop is similar to that in a bulb crop (Kalhapure *et al.*, 2014; Pannacci *et al.*, 2020). Very low levels of weed infestation can have a serious economic effect on seed quality since contamination at very low levels with seed of some weed species, such as Johnson grass, witchweed and morning glory, can render a seed lot unsaleable in some countries. Some weeds have seeds that are very difficult to remove using seed-cleaning equipment, since they are so similar to onion seed in size, shape and colour.

Hand weeding at flowering and later can be more damaging than beneficial; so it is very important that the field be clean before flowering (Kolse *et al.*, 2010). Chemical control with repeated applications of low levels of oxyfluorfen to burn up dicot foliage, when applied below the onion inflorescence, is an option since this does not damage the onion plant.

The same diseases that affect the bulb-onion crop also affect the seed crop, but since the seed crop is often in the field longer than a bulb crop, and due to its denser foliage, it is even more prone to attack by foliar diseases (Gonzalez *et al.*, 2011). Jangwad *et al.* (2021) found that farmers evaluated diseases and pests to be the largest limiting factor in onion seed production.

Purple blotch caused by *Alternaria* and/or *Stemphylium* is a common problem in the tropics. A number of studies have addressed its chemical control in seed production (Islam *et al.*, 2020; Paneru *et al.*, 2020; Sofy *et al.*, 2020).

Botrytis spp. can attack the flower stalk just below the umbel. Abo-Elyousr *et al.* (2020) used molecular markers to identify the various *Botrytis* spp. and strains that did this and looked at both biological and chemical control. They found that while some *B. subtilis* strains reduced infection by 60%, Ridomil Gold gave 100% control. Iris yellow spot virus (ISYV) and its vector, thrips, have been under intense investigation for the last 25 years. While the incidence of ISYV has increased greatly over this period, the ability to control thrips with chemicals has decreased. It was found that the waxy foliage varieties with a bluish cast found on many varieties from higher-rainfall regions are more attractive to thrips, while the semi-glossy varieties with a green colour that are more suited to the deserts are less attractive. This phenomenon is compensated by the fact that rains reduce thrips infestation. The thrips are not often given a choice. But seed growers, who are usually located in desert regions, should be aware that the waxy varieties need more attention than the semi-glossy ones. ISYV is a more serious problem for seed growers than for bulb growers for the same reasons as mentioned previously (Gent *et al.*, 2006).

3.10. Hormone treatments

Onion seed fields can be treated with gibberellins to speed up and synchronize inflorescence emergence and flowering (Ruggeri and Branca, 1994; Geetharani *et al.*, 2008; Khatun *et al.*, 2020; Lokesh *et al.*, 2020). This is sometimes done when the vernalization period is marginally cold enough, which can result in less uniformity of flowering date and a lower percentage of flowering, although GA has not been found to substitute for vernalization. Cytokinins and gibberellin antagonists can be applied to dwarf the scape and reduce lodging (Ashdafuzzaman *et al.*, 2009; Kamboj *et al.*, 2017).

3.11. Pollination

The biology of onion seed pollination was reviewed by Devi *et al.* (2015b). Pollinator behaviour in hybrid cauliflower seed production was studied by Gagic *et al.* (2020). The fact that when external sources of insects are excluded onion seed production is very low was reinforced by the work of Rasekh *et al.* (2013) and Korichi *et al.* (2021).

Small production fields neighboured by uncultivated land can get good insect pollination from many insects (Chandel *et al.*, 2002; Kalmath, 2002; Walker *et al.*, 2011; Kavitha and Reddy, 2019; Korichi *et al.*, 2021; Saleh *et al.*, 2021). Large fields surrounded by modern agriculture rarely have sufficient native pollinators to do a good job. Honeybees are the most common pollinators used for onion seed production, although carrion and house flies are often used in cages alone or as supplements to honey bees. Beehives are usually placed in onion seed production fields at the rate of about 5–10/ ha. They are generally added gradually rather than all at once to take advantage of the behaviour of naive nectar gatherers. Soto *et al.* (2021) investigated the sugar profile of onion nectar over the flowering period and compared that to bee visits. They found a strong positive correlation with fructose and glucose and a strong negative correlation with sucrose. They recommended initiating hive placement when 50% of plants had at least one open flower with staggered supplementation after that.

Nectar is a natural attractant to all pollinators. Any agronomic practices that promote nectar production increase the attractiveness of the crop to pollinators (Gillespie *et al.*, 2015; Kavitha and Reddy, 2019). There is a genetic component to the attractiveness of onion nectar that breeders should be aware of, especially concerning the relative attractiveness of male and female lines (Kavitha and Reddy, 2015; Soto *et al.*, 2015, 2018). Bee attractants have been offered for application to onion flowers for decades. Few scientific investigations have found them very effective (Mayer *et al.*, 1989; Kalmath, 2002; Abdelghafar *et al.*, 2016; More *et al.*, 2020).

Honeybee pollination, while it is the best supplement to natural pollinators, is quite problematic. Onions, specifically, are not the most attractive crop to bees and the bees can often be lured away to pollinate other crops (up to 5 km). Bees do not intentionally gather onion pollen. This forces them to find other sources of pollen for food unless supplied by the beekeeper.

Honeybee-keeping is becoming more and more difficult around the world, especially for the most docile strains. Inspecting hives upon initial introduction to onion seed fields to ensure that they reach minimum standards greatly improves their effectiveness (Gabai *et al.*, 2018). This document from the ISF goes into great detail about how to optimize pollination. Many countries have a beekeeper's association that can provide inspection services for a fee.

Seed production in cages is often used for breeding and variety maintenance purposes to obtain the isolation required. Seed yield in cages is usually lower than in the open field. Caged seed production often uses blowflies or house flies instead of or with honeybees (Munawar *et al.*, 2011). Devi *et al.* (2015a) reported on the effect of bee population size on seed yield in cages.

4. Seed Harvest

Onions, unlike most crops, have not been selected for retaining their seed and shatter quite easily when they ripen on the plant. To prevent this, the umbels are harvested when the first seeds are ripe with a long portion of the stalk attached. This allows the unripe fruits to continue to receive moisture, continue to photosynthesize and ripen immature seed. They are placed on plastic sheeting, threshing floors or in ventilated bags and allowed to ripen and dry. The seed usually takes about a month to be ready to thresh. In many locations, they are dried under a roof so that they will not be wet by rain nor overheated by the sun. They are turned every day or so to prevent rotting (Voss *et al.*, 2013).

Onion seed is usually harvested by hand when the first 10% of umbels have some black seed showing. This is done when it is expected that it will take a week or more to harvest the field. More and more, mechanical harvesting is being used due to the cost and limited availability of labour at harvest time. If this means that the field can be harvested in a day or two rather than a week, the seed can be riper when harvesting starts. This means that more seed will be shattered initially, but that more of the seed will reach maturity during the ripening and drying process. The effects of harvest date on seed yield and quality were investigated by Spurr *et al.* (2002). They looked at umbel moisture content as a measure of seed maturity and for deciding upon harvest date. Badawi *et al.* (2010) found no significant difference in hand-harvested seed yield when comparing harvesting on 150, 157 and 164 days after planting.

4.1. Seed yield

Seed yields vary greatly over the world and even locally with one variety over years (Voss *et al.*, 2013). This greatly complicates planning and inventory management. It also makes the price of onion seed quite high since the risk is reflected in the price. Outstanding yields of over 1000 kg ha^{-1} occur most commonly in seed-to-seed production and with agriculturally sophisticated growers where drip-irrigation and chemical pest, disease and weed control can be used in combination with high stands. This should not discourage those with more limited resources, since lower yields may be linked to lower input costs (bulbs, chemicals, fertilizer, labour, mechanization etc.) (Aklilu, 1994). The important factor is the cost/kg and not kg/ha.

Lots produced in low-yielding years are commonly associated with low 1000 seed weight, poor germination, poor shelf life and more off-types. This compounds the losses due to reduced seed yield. Many investigations of seed yield also look at the effect of the variable in question on seed quality, particularly germination. Sometimes the best yields correlate with the best germination.

4.2. Quality control

Once the seed is cleaned, it needs to be evaluated for quality; 1000 seed weight, germination under standard conditions, contamination with foreign material, noxious weeds and, more and more, bacteria (Walcott *et al.*, 2002; Humeau *et al.*, 2006), and fungi that can cause onion diseases. Germination under extreme temperatures may be evaluated to see how the seed will perform under especially cold and extremely hot conditions as well as checking how quickly seedlings emerge. These responses may be referred to as vigour. After evaluation, the seed may be re-cleaned, sized, blended with other lots, treated, pretreated, coated and/or pelleted before packaging. Tajbakhsh *et al.* (2004) reported that abnormal seedlings with stunted roots found in germination tests could be ameliorated with various priming treatments. Rodo and Marcos-Filho (2003) investigated a new method of evaluating seed vigour and a rapid test for seed ageing.

4.3. Seed storage

Onion seeds are notorious for having poor shelf life (Nikus and Mulugeta, 2010). They usually do not germinate well after one year if they are kept in room conditions. Low and even temperature, low oxygen and low humidity all contribute to good shelf life. A good rule of thumb is that if the storage temperature in degrees F and the relative humidity add up to less than 100, conditions are good and the seed should lose its germ slow enough to last for several years once they have been dried to 6% moisture. Vacuum packing seed greatly boosts shelf life. Under low-input conditions, well-dried seed placed in a sealed container with a lighted candle to consume the oxygen can give good results when stored in the dark at room temperature.

4.4. Climate change as it affects onion seed production

Onion seed production has always been a risky business. With the change in the climate, we are encountering more variable conditions and, in general, slightly warmer conditions. Higher variability increases the chances of encountering unfavourable ones. Higher temperature, especially during winter vernalization, can reduce flowering percentage.

5. Conclusions

A number of improvements in onion seed production have been achieved. Good beehive preparation and inspection have shown their worth. Many handbooks – guides to local seed production have been published expanding professional onion seed production in many countries. How genetics interacts with nectar quality, vernalization and daylength response for seed production is better understood. Many aspects of onion seed production have become more challenging including the production of quality bee colonies and climatic variation. The control of thrips and ISYV continues to be more difficult as the years go by.

References

Abdelghafar, M.S., Al-Abd, M.T., Helaly, A.A. and Rashwan, A.M. (2016) Foliar application of lithovit and rose water as factor for increasing onion seed production. *Natural Sciences* 14(3), 53–61.

Abo-Elyousr, K.A., Alamri, S.A., Hussein, M.M., Hassan, M.A., Abd El-Fatah, B.E. and Hashem, M. (2020) Molecular disparities among Botrytis species involved in onion umbel blight disease and its management using Bacillus subtilis PHYS7. *Egyptian Journal of Biological Pest Control* 30(1), pp. 1–12.

Abrha, H., Gebretsadik, A. and Tesfay, G. (2020) Evaluation of different onion seed production methods for improving production and productivity of onion in Southern Tigray. *Evaluation – Journal of Natural Sciences Research* 10(9) (ISSN 2225-0921).

Adagale, S.V., Masalkar, S.D. and Pandure, B.S. (2009) Effect of integrated nutrient management on growth and yield of onion seed production. *The Asian Journal of Horticulture* 4(2), 484–487.

Aklilu, S. (1994) Onion research and production in Ethiopia. *Acta Horticulturae* 433, 95–98.

Ali, M.A., Hossain, M.M., Zakaria, M., Naznin, A. and Islam, M.M. (2015) Effect of bulb size on quality seed production of onion in Bangladesh. *International Journal of Agronomy and Agricultural Research* 6(4), 174–180.

Ali, M.K., Alam, M.F., Alam, M.N., Islam, M.S. and Khandaker, S.M.A.T. (2007) Effect of nitrogen and potassium level on yield and quality seed production of onion. *Journal of Applied Sciences Research* 3(12), 1889–1899.

Ali, M.K., Alam, M.N., Islam, M.S., Islam, M.K. and Baree, M.A. (2008) Effect of cowdung at different level of phosphorus on growth, yield and quality seed production of onion. *Research Journal of Agriculture and Biological Sciences* 4(1).

Ali, N., Baloch, M.A. and Hussain, S.A. (1998) Study on the effects of planting space and bulb size on seed. *Sarhad Journal of Agriculture* (Pakistan)14, 563–568.

Amare, G., Mohammed, W. and Tana, T. (2020) Effect of plant spacing and Np fertilizer levels on growth, seed yield and quality of onion (L.) at Shewa Robit, Northern Ethiopia. *The Open Biotechnology Journal* 14(1).

Ami, E.J., Islam, M.T. and Farooque, A.M. (2013) Effect of vernalization on seed production of onion. *Agriculture Forestry & Fishing* 2(6), 212–217.

Anisuzzaman, M., Ashrafuzzaman, M., Ismail, M.R., Uddin, M.K. and Rahim, M.A. (2009) Planting time and mulching effect on onion development and seed production. *African Journal of Biotechnology* 8(3).

Ashdafuzzaman, M., Millat, M.N., Ismail, M.R., Uddin, M.K., Shahidullah, S.M. *et al.* (2009) Paclobutrazol and bulb size effect on onion seed production. *International Journal of Agriculture And Biology* 11(3), 245–250.

Asaduzzaman, M.D., Hasan, M.M., Hasan, M.M., Moniruzzaman, M.D. and Howlander, M.H.K. (2012a) Effect of bulb size and plant spacing on seed production of onion (*Allium cepa* L.). *Bangladesh Journal of Agricultural Research* 37(3), 405–414.

Asaduzzaman, M.D., Hasan, M.M. and Moniruzzaman, M. (2012b) Quality seed production of onion (Allium cepa L.): an integrated approach of bulb size and plant spacing. *Indian Journal of Agricultural Research* 50(1), 119–128.

Ashagrie, T., Belew, D. and Nebiyu, A. (2021) Influence of planting date and bulb size on yield and quality of onion (*Allium cepa* L.) seed production. *Cogent Food & Agriculture* 7(1), 1908656.

Augagneur, M., Brun, L., Laurent, E., Dorand, P. and Conseil, M. (2021) *Produire des semences en agriculture biologique-Oignon*. FNAMS, Loire-Authion, ITAB, Paris, p. 1–6.

Badawi, M.A., Seadh, S.E., El-Emery, M.I. and Shalaby, A.E.M. (2010) Onion seed yield and its quality as influenced by storage methods, mother bulb size and harvesting time. *International Journal of Plant Production* 1(2), 239–249.

Bhonde, S.R., Mishra, V.K. and Chougule, A.B. (1996) Effect of frequency of irrigation and nitrogen levels on yield and quality of onion seed variety Agrifound Light Red. *National Horticultural Research and Development Foundation News Letter* 16(3), 4–7.

Branca, F. and Ruggeri, A. (1994). Reproductive response of onion seed plants to photothermal conditions. *Acta Horticulturae* 362, 25–34.

Brewster, J.L. (2008) Onions and Other Vegetable Alliums, 2nd edn. *Crop Production Science in Horticulture* 15, pp.1–432.

Chandel, R.S., Thakur, R.K., Bhardwaj, N.R. and Pathania, N. (2002) Onion seed crop pollination: a missing dimension in mountain horticulture. *Acta Horticulturae* 631, 79–86.

Das, T., Mandal, J. and Mohanta, S. (2020) Influence of mother bulb weight and spacing on quality seed yield of onion. *Indian Journal of Ecology* 47(4), 1130–1134.

Das, B.C., Tripathy, P., Sahu, G.S., Das, A.K., Swain, S.K. *et al.* (2021) Effect of planting dates and varieties on seed yielding attributes of onion (*Allium cepa* Linn.) under west central table land zone of Odisha. *Journal of Pharmaceutical Innovation* 10(7), 151–154. ISSN 2277-7695.

Devi, S., Gulati, R., Tehri, K. and Poonia, A. (2015a) Optimum population of *A. mellifera* required for seed production in onion. *Bioinfolet* 12(3a), 593–596.

Devi, S., Gulati, R., Tehri, K. and Poonia, A. (2015b) The pollination biology of onion (*Allium cepa* L.) - a Review. *Agric Reviews* 36(1), 1–13.

Dingre, S.K., Pawar, D.D. and Kadam, K.G. (2012) Productivity, water use and quality of onion (*Allium cepa*) seed production under different irrigation scheduling through drip. *Indian Journal of Agronomy* 57(2), 186–190.

Edgerton, C.W. (1921) *Onion Diseases and Onion Seed Production* (No. 182). LA AES BULL.

El Abas, S.I., Ali, A.M., Mohamed, O.E. and Nourai, A.H. (2015) Effects of nitrogen fertilization and bulb spacing on'Saggai Red'onion seed production in Berber area of River Nile State, Sudan. *Acta Horticulturae* 1143, 235–244.

Fufa, N., Gebretensay, F., Tsagaye, D., Fikre, D., Wegayehu, G. *et al.* (2021) Evaluation of seed production and quality performance of onion (*Allium cepa* L.) varieties at Kulumsa in Arsi Zone South East Ethiopia. *Journal of Agriculture and Crops* 7(1), 33–38. ISSN 241-6381.

Gabai, A., Vaissière, B.E., Blacquiere, T., Freitas, B.M., Allsopp, M. *et al.* (2018) Protocol for using Pollinators in Hybrid Seed Production: An Outline for Improving Pollinator Effectiveness. *International Seed Federation.*

Gagic, V., Kirkland, L., Kendall, L.K., Jones, J., Kirkland, J. *et al.* (2020) Understanding pollinator foraging behaviour and transition rates between flowers is important to maximize seed set in hybrid crops. *Apidologie* 1–12.

Geetharani, P., Manivannan, M.I. and Ponnuswamy, A.S. (2008) Seed production of onion as influenced by the application of growth regulators and nutrients. *The Asian Journal of Horticulture* 3(2), 301–303.

Gent, D.H., du Toit, L.J., Fichtner, S.F., Mohan, S.K., Pappu, H.R. *et al.* (2006) Iris yellow spot virus: an emerging threat to onion bulb and seed production. *Plant Disease* 90(12), 1468–1480.

Ghosh, D.K., Biswas, L.B., Das, S., Kartick, C.S. and Bandyopadhyay, A. (2018) Minimizing mother bulb requirement through mechanical option: A cost friendly approach of onion seed production. *International Journal of Current Microbiology and Applied Sciences* 7(7), 2646–2655.

Gillespie, S., Long, R., Seitz, N. and Williams, N. (2014) Insecticide use in hybrid onion seed production affects pre-and postpollination processes. *Journal of Economic Entomology* 107(1), 29–37.

Gillespie, S., Long, R. and Williams, N. (2015) Indirect effects of field management on pollination service and seed set in hybrid onion seed production. *Journal of Economic Entomology* 108(6), 2511–2517.

González, P.H., Colnago, P., Peluffo, S., Idiarte, H.G., Zipitría, J. *et al.* (2011) Quantitative studies on downy mildew (*Peronospora destructor* Berk. Casp.) affecting onion seed production in southern Uruguay. *European Journal of Plant Pathology* 129(2), 303–314.

Green, J.H. (1972) The influence of bulb size, bulb cutting and separation of axillary shoots on seed production of onion (*Allium cepa* L.). *Journal of Horticultural Sciences* 47(3), 365–368.

Haile, A., Tesfaye, B. and Worku, W. (2017) Seed yield of onion (*Allium cepa* L.) as affected by bulb size and intra-row spacing. *African Journal of Agricultural Research* 12(12), 987–996.

Haque, M.R., Robbani, M., Hasan, M.M., Asaduzzaman, M., Hasan, M.M. *et al.* (2014) Zinc and boron affect yield and quality of onion (*Allium cepa* L.) seed. *International Journal of Vegetable Science* 20(2), 131–140.

Hayat, S., Pichtel, J., Faizan, M. and Fariduddin, Q. (eds) (2020) *Sustainable Agriculture Reviews 41: Nanotechnology for Plant Growth and Development*, vol. 41. Springer Nature.

Howlader, M.M., Yousuf, M.N., Kamal, M.M., Khatun, M.R. and Bashar, H.M.K. (2012) Effect of phosphorus, potassium and boron on onion seed production. *International Journal of Agricultural Sustainability* 8(7), 28–30.

Humeau, L., Roumagnac, P., Picard, Y., Robène-Soustrade, I., Chiroleu, F. *et al.* (2006) Quantitative and molecular epidemiology of bacterial blight of onion in seed production fields. *Phytopathology* 96(12), 1345–1354.

Islam, M.M., Begum, F., Nahar, N., Habiba, U.A. and Fakruzzaman, K.M. (2020) In-vivo management of purple blotch of onion caused by *Alternaria porri* (Ellis) Cif. through fungicides. *American Journal of Plant Sciences* 11(11), 1847–1859.

Jangwad, N.P., Gaware, K.M., Kale, N.M., Bhople, P.P., Salame, S.P. *et al.* (2021) Constraints experienced by the onion seed producers regarding the onion seed production in Akola district of Maharashtra state. Journal of Pharmaceutical Innovation. *TPI* 10(2), 72–75 *ISSN* 2277-7695.

Johan van der Westhuizen and Sons (PTY) Ltd. Available at: https://www.jwseeds.co.za/downloads_docs/onionseed.pdf (accessed 15 May 2020).

Jones, H.A. and Mann, L.K. (1963) *Onions and Their Allies: Botany, Cultivation and Utilization*. Leonard Hill, London and Interscience, New York.

Jones, H.A. and Mann, L.K. (1964) Onions and their allies. *Soil Science* 98(1), 68.

Kalhapure, A., Shete, B. and Dhonde, M. (2014) Weed management in onion by pre-planting and post-emergence herbicides for seed production. *Indian Journal of Weed Science* 46(2), 142–145.

Kalmath, B.S. (2002) Impact of bee pollination on onion seed production. Doctoral dissertation, University of Agricultural Science, Dharwad, India.

Kamboj, N.V., Batra, V.K., Vilas, C.A. and Sharma, P.K. (2017) Effect of planting dates and paclobutrazol on yield and quality of onion (*Allium cepa* L.) seed. *International Journal of Pure and Applied Bioscience* 5(1), 417–424.

Kavitha, S.J. and Reddy, P.R. (2019) Influence of floral nectar quantity and quality on honey bee foraging activity in onion. *Journal of Entomology and Zoology Studies* 7(3), 92–96.

Khang, V.T., Patil, H.M. and Gudedhe, N.N. (2011) Effect of integrated nutrient management on onion yield and soil properties in soybean–onion cropping sequence. *Omonrice* 18, 112–120.

Khatun, L., Karim, M.R., Talukder, F.U. and Rahman, M.S. (2020) Combined effects of vernalization and gibberellic acid on quality seed production of summer onion (*Allium cepa* L.). *Agricultural Sciences* 2(2), 148–148. ISSN 2690-4799.

Khokhar, K.M. (2008) Effect of temperature and photoperiod on the incidence of bulbing and bolting in seedlings of onion cultivars of diverse origin. *The Journal of Horticultural Science and Biotechnology* 83(4), 488–496.

Khokhar, K.M. (2009) Effect of set-size and storage temperature on bolting, bulbing and seed yield in two onion cultivars. *Scientia Horticulturae* 122(2), 187–194.

Khokhar, K.M. (2014) Flowering and seed development in onion: a review. *Open Access Library Journal* 1(7), 1–13.

Khokhar, K.M. (2019) *Onion an Ancient Crop and Modern Practices: A Review*. Noor Publishing, Chisinau, Moldova.

Khokhar, K.M., Hadley, P. and Pearson, S. (2007a) Effect of cold temperature durations of onion sets in store on the incidence of bolting, bulbing and seed yield. *Scientia Horticulturae* 112(1), 16–22.

Khokhar, K.M., Hadley, P. and Pearson, S. (2007b) Effect of photoperiod and temperature on inflorescence appearance and subsequent development towards flowering in onion raised from sets. *Scientia Horticulturae* 112(1), 9–15.

Kimani, P.M., Kariuki, J.L.W., Peters, R. and Rabinowitch, H.D. (1993) Potential of onion seed production in a tropical environment. *Scientia Horticulturae* 358, 341–348.

Kolse, R.H., Gaikwad, C.B., Jadhav, J.D. and Yadav, S.T. (2010) Efficiency of chemical weed control methods in onion seed production for controlling weeds and its effect on yield. *International Journal of Plant Protection* 3(1), 55–59.

Korichi, Y., Malika, A.S., Karima, K.G. and Hassina, I. (2021) Foraging bees of the onion (*Allium cepa* L.) and their impact on seed production in Tizi-Ouzou area (Algeria). *Journal of Apicultural Research* 1–8.

Lal, M. and Kanwar, H.S. (2017) Onion seed production: a challenge in northwestern Himalayas. *AgricIN-TERNATIONAL* 4(1), 31–34.

Laware, S.L. and Raskar, S. (2014) Influence of zinc oxide nanoparticles on growth, flowering and seed productivity in onion. *International Journal of Current Microbiology and Applied Sciences* 3(7), 874–881.

Lazko, V.E. and Yakimova, O.V. (2020) Planting depth-effective agricultural practice preventing lodging of seedstalks of onion seeds. In: *E3S Web of Conferences*, vol. 224. EDP Sciences.

Lazko, V. and Yakimova, O. (2021) Influence of the orientation of uterine bulbs during planting on the seed yield of winter onion variety Ellan in a two-yield direct crop. In: *E3S Web of Conferences*, vol. 285. EDP Sci, p. 02040.

Lemma, D. (1998) Seed production guideline for tomatoes onion and hot pepper. *Ethiopian Institute of Agricultural Research* 23.

Limeneh, D.F. (2021) Review on production status of onion seed yield, nutrient uptake and use efficiency of nitrogen and phosphorus fertilizations in Ethiopia. *International Journal of Agriculture Innovations and Research* 9. ISSN 2319-1473.

Lokesh, K., Veeregowda, R. and Shekar, G.C. (2020) Seed production potential of long day and intermediate group of onion genotypes as affected by period of cold treatment and growth regulator. *Journal of Pharmaceutical Innovation* 9(6), 525–527. ISSN 2277-7693.

Long, R. and Morandin, L. (2011) Low hybrid onion seed yields relate to honey bee visits and insecticide use. *California Agriculture* 65(3), 155–159.

Mansoor, M., Zaman, K., Jamil, M., Ali, R., Waheed, M. *et al.* (2019) Maximizing onion seed production through foliar application of zinc and boron. *Biological Sciences-PJSIR* 62(1), 1–7.

Mayer, D.F., Britt, R.L. and Lunden, J.D. (1989) Evaluation of BeeScent as a honey bee attractant. *The American Bee Journal* 129, 41–42.

More, M.R., More, S.A., Tamboli, N.D. and Kulkarni, S.R. (2020) Effect of sprays of bee attractants on qualitative and quantitative yield parameters on seed onion crop (*Allium cepa* L.). *Journal of Entomology and Zoology Studies* 8, 1513–1516, ISSN: 2320-7078.

Msika, R.L., Jackson, J.E. and Currah, L. (1994) Onion seed production trials in the highveld of Zimbabwe. *Acta Horticulturae* 433, 337–346.

Munawar, M.S., Raja, S., Niaz, S. and Sarwar, G. (2011) Comparative performance of honeybees (*Apis mellifera* L.) and blow flies (*Phormia terronovae*) in onion (*Allium cepa* L.) seed setting. *Journal of Agricultural Research* 49(1), 49–56.

Nikus, O. and Mulugeta, F. (2010) *Onion Seed Production Techniques: A Manual for Extension Agents and Seed Producers*. FAO Crop Diversification and Marketing Development Project, Asella, Ethiopia.

Paneru, N., Adhikari, P. and Tandan, P. (2020) Management of purple blotch complex of onion (*Allium cepa* Cv Red Creole) under field condition in Rukum-West, Nepal. *Malaysian Journal of Sustainable Agriculture* 4(2), 71–74. ISSN 2521-294X.

Pannacci, E., Farneselli, M., Guiducci, M. and Tei, F. (2020) Mechanical weed control in onion seed production. *Crop Protection* 105221.

Pathak, C.S. (2000) Hybrid seed production in onion. *Journal of New Seeds* 1(3-4), 89–108.

Peluffo, S., González Idiarte, H., Borges, A., Arboleya, J. and Galván, G.A. (2016) Onion sets as planting material for seed production of three cultivars in Uruguay. *Seed Science and Technology* 44(3), 500–513.

Rabinowitch, H.D. and Brewster, J.L. (1990) *Onions and Allied Crops: Biochemistry Food Science Minor Crops*, vol. 3. CRC Press, Boca Raton, Florida.

Rabinowitch, H.D. and Currah, L. (eds) (2002) *Allium Crop Science: Recent Advances*. CAB International, Wallingford, UK.

Rasekh, A.M., Sadeghi, N.H. and Hosseini, M. (2013) Effects of insect pollinators on onion seed production quality and quantity. *Journal of Crop Protection* 2(4), 395–402.

Rodo, A.B. and Marcos-Filho, J. (2003) Onion seed vigor in relation to plant growth and yield. *Horticultura Brasileira* 21(2), 220–226.

Rosliani, R., Prathama, M., Sulastiningsih, N.W.H., Hermanto, C. and Yufdy, M.P. (2021) Flowering and yield of true shallot seed from bulb and different seedling age vernalized at low temperature. In: *IOP Conference Series: Earth and Environmental Sciences* 752(1), 012045. IOP Publishing, Bristol, UK.

Ruggeri, A. and Branca, F. (1994) Sowing date and GA3 in onion seed production. *Acta Horticulturae* 362, 35–42.

Saleh, M., Bashir, M.A., Khan, K.A., Mahmood, R., Sarwar, G. *et al.* (2021) Onion flowers anthesis and insect pollinators preferences on onion (*Allium cepa* L.) Crop. *Fresenius Environmental Bulletin* 30(3), 2580–2585.

Sofy, S.H., Abd-El-Moneim, L.M. and Hassan, M. (2020) Effect of potassium salts on onion purple blotch incidence and some physiological and yield parameters in onion seed plants. *Fayoum Journal of Agricultural Research and Development* 34(2), 158–168.

Soto, V.C., Maldonado, I.B., Jofré, V.P., Galmarini, C.R. and Silva, M.F. (2015) Direct analysis of nectar and floral volatile organic compounds in hybrid onions by HS-SPME/GC–MS: relationship with pollination and seed production. *Microchemical Journal* 122, 110–118.

Soto, V.C., Caselles, C.A., Silva, M.F. and Galmarini, C.R. (2018) Onion hybrid seed production: relation with nectar composition and flower traits. *Journal of Economic Entomology* 111(3), 1023–1029.

Soto, V.C., Gatica, I.J.G. and Galmarini, C.R. (2021) Onion (*Allium cepa* L.) hybrid seed production: Sugar content variation during the flowering period. *Crop Science* 61(5), 3529–3537.

Spurr, C.J., Fulton, D.A., Brown, P.H. and Clark, R.J. (2002) Changes in seed yield and quality with maturity in onion (*Allium cepa* L., cv. 'Early Cream Gold'). *Journal of Agronomy and Crop Science* 188(4), 275–280.

Tajbakhsh, M., Brown, P.H., Gracie, A.J., Spurr, C.J., Donovan, N. *et al.* (2004) Mitigation of stunted root abnormality in onion (*Allium cepa* L.) using seed priming treatments. *Seed Science and Technology* 32(3), 683–692.

Tarpaga, M.W.V., Albert, R. and Zoumbiessé, T. (2011) Effects of seasons of bulb and seed production on the early bolting of onion (*Allium cepa* L.) cv. Violet de Galmi. *Journal of Applied Biosciences* 40, 2652–2658.

Tekle, G., Belay, F. and Chernet, S. (2020) Effect of Different Planting Time on Seed Yield and Quality of Onion (*Allium cepa* L.) in Tigrai Region, Northern Ethiopia. *International Journal of Agriculture and Biosciences* 9(5), 216–220. *ISSN* 2306-3599.

Tesfaye, M., Belew, D., Dessalegn, Y. and Shumye, G. (2018) Effect of planting time on growth, yield components, seed yield and quality of onion (*Allium cepa* L.) at Tehuledere district, northeastern Ethiopia. *Agriculture & Food Security* 7(1), 28.

University of Arizona (2006) Available at: https://cals.arizona.edu/fps/sites/cals.arizona.edu.fps/files/cotw/Onion_Seed.pdf (accessed December 2020).

Voss, R., Murray, M., Bradford, K., Mayberry, K. and Miller, I. (2013) *Onion Seed Production in California*. UCANR Publications, Davis, California.

Wahyuni, A.N. and Padang, I.S. (2020) The growth and yield performance of true shallot seed production in Central Sulawesi, Indonesia. *Asian Journal of Agriculture* 4(1). ISSN 2580-4537.

Walcott, R.R., Gitaitis, R.D., Castro, A.C., Sanders, F.H. Jr and Diaz-Perez, J.C. (2002) Natural infestation of onion seed by *Pantoea ananatis*, causal agent of center rot. *Plant Disease* 86(2), 106–111.

Walker, M.K., Howlett, B.G., Wallace, A.R., McCallum, J.A. and Teulon, D.A.J. (2011) The diversity and abundance of small arthropods in onion, *Allium cepa*, seed crops, and their potential role in pollination. *Journal of Insect Science* 11(1), 98.

13 Crop Protection

Edivânio R. Araújo[1]*, Renata S. Resende[1] and Mirtes F. Lima[2]
[1]*Epagri – Estação Experimental de Ituporanga, Santa Catarina, Brazil;*
[2]*Embrapa Hortaliças, Brasília, Brazil*

1. Introduction

The control of plant disease can be considered the ultimate goal of plant pathology. However, to achieve this goal it is necessary to have a comprehensive and accurate knowledge of etiology, symptomatology, epidemiology and other topics related to disease management. In addition, the location of the pathogen inside the plant, aerial part or roots, also determines the best combat strategy.

This chapter focuses on the control of diseases caused by fungi, oomycetes, bacteria and viruses. For additional information on insect pests and weeds, see Chapter 11. The main objective of this chapter is to present current strategies for the management of important diseases that affect the edible alliums.

2. Management of Foliar Diseases Caused by Fungi and Oomycetes

2.1. Botrytis leaf blight (*Botrytis squamosa* J.C Walker)

Botrytis leaf blight (BLB) is the most important disease caused by the genus *Botrytis*, especially in the onion crop (*Allium cepa* L.). There are reports on low levels of infection by *B. porri* (Zhang *et al.*, 2009) and *B. sinoallii* (Zhang *et al.*, 2010) in association with leaf blight in garlic (*Allium sativum* L.). Bulb onion cultivars vary in susceptibility to *Botrytis* (Araújo *et al.*, 2018), and Van der Meer and De Vries (1990) and De Vries *et al.* (1992) reported on a *Bs* gene in *Allium roylei*, which confers total resistance to *B. squamosa*. Yet, no resistant commercial cultivars have been released and management of BLB is based mainly on fungicides control.

Van der Heyden *et al.* (2012) proposed that for better management of BLB, fungicides application will start with the first detection of airborne *B. squamosa* conidia. This recommendation demonstrates how difficult it is to manage this disease and the consequent requirement for frequent fungicide sprays. Frequent fungicides applications enforce selection pressures with the consequent selection of resistant pathogens to the active ingredients. Hence, tolerance of *B. squamosa* to iprodione (group 2 FRAC) has been observed both *in vitro* (Carisse and Tremblay, 2007) and *in situ* (Araújo *et al*, 2018). Unfortunately, efforts to search for control alternatives, such as the use of biological control agents, have not been effective so far (Carisse *et al.*, 2011).

To rationalize the number of applications and quantity of fungicides involved, several

*Address for correspondence: edivanioaraujo@epagri.sc.gov.br

 Edible Alliums: Botany, Production and Uses (eds H.D. Rabinowitch and B. Thomas)
DOI: 10.1079/9781789249996.0013

forecasting systems/disease management systems have been developed, such as weather-based prediction models, e.g. BOTCAST (Sutton *et al.*, 1986), PREDICTOR (Lacy and Pontius, 1983) and BLIGH-ALERT (Vincelli and Lorbeer, 1989). Their accuracy thus resulted in a marked reduction in both fungicides quantity and number of treatments, and thus in economic saving and environmental sustainability.

Table 13.1 lists some active ingredients tested for BLB control and their respective effectiveness. It is noted that several of the active ingredients tested so far for BLB control have not been efficient, showing a broad insensitivity of the pathogen.

2.2. Stemphylium leaf blight (*Stemphylium vesicarium* (Wallr.) Simmons) and purple blotch (*Alternaria porri* (Ellis) Cif.)

Stemphylium leaf blight (SLB), caused by *S. vesicarium*, recently emerged in the United States and Canada where the pathogen's tolerance to several fungicides, such as quinone outside inhibitor fungicides – QoI (azoxystrobin and pyraclostrobin due to a mutation in the *cytochrome b* gene (G143A), FRAC group 11), carboxamide (boscalid, FRAC group 7) and anilinopyrimidine (cyprodinil and pyrimethanil, FRAC group 9), causes great concerns (Hay *et al.*, 2019). Likewise, in Ontario, Canada, 90% of the strains evaluated were resistant to azoxystrobin, and 57% to pyrimethanil (Stricker *et al.*, 2020). It is not known, however, whether these resistance acquisitions are associated with loss of fitness in those strains. The high tolerance rate causes concerns since no resistance to *S. vesicarium* is known in bulb onion.

SLB epidemics have recently been reported in New Zealand (Wright *et al.*, 2019) and India, where SLB has is considered the main fungal leaf disease (Hassan *et al.*, 2020). Interestingly, in Brazil, especially in onions, *Stemphylium* sp. is usually described as a secondary agent that colonizes lesions developed by other pathogens (downy mildew, for example), and thus is considered a less important disease.

Some promising chemical control alternatives such as acibenzolar-*S*-methyl and salicylic acid reduced the severity of SLB in greenhouse conditions (Hussein *et al.*, 2007), and control by biological agents was reported. The To strain of *Trichoderma asperellum* reduced mycelial growth and conidia production *in vitro*, as well as reduced SLB severity in onion plants (Zapata-Sarmiento *et al.*, 2020). The integration of alternative

Table 13.1. Efficiency of some active ingredients in controlling botrytis leaf blight of onion (*Botrytis squamosa*) according to research results.

Active ingredient	FRAC group	Control efficiency	References
iprodione + pyrimethanil	2 + 9	Poor	Dutta *et al.* (2020b); Hoepting *et al.* (2020)
Bacillus subtilis, strain QST 713	BM02	Poor	Araújo *et al.* (2018)
procyimidone	2	Poor	Araújo *et al.* (2018)
iprodione	2	Poor	Carisse and Tremblay (2007); Araújo *et al.* (2018)
pyrimethanil	9	Poor	Dutta *et al.* (2020b); Hoepting *et al.* (2020); Araújo *et al.* (2018)
azoxystrobin + difenoconazole	11 + 3	Poor	Dutta *et al.* (2020b); Hoepting *et al.* (2020)
azoxystrobin + difenoconazole	11 + 3	Medium	Araújo *et al.* (2018)
fluazinam	29	High	Araújo *et al.* (2018)
fluopyram + pyrimethanil	7 + 9	High	Dutta *et al.* (2020b); Hoepting *et al.* (2020)
fluopyram + tebuconazole	7 + 3	High	Dutta *et al.* (2020b); Hoepting *et al.* (2020)
trifloxystrobin + tebuconazole	11 + 3	High	Araújo *et al.* (2018)
boscalid + kresoxim-methyl	7 + 11	High	Araújo *et al.* (2018)

management methods of SLB can replace the older, currently inefficient, fungicides.

Alliaceous crops vary in susceptibility to *A. porri* (Nanda *et al.*, 2016); however, currently, no resistant commercial cultivars were reported.

Unlike *S. vesicarium*, there are no reports of *A. porri* strains with long-lasting tolerance to fungicides. In general, mixtures of strobilurins and triazoles have shown good control of purple blotch (Veeraghanti *et al.*, 2017). Additionally, biocontrol agents and other molecules have potential in the management of purple blotch disease. In India, soil application of silicic acid and silicon reduced purple blotch severity in the onion by up to 45% (Haroon *et al.*, 2020), and sprays of salicylic acid (250 mg L^{-1}) significantly reduced its incidence in the field (Pradhan *et al.*, 2018). Interestingly, elemental sulphur associated with *Trichoderma* spp. also controlled purple blotch of onion and resulted in greater yield than fungicide treatment (Bayoumi *et al.*, 2019).

2.3. Downy mildew (*Peronospora destructor* (Berk.) Caspary)

Fungicides and oomicides are the main means of downy mildew control in onion. A dominant gene (*Pd*) for complete resistance to *P. destructor* was identified in *A. roylei* (Kofoet *et al.*, 1990; Van der Meer and de Vries, 1990). Yet, a strong linkage to a region in the *A. roylei* genome is lethal in *A. cepa*, hence the introgression of this gene in commercial cultivars is rather difficult (Scholten *et al.*, 2007). Using genomic tools, Scholten *et al.* (2007) resulted in the breeding of a few cultivars containing the *Pd* gene (e.g. F_1 Santero), yet genetic control as a management tool for downy mildew is still incipient and insufficient. Additionally, quantitative genetic resistance to *P. destructor* shows promise for use as part of downy mildew integrated management (Arias *et al.*, 2020).

To rationalize the control of downy mildew in onion, several warning systems have been developed. The first forecasting system was the Canadian DOWNCAST (Jesperson and Suton, 1987), which combines temperature, relative humidity and precipitation to predict sporulation and infection by the pathogen. An improved model included additional parameters, such as leaf wetness (Visser, 1998); and, based on climatic parameters, Battilani *et al.* (1996) developed the ONIMIL model. The success of both models depends on the application of systemic fungicides as the infection conditions could prevail at the time of decision making. Friedrich *et al.* (2003) thus developed the ZWIPERO model capable of predicting (for up to four days) the risk of *P. destructor* sporulation and infection, thus enabling management with both systemic and protective fungicides. Gilles *et al.* (2004) developed the MILLIONCAST model based on temperature and relative humidity. Evaluation tests revealed that it is more precise in sporulation predictions than both DOWNCAST and ONIMIL. Araújo *et al.* (2017) validated a warning system for south Brazil based on a modified Wallin table (Wallin, 1962) for potato late blight. This version considers different combinations of temperature and relative humidity as risk factors, concluding an up-to-20% decrease in the number of sprays with systemic fungicides, without compromising the health and productivity of the onion crop.

Despite the development of increasingly accurate mathematical and/or empirical predictive models of the *P. destructor* sporulation and infection, little information is available on the practical application of these models, as data from regions with highly favourable environment for the spread of the pathogen, associated with a high concentration of inoculum, complicate the application and interpretation of the results.

Determining the efficiency of fungicides' active ingredients is as important as knowing the warning systems, in controlling the disease. Only a few chemicals effectively control downy mildew (Table 13.2). These include metalaxyl-M and mancozeb (Araújo *et al.*, 2020), oxathiapiprolin (Dutta *et al.*, 2019, 2020a) and propineb (Araújo and Resende, 2020).

Interestingly, despite the ongoing use of metalaxyl-M (mefenoxam) for several years for downy mildew control, no record of *P. destructor* resistance has been reported, probably due to fitness, as reported for *Plasmopara viticola* (the causal agent of grape downy mildew) in Brazil, where strains resistant to azoxystrobin tend to disappear in an environment free of application of the active ingredient (Sierotzki *et al.*, 2008). The prevalence of fungicide-susceptible strains in the absence of azoxystrobin indicates loss of fitness in resistant mutants. On the other hand, mefenoxam-resistant strains of *Phytophthora* species are equally, or even more, aggressive

Table 13.2. Efficiency of some active ingredients in controlling downy mildew of onion (*Peronospora destructor*) according to research results.

Active ingredient	FRAC group	Control efficiency	References
copper oxychloride	M01	Poor	Ahmad and Khan (2000)
dimethomorph + mancozeb	40 + M03	Poor	Abkhoo (2012)
mono and di potassium phosphonate	–	Poor	Abkhoo (2012)
propamocarb + fluopicolide	28 + 43	Poor	Araújo *et al.* (2020)
cymoxanil + famoxadone	28 + 11	Poor	Araújo *et al.* (2020)
Bacillus subtilis, strain QST 713	BM02	Poor	Araújo *et al.* (2020)
propiconazole	3	Poor	Dutta (2019); Dutta *et al.* (2020a)
acibenzolar-*S*-methyl	P01	Poor	Wordell Filho *et al.* (2007); Araújo *et al.* (2017)
calcium phosphite	P07	Poor	Araújo *et al.* (2017)
potassium phosphite	P07	Poor	Wordell Filho *et al.* (2007); Araújo *et al.* (2017)
chlorothalonil	M05	Poor	Dutta (2019); Dutta *et al.* (2020a)
fluazinam	29	Medium	Dutta (2019); Araújo *et al.* (2020); Dutta *et al.* (2020a)
azoxystrobin + difenoconazole	11 + 3	Medium	Dutta (2019); Araújo *et al.* (2020); Dutta *et al.* (2020a)
cymoxanil + famoxadone	27 + 11	Medium	Dutta (2019); Dutta *et al.* (2020a)
metalaxyl-M + mancozeb	4 + M03	High	Develash and Sugha (1997); Raziq *et al.* (2008); Surviliene *et al.* (2008); Ahmad and Khan (2000); Araújo *et al.* (2020)
mancozeb	M03	High	Ahmad and Khan (2000); Araújo *et al.* (2020)
propineb	M03	High	Raziq *et al.* (2008); Ahmad and Khan (2000); Araújo *et al.* (2020)
propamocarb + fluopicolide	28 + 43	High	Abkhoo (2012)
thiophanate methyl	1	High	Raziq *et al.* (2008)
metalaxyl-M + chlorothalonil	4 + M05	High	Dutta (2019); Araújo *et al.* (2020); Dutta *et al.* (2020a)
oxathiapiprolin + chlorothalonil	49 + M05	High	Dutta (2019); Dutta *et al.* (2020a)
oxathiapiprolin + mandipropamid	49 + 40	High	Dutta (2019); Dutta *et al.* (2020a)

compared to susceptible strains (Café-Filho and Ristaino, 2008; Chapara *et al.*, 2011). No published data on *P. destructor* strains and fitness are available. Also, there are no reports of biological control agents used satisfactorily in the management of the disease.

Basic research studies, such as the sequencing of *P. destructor* by Natesan *et al.* (2020) open possibilities for the management of downy mildew, such as the use of dsRNA applied exogenously.

3. Management of Diseases Caused by Soilborne Fungi

The management of plant diseases caused by soil-dwelling pathogens is challenging. Plant pathologists still find it difficult to get an accurate answer, if any, when farmers ask about this type of problem in their area. In fact, the standard of best practices includes a series of preventive actions aimed at mitigating the problem. But we still see significant or total field losses. For edible alliums, this is no different. Below, we will list three important diseases that still challenge farmers and researchers constantly.

3.1. White rot (*Stromatinia cepivora* (Berk.) Whetzel)

Although active ingredients such as boscalid and tebuconazole have shown satisfactory results in controlled experimental trials, chemical control of white rot has not been effective in the field. The use of 78% diallyl disulphide injected into the soil, with the aim of stimulating the germination of sclerotia in the absence of the plant, has shown promising results (Villalta *et al.*, 2012). As they form sclerotia that survive for many years in the soil, management strategies such as crop rotation may not have the expected effectiveness. Looking for more recent studies, we observed that the main lines of research are based on biological control, which seems to be the most viable alternative to mitigate the losses resulting from the disease. However, we know that plant pathology has not yet reached the level of success that entomology has achieved with the use of biological control. Of course, the complexity of pathosystems, the interaction with other soil micro-organisms, among other factors, makes this tool more complex in the use of plant disease control.

Trichoderma harzianum and *T. koningii* strains, formulated on wheat bran powder and applied to soil, were effective in controlling white rot of onion under greenhouse and field conditions in Egypt (Elshahawy *et al.*, 2017a). Also in Egypt, Elshahawy *et al.* (2017b) reported that sclerotial mycoparasite fungi such as *Chaetomium globosum* (Chg6) Kunze, *Clonostachys rosea* (Crl2) (Link) Schroers and *Penicillium oxalicum* (Po9) were able to reduce the incidence of white rot of onions by around 80% when compared to the control plots. In addition to fungi, antagonistic bacteria have demonstrated control potential. Simultaneous application of *Pseudomonas fluorescens*, *Bacillus subtilis* and *Bacillus pumilus* strains reduced the incidence of white rot of onion and garlic, in field experiments, by 75.6% and 40.4%, respectively (Elshahawy *et al.*, 2018). However, these bacteria were less efficient than the fungicide tebuconazole. Endophytic strains of *Bacillus siamensis* and *Streptomyces setonii*, obtained from healthy garlic bulbs, were able to reduce the incidence of garlic white rot by up to 66% under controlled conditions (Wang *et al.*, 2019).

3.2. Pink root (*Setophoma terrestris* = syn. *Phoma terrestris*; *Pyrenochaeta terrestris*)

Despite the importance of the pink root for edible alliums, there are few studies on the management of this disease. In fact, there are no recent records of extremely destructive pink root epidemics which can distract from the relevance of this disease. Methods such as solarization (Katan *et al.*, 1980) and soil fumigation (Hartz *et al.*, 1989) have already demonstrated high effectiveness in controlling the pathogen. However, with the demand for a more sustainable agriculture, these methods have become obsolete since it can eliminate much of the soil microbial community.

Thus, as is the case for most diseases caused by soilborne pathogens, management strategies are based on biological control and genetic resistance. Inheritance studies have shown that resistance to *S. terrestris* is linked to a common chromosomal region and co-dominance conditioned by a single recessive locus (Marzu *et al.*, 2018). In practice, cultivars demonstrate different levels of resistance, something like quantitative resistance (Esfahani and Pour, 2008). However, there is still no commercial cultivar that can be used in soils with high pathogen infestation.

Biological control practices are scarce. Noteworthy is the use of an Argentinian strain of *Bacillus subtilis*, which has demonstrated significant effect in inhibiting growth of the pathogen *in vitro* and in reducing the incidence and severity of the disease under field conditions (Sayago *et al.*, 2020).

3.3. Basal rot (*Fusarium* spp.)

Basal rot of edible alliums is associated with different *Fusarium* species. Besides *Fusarium*, *Rhizoctonia* and *Pythium* species can also cause damping-off in seedlings. This fact makes the management of the disease extremely complex due to the genetic diversity of causal agents. It is important to highlight that the occurrence of maggots in onions (*Delia* spp.) can increase the incidence of wilt and reduce plant stand due to the attack of *Fusarium* spp. (Everts *et al.*, 1985). To learn more about maggots management, see Chapter 11. As reported for pink root, soil fumigation has a good control of basal rot, even surpassing chemical control (Sumner *et al.*, 1997). However, it is not an environmentally friendly method. Carbendazim and benomyl are among the active ingredients most used in the management of basal rot. However, chemical control may not promote adequate disease control (Haapalainen *et al.*, 2016). Crop rotation, a recommended practice for managing various diseases caused by soilborne pathogens, can be far less effective than expected. In addition to surviving for long periods in the soil, there are reports of *Fusarium* spp. infecting hosts of different botanical families, such as maize, wheat, potato and sunflower (Molinero-Ruiz *et al.*, 2011). This puts the farmer in a deadlock. Again, the most adopted strategies are genetic resistance and biological control. The commercial cultivars used do not have complete resistance to *Fusarium* spp. There are different levels of susceptibility among cultivars (Galván *et al.*, 2008; Taylor *et al.*, 2013). However, the phenotypic and genotypic variability of different *Fusarium* species, the different edaphoclimatic conditions in which the onion is cultivated, and the different inoculum pressures of each soil are still a challenge for genetic improvement.

The prospects for biological control of basal rot use the most traditionally known biocontrol agents: *Trichoderma* spp.; *Pseudomonas* spp. and *Bacillus* spp. The use of *T. harzianum* or *T. viride* associated with *Pseudomonas* spp. reduced the incidence of basal rot of onion under field conditions (Rajendran and Ranganathan, 1996; Malathi, 2015). Volatile compounds produced by *Bacillus* species inhibited the *in vitro* mycelial growth of *F. oxysporum* which causes wilt of onion (Sharifi Tehrani and Ramezani, 2003).

4. Management of Bacterial Diseases

The most common bacterial diseases in onions are listed in Table 13.3.

Recently, species from genera *Serratia* sp., *Klebsiella* sp., *Lactobacillus* sp., *Leuconostoc* sp. and *Rahnella* sp. have been described as onion bulb pathogens (Bonasera *et al.*, 2017; Asselin *et al.*, 2019). The co-occurrence of multiple species of pathogenic bacteria in symptomatic bulbs has been commonly reported, thus diagnosis confusion and complicated management of these diseases are common.

Onion cultivars vary in susceptibility to bacterial bulb rot, but resistant cultivars are not commercially available. Hence, severity differences in Enterobacter bulb decay (*E. cloacae*) and slippery skin (*B. gladioli* pv. *alliicola*) have been reported among onion cultivars, but none of the tested cultivars has shown a high level of resistance (Schroeder and Du Toit, 2010; Velez *et al.*, 2020). *Allium* germplasm evaluation for resistance to *Pseudomonas marginalis* and *P. viridiflava* showed high susceptibility to both pathogens while other *Allium* species were resistant (Wright and Grant, 1998), and Ko *et al.* (2002) reported that onion cultivars with high dry matter contents and small bulbs store better than cultivars with big bulbs and low dry matter, and that average bulb fresh weight was positively correlated with the incidence of storage diseases. They concluded that these genotypes could serve as donors for improved onion storability.

Copper bactericides are recommended as part of integrated management of both bulb rot and foliar bacterial diseases of alliums. A preventive programme with copper fixed formulations, sprayed alone or in combination with an ethylene bisdithiocarbamate (EBDC) fungicide such as mancozeb, has been successfully used to suppress Xanthomonas leaf blight in onions (Schwartz and Otto, 1998). In general, copper is a powerful tool for controlling bacterial disease,

Table 13.3. Main bacterial diseases in bulb onion (Schwartz and Mohan 2008).

Disease	Bacterial name	Disease	Bacterial name
Leaf blight	*Xanthomonas axonopodis* pv. *allii*	Centre rot	*Pantoea ananatis*; *P. allii;* and *P. agglomerans*
Bacterial stalk and leaf necrosis	*Pantoea agglomerans*	Slippery skin	*Burkholderia gladioli* pv. *alliicola*
Leaf streak and bulb rot	*Pseudomonas viridiflava*	Sour skin	*Burkholderia cepacia* complex
Soft rots	*Pseudomonas marginalis* pv. *marginalis;* and *Pectobacterium carotovorum* ssp. *carotovorum*	Bulb decay	*Enterobacter cloacae*

but frequent applications led to the evolution of resistance in plant pathogen populations (Sundin *et al.*, 2016). Indeed, copper-tolerant strains of *Pantoea ananatis* and *P. agglomerans* have been isolated from onion plants (Nischwitz *et al.*, 2007; Tho *et al.*, 2019). Moreover, EBDC fungicides have been classified as possibly carcinogenic to humans and there is regulatory pressure to reduce their use.

The adoption of alternative strategies for the management of bacteriosis is of utmost importance. Gent and Schwartz (2005) have shown that the integration of alternative products such as acibenzolar-*S*-methyl, a plant activator, and biological control agents is an effective management strategy for Xanthomonas leaf blight, thus reducing the need for conventional copper bactericide and EBDC fungicides. Four weekly applications of acibenzolar-*S*-methyl reduced severity of Xanthomonas leaf blight as, or more, effectively than 9–12 weekly applications of copper hydroxide or copper hydroxide-mancozeb. Combining bacteriophages with acibenzolar-*S*-methyl effectively controlled Xanthomonas leaf blight (Lang *et al.*, 2007), but despite the absence of symptoms, ten weekly applications of acibenzolar-*S*-methyl resulted in reduced bulb yield.

Interestingly, copper alone did not provide effective protection from *P. agglomerans* foliar bacterial diseases in onion (Grode *et al.*, 2019). However, insecticides' control of onion thrips led to the control of bacterial stalk and leaf necrosis disease caused by *P. agglomerans*, possibly since onion thrips transmit *P. agglomerans* and *P. ananatis* (Gitaitis *et al.*, 2003, Dutta *et al.*, 2014). Indeed, a positive correlation between the number of thrips per plant and bacterial stalk and leaf necrosis incidence in onion fields has been determined (Grode *et al.*, 2019).

In contrast to foliar bacterial diseases, only limited data on the efficacy of bactericides and other chemicals in reducing bacterial bulb rot is available, possibly due to the lack of knowledge of the exact onset of the infection, since foliar infections can lead to bulb infection (Carr *et al.*, 2013). Since bactericides have no curative or systemic activities, treatment must be applied preventively to be effective.

The management of bacterial diseases in alliums depends primarily on cultural strategies. Among these strategies, managing soil fertility, managing water availability to avoid/limit excessively wet necks, rapid and thorough curing, bulbs harvest (for storage) at the appropriate crop maturity, and protection from mechanical damages and bulbs wounding during harvest and sorting seem to considerably reduce the damages caused by bacterial diseases.

Excessive nitrogen fertilization may delay plant maturation and keep onion necks thicker and succulent, thus increasing the plants' susceptibility to bacterial infection, and bacterial bulb rots in store (Wright, 1993), and reduce bulb quality (Wright *et al.*, 1993; Diaz-Perez *et al.*, 2003), but then Leach *et al.* (2020) reported that bacterial rot differed between years and that N applications had no effect (Leach *et al.*, 2017). Furthermore, foliar nitrogen was negatively related to the incidence of bacterial rots of onion, while high early-season levels of soil nitrate were associated with the incidence of bacterial rots (Pfeufer and Gugino, 2018). They suggested a greater complexity for N fertility recommendations, as timing and form of N fertilizer and seasonal application rates may play a considerable role in bacterial disease development.

The irrigation method has a substantial impact on the incidence of sour skin, caused by *B. cepacia*, especially when applied from bulbing onset to harvest. Hence, overhead sprinkler irrigation provides a favourable environment for sour skin rot, whereas furrow or drip-irrigation resulted in almost complete control of the disease (Teviotdale *et al.*, 1989). In addition, the incidence of bacterial bulb rot of onion in store is greater in experimental plots that received supplementary water during the field-curing than in those that did not receive additional water (Wright *et al.*, 1993). Since contaminated irrigation water can potentially spread bacteria, irrigation with recycled, or run-off, water should be avoided (Schwartz and Mohan, 2008).

Curing, temperature and its duration are important factors in bacterial bulb infection and disease progress. Hence, the incidence of bulb rot in storage, caused by *E. cloacae*, *B. cepacia* and *B. gladioli* pv. *allii*, was reduced when onion bulbs were cured at temperatures <35°C for two v. 14 days (Schroeder and du Toit, 2010; Schroeder *et al.*, 2012), while the incidence of bulb rot decreased for *Pantoea* spp. at temperatures >35°C (Vahling-Armstrong *et al.*, 2016). Proper storage environment is essential to maintain onion quality. Storing onions at 0°C with a relative humidity of 65–75% has been recommended for long-term storage.

5. Management of Diseases Caused by Viruses

Multiple virus infections causing the known 'viral complex' of garlic are induced mainly by species in three main genera *Potyvirus*, *Carlavirus* and *Allexivirus* (Mituti *et al.*, 2011, 2015; Oliveira *et al.*, 2014) whose occurrence has been frequently reported (Fajardo *et al.*, 2001; Mituti *et al.*, 2014; Araújo *et al.*, 2018). Viruses are among the principal causes of degeneracy of the vegetatively propagated garlic cultivars. Their management in onion and especially in clonally propagated garlic is challenging, as the propagation system facilitates virus accumulation and dissemination, with the consequent drastic reduction in vegetative vigour (Oliveira *et al.*, 2014; Marodin *et al.*, 2019) and heavy losses in quality and yield (up to 78%) (Davis, 1995; Lunello *et al.*, 2007).

Latent viral infection in garlic is common in most cultivars (Davis, 1995), therefore diagnosis based on symptom recognition is difficult. The majority of the viruses identified in these complexes are spread globally due to international commerce in propagules and produce with inadequate phytosanitary inspections (Silva *et al.*, 2019). Considering the significant losses induced by virus infection, the main control measures for viral diseases in garlic and onions are discussed below (and see Chapter 11, this volume).

5.1. Production of virus-free stocks

As garlic is produced solely by vegetative propagation, obtaining virus-free stock is highly recommended to plant commercial fields, annually, as plants are rapidly reinfected by viruliferous insect vectors during the production season. Obtaining virus-free propagules involves a variety of techniques including meristem tip culture, with or without thermotherapy (Conci and Nome, 1991; Torres *et al.*, 2000; Pramesh and Baranwal, 2015; Vieira *et al.*, 2015; Marodin *et al.*, 2019) or other treatments (Chapter 11, this volume). Regenerated plantlets are tested or indexed for the crop's main viruses by ELISA (enzyme-linked immunosorbent assay) and/or RT-PCR (reverse transcription and polymerase chain reaction). Meristem tip culture method efficiently disposes of potiviruses, followed by carlaviruses; while allexiviruses are considered more difficult to eliminate (Davis, 1995). Garlic growers obtain a virus-free stock and multiply it in isolated territories or in a screenhouse protected from insects.

Plants raised from virus-free propagules are 14–32% more productive than standard control (Bereda and Paduch-Cichal, 2016), and field trials indicated a faster plant emergence increased share of marketable bulbs, and higher mass of bulbs and cloves in virus-free plants compared to the standard plants (Marodin *et al.*, 2019).

For onions, with the exception of Iris yellow spot orthotospovirus – IYSV (Kritzman *et al.*, 2001; Mullis *et al.*, 2004), most viruses remain in bulbs, sets and volunteer onions produced by the infected plants, and therefrom spread in the following production field. Unlike the clonally propagated garlic, onions are raised from seeds, seedlings, sets and bulbs. Most viruses (e.g. Onion yellow dwarf virus – OYDV (Davis, 1995; Manglli *et al.*, 2020) and IYSV (Kritzman *et al.*, 2001)) are not transmitted by onion seed, hence starting with true seeds guarantees a clean start.

5.2. Eradication of alternative and volunteer plants

Alternate hosts and volunteer garlic and onion plants may serve as an insect and virus wide or narrow reservoir, thus a potential source of next growing-season infection. Eradication of these hosts reduces these sources and consequently the risks of infection. OYDV has a host range mainly but not exclusively limited to alliaceous species (Davis, 1995), as *Cynanchum acutum* was reported a OYDV host in Egypt (El-Wahab *et al.*, 2009) and 50% and 56% rates of OYDV infection were recorded. In Colorado, IYSV-infected onion volunteers were identified in fields of dry bean, alfalfa, field corn and carrot (Gent *et al.*, 2004).

In addition, the polyphagous onion thrips (*Thrips tabaci*) thrives on a wide range of hosts including many weeds (Gent *et al.*, 2006), hence the importance of elimination of alternate hosts and volunteer plants from the farmed areas and their surroundings.

5.3. Chemical control

For non-persistent, non-circulative aphid-transmitted viruses (OYDV; Leek yellow stripe virus – LYSV, and more), insecticides have little or even no effect on virus dissemination. Virus transmission by aphids takes seconds, and thereafter they move to feed from another plant (Davis, 1995), and chemical control has very low efficiency due to the short contact between the insect and the host's tissues.

Chemical control of *Thrips tabaci* is largely employed to reduce the persistent-propagative manner of IYSV dissemination in onion. *Thrips* control in onion, however, is complex due to the fast reproduction and wide host range (Gent *et al.*, 2006), and the development of resistance to insecticides (Shelton *et al.*, 2003; Allen *et al.*, 2005).

5.4. Other control measures

Farming where virus infestation is low is extremely important and new production fields should be as distant as possible from virus-infected existing fields or those with volunteers (Davis, 1995). Manglli *et al.* (2020) reported that OYDV infection was close to 0% when seedbeds were situated about 1 km away from onion fields. Early elimination of garlic and onion plants infected with viruses ('roguing') minimizes the virus source in the production field.

Screening of onion germplasm in search of genotypes resistant/tolerant to IYSV and its insect vector resulted in onion lines with improved tolerance to onion thrips after two generation selections, most showing lower thrips and IYSV severity (Singh and Cramer, 2019).

For OYDV, regional crop rotation to free it for a period of time of *Allium* spp. can interrupt the disease cycle (Davis, 1995). Effective management of viruses must involve diverse components such as host plant resistance, cultural practices and control of thrips.

Besides being the main pest of onion worldwide, thrips is associated with several diseases, either as a vector (IYSV, for example) or as a symptom enhancer due to its feeding. Our experience has shown that without an efficient control of thrips, it is quite difficult to keep crops healthy. Thus, a joint effort by plant pathologists and entomologists is needed to solve this complex problem.

6. Future Prospects and Summary

Given the various difficulties we face to efficiently manage the diseases of edible alliums, some medium-/long-term perspectives appear on the horizon. In general, the development of new molecules/active ingredients takes a lot of time and investment. In addition, these investments originating from private companies are mostly destined for disease management of commodities, since it is the crops that move large amounts of money. Rarely, active ingredients are developed specifically for use in vegetables. What happens is the use of active ingredients that are already used in cultures of greater economic expression. Hence the importance of public investment in research on management of vegetable diseases.

New fungicide formulations, such as the encapsulation of fluazinam with ethyl cellulose polymer, may extend the period in which the active ingredient protects the plant against attack by pathogens (Liu *et al.*, 2021). This may be an alternative in reducing the high number of sprays that disease control in edible alliums currently demands. The use of nanoparticles has emerged as a promising technology for controlling plant diseases. Silver nanoparticles were effective in inhibiting the germination of *Stromatinia cepivora* sclerotia *in vitro* and *in planta* (Darwesh and Elshahawy, 2021).

Another weapon in the fight against phytopathogens is the application of exogenous dsRNA. Although studies using this technology in alliums are rare, Wang *et al.* (2016) demonstrated that the application of dsRNA in onions has a potential protective action against *Botrytis cinerea*. Although the use of exogenous dsRNA in the control of plant diseases is extremely exciting, some aspects are still limiting the extensive use of this technology, such as: cost of production of RNA double strands, stability of molecules for field application, and legal and regulatory issues (would it be a type of transgenic?).

It is clear that overcoming the challenges in the control of allium diseases will require a multi-disciplinary effort, which will involve phytopathologists, geneticists, chemists, physicists, mathematicians and countless other professionals.

In summary, the best management strategy is linked to the complexity of each pathosystem. Chemical control requires the use of a variety of active ingredients, but presents the risk of selection pressure for resistant pathogen populations. Biological control is an environmentally friendly method, but still highly dependent on availability of predators, weather, soil and other variations for its success. Genetic control is the most promising approach, but sources for resistance are scarce, and cross-pollination together with a biannual life cycle requires a very long time for traditional breeding to be successful. Meanwhile, cultural control depends on the technological level of the farmer, among others. Each method has its advantages and disadvantages. Fig. 13.1 presents the level of dependence of the diseases presented in this chapter to current control strategies.

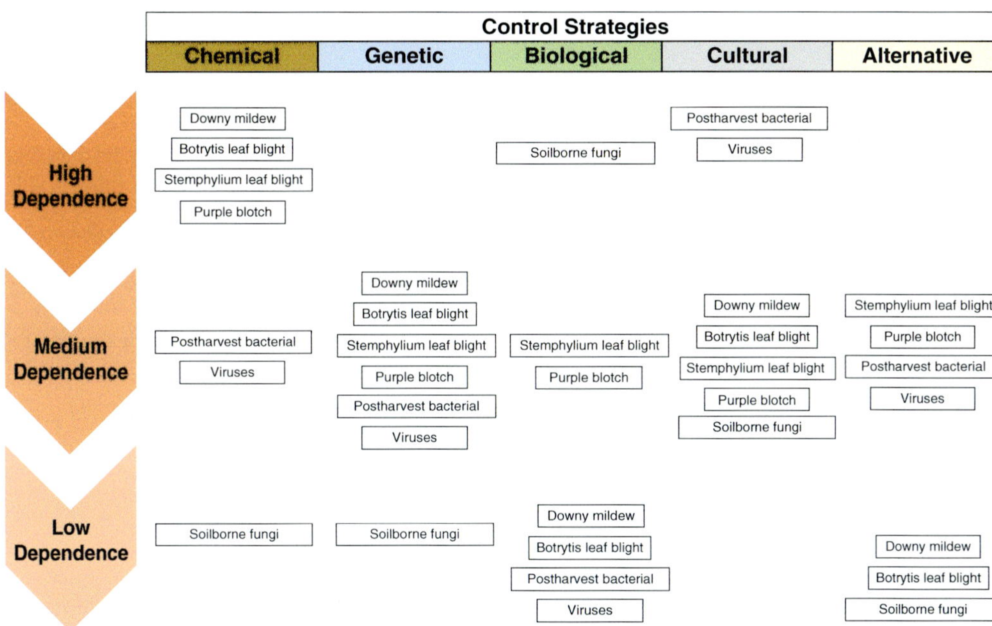

Fig. 13.1. Current level of dependence on different strategies to control diseases of edible alliums.

References

Abkhoo, J. (2012) Efficacy of different fungicides for the control of downy mildew of onion. *Open Access Scientific Reports* 1(6). DOI: 10.4172/scientificreports.331.

Ahmad, S. and Khan, H. (2000) Effect of fungicide synergy on downy mildew control in onions. *Pakistan Journal of Biological Sciences* 3, 1042–1043. DOI: 10.3923/pjbs.2000.1042.1043.

Allen, J.K.M., Scott-Dupree, C.D., Tolman, J.H. and Harris, C.R. (2005) Resistance of *Thrips tabaci* to pyrethroid and organophosphorus insecticides in Ontario, Canada. *Pest Management Science* 61, 809–815. DOI: 10.1002/ps.1068.

Araújo, E.R. and Resende, R.S. (2020) Propineb efficiency in controlling downy mildew of onion under field conditions. *Australasian Plant Disease Notes* 15, 34. DOI: 10.1007/s13314-020-00403-x.

Araújo, E.R., Alves, D.P. and Knoth, J.R. (2017a) Weather-based decision support reduces the fungicide spraying to control onion downy mildew. *Crop Protection* 92, 89–92. DOI: 10.1016/j.cropro.2016.10.022.

Araújo, E.R., Gonçalves, P.A.S. and Alves, D.P. (2017b) Acibenzolar-*S*-methyl, and potassium and calcium phosphites are not effective to control downy mildew of onion in Brazil. *Australasian Plant Disease Notes* 12, 30. DOI: 10.1007/s13314-017-0255-4.

Araújo, E.R., Alves, D.P. and Higashikawa, F.S. (2018a) Cultivar resistance and chemical, biological and fertilizer treatments for controlling Botrytis leaf blight of onion. *Tropical Plant Pathology* 43, 160–164. DOI: 10.1007/s40858-017-0193-x.

Araújo, E.R., Higashikawa, F.S. and Lima, M.F. (2018b) First report of *Onion yellow dwarf virus* and *Allexivirus* associated with noble garlic in Itajai Valley, Santa Catarina State, Brazil. *Summa Phytopathologica* 44, 195–196. DOI: 10.1590/0100-5405/178028.

Araújo, E.R., Resende, R.S., Alves, D.P. and Higashikawa, F.S. (2020) Field efficacy of fungicides to control downy mildew of onion. *European Journal of Plant Pathology* 156, 305–309. DOI: 10.1007/s10658-019-01874-0.

Arias, M., Curbelo, N., Rabelino, P.G., Vicente, E., Giménez, G. *et al.* (2020) Inheritance of resistance against *Peronospora destructor* in onion cv. 'Regia'. *Australian Journal of Crop Science* 14, 1999–2009. DOI: 10.21475/ajcs.20.14.12.2883.

Asselin, J., Eikemo, H., Perminow, J., Nordskog, B., Brurberg, M. *et al.* (2019) *Rahnella* spp. are commonly isolated from onion (*Allium cepa*) bulbs and are weakly pathogenic. *Journal of Applied Microbiology* 127, 812–824. DOI: 10.1111/jam.14340.

Battilani, P., Rossi, V., Racca, P., and Giosuè, S. (1996) ONIMIL, a forecaster for primary infection of downy mildew of onion. *Bulletin OEPP/EPPO* 26, 567–576. DOI: 10.1111/j.1365-2338.1996.tb01499.x.

Bayoumi, Y., Taha, N., Shalaby, T., Alshaal, T. and El-Ramady, H. (2019) Sulfur promotes biocontrol of purple blotch disease via *Trichoderma* spp. and enhances the growth, yield and quality of onion. *Applied Soil Ecology* 134, 15–24. DOI: 10.1016/j.apsoil.2018.10.011.

Bereda, M. and Paduch-Cichal, E. (2016) Allexiviruses – pathogens of garlic plants. *Progress in Plant Protection* 56, 302–311. DOI: 10.14199/ppp-2016-049.

Bonasera, J.M., Asselin, J.E. and Beer, S.V. (2017) Lactic acid bacteria cause a leaf blight and bulb decay of onion (*Allium cepa*). *Plant Disease* 101, 29–33. DOI: 10.1094/PDIS-06-16-0860-RE.

Café-Filho, A.C. and Ristaino, J.B. (2008) Fitness of isolates of *Phytophthora capsici* resistant to mefenoxam from squash and pepper fields in North Carolina. *Plant Disease* 92, 1439 1443. DOI: 10.1094/PDIS-92-10-1439.

Carisse, O. and Tremblay, D.M. (2007) Incidence and significance of iprodione-insensitive isolates of *Botrytis squamosa*. *Plant Disease* 91, 41–46. DOI: 10.1094/PD-91-0041.

Carisse, O., Tremblay, D.M., McDonald, M.R., Brodeur, L. and McRoberts, N. (2011) Management of botrytis leaf blight of onion: the Québec experience of 20 years of continual improvement. *Plant Disease* 95, 504–514. DOI: 10.1094/PDIS-11-10-0797.

Carr, E.A., Zaid, A.M., Bonasera, J.M., Lorbeer, J.W. and Beer, S.W. (2013) Infection of onion leaves by *Pantoea ananatis* leads to bulb infection. *Plant Disease* 97, 1524–1528. DOI: 10.1094/PDIS-06-12-0597-RE.

Chapara, V., Taylor, R.J., Pasche, J.S. and Gudmestad, N.C. (2011) Competitive parasitic fitness of mefenoxam-sensitve and -resistant isolates of *Phytophthora erythroseptica* under fungicide selection pressure. *Plant Disease* 95, 691–696. DOI: 10.1094/PDIS-10-10-0730.

Conci, V.C. and Nome, S.F. (1991) Virus free garlic (*Allium sativum* L.) Plants obtained by thermotherapy and meristem tip culture. *Journal of Phytopathology* 132, 186–192. DOI: 10.1111/j.1439-0434.1991.tb00111.x.

Darwesh, O.M. and Elshahawy, I.E. (2021) Silver nanoparticles inactivate sclerotial formation in controlling white rot disease in onion and garlic caused by the soil borne fungus *Stromatinia cepivora*. *European Journal of Plant Pathology* 160, 917–934. DOI: 10.1007/s10658-021-02296-7.

Davis, R.M. (1995) Diseases caused by viruses and mycoplasmalike organisms. In: Schwartz, H.F. and Mohan, S.K. (eds) *Compendium of Onion and Garlic Diseases*. APS Press, St. Paul, Minnesota.

De Vries, J., Wietsma, W. and De Vries, T. (1992) Introgression of leaf blight resistance from *Allium roylei* Stearn into onion (*A. cepa* L.). *Euphytica* 62, 127–133. DOI: 10.1007/BF00037938.

Develash, R.K. and Sugha, S.K. (1997) Management of downy mildew (*Peronospora destructor*) of onion (*Allium cepa*). *Crop Protection* 16, 63–67. DOI: 10.1016/S0261-2194(96)00056-7.

Diaz-Perez, J.C., Purvis, A.C. and Paulk, J.T. (2003) Bolting, yield, and bulb decay of sweet onion as affected by nitrogen fertilization. *Journal of the American Society for Horticultural Science* 128, 144–149. DOI: 10.21273/JASHS.128.1.0144.

Dutta, B. (2019) Evaluation of fungicides to manage downy mildew in Toombs County, Georgia, 2018. *Plant Disease Management Reports* 13, V019. DOI: 10.1094/PDMR13.

Dutta, B., Barman, A.K., Srinivasan, R., Avci, U., Ullman, D.E. *et al.* (2014) Transmission of *Pantoea ananatis* and *P. agglomerans*, causal agents of center rot of onion (*Allium cepa*), by onion thrips (*Thrips tabaci*) through feces. *Phytopathology* 104, 812–819. DOI: 10.1094/PHYTO-07-13-0199-R.

Dutta, B., Edenfield, J. and Williams, Z. (2020a) Evaluation of fungicide programs for post-infection curative activity on onion downy mildew in Toombs County, Georgia, 2019. *Plant Disease Management Reports* 14, V019. DOI: 10.1094/PDMR14.

Dutta, B., Foster, M.J. and Donahoo, W. (2020b) Evaluation of fungicides to manage Botrytis leaf blight in Georgia, 2019. *Plant Disease Management Reports* 14, V018. DOI: 10.1094/PDMR14.

Elshahawy, I.E., Saied, N., Abd-El-Kareem, F. and Morsy, A.A. (2017a) Biocontrol of onion white rot by application of *Trichoderma* species formulated on wheat bran powder. *Archives of Phytopathology and Plant Protection* 50, 150–166. DOI: 10.1080/03235408.2016.1276423.

Elshahawy, I.E., Saied, N., Abd-El-Kareem, F., and Morsy, A.A. (2017b) Field application of sclerotial mycoparasites as biocontrol agents to *Stromatinia cepivora*, the cause of onion white rot. *Journal of Plant Pathology* 99, 391–401. DOI: 10.4454/jpp.v99i2.3888.

Elshahawy, I.E., Saied, N., Abd-El-Kareem, F. and Morsy, A.A. (2018) Field application of selected bacterial strains and their combinations for controlling onion and garlic white rot disease caused by *Stromatinia cepivora*. *Journal of Plant Pathology* 100, 493–503. DOI: 10.1007/s42161-018-0113-z.

El-Wahab, A.A., Elnagar, S. and El-Sheikh, M.K. (2009) Incidence of aphid-borne *Onion yellow dwarf virus* (OYDV) in alliaceae crops and associated weeds in Egypt. In: *4th Conference on Recent Technologies in Agriculture*, pp. 21–33.

Esfahani, M.N. and Pour, B.A. (2008) Differences in resistance in onion cultivars to pink root rot disease in Iran. *Journal of General Plant Pathology* 74, 46–52. DOI: 10.1007/s10327-007-0070-4.

Everts, K.L., Schwartz, H.F., Epsky, N.D. and Capinera, J.L. (1985) Effects of maggots and wounding on occurrence of Fusarium basal rot of onions in Colorado. *Plant Disease* 69, 878–882. DOI: 10.1094/PD-69-878.

Fajardo, T.V.M., Nishijima, M., Buso, J.A., Torres, A.C., Ávila, A.C. *et al.* (2001) Garlic viral complex: identification of *Potyviruses* and *Carlavirus* in Central Brazil. *Fitopatologia Brasileira* 26, 619–626.

Friedrich, S., Leinhos, G.M.E. and Löpmeier, F.J. (2003) Development of ZWIPERO, a model forecasting sporulation and infection periods of onion downy mildew based on meteorological data. *European Journal of Plant Pathology* 109, 35–45. DOI: 10.1023/A:1022024612563.

Galván, G.A., Koning-Boucoiran, C.F.S., Koopman, W.J.M., Burger-Meijer, K., González, P.H. *et al.* (2008) Genetic variation among *Fusarium* isolates from onion, and resistance to *Fusarium* basal rot in related *Allium* species. *European Journal of Plant Pathology* 121, 499–512. DOI: 10.1007/s10658-008-9270-9.

Gent, D.H. and Schwartz, H.F. (2005) Management of Xanthomonas leaf blight of onion with a plant activator, biological control agents, and copper bactericides. *Plant Disease* 89, 631–639. DOI: 10.1094/PD-89-0631.

Gent, D.H., Schwartz, H.F. and Khosla, R. (2004) Managing *Iris yellow spot virus* of onion with cultural practices, host genotype, and novel chemical treatments. (Abstr.) *Phytopathology* 94, S34.

Gent, D.H., du Toit, L., Fichtner, S.F., Krishna Mohan, S., Pappu, H.R. *et al.* (2006) *Iris yellow spot virus*: an emerging threat to onion bulb and seed production. *Plant Disease* 90, 1468–1479. DOI: 10.1094/PD-90-1468.

Gilles, T., Phelps, K., Clarkson, J.P. and Kennedy, R. (2004) Development of MILIONCAST, an improved model for predicting downy mildew sporulation on onions. *Plant Disease* 88, 695–702. DOI: 10.1094/PDIS.2004.88.7.695.

Gitaitis, R.D., Walcott, R.R., Wells, M.L., Perez, J. and Sanders, F.H. (2003) Transmission of *Pantoea ananatis*, causal agent of center rot of onion, by tobacco thrips, *Frankliniella fusca*. *Plant Disease* 87, 675–678. DOI: 10.1094/PDIS.2003.87.6.675.

Grode, A.S., Brisco-McCann, E., Wiriyajitsomboon, P., Hausbeck, M. and Szendrei, Z. (2019) Managing onion thrips can limit bacterial stalk and leaf necrosis in Michigan onion fields. *Plant Disease* 103, 938–943. DOI: 10.1094/PDIS-07-18-1271-RE.

Haapalainen, M., Latvala, S., Kuivainen, E., Qiu, Y., Segerstedt, M. *et al.* (2016) *Fusarium oxysporum*, *F. proliferatum* and *F. redolens* associated with basal rot of onion in Finland. *Plant Pathology* 65, 1310–1320. DOI: 10.1111/ppa.12521.

Hancock, J. and Lorbeer, J. (1963) Pathogenesis of *Botrytis cinerea*, *B. squamosa*, and *B. allii* on onion leaves. *Phytopathology* 53, 669–673.

Haroon, M., Bhat, A.S., Prakash, N.B., Rangaswamy, K.T. and Lingaiah, H.B. (2020) Effect of silicon on incidence and severity of purple blotch disease (*Alternaria porri*) (Ellis) Cif.) in onion (*Allium cepa* L.). *International Journal of Current Microbiology and Applied Sciences* 9, 429–439. DOI: 10.20546 /ijcmas.2020.902.053.

Hartz, T.K., Bogle, C.R., Bender, D.A. and Avila, F.A. (1989) Control of pink root disease in onion using solarization and fumigation. *Journal of the American Society for Horticultural Science* 114, 587–590.

Hassan, M., Yousuf, V., Shah, T.A., Bhat, N.A., Bhat, Z.A. *et al.* (2020) Stemphylium blight of onion: a review. *Agricultural Reviews* 41, 51–58. DOI: 10.18805/ag.R-1902.

Hay, F.S., Sharma, S., Hoepting, C., Strickland, D., Luong, K. *et al.* (2019) Emergence of Stemphylium leaf blight of onion in New York associated with fungicide resistance. *Plant Disease* 103, 3083–3092. DOI: 10.1094/PDIS-03-19-0676-RE.

Hoepting, C.A., Vande Brake, S.K. and van der Heide, E.R. (2020) Efficacy of fungicide treatments for control of Stemphylium leaf blight and Botrytis leaf blight in onion, Elba, 2019. *Plant Disease Management Reports* 14, V220. DOI: 10.1094/PDMR14.

Hussein, M.A.M., Hassan, M.H.A., Allam, A.D.A. and Abo-Elyousr, K.A.M. (2007) Management of Stemphylium blight of onion by using biological agents and resistance inducers. *Egyptian Journal of Phytopathology* 35, 49–60.

Jesperson, G.D. and Suton, J.C. (1987) Evaluation of a forecaster for downy mildew of onion (*Allium cepa* L.). *Crop Protection* 6, 95–103. DOI: 10.1016/0261-2194(87)90106-2.

Katan, J., Rotem, I., Finkel, Y. and Daniel, J. (1980) Solar heating of the soil for the control of pink root and other soilborne diseases in onions. *Phytoparasitica* 8, 39–50. DOI: 10.1007/BF02986234.

Ko, S., Chang, W., Wang, J., Cherng, S. and Shanmugasundaram, S. (2002) Storage variability among short-day onion cultivars under high temperature and high relative humidity, and its relationship with disease incidence and bulb characteristics. *Journal of the American Society for Horticultural Science* 127, 848–854. DOI: 10.21273/JASHS.127.5.848.

Kofoet, A., Kik, C., Wietsma, W.A. and de Vries, J.N. (1990) Inheritance of resistance to downy mildew (*Peronospora destructor* [Berk.] Casp.) from *Allium roylei* Stearn in the backcross *Allium cepa* L. x (*A. roylei* x *A. cepa*). *Plant Breeding* 105, 144–149. DOI: 10.1111/j.1439-0523.1990.tb00467.x.

Kritzman, A., Lampel, M., Raccah, B. and Gera, A. (2001) Distribution and transmission of *Iris yellow spot virus*. *Plant Disease* 85, 838–842. DOI: 10.1094/PDIS.2001.85.8.838.

Lacy, M.L. and Pontius, G.A. (1983) Prediction of weather-mediated release of conidia of *Botrytis squamosa* from onion leaves in the fields. *Phytopathology* 73, 670–676. DOI: 10.1094/Phyto-73-670.

Lang, J.M., Gent, D.H. and Schwartz, H.F. (2007) Management of Xanthomonas leaf blight of onion with bacteriophages and a plant activator. *Plant Disease* 91, 871–878. DOI: 10.1094/PDIS-91-7-0871.

Leach, A., Reiners, S., Fuchs, M. and Nault, B.A. (2017) Evaluating integrated pest management tactics for onion thrips and pathogens they transmit to onion. *Agriculture, Ecosystems & Environment* 250, 89–101. DOI: 10.1016/j.agee.2017.08.031.

Leach, A., Reiners, S. and Nault, B.A. (2020) Challenges in integrated pest management: a case study of onion thrips and bacterial bulb rot in onion. *Crop Protection* 133, 105123. DOI: 10.1016/j.cropro. 2020.105123.

Liu, Q., Liu, P., Xu, Y., Wang, B., Liu, P. *et al.* (2021) Encapsulation of fluazinam to extend efficacy duration in controlling *Botrytis cinerea* on cucumber. *Pest Management Science* 77, 2836–2842. DOI: 10.1002/ ps.6318.

Lunello, P., Di Rienzo, J. and Conci, V.C. (2007) Yield loss in garlic caused by Leek yellow stripe virus Argentinean isolate. *Plant Disease* 9, 153–158. DOI: 10.1094/PDIS-91-2-0153.

Malathi, S. (2015) Biological control of onion basal rot caused by *Fusarium oxysporum* f. sp. *cepae*. *Asian Journal of Biological Sciences* 10, 21–26. DOI: 10.15740/HAS/AJBS/10.1/21-26.

Manglli, A., Tomassoli, L., Tiberini, A., Agosteo, G.E., Fontana, A. *et al.* (2020) A survey on the infection of *Onion yellow dwarf virus* and *Iris yellow spot tospovirus* in seed and bulb productions systems of onion in Calabria, Italy. *European Journal of Plant Pathology* 156, 767–778. DOI: 10.1007/s10658-019-01927-4.

Marodin, J.C., Resende, F.V., Gabriel, A., Souza, R.J., Resende, J.T.V. *et al.* (2019) Agronomic performance of both virus-infected and virus-free garlic with different seed bulbs and clove sizes. *Pesquisa Agropecuária Brasileira* 54, 01448. DOI: 10.1590/s1678-3921.pab2019.v54.01448.

Marzu, J.C., Straley, E. and Havey, M.J. (2018) Genetic analyses and mapping of pink-root resistance in onion. *Journal of the American Society for Horticultural Science* 143, 503–507. DOI: 10.21273/ JASHS04509-18.

Mituti, T., Marubayashi, J.M., Moura, M.F., Krause-Sakate, R. and Pavan, M.A. (2011) First report of *Shallot latent virus* in garlic in Brazil. *Plant Disease* 95, 227. DOI: 10.1094/PDIS-08-10-0599.

Mituti, T., Oliveira, M.L., Nardini, J.P.C., Moura, M.F., Hoffmann, M.I.M. *et al.* (2014) Bulbilhos aéreos de alho, provenientes de escapes florais, são infectados por virus. *Summa Phytopathologica* 40, 368–370. DOI: 10.1590/0100-5405/1971.

Mituti, T., Moura, M.F., Marubayashi, J.M., Oliveira, M.L., Imaizumi, V.M. *et al.* (2015) Survey of viruses belonging to different genera and species in noble garlic in Brazil. *Scientia Agricola* 72, 278–281. DOI: 10.1590/0103-9016-2014-0168.

Molinero-Ruiz, L., Rubio-Pérez, E., González-Domínguez, E. and Basallote-Ureba, M.J. (2011) Alternative hosts for *Fusarium* spp. causing crown and root rot of asparagus in Spain. *Journal of Phytopathology* 159, 114–116. DOI: 10.1111/j.1439-0434.2010.01723.x.

Mullis, S.W., Langston, D.B., Gitaitis, R.D., Sherwood, J.L., Csinos, A.C. *et al.* (2004). First report of Vidalia onion (*Allium cepa*) naturally infected with *Tomato spotted wilt virus* and *Iris yellow spot virus* (family *Bunyaviridae*, genus *Tospovirus*) in Georgia. *Plant Disease* 88, 1285. DOI: 10.1094/PDIS.2004.88.11.1285B.

Nanda, S., Chand, S.K., Mandal, P., Tripathy, P. and Joshi, R.K. (2016) Identification of novel source of resistance and differential response of *Allium* genotypes to purple blotch pathogen, *Alternaria porri* (Ellis) Ciferri. *The Plant Pathology Journal* 32, 519–527. DOI: 10.5423/PPJ.OA.02.2016.0034.

Natesan, K., Park, J.Y., Kim, C.W., Park, D.S. and Kwon, Y.S. (2020) High-quality genome assembly of *Peronospora destructor*, the causal agent of onion downy mildew. *Molecular Plant-Microbe Interactions* 33, 718–720. DOI: 10.1094/MPMI-10-19-0280-A.

Nischwitz, C., Gitaitis, R., Sanders, H., Langston, D., Mullinix, B. *et al.* (2007) Use of fatty acid methyl ester profiles to compare copper-tolerant and copper-sensitive strains of *Pantoea ananatis*. *Phytopathology* 97, 1298–1304. DOI: 10.1094/PHYTO-97-10-1298.

Oliveira, M.L., Marchi, B.R., Mituti, T., Pavan, M.A. and Krause-Sakate, R. (2014) Identification and sequence analysis of five allexiviruses species infecting garlic crops in Brazil. *Tropical Plant Pathology* 39, 483–489. DOI: 10.1590/S1982-56762014000600011.

Palti, J. (1989) Epidemiology, predicton and control of onion downy mildew caused by *Peronospora destructor*. *Phytoparasitca* 17, 31–48. DOI: 10.1007/BF02979603.

Pfeufer, E.E. and Gugino, B.K. (2018) Environmental and management factors associated with bacterial diseases of onion in Pennsylvania. *Plant Disease* 102, 2205–2211. DOI: 10.1094/PDIS-11-17-1703-RE.

Pradhan, M., Tripathy, P., Mandal, P. and Sahoo, B.B. (2018) Effect of salicylic acid (SA) on incidence of purple blotch (*Alternaria porri*) in onion (*Allium cepa* L.). *Journal of Allium Research* 1, 32–36.

Pramesh, D. and Baranwal, V.K. (2015) Production of virus-free garlic (*Allium sativum* l.) through meristem tip culture after solar or hot air treatment of cloves. *The Journal of Horticultural Science and Biotechnology* 90, 180–186. DOI: 10.1080/14620316.2015.11513170.

Rajendran, K. and Ranganathan, K. (1996) Biological control of onion basal rot (*Fusarium oxysporum* f. sp. *cepae*) by combined application of fungal and bacterial antagonists. *Journal of Biological Control* 10, 97–102.

Raziq, F., Alam, I., Naz, I. and Khan, H. (2008) Evaluation of fungicides for controlling downy mildew of onion under field conditions. *Sarhad Journal of Agriculture* 24, 85–91.

Sayago, P., Juncosa, F., Albarracín Orio, G.A., Luna, D.F., Molina, G. *et al.* (2020) *Bacillus subtilis* ALBA01 alleviates onion pink root by antagonizing the pathogen *Setophoma terrestris* and allowing physiological status maintenance. *European Journal of Plant Pathology* 157, 509–519. DOI: 10.1007/s10658-020-02012-x.

Scholten, O.E., van Heusden, A.W., Khrustaleva, L.I., Burger-Meijer, K., Mank, R.A. *et al.* (2007) The long and winding road leading to the successful introgression of downy mildew resistance into onion. *Euphytica* 156, 345–353. DOI: 10.1007/s10681-007-9383-9.

Schroeder, B.K. and du Toit, L.J. (2010) Effects of postharvest onion curing parameters on Enterobacter bulb decay in storage. *Plant Disease* 94, 1425–1430. DOI: 10.1094/PDIS-06-10-0457.

Schroeder, B.K., Humann, J.L. and du Toit, L.J. (2012) Effects of postharvest curing parameters on the development of sour skin and slippery skin in storage. *Plant Disease* 96, 1548–1555. DOI: 10.1094/PDIS-02-12-0117-RE.

Schwartz, H.F. and Mohan, S.K. (2008) *Compendium of Onion and Garlic Diseases and Pests*, 2nd edn. APS Press, St. Paul, Minnesota.

Schwartz, H.F. and Otto, K.J. (1998) Onion bacterial disease management in Colorado. In: *Proceedings of the 1998 National Onion (and other Allium) Research Conference, Sacramento*, California, pp. 214–218.

Sharifi Tehrani, A. and Ramezani, M. (2003) Biological control of *Fusarium oxysporum*, the causal agent of onion wilt by antagonistic bacteria. *Communications in Agricultural and Applied Biological Sciences* 68, 543–547.

Shelton, A.M., Nault, B.A., Plate, J. and Zhao, J.Z. (2003) Regional and temporal variation in susceptibility to lambda-cyhalothrin in onion thrips, *Thrips tabaci* (Thysanoptera: Thripidae), in onion fields in New York. *Journal of Economic Entomology* 96, 1843–1848. DOI: 10.1093/jee/96.6.1843.

Sierotzki, H., Kraus, N., Pepin, S., Ferandes, N., Fernandes, N. *et al.* (2008) Dynamics of QoI resistance in *Plasmopara vitícola*. In: Dehne, H.W., Deising, H.B., Gisi, U., Kuck, K.H., Russell, P.E. *et al.* (eds) *Modern Fungicides and Antifungal Compounds* V. DPG Selbstverlag, Braunschweig, Germany, pp. 151–157.

Silva, L.A., Oliveira, A.S., Melo, F.L., Ardisson-Araújo, D.M.P., Resende, F.V. *et al.* (2019) A new virus found in garlic virus complex is a member of possible novel genus of the family *Betaflexiviridae* (order *Tymovirales*). *PeerJ* 7, e6285. DOI: 10.7717/peerj.6285.

Singh, N. and Cramer, C.S. (2019) Improved tolerance for onion thrips and iris yellow spot in onion plant introductions after two selection cycles. *Horticulturae* 5, 18. DOI: 10.3390/horticulturae5010018.

Stricker, S.M., Tayviah, C.S., Gossen, B.D. and McDonald, M.R. (2020) Fungicide efficacy and timing for the management of *Stemphylium vesicarium* on onion. *Canadian Journal of Plant Pathology*. DOI: 10.1080/07060661.2020.1804461.

Sumner, D.R., Gitaitis, R.D., Gay, J.D., Smittle, D.A., Maw, B.W. *et al.* (1997) Control of soilborne pathogenic fungi in fields of sweet onion. *Plant Disease* 81, 885–891. DOI: 10.1094/PDIS.1997.81.8.885.

Sundin, G.W., Castiblanco, L.F., Yuan, X., Zeng, Q. and Yang, C.H. (2016) Bacterial disease management: challenges, experience, innovation and future prospects. *Molecular Plant Pathology* 17, 1506–1518. DOI: 10.1111/mpp.12436.

Surviliene, E., Valiuskaite, A. and Raudonis, L. (2008) The effect of fungicides on the development of downy mildew of onions. *Zemdirbyste-Agriculture* 95, 171–179.

Sutton, J.C., James, T.D.W. and Rowell, P.M. (1986) Botcast: a forecasting system to time the initial fungicide spray for managing Botrytis leaf blight of onions. *Agriculture, Ecosystems & Environment* 18, 123–143. DOI: 10.1016/0167-8809(86)90136-2.

Taylor, A., Vagany, V., Barbara, D.J., Thomas, B., Pink, D.A.C. *et al.* (2013) Identification of differential resistance to six *Fusarium oxysporum* f. sp. *cepae* isolates in commercial onion cultivars through the development of a rapid seedling assay. *Plant Pathology* 62, 103–111. DOI: 10.1111/j.1365-3059.2012.02624.x.

Teviotdale, B.L., Davis, R.M., Guerard, J.P. and Harper, D.H. (1989) Effect of irrigation management on sour skin of onion. *Plant Disease* 73, 819–822. DOI: 10.1094/PD-73-0819.

Tho, K.E., Brisco-McCann, E., Wiriyajitsomboon, P., Sundin, G. and Hausbeck, M.K. (2019) Bacteria associated with onion foliage in Michigan and their copper sensitivity. *Plant Health Progress* 20, 170–177. DOI: 10.1094/PHP-03-19-0022-RS.

Torres, A.C., Fajardo, T.V., Dusi, A.N., Resende, R.O. and Buso, J.A. (2000) Shoot tip culture and thermotherapy for recovering virus-free plants of garlic. *Horticultura Brasileira* 18, 192–195. DOI: 10.1590/S0102-05362000000300010.

Vahling-Armstrong, C., Dung, J.K.S., Humann, J.L. and Schroeder, B.K. (2016) Effects of postharvest onion curing parameters on bulb rot caused by *Pantoea agglomerans*, *Pantoea ananatis* and *Pantoea allii* in storage. *Plant Pathology* 65, 536–544. DOI: 10.1111/ppa.12438.

Van der Heyden, H., Carisse, O. and Brodeur, L. (2012) Comparison of monitoring based indicators for initiating fungicide spray programs to control Botrytis leaf blight of onion. *Crop Protection* 33, 21–28. DOI: 10.1016/j.cropro.2011.11.008.

Van der Meer, Q. and De Vries, J. (1990) An interspecific cross between *Allium roylei* Stearn and *Allium cepa* L., and its backcross to *A. cepa*. *Euphytica* 47, 29–31. DOI: 10.1007/BF00040359.

Veeraghanti, K.S., Naik, B.G. and Hegde, K.T. (2017) Management of purple blotch disease of onion under field condition. *Journal of Pharmacognosy and Phytochemistry* 6, 1768–1769.

Velez, L.S., Silva, A.M.F., Santos, C.A.F., Assunção, E.F., Silva, M.S. *et al.* (2020) Evaluation of onion genotypes to slippery skin caused by *Burkholderia gladioli* pv. *alliicola*. *Horticultura Brasileira* 38, 350–355. DOI: 10.1590/s0102-053620200402.

Vieira, R.L., Silva, A.L., Zaffari, G.R., Steinmacher, D.A., Fraga, H.P.F. *et al.* (2015) Efficient elimination of virus complex from garlic (*Allium sativum* L.) by cryotherapy of shoot tips. *Acta Physiologiae Plantarum* 37, 1733. DOI: 10.1007/s11738-014-1733-3.

Villalta, O.N., Wite, D., Porter, I.J., McLean, K.L., Stewart, A. *et al.* (2012) Integrated control of onion white rot on spring onions using diallyl disulphide, fungicides and biocontrols. *Acta Horticulturae* 944, 63–71. DOI: 10.17660/ActaHortic.2012.944.8.

Vincelli, P.C. and Lorbeer, J.W. (1989) BLIGHT-ALERT: a weather-base predictive system for timing fungicide applications on onion before infection periods of *Botrytis squamosa*. *Phytopathology* 79, 493–498. DOI: 10.1094/Phyto-79-493.

Visser, C.L.M. (1998) Development of a downy mildew advisory model based on downcast. *European Journal of Plant Pathology* 104, 933–943. DOI: 10.1023/A:1008656122629.

Wallin, J.R. (1962) Summary of recent progress in predicting late blight epidemics in United States and Canada. *American Journal of Potato* 39, 306–312. DOI: 10.1007/ BF02862155.

Wang, J., Shi, L., Wang, D., Li, L., Loake, G.J. *et al.* (2019) White rot disease protection and growth promotion of garlic (*Allium sativum*) by endophytic bacteria. *Plant Pathology* 68, 1543–1554. DOI: 10.1111/ ppa.13066.

Wang, M., Weiberg, A., Lin, F.M., Thomma, B.P.H.J., Huang, H.D. *et al.* (2016) Bidirectional cross-kingdom RNAi and fungal uptake of external RNAs confer plant protection. *Nature Plants* 2, 16151. DOI: 10.1038/nplants.2016.151.

Wordell Filho, J.A., Martins, D.A. and Stadnik, M.J. (2007) Aplicação foliar de tratamentos para o controle do míldio e da podridão-de-escamas de bulbos de cebola. *Horticultura Brasileira* 25, 544–549. DOI: 10.1590/S0102-05362007000400010.

Wright, P.J. (1993) Effects of nitrogen fertilizer, plant maturity at lifting, and water during field curing on the incidence of bacterial soft rot of onions in store. *New Zealand Journal of Crop and Horticultural Science* 21, 377–381. DOI: 10.1080/01140671.1993.9513796.

Wright, P.J. and Grant, D.G. (1998) Evaluation of *Allium* germplasm for susceptibility to foliage bacterial soft rot caused by *Pseudomonas marginalis* and *Pseudomonas viridiflava*. *New Zealand Journal of Crop and Horticultural Science* 26, 17–21. DOI: 10.1080/01140671.1998.9514034.

Wright, P.J., Hale, C.N. and Fullerton, R.A. (1993) Effect of husbandry practices and water applications during field curing on the incidence of bacterial soft rot of onions in store. *New Zealand Journal of Crop and Horticultural Science* 21, 161–164. DOI: 10.1080/01140671.1993.9513762.

Wright, P.J., Searle, B., Tyson, J.L. and Mellow, K.D. (2019) The current outbreak of stemphylium leaf blight of onion in New Zealand – identification of cause and review of possible risk factors associated with the disease. *New Zealand Plant Protection* 72, 10–20. DOI: 10.30843/nzpp.2019.72.254.

Zapata-Sarmiento, D.H., Palacios-Pala, E.F., Rodríguez-Hernández, A.A., Melchor, D.L.M., Rodríguez-Monroy, M. *et al.* (2020) *Trichoderma asperellum*, a potential biological control agent of *Stemphylium vesicarium*, on onion (*Allium cepa* L.). *Biological Control* 140, 104105. DOI: 10.1016/j.biocontrol. 2019.104105.

Zhang, J., Li, G.Q. and Jiang, D.H. (2009) First report of garlic leaf blight caused by *Botrytis porri* in China. *Plant Disease* 93, 1216–1216. DOI: 10.1094/PDIS-93-11-1216B.

Zhang, J., Zhang, L., Li, G.Q., Yang, L., Jiang, D-H. *et al.* (2010) *Botrytis sinoallii*: a new species of the grey mould pathogen on *Allium* crops in China. *Mycoscience* 51, 421–431. DOI: 10.1007/S10267-010-0057-4.

14 Trait Genetics

Kiran Khandagale[1]*, Dhananjay Shirsat[2], Avinash Ade[1], Ashwini Benke[2], Pranjali Gedam[2], Pragati Randive[2], Suresh Gawande[2] and Major Singh[2]
[1]Savitribai Phule Pune University, Pune, India; [2]ICAR-Directorate of Onion and Garlic Research, Rajgurunagar, India

1. Introduction

Alliums are not considered breeder-friendly crops due to their complex nature, breeding behaviour and large genome. The study of trait genetics and breeding of *Allium* crops is challenging due to some inherent characteristics such as a high degree of cross-pollination, inbreeding depression and biannual life cycle in many alliaceous crops. Hence, despite their nutritional and economic importance, only limited genetic and genomic information is available. The challenging task of working with large genome size (10–30 pg/IC) (Ricroch *et al.*, 2005) leads to difficulty in sequencing, limited markers availability and, eventually, hampered mapping and genomics-assisted breeding in these crops. The present chapter reviews genetic studies of important traits in alliums.

2. Bulb Colour

The bulb onion colour influences consumer preference across the world. White, yellow, red, pink, chartreuse and gold colours are available in different parts of the world. This trait is governed by a complex genetic mechanism involving multiple genes of the flavonoid biosynthesis metabolic pathway (Khandagale and Gawande, 2019). Early in the 20th century, researchers reported that yellow and red were dominant over white bulb colour (Tscreryae, 1916; Meunissier, 1918). Rieman (1931) reported that the incomplete dominance of factor *I* was responsible for the production of the white bulb. Later, researchers reported that the interaction of the following genes determine the expression of this trait (Clarke *et al.*, 1944; El-Shafie and Davis, 1967). The dominant colour inhibitory factor (*I*) is responsible for white bulb color. Homozygous dominant and heterozygous factor (*C*) develops colour while the recessive (cc) condition produces white bulbs irrespective of other factors and is, therefore, called a basic colour factor. Locus *G* in homozygous recessive *(gg)* condition gives chartreuse colour, otherwise leading to golden yellow bulbs. *L* and *R* are complementary factors that are involved in imparting red colour to bulbs (El-Shafie and Davis, 1967). Advances in molecular techniques revealed that structural genes from the flavonoid biosynthesis pathway – chalcone synthase (CHS), dihydroflavonol 4-reductase (DFR) and anthocyanidin synthase (ANS) etc. – have major roles in onion bulb colour. Studies on

*Address for correspondence: kirankhandagale253@gmail.com

 Edible Alliums: Botany, Production and Uses (eds H.D. Rabinowitch and B. Thomas)
DOI: 10.1079/9781789249996.0014

gene expression and allelic diversity in varied genetic backgrounds revealed that *C*, *R* and *L* loci correspond to *CHS*, *DFR* and *ANS* genes, respectively (Khandagale and Gawande, 2019). Further, these genes were assigned to particular chromosomal locations with the help of alien chromosome addition techniques (Masuzaki *et al.*, 2006).

Expression analyses, gene cloning and sequencing of candidate genes of flavonoid pathway revealed several alleles of these genes and their genetic control over bulb colour and biosynthesis of flavonoids (Khandagale and Gawande, 2019). Hence, mutations or inactivity of the *DFR* gene leads to loss of function, and yellow colour develops rather than red. Currently, ten different *DFR* alleles were reported to have variations in the intron, exon and promoter regions of the gene, and molecular markers have been developed to identify or select these mutations in the population (Kim *et al.*, 2004a, 2005a, 2009, 2015a, 2017; Park *et al.*, 2013; Song *et al.*, 2014). Pink bulb colour is recessive as evidenced from the genetic analyses under the influence of expression of the *ANS* gene. Five alleles of the *ANS* gene were reported with mutations in promoter as well as coding regions (Kim *et al.*, 2004b, 2005b, 2006, 2015b, 2016a). Some of the mutations resulted in a change of a critical amino acid at the enzyme active site with the consequent effect on its activity (Kim *et al.*, 2005c; Kim *et al.*, 2016a). Similarly, the non-functionality of the *CHI* gene resulted in a golden-coloured bulb (Kim *et al.*, 2004c) (details in Table 14.1).

Binary quantitative genetic analyses of anthocyanin concentrations in segregating haploid progenies of cross yellow (OH1) and red (5225) inbreds suggest that a region on chromosome 7 and two regions linked in the repulsion phase on chromosome 4 interact in imparting bulb colour. Further, chromosome 7 corresponded to the *DFR* gene (*R* locus), and two loci on chromosome 4 may correspond to the L_1 and L_2 loci (Duangjit *et al.*, 2014). Genetic linkage mapping using genotyping-by-sequencing also revealed one major QTL, *qAS7.1*, for anthocyanin synthesis and two significant QTLs, *qAC4.1* and *qAC4.2*, for the anthocyanin quantity (Duangjit *et al.*, 2014; Choi *et al.*, 2020). Tsukazaki *et al.* (2017) reported two QTLs for pigmentation of bunching onion pseudostem on linkage groups chromosome 4a and chromosome 5a-2, among them *qPig5a-2*, found to be major effect QTL. Similarly, Havey (2020) mapped chartreuse bulb colour (*G* locus) on chromosome 7.

RNAseq was also performed for the elucidation of bulb pigmentation. Baek *et al.* (2017) reported 97 genes of flavonoid pathway showing differential expression in yellow and white onion and identified marker linked to *C* locus. Further, Zhang *et al.* (2018) identified 16 differentially expressed genes that might play a key role in bulb colour development and flavonoid biosynthesis. RNAseq and BSA analyses revealed a non-autonomous DNA transposon insertion (577 bp) in the gene of bHLH transcription factor resulted in white bulb colour. It showed perfect linkage with the *C* locus and thus it might be a causal mutation for this locus (Jo and Kim, 2020).

3. Male Sterility

Male sterility is characterized by the inability to produce viable pollen without compromising female fertility (Jones and Clarke, 1943). This trait plays a major role in reducing cost and labour in hybrid seed production. Hybrid varieties of onion are available in American and European markets but open-pollinated varieties are still dominant in major onion-growing areas like India.

Mitochondrial genomes are dynamic and undergo recombination and genome rearrangements through short repeats which leads to plant mtDNA evolution as well as male sterility through the generation of chimeric open-reading frames (ORFs) (Satoh *et al.*, 2004; Kmiec *et al.*, 2006). Two major types of CMS cytoplasm have been reported in onion – CMS-S and CMS-T (Jones and Emsweller, 1936; Berninger 1965). The fertility of CMS-S lines is restored by a single restoration factor (*Rf*) gene called *Ms* locus (Jones and Clarke, 1943) whereas three independent *Rf* loci (A, B, C) are required to restore the fertility of the CMS-T system (Schweisguth, 1973). Due to stability and simple inheritance, the CMS-S system is generally preferred over CMS-T in commercial hybrid seed production in onion (Havey, 2000). The molecular genetics and sequencing of the mitochondrial genome identified *cytochrome C oxidase 1* (*cox1*) as part of chimeric *orf725*, which is

Table 14.1. Multiple alleles of structural genes in flavonoid pathway responsible for bulb colour variation.

Allele	Mutation details	Reference
Alleles of DFR gene		
DFR-A^{PS}	A premature stop codon and 20 bp deletion in the promoter region	Kim *et al.* (2009a)
DFR-A^{DEL}	Complete deletion of the DFR-A gene	Kim *et al.* (2009a)
DFR-A^{PD}	171 bp deletion in the promoter region	Song *et al.* (2014)
DFR-A$^{5'DEL}$	518 bp deletion in exons 1 and 2	Song *et al.* (2014)
DFR-A^{GT}	2 bp insertion led to premature stop codon	Song *et al.* (2014)
DFR-A^{2AT}	4 bp insertion led to premature stop codon	Song *et al.* (2014)
DFR-A^{K48N}	1 bp substitution	Song *et al.* (2014)
DFR-A^{TTA}	Insertion of a leucine near the catalytic site	Song *et al.* (2014)
DFR-A^{PS2}	Stop codon in the 6th exon	Kim *et al.* (2017)
DFR-A^{DTP}	4615 bp DNA transposon between exon 1 and 2	Kim *et al.* (2015b)
DFR-A^{LTR}	5070 bp long terminal repeat (LTR) retrotransposon in exon 3	Kim *et al.* (2015b)
DFR-A^{TRN}	5073 bp DNA transposon insertion in exon 5	Kim *et al.* (2015)
Alleles of ANS gene		
390-bp insertion	390 bp insertion in the promoter region	Kim *et al.* (2005c)
ANS^{G229R}	A point mutation led to an amino acid change of a glycine to an arginine at residue 229	Kim *et al.* (2005b)
ANS^{-h1}	A 400 bp insertion at 632 bp upstream from the putative transcription start site	Kim *et al.* (2006)
ANS^{PS}	4 bp insertion in exon 1 resulted in a premature stop codon	Kim *et al.* (2016a)
ANS^{S188L}	A critical amino acid change from serine into leucine	Kim *et al.* (2016a)
ANS^{P}	A non-autonomous DNA transposon of 6258 bp was inserted in the promoter region.	Kim *et al.* (2015b)

responsible for CMS in onion (Kim *et al.*, 2016b; Kim *et al.*, 2019). The chimera of other genes such as *atp6*, *orfA501*and *atp9* are also involved in onion male sterility (Engelke *et al.*, 2003). Further, the expression pattern of *cox1* and *orfA501* differed in male-sterile lines with CMS-T and CMS-S systems indicating their relationship with onion CMS (Kim *et al.*, 2009a). Additionally, five genes linked with male sterility were identified using transcriptomics at the tetrad stage, among them *atp9* and *cox1* from the cytoplasmic genome and *SERK1*, *AG* and *AMS* from the nuclear genome (Yuan *et al.*, 2018). Similarly, RNAseq of male-sterile and maintainer lines of *A. fistulosum* revealed several thousand differentially expressed genes, some of them known for their association with male sterility – *F-type ATPase*, *NADH dehydrogenase* and *cytochrome c oxidase* (Liu *et al.*, 2016). Shemesh-Mayer *et al.* (2013) provided the first detailed study of male gametophyte development in garlic and observed the presence of three types of plant-sterility mechanisms: complete sterility, male sterility and environmentally induced male sterility. Further, Shemesh-Mayer *et al.* (2015a) also suggested that the sterility in garlic is governed by genetic factors as it was not restored by temperature changes. RNAseq and proteomic studies indicated the energy deficiency caused due to respiratory restrictions or non-regulated programmed cell death led to the pollen abortion in the male-sterile genotype (Shemesh-Mayer *et al.*, 2015b).

The genes for the restorer of fertility factor (*Rf*) have been identified in several species and are found to be a member of the pentatricopeptide repeat family of proteins (PPR). These proteins were characterized by the presence of a 35 amino acids conserved motif (Chen and Liu, 2014; Islam *et al.*, 2014). Transcriptome analyses of the unopened flowers of male-sterile and fertile plants revealed 483 transcripts containing the PPR motif. Among them, 41 were closely related to *Rf* genes but none of the *Rf*-like genes showed any linkage with *Ms* locus. The integrated study using BSA and RNAseq led to the identification of the best candidate gene for fertility restoration in onion as *AcPMS1* (Kim *et al.*, 2015a).

This gene is involved in DNA mismatch repair. Such DNA mismatch repair genes were known to be involved in fertility restoration in other species also (Li *et al.*, 2009; Chen and Liu, 2014).

Due to the biannual nature of onion, it takes 4–8 years to identify the different cytotypes in male-sterile and maintainer lines. Fortunately, researchers have developed several molecular markers linked to the *Ms* locus as well as S & T cytotypes (Table 14.2), for both cytoplasmic and nuclear genomes.

4. Pungency

The pungent flavour in alliums forms when alk(en)yl cysteine sulfoxide (ACSO) precursors like propenyl-, propyl- and methyl-cysteine sulfoxides react with allinase enzyme during physical injury caused by cutting, which leads to the formation of highly unstable sulfenic acid products (Randle and Lancaster, 2002; Chapter 16 this book). 1-propenyl sulphenic acid acted upon by lachrymatory factor synthase to produce lachrymatory factor which led to tearing (Brodnitz and Pascale, 1971; Imai *et al.*, 2002). Pungent flavour intensity is affected by sulphur fertilization and physiological partitioning of sulphate and organic sulphur compounds (Randle, 1992; Randle *et al.*, 1999; Chapter 11, this volume).

Pungency is a genotype-specific characteristic and total soluble solid content is positively correlated with pungency (Foskett and Peterson, 1950; Simon, 1995). Pal and Singh (1987) studied onion genetics of pungency and found that pungency (expressed as pyruvic acid) is mainly regulated by additive gene effect and thus pungency could be improved by heterosis breeding. Yoo *et al.* (2006) studied the effect of cultivars, environment and bulb size on variation in pungency and reported that 81.3% of the variation was due to genetic differences. The first marker-based QTL pungency mapping was performed by Galmarini *et al.* (2001) in the cross between 'BYG15-23' and 'AC43' and revealed that region on chromosome 5 influences the trait. Molecular marker-based linkage mapping of pungency in the cross of pungent onion (W202A) × mild onion (Texas Grano 438) revealed that genes in the sulphate assimilation pathway such as plastidic ferredoxin-sulfite reductase (SiR) and plastidic ATP sulfurylase (ATPS) are very tightly linked on chromosome 3. Further, single-locus analyses showed that a marker linked to candidate genes controls a major portion of the pungency variation. The close linkage between *SiR* and *ATPS* genes indicates mutation in one of the genes may cause the variation. The marker used in this study could be used for selecting lower-pungency phenotypes in the population (McCallum *et al.*, 2007).

Mild pungency is one of the desirable traits of bunching onion, where the broad- (0.97) and narrow-sense (0.56) heritability of pungency were estimated in eight inbred lines (Kojima *et al.*, 2004). Further, Tsukazaki *et al.* (2012) mapped a major QTL controlling pseudostem pungency in bunching onion on chromosome 2a. Marker 'AFAT04B03' can be effectively used in the selection of low-pungent genotypes in the breeding programme. They also released low-pungent

Table 14.2. Markers for identification of male-sterile and restorer lines.

Marker description	Species	Reference
SCAR marker for identification of S and N cytoplasm	*A. fistulosum* L.	Gao *et al.* (2015)
SCAR markers co-segregated with the dominant *Ms* and recessive *ms* allele	*A. cepa* L.	Yang *et al.* (2013)
PCR-based markers linked to the *Ms* locus	*A. cepa* L.	Bang *et al.* (2011)
Co-dominant marker in linkage disequilibrium with *Ms* locus	*A. cepa* L.	Kim (2014)
PCR-based markers for CMS-(S), CMS-(T), (N)-cytoplasm	*A. cepa* L.	Engelke *et al.* (2003)
A marker based on the copy numbers of *orf725* and *coxI* was developed for distinguishing three cytoplasm types by single PCR	*A. cepa* L.	Kim *et al.* (2009b)
Co-dominant CAPS marker associated with *Ms* locus	*A. cepa* L.	Bang *et al.* (2013)
High-resolution melting (HRM) markers were developed for both cytoplasm identification and *Ms* locus.	*A. cepa* L.	Kim and Kim (2019)

pseudostem cultivar 'Fuyuwarabe' in Japan (Wako *et al.*, 2010). Yoo *et al.* (2020) developed sweet onion (low-pungent) by using recurrent selection. They started with pungencies of 3–4 mM pyruvic acid and after the fourth recurrent selections achieved bulb mean pungency of 1.5 mM, thus low-pungent lines could be developed in ten years with the availability of low-pungency parent lines, enough manpower and strong selection pressure.

Genes involved in pungency in alliums have been identified and studied (McCallum *et al.*, 2011; McManus *et al.*, 2012; Kamenetsky *et al.*, 2015; Sayadi *et al.*, 2020). Several sulphur assimilation-related genes were assigned to their chromosomal locations using monosomic addition lines (Masamura *et al.*, 2011). Candidate genes such as *O-acetyl-L-serine (thiol)-lyase*, *APS Reductase*, *high-affinity sulphate transporter homolog* (*HAST*) and *ATP sulfurylase* in sulphur assimilation were studied by McCallum *et al.* (2011) and variation in their expression might be playing a role in the pungency of onion. Sulphite reductase (SiR) is a key enzyme in sulphur assimilation in several plants. Functional studies revealed that the onion contains one functional and one pseudogene *SiR* gene. The activity of this gene depends on the sulphur availability and plays a key role in S-assimilation in onion (McManus *et al.*, 2012).

Allicin is an organosulphur compound in garlic that has several medicinal and nutraceutical properties and imparts tolerance to stress (Chapter 16, this volume). Allicin is catalysed by alliinase from cysteine sulfoxides in alliums. Sequence analysis of the *alliinase* gene revealed its multi-gene nature with the presence of sequence variants. Bolting-type garlic cultivars showed higher variability than non-bolting (Ovesná *et al.*, 2015). Recently, Sayadi *et al.* (2020) sequenced the allinase gene from *Allium* species which revealed novel allinase alleles, which express in roots and bulbs but not in leaves. Thus the presence of isoforms and variation in spatial expression might be playing a role in different allicin content in these species. Garlic became the first *Allium* species whose genome is sequenced, being 16.24 GB in size with 57,561 coding genes (Sun *et al.*, 2020). They predicted 60 genes for allinase in garlic of which 38 were reported to be expressed in different tissues. Among these, four genes having high similarity and expression in bulbs indicate functional redundancy to enhance allicin content in bulbs.

5. Bolting

Bolting is a premature flowering in the first production season before bulb maturation and it is considered an undesirable trait in the bulb onion crop but this character is widely used for breeding and seed production. Despite its economic importance, the genetics of bolting in alliums is little studied. Hyun *et al.* (2009) first studied bolting time in segregating F_2 onion population between late-bolting-type cultivar (MOS8) and very early-bolting-type cultivar (Guikum). They revealed that late bolting was a dominant trait. Proteomics study revealed that the differential protein spot between late- and early-bolting genotypes showed similarities to heterochromatin protein 1 (Hyun *et al.*, 2009), thus indicating that floral genetic pathways governed by several loci might play a role in the regulation of bolting in onion and suggested the future need of QTL mapping and genomics for a better understanding of this trait. Baldwin *et al.* (2014) first reported QTLs for the adaptive trait in bulb onion. The amplicon sequencing of ten candidate genes in eight different populations was performed but differentiation was found only in two candidate genes (*AcFT2* and *AcSOC1*). Further QTL analysis of bolting in Nasik Red × CUDH2150 F_2 population showed that QTLs for bolting were present on chromosomes 1, 3 and 6. QTL on chromosome 1 (*AcBlt1*) was consistently evident in bolting susceptibility in two different environments. Bunching onion (*Allium fistulosum* L.) is an important vegetable crop (Chapter 7, this volume). Bolting severely affects its quality and production during the spring and early summer and the scape's toughness in bolted plants makes them inedible. Thus, late bolting is an important trait in *A. fistulosum*. QTL mapping in two different $F_{2:3}$ populations originated from the crosses between ever-bolting, early-bolting and late-bolting genotypes revealed that QTL on chromosome 2 (chromosome 2a) was having a major effect and is linked to the same marker in both populations, thus indicating its significant influence on bolting time in *A. fistulosum* (Wako *et al.*, 2016).

6. Disease and Pest Resistance

Allium crops are attacked by several fungal, bacterial and viral pathogens and a variety of insect pests. Therefore, resistance to disease and pests is an important trait for breeding. The resistance to purple blotch in onion was reported to be controlled by additive and non-additive gene effects (Evoor *et al.*, 2010), and a single dominant qualitative gene (Abubaker and Ado, 2008). Further, linkage studies in short-day onion F_2 population revealed that a single dominant gene imparts resistance to *Alternaria porri*, and RAPD marker associated with purple blotch resistance was discovered (Ganesh and Veeregowda, 2007). Chand *et al.* (2018) first mapped the purple blotch resistance loci in onion using F_1, F_2 and BC_1 populations of Arka Kalyan × Agrifound Rose and reported that resistance was governed by a single dominant gene (*ApR1*). The SSR marker (AcSSR7) and STS marker (ApR-450) were found tightly linked to *ApR1* and thus could be used in marker-assisted breeding in onion.

Fusarium basal rot (FBR) caused by *Fusarium oxysporum* f. sp. *cepae* (FOC) is one of the important diseases of onion and other alliums at the global level. Genetics of FBR resistance involves two partially dominant genes (Bacher *et al.*, 1989), cytoplasmic genes, single dominant gene (Tsutsui, 1991) and multiple genes (Holz and Knox-Davies, 1985). Galván *et al.* (2008) studied the genetic diversity in different *Fusarium* isolates and screened six *Allium* species for FBR resistance. *A. fistulosum*, *A. roylei* and *A. galanthum* were found to be resistant and thus could be exploited for introgression of fusarium resistance into the onion. Diversity plays a key role in the selection of beneficial traits. Taylor *et al.* (2019) assembled a diverse set of onions and identified fusarium-resistant genotypes and linked SNP markers with seedling vigour and FBR resistance. Similarly, screening for resistance in garlic selected accession CBT-As153 as resistant to FBR (Rout *et al.*, 2014). Further qPCR analyses of nucleotide-binding sites (NBS) encoding resistance genes revealed that *AsRGA29* showed 20-fold higher gene expression in CBT-As153 compared to other resistant wild alliums. *AsRGA29* was also found to be induced upon infection by FOC as well as priming with defence-signalling hormones (Rout *et al.*, 2014). FBR-resistant varieties have been developed in long-day onions but short-day onions are susceptible (Cramer, 2000).

Downy mildew caused by *Peronospora destructor* is the most devastating disease of alliums and causes higher damages in cool and moist weather. Warid and Tims (1952) reported that resistance was governed by dominant alleles at two genes. Due to the limitation of resistance source in cultivated Alliums, however, researchers mined the wild germplasm in search of a resistance source. Interspecific crosses between *A. roylei* × *A. cepa*, revealed that downy mildew resistance is controlled by a single, dominant locus (Kofoet *et al.*, 1990) and two dominant, weakly linked nuclear genes in *A. roylei* (De Vries *et al.*, 1992). The first successful introgression of resistance from wild *Allium* to onion was achieved by introgression of resistance to downy mildew from *A. roylei*. It became possible by breaking the link between a lethal recessive allele and the resistance allele due to a rare crossing over (Scholten *et al.*, 2007). Similarly, a few wild relatives were also identified as the source of resistance against fusarium basal rot, purple blotch and anthracnose (Malik *et al.*, 2020).

Pink root is a major soilborne disease of onion caused by *Phoma terrestris*, which significantly reduces the quality and bulb yield in onion (Coleman and Ellerbrock, 1992). The development of the resistant cultivars to the pink root disease is more practical than the soil treatment (Rengwalska and Simon, 1986). Marzu *et al.* (2018) had developed segregating families from different sources of pink root resistance and evaluated them by screening. The severity of PR in two segregating families from the same sources of resistance were mapped to one position on chromosome 4 and found to be co-dominantly inherited in onion.

Neck rot caused by *Botrytis alli* causes heavy storage losses in bulb onion. Its infection is favoured by harvesting early before bulb maturity and inadequate curing of the bulb, also the cool temperature and high humidity in storage (Lin *et al.*, 1995). Van der Meer *et al.* (1970) demonstrated different methods for testing resistance of *Allium* species to *Botrytis alli* Munn. It was reported that the pathogen behaves as a wound parasite and its pathogenicity is rather low. The resistance to the neck rot is a qualitative character; however, inbreeding and selection can reduce the disease incidence (Vik and Aastveit, 1984; Lin *et al.*, 1995).

Epicuticular wax plays an important role in imparting resistance to pests and diseases in plants. This particularly acts as the first line of defence in biotic stress and it varies with the structure, composition and thickness of the wax layer (Dhanyalakshmi *et al.*, 2019). In onion, several studies show that glossy or semi-glossy genotypes are less damaged by thrips than waxy genotypes (Molenaar, 1984; Damon *et al.*, 2014; Munaiz *et al.*, 2020). Molenaar (1984) reported that the non-glossy foliage trait was dominant over glossy foliage and was in epistatic interaction with genes imparting glossy scapes. Transcriptome analyses of waxy and glossy welsh onion identified several potential genes for wax cuticle formation and defence (Liu *et al.*, 2014).

7. Summary

A genetic understanding of important traits in *Allium* crops is difficult due to characters like large genome size and inbreeding depression. An understanding of key traits has been developed in onion, but in other alliums studies are limited. Double haploids in onion varieties help analysing the genetics behaviour of several traits and overcome the problem of inbreeding depression. Transcriptomics, proteomics and metabolomics have helped to reveal the genetics behind complex traits such as flowering, bulbing, biotic and abiotic stress. Overall, genetic studies in alliums are limited and co-ordinated efforts are needed for future investigation.

References

Abubakar, L. and Ado, S.G. (2008) Heterosis of purple blotch (*Alternaria porri* (Ellis) Cif.) resistance, yield and earliness in tropical onions (*Allium cepa* L.). *Euphytica* 164(1), 63–74. DOI: 10.1007/s10681-007-9639-4.

Bacher, J.W., Pan, S. and Ewart, L. (1989) Inheritance of resistance to *Fusarium oxysporum* f.sp. *cepae* in cultivated onions. In: Jensen, L. (ed.) *Proceedings of the National Onion Research Conference*, Boise, Idaho, pp. 85–91.

Baek, G., Kim, C.W. and Kim, S. (2017) Development of a molecular marker tightly linked to the C locus conferring a white bulb color in onion (*Allium cepa* L.) using bulked segregant analysis and RNA-Seq. *Molecular Breeding* 37(8), 1–13. DOI: 10.1007/s11032-017-0697-6.

Baldwin, S., Revanna, R., Pither-Joyce, M., Shaw, M., Wright, K. *et al.* (2014) Genetic analyses of bolting in bulb onion (*Allium cepa* L.). *Theoretical and Applied Genetics* 127(3), 535–547. DOI: 10.1007/s00122-013-2232-4.

Bang, H., Cho, D.Y., Yoo, K.S., Yoon, M.K., Patil, B.S. *et al.* (2011) Development of simple PCR-based markers linked to the Ms locus, a restorer-of-fertility gene in onion (*Allium cepa* L.). *Euphytica* 179(3), 439–449. DOI: 10.1007/s10681-010-0342-5.

Bang, H., Kim, S., Park, S.O., Yoo, K.S., and Patil, B.S. (2013) Development of a codominant CAPS marker linked to the Ms locus controlling fertility restoration in onion (*Allium cepa* L.). *Scientia Horticulturae* 153, 42–49. DOI: 10.1016/j.scienta.2013.01.020.

Berninger, E. (1965) Contribution a l'étude de la stérilité male de l'oignon (*Allium cepa*). *Annal Amelior Plant* 15, 183–199.

Brodnitz, M.H. and Pascale, J.V. (1971) Thiopropanal S-oxide: a lachrymatory factor in onions. *Journal of Agricultural and Food Chemistry* 19(2), 269–272.

Chand, S.K., Nanda, S. and Joshi, R.K. (2018) Genetics and molecular mapping of a novel purple blotch-resistant gene ApR1 in onion (*Allium cepa* L.) using STS and SSR markers. *Molecular Breeding* 38(9), 1–13. DOI: 10.1007/s11032-018-0864-4.

Chen, L. and Liu, Y.G. (2014) Male sterility and fertility restoration in crops. *Annual Review of Plant Biology* 65, 579–606. DOI: 10.1146/annurev-arplant-050213-040119.

Choi, Y., Kim, S. and Lee, J. (2020) Construction of an onion (*Allium cepa* L.) genetic linkage map using genotyping-by-sequencing analysis with a reference gene set and identification of QTLs controlling anthocyanin synthesis and content. *Plants* 9(5), 616. DOI: 10.3390/plants9050616.

Clarke, A.E., Jones, H.A. and Little, T.M. (1944) Inheritance of bulb color in the onion. *Genetics* 29(6), 569.

Coleman, P.M., Ellerbrock, L.A. and Lorbeer, J.W. (1992) Interaction of *Phoma terrestris* and soil moisture level on yield of two onion cultivars differentially susceptible to pink root. *Plant Disease* 76(12), 1213–1216.

Cramer, C.S. (2000) Breeding and genetics of Fusarium basal rot resistance in onion. *Euphytica* 115(3), 159–166. DOI: 10.1023/A:1004071907642.

Damon, S.J., Groves, R.L. and Havey, M.J. (2014) Variation for epicuticular waxes on onion foliage and impacts on numbers of onion thrips. *Journal of the American Society for Horticultural Science* 139(4), 495–501. DOI: 10.21273/JASHS.139.4.495.

De Vries, J.N., Wietsma, W.A. and Jongerius, M.C. (1992) Linkage of downy mildew resistance genes Pd 1 and Pd 2 from *Allium roylei* Stearn in progeny of its interspecific hybrid with onion (*A. cepa* L.). *Euphytica*, 64(1), 131–137. DOI: 10.1007/BF00023546.

Dhanyalakshmi, K.H., Soolanayakanahally, R.Y., Rahman, T., Tanino, K.K. and Nataraja, K.N. (2019) Leaf cuticular wax, a trait for multiple stress resistance in crop plants. In: *Abiotic Biotic Stress Plants*. IntechOpen, London, UK. Available at: https://doi.org/10.5772/intechopen.84565 (accessed 24 July 2022).

Duangjit, J., Welsh, K., Wise, M.L., Bohanec, B. and Havey, M.J. (2014) Genetic analyses of anthocyanin concentrations and intensity of red bulb color among segregating haploid progenies of onion. *Molecular Breeding* 34(1), 75–85. DOI: 10.1007/s11032-014-0018-2.

El-Shafie, M. and Davis, G.N. (1967) Inheritance of bulb color in the onion (*Allium cepa* L.). *Hilgardia* 38(17), 607–622. DOI: 10.3733/hilg.v38n17p607.

Engelke, T., Terefe, D. and Tatlioglu, T. (2003) A PCR-based marker system monitoring CMS-(S), CMS-(T) and (N)-cytoplasm in the onion (*Allium cepa* L.). *Theoretical and Applied Genetics* 107(1), 162–167.

Evoor, S., Veeregowda, R. and Ganeshan, G. (2010) Genetics of resistance to purple blotch disease in onion (*Allium cepa* L.). *Karnataka Journal of Agricultural Sciences* 20(4).

Foskett, R.L. and Peterson, C.E. (1950) Relation of dry matter content to storage quality in some onion varieties and hybrids. *Proceedings of the American Society for Horticultural Science*, vol. 55, pp. 314–318.

Galmarini, C.R., Goldman, I.L. and Havey, M.J. (2001) Genetic analyses of correlated solids, flavor, and health-enhancing traits in onion (*Allium cepa* L.). *Molecular Genetics and Genomics* 265, 543–551.

Galván, G.A., Koning-Boucoiran, C.F., Koopman, W.J., Burger-Meijer, K., González, P.H. *et al.* (2008) Genetic variation among *Fusarium* isolates from onion, and resistance to Fusarium basal rot in related *Allium* species. *European Journal of Plant Pathology* 121(4), 499–512.

Ganesh, C.T. and Veeregowda, R. (2007) Marker assisted selection for purple blotch disease resistance in onion (*Allium cepa* L.). *Recent Trends in Horticultural Biotechnology, Vol. II.* ICAR National Symposium on Biotechnological Interventions for Improvement of Horticultural Crops: Issues and Strategies. Vellanikkara, India. New India Publishing Agency, New Delhi, pp. 505–514.

Gao, L.M., Chen, Y.Q., Huo, Y.M., Dong, F., Yang, Y.Y. *et al.* (2015) Development of SCAR markers to distinguish male-sterile and normal cytoplasm in bunching onion (*Allium fistulosum* L.). *The Journal of Horticultural Science and Biotechnology* 90(1), 57–62. DOI: 10.1080/14620316.2015.11513153.

Havey, M.J. (2000) Diversity among male-sterility-inducing and male-fertile cytoplasms of onion. *Theoretical and Applied Genetics* 101(5–6), 778–782.

Havey, M.J. (2020) Genetic mapping of chartreuse bulb color in onion. *Journal of the American Society for Horticultural Science* 145(2), 110–119.

Holz, G. and Knox-Davies, P.S. (1985) Production of pectic enzymes by *Fusarium oxysporum* f. sp. *cepae* and its involvement in onion bulb rot. *Journal of Phytopathology* 112(1), 69–80.

Hyun, D.Y., Kim, O.T., Bang, K.H., Kim, Y.C., Yoo, N.H. *et al.* (2009) Genetic and molecular studies for regulation of bolting time of onion (*Allium cepa* L.). *Journal of Plant Biology* 52(6), 602.

Imai, S., Tsuge, N., Tomotake, M., Nagatome, Y., Sawada, H. *et al.* (2002) An onion enzyme that makes the eyes water. *Nature* 419(6908), 685–685. DOI: 10.1038/419685a.

Islam, M.S., Studer, B., Moller, I.M. and Asp, T. (2014) Genetics and biology of cytoplasmic male sterility and its applications in forage and turf grass breeding. *Plant Breeding* 133(3), 299–312.

Jo, C. and Kim, S. (2020) Transposition of a non-autonomous DNA transposon in the gene coding for a bHLH transcription factor results in a white bulb color of onions (*Allium cepa* L.). *Theoretical and Applied Genetics* 133(1), 317–328. DOI: 10.1007/s00122-019-03460-8.

Jones, H.A. and Clarke, A. (1943) Inheritance of male sterility in the onion and the production of hybrid seed. *Proceedings of the American Society for Horticultural Science* 43, 189–194.

Jones, H. and Emsweller, S.L. (1936) A male sterile onion. *Proceedings of the American Society for Horticultural Science* 34, 582–585.

Kamenetsky, R., Faigenboim, A., Mayer, E.S., Michael, T.B., Gershberg, C. *et al.* (2015) Integrated transcriptome catalogue and organ-specific profiling of gene expression in fertile garlic (*Allium sativum* L.). *BMC Genomics* 16(1), 1–16.

Khandagale, K. and Gawande, S. (2019) Genetics of bulb colour variation and flavonoids in onion. *The Journal of Horticultural Science and Biotechnology* 94(4), 522–532. DOI: 10.1080/14620316.2018.1543558.

Kim, B. and Kim, S. (2019) Identification of a variant of CMS-T cytoplasm and development of high resolution melting markers for distinguishing cytoplasm types and genotyping a restorer-of-fertility locus in onion (*Allium cepa* L.). *Euphytica* 215(10), 1–11. DOI: 10.1007/s10681-019-2492-4.

Kim, B., Kim, K., Yang, T.J. and Kim, S. (2016b) Completion of the mitochondrial genome sequence of onion (*Allium cepa* L.) containing the CMS-S male-sterile cytoplasm and identification of an independent event of the ccmF N gene split. *Current Genetics* 62(4), 873–885. DOI: 10.1007/s00294-016-0595-1.

Kim, B., Cho, Y. and Kim, S. (2017) Identification of a novel *DFR-A* mutant allele determining the bulb color difference between red and yellow onions (*Allium cepa* L.). *Plant Breeding and Biotechnology* 5(1), 45–53. DOI: 10.9787/PBB.2017.5.1.45.

Kim, B., Yang, T.J. and Kim, S. (2019) Identification of a gene responsible for cytoplasmic male-sterility in onions (*Allium cepa* L.) using comparative analysis of mitochondrial genome sequences of two recently diverged cytoplasms. *Theoretical and Applied Genetics* 132(2), 313–322. DOI: 10.1007/s00122-018-3218-z.

Kim, E.Y., Kim, C.W. and Kim, S. (2016a) Identification of two novel mutant ANS alleles responsible for inactivation of anthocyanidin synthase and failure of anthocyanin production in onion (*Allium cepa* L.). *Euphytica* 212(3), 427–437. DOI: 10.1007/s10681-016-1774-3.

Kim, S. (2014) A codominant molecular marker in linkage disequilibrium with a restorer-of-fertility gene (Ms) and its application in reevaluation of inheritance of fertility restoration in onions. *Molecular Breeding* 34(3), 769–778. DOI: 10.1007/s11032-014-0073-8.

Kim, S., Binzel, M.L., Park, S., Yoo, K.S. and Pike, L. (2004a) Inactivation of DFR (Dihydroflavonol 4-reductase) gene transcription results in blockage of anthocyanin production in yellow onions (*Allium cepa*). *HortScience* 39(4), 889C–889. DOI: 10.1023/B:MOLB.0000047770.92977.04.

Kim, S., Binzel, M.L., Yoo, K.S., Park, S. and Pike, L.M. (2004b) Pink (P), a new locus responsible for a pink trait in onions (*Allium cepa*) resulting from natural mutations of anthocyanidin synthase. *Molecular Genetics and Genomics* 272(1), 18–27. DOI: 10.1007/s00438-004-1041-5.

Kim, S., Jones, R., Yoo, K.S. and Pike, L.M. (2004c) Gold color in onions (*Allium cepa*): a natural mutation of the chalcone isomerase gene resulting in a premature stop codon. *Molecular Genetics and Genomics* 272(4), 411–419. DOI: 10.1007/s00438-004-1076-7.

Kim, S., Jones, R., Yoo, K.S. and Pike, L.M. (2005a) The L locus, one of complementary genes required for anthocyanin production in onions (*Allium cepa*), encodes anthocyanidin synthase. *Theoretical and Applied Genetics* 111(1), 120–127. DOI: 10.1007/s00122-005-2000-1.

Kim, S., Yoo, K.S. and Pike, L.M. (2005b) Development of a PCR-based marker utilizing a deletion mutation in the dihydroflavonol 4-reductase (DFR) gene responsible for the lack of anthocyanin production in yellow onions (*Allium cepa*). *Theoretical and Applied Genetics* 110(3), 588–595. DOI: 10.1007/s00122-004-1882-7.

Kim, S., Yoo, K.S. and Pike, L.M. (2005c) Development of a codominant PCR-based marker for allelic selection of the pink trait in onions (*Allium cepa*), based on the insertion mutation in the promoter of the anthocyanidin synthase gene. *Theoretical and Applied Genetics* 110(3), 573–578. DOI: 10.1007/s00122-004-1875-6.

Kim, S., Bang, H., Yoo, K.S. and Pike, L.M. (2006) Identification of the fourth allele of the ANS (anthocyanidin synthase) gene and its effect on red color intensity in onions (*Allium cepa*). *Euphytica*, 149(1), 45–51. DOI: 10.1007/s10681-005-9053-8.

Kim, S., Baek, D., Cho, D.Y., Lee, E.T. and Yoon, M.K. (2009a) Identification of two novel inactive DFR-A alleles responsible for failure to produce anthocyanin and development of a simple PCR-based molecular marker for bulb color selection in onion (*Allium cepa* L.). *Theoretical and Applied Genetics* 118(7), 1391–1399. DOI: 10.1007/s00122-009-0989-2.

Kim, S., Lee, E.T., Cho, D.Y., Han, T., Bang, H. *et al.* (2009b) Identification of a novel chimeric gene, *orf725*, and its use in development of a molecular marker for distinguishing among three cytoplasm types in onion (*Allium cepa* L.). *Theoretical and Applied Genetics* 118(3), 433–441. DOI: 10.1007/s00122-008-0909-x.

Kim, S., Kim, C.W., Park, M. and Choi, D. (2015a) Identification of candidate genes associated with fertility restoration of cytoplasmic male-sterility in onion (*Allium cepa* L.) using a combination of bulked segregant analysis and RNA-seq. *Theoretical and Applied Genetics* 128(11), 2289–2299. DOI: 10.1007/s00122-015-2584-z.

Kim, S., Park, J.Y. and Yang, T.J. (2015b) Characterization of three active transposable elements recently inserted in three independent DFR-A alleles and one high-copy DNA transposon isolated from the

Pink allele of the ANS gene in onion (*Allium cepa* L.). *Molecular Genetics and Genomics* 290(3), 1027–1037. DOI: 10.1007/s00438-014-0973-7.

Kmiec, B., Woloszynska, M. and Janska, H. (2006) Heteroplasmy as a common state of mitochondrial genetic information in plants and animals. *Current Genetics* 50(3), 149–159. DOI: 10.1007/s00294-006-0082-1.

Kofoet, A., Kik, C., Wietsma, W.A. and De Vries, J.N. (1990) Inheritance of resistance to downy mildew (*Peronospora destructor* [Berk.] Casp.) from *Allium roylei* Stearn in the backcross *Allium cepa* L. × (*A. roylei* × *A. cepa*). *Plant Breeding* 105(2), 144–149. DOI: 10.1111/j.1439-0523.1990.tb00467.x.

Kojima, A., Ando, T., Tsukazaki, H., Yamashita, K. and Ohara, T. (2004) Diallel analyses of leaf blade length-width ratio and pyruvic acid production in bunching onion. *Breed Rescue* 6(suppl 1), 40.

Li, L., Dion, E., Richard, G., Domingue, O., Jean, M. *et al.* (2009) The Arabidopsis DNA mismatch repair gene *PMS1* restricts somatic recombination between homeologous sequences. *Plant Molecular Biology* 69(6), 675–684. DOI: 10.1007/s11103-008-9447-9.

Lin, M.W., Watson, J.F. and Baggett, J.R. (1995) Inheritance of resistance to neck-rot disease incited by *Botrytis allii* in bulb onions. *Journal of the American Society for Horticultural Science* 120(2), 297–299. DOI: 10.21273/JASHS.120.2.297.

Liu, Q., Wen, C., Zhao, H., Zhang, L., Wang, J. *et al.* (2014) RNA-Seq reveals leaf cuticular wax-related genes in Welsh onion. *PLOS ONE* 9(11), e113290. DOI: 10.1371/journal.pone.0113290.

Liu, Q., Lan, Y., Wen, C., Zhao, H., Wang, J. *et al.* (2016) Transcriptome sequencing analyses between the cytoplasmic male sterile line and its maintainer line in welsh onion (*Allium fistulosum* L.). *International Journal of Molecular Sciences* 17(7), 1058. DOI: 10.3390/ijms17071058.

Malik, G., Dhatt, A.S. and Malik, A.A. (2020) A review of genetic understanding and amelioration of edible Allium species. *Food Reviews International* 1–32. DOI: 10.1080/87559129.2019.1709202.

Marzu, J.C., Straley, E. and Havey, M.J. (2018) Genetic analyses and mapping of pink-root resistance in onion. *Journal of the American Society for Horticultural Science* 143(6), 503–507. DOI: 10.21273/JASHS04509-18

Masamura, N., Yaguchi, S., Ono, Y., Nakajima, T., Masuzaki *et al.* (2011) Characterization of amino acid and S-alk (en) yl-L-cysteine sulfoxide production in Japanese bunching onion carrying an extra chromosome of shallot. *Journal of the Japanese Society for Horticultural Science* 80(3), 322–333. DOI: 10.2503/jjshs1.80.322.

Masuzaki, S.I., Shigyo, M. and Yamauchi, N. (2006) Complete assignment of structural genes involved in flavonoid biosynthesis influencing bulb color to individual chromosomes of the shallot (*Allium cepa* L.). *Genes & Genetic Systems* 81(4), 255–263. DOI: 10.1266/ggs.81.255.

McCallum, J., Pither-Joyce, M., Shaw, M., Kenel, F., Davis, S. *et al.* (2007) Genetic mapping of sulfur assimilation genes reveals a QTL for onion bulb pungency. *Theoretical and Applied Genetics* 114(5), 815–822. DOI: 10.1007/s00122-006-0479-8.

McCallum, J., Thomas, L., Shaw, M., Pither-Joyce, M., Leung, S. *et al.* (2011) Genotypic variation in the sulfur assimilation and metabolism of onion (*Allium cepa* L.) I. Plant composition and transcript accumulation. *Phytochemistry* 72(9), 882–887. DOI: 10.1016/j.phytochem.2011.03.006.

McManus, M.T., Joshi, S., Searle, B., Pither-Joyce, M., Shaw, M. *et al.* (2012) Genotypic variation in sulfur assimilation and metabolism of onion (*Allium cepa* L.) III. Characterization of sulfite reductase. *Phytochemistry* 83, 34–42. DOI: 10.1016/j.phytochem.2012.07.028.

Meunissier, A. (1918) Expériences génétiques faites a verriéres. *Bulletin de la Société Nationale d'Acclimatation de France* 65, 81–90.

Molenaar, N.D. (1984) Genetics, thrips (*Thrips tabaci* L.) resistance and epicuticular wax characteristics of nonglossy and glossy onions (*Allium cepa* L.). *Dissertation Abstracts International. B, Sciences and Engineering*, 45(4).

Munaiz, E.D., Groves, R.L. and Havey, M.J. (2020) Amounts and types of epicuticular leaf waxes among onion accessions selected for reduced damage by onion thrips. *Journal of the American Society for Horticultural Science* 145(1), 30–35. DOI: 10.21273/JASHS04773-19.

Ovesná, J., Mitrova, K. and Kucera, L. (2015) Garlic (*A. sativum* L.) alliinase gene family polymorphism reflects bolting types and cysteine sulphoxides content. *BMC Genomics* 16(1), 1–10. DOI: 10.1186/s12863-015-0214-z.

Pal, N. and Singh, N. (1987) Analysis of genetic architecture for pungency (pyruvic acid) in onion (*Allium cepa* L.). *Current Science* 56(14), 719–720.

Park, J., Cho, D.Y., Moon, J.S., Yoon, M.K. and Kim, S. (2013) Development of functional markers for detection of inactive DFR-A alleles responsible for failure of anthocyanin production in onions (*Allium cepa* L.). *Horticultural Science and Technology* 31(1), 72–79. DOI: 10.7235/hort.2013.12131.

Randle, W.M. (1992) Onion germplasm interacts with sulfur fertility for plant sulfur utilization and bulb pungency. *Euphytica* 59(2), 151–156. DOI: 10.1007/BF00041267.

Randle, W.M. and Lancaster, J.E. (2002) Sulphur compounds in Alliums in relation to flavour quality. In: Rabinowitch, H. and Currah, L. (eds) *Allium Crop Science: Recent Advances*. CABI Publishing, Wallingford, UK, pp. 1–62. DOI: 10.1079/9780851995106.0329.

Randle, W.M., Kopsell, D.E., Kopsell, D.A. and Snyder, R.L. (1999) Total sulfur and sulfate accumulation in onion is affected by sulfur fertility. *Journal of Plant Nutrition* 22(1), 45–51. DOI: 10.1080/01904169909365605.

Rengwalska, M.M. and Simon, P.W. (1986) Laboratory evaluation of pink root and Fusarium basal rot resistance in garlic. *Plant Disease* 70(7), 670–672. Available at: http://pascal-francis.inist.fr/vibad/index.php?action=getRecordDetail&idt=8764749 (accessed 24 July 2022).

Ricroch, A., Yockteng, R., Brown, S.C. and Nadot, S. (2005) Evolution of genome size across some cultivated Allium species. *Genome* 48(3), 511–520. DOI: 10.1139/g05-017.

Rieman, G.H. (1931) Genetic factors for pigmentation in the onion and their relation to disease resistance. *Indian Journal of Agricultural Research* 42, 251–278.

Rout, E., Nanda, S., Nayak, S. and Joshi, R.K. (2014) Molecular characterization of NBS encoding resistance genes and induction analysis of a putative candidate gene linked to Fusarium basal rot resistance in *Allium sativum*. *Physiological and Molecular Plant Pathology* 85, 15–24. DOI: 10.1016/j.pmpp.2013.11.003.

Satoh, M., Kubo, T., Nishizawa, S., Estiati, A., Itchoda, N. *et al.* (2004) The cytoplasmic male-sterile type and normal type mitochondrial genomes of sugar beet share the same complement of genes of known function but differ in the content of expressed ORFs. *Molecular Genetics and Genomics* 272(3), 247–256. DOI: 10.1007/s00438-004-1058-9.

Sayadi, V., Karimzadeh, G., Rashidi Monfared, S. and Naghavi, M.R. (2020) Identification and expression analysis of S-alk (en) yl-L-cysteine sulfoxide lyase isoform genes and determination of allicin contents in Allium species. *PLOS One* 15(2), e0228747. DOI: 10.1371/journal.pone.0228747.

Scholten, O.E., Van Heusden, A.W., Khrustaleva, L.I., Burger-Meijer, K., Mank, R.A. *et al.* (2007) The long and winding road leading to the successful introgression of downy mildew resistance into onion. *Euphytica* 156(3), 345–353. DOI: 10.1007/s10681-007-9383-9.

Schweisguth, B. (1973) Étude d'un nouveau type de stérilité male chez l'oignon, *Allium cepa* L. *Annal Amelior Plant* 23, 221–233.

Shemesh-Mayer, E., Winiarczyk, K., Błaszczyk, L., Kosmala, A., Rabinowitch, H.D. *et al.* (2013) Male gametogenesis and sterility in garlic (*Allium sativum* L.): barriers on the way to fertilization and seed production. *Planta* 237(1), 103–120. DOI: 10.1007/s00425-012-1748-1.

Shemesh-Mayer, E., Ben-Michael, T., Kimhi, S., Forer, I., Rabinowitch, H.D. *et al.* (2015a) Effects of different temperature regimes on flower development, microsporogenesis and fertility in bolting garlic (*Allium sativum*). *Functional Plant Biology* 42(6), 514–526.

Shemesh-Mayer, E., Ben-Michael, T., Rotem, N., Rabinowitch, H.D., Doron-Faigenboim, A. *et al.* (2015b) Garlic (*Allium sativum* L.) fertility: transcriptome and proteome analyses provide insight into flower and pollen development. *Frontiers in Plant Science* 6, 271. DOI: 10.3389/fpls.2015.00271.

Simon, P.W. (1995) Genetic analysis of pungency and soluble solids in long-storage onions. *Euphytica* 82(1), 1–8. DOI: 10.1007/BF00028703.

Song, S., Kim, C.W., Moon, J.S. and Kim, S. (2014) At least nine independent natural mutations of the *DFR-A* gene are responsible for appearance of yellow onions (*Allium cepa* L.) from red progenitors. *Molecular Breeding* 33(1), 173–186. DOI: 10.1007/s11032-013-9942-9.

Sun, X., Zhu, S., Li, N., Cheng, Y., Zhao, J. *et al.* (2020) A chromosome-level genome assembly of garlic (*Allium sativum*) provides insights into genome evolution and allicin biosynthesis. *Molecular Plant* 13(9), 1328–1339. DOI: 10.1016/j.molp.2020.07.019.

Taylor, A., Teakle, G.R., Walley, P.G., Finch-Savage, W.E., Jackson, A.C. *et al.* (2019) Assembly and characterisation of a unique onion diversity set identifies resistance to Fusarium basal rot and improved seedling vigour. *Theoretical and Applied Genetics* 132(12), 3245–3264. DOI: 10.1007/s00122-019-03422-0.

Tscreryae, K. (1916) Uber den gegenwartigen Stand der Gemiisenzuchtung. *Z. Zucht* 4, 65–104.

Tsukazaki, H., Yaguchi, S., Yamashita, K.I., Shigyo, M., Kojima, A. *et al.* (2012) QTL analysis for pseudostem pungency in bunching onion (*Allium fistulosum*). *Molecular Breeding* 30(4), 1689–1698. DOI: 10.1007/s11032-012-9752-5.

Tsukazaki, H., Yaguchi, S., Yamashita, K.I. and Wako, T. (2017) QTL analysis of morphological traits and pseudostem pigmentation in bunching onion (*Allium fistulosum*). *Euphytica* 213(7), 1–10. DOI: 10.1007/s10681-017-1944-y.

Tsutsui, K. (1991) Inheritance of resistance to Fusarium oxysporum in onion. Doctoral dissertation. University of Wisconsin-Madison, Madison, Wisconsin.

Van der Meer, Q.P., Van Bennekom, J.L. and Van Der Giessen, A.C. (1970) Testing onions (*Allium cepa* L.) and other Allium species for resistance to *Botrytis allii* Munn. *Euphytica* 19(2), 152–162.

Vik, J. and Aastveit, K. (1984) Breeding for resistance to neck rot (*Botrytis allii*) in onions (*Allium cepa*). *Euphytica* 33, 257–262.

Wako, T., Kojima, A., Yamashita, K., Tsukazaki, H., Ohara, T. *et al.* (2010) Breeding and characteristics of a bunching onion (*Allium fistulosum* L.) cultivar 'Fuyuwarabe' with short leaf sheath. *Horticulture Research* 9(3), 279–285.

Wako, T., Tsukazaki, H., Yaguchi, S., Yamashita, K.I., Ito, S.I. *et al.* (2016) Mapping of quantitative trait loci for bolting time in bunching onion (*Allium fistulosum* L.). *Euphytica* 209(2), 537–546. DOI: 10.1007/s10681-016-1686-2.

Warid, W.A. and Tims, E.C. (1952) Studies on the inheritance of resistance to downy mildew studies in onion incited by Peronospora destructor. *Phytopathology* 42, 22.

Yang, Y.Y., Huo, Y.M., Miao, J., Liu, B.J., Kong, S.P. *et al.* (2013) Identification of two SCAR markers co-segregated with the dominant Ms and recessive ms alleles in onion (*Allium cepa* L.). *Euphytica* 190(2), 267–277. DOI: 10.1007/s10681-012-0842-6.

Yoo, K.S., Pike, L., Crosby, K., Jones, R. and Leskovar, D. (2006) Differences in onion pungency due to cultivars, growth environment, and bulb sizes. *Scientia Horticulturae* 110(2), 144–149. DOI: 10.1016/j.scienta.2006.07.006.

Yoo, K.S., Pike, L.M., Patil, B.S. and Lee, E.J. (2020) Developing sweet onions by recurrent selection in a short-day onion breeding program. *Scientia Horticulturae* 266, 109269. DOI: 10.1016/j.scienta.2020.109269.

Yuan, Q., Song, C., Gao, L., Zhang, H., Yang, C. *et al.* (2018) Transcriptome de novo assembly and analysis of differentially expressed genes related to cytoplasmic male sterility in onion. *Plant Physiology and Biochemistry* 125, 35–44. DOI: 10.1016/j.plaphy.2018.01.015.

Zhang, C., Li, X., Zhan, Z., Cao, L., Zeng, A. *et al.* (2018) Transcriptome sequencing and metabolism analysis reveals the role of cyanidin metabolism in dark-red onion (*Allium cepa* L.) bulbs. *Scientific Reports* 8(1), 1–10. DOI: 10.1038/s41598-018-32472-5

15 Nuclear and Cytoplasmic Genomics

Mostafa Abdelrahman[1,2] and Masayoshi Shigyo[3*]
[1]Botany Department, Faculty of Science, Aswan University, Egypt; [2]Faculty of Science, Galala University, Elsokhna, Egypt; [3]Laboratory of Vegetable Crop Science, College of Agriculture, Graduate School of Sciences and Technology for Innovation, Yamaguchi University, Japan

1. Introduction

The genus *Allium* contains many economically important crops that have been cultivated globally as spices, condiments and vegetables (Abdelrahman *et al.*, 2016). The genus comprises ~1000 species (Govaerts *et al.*, 2020), of which onion (*A. cepa* L. Common onion group), shallot (*A. cepa* L. Aggregatum group), Japanese bunching onion (*A. fistulosum* L.) and leek (*A. ampeloprasum* L. Leek group) are widely cultivated and exhibit great diversity in morphological characters, particularly in life form, with rhizomes and bulbs (Abdelrahman *et al.*, 2017). Additionally, *Allium* species showed considerable variation in various cytogenetic characters such as chromosome numbers (x=7, 8, and 9), ploidy level (varying from 2x to 16x), and genome size. Specifically, Alliums are known to possess one of the largest genomes ranging from 18.1 picograms (pg) in *A. triquetrum* L. to 31.5 pg in *A. ursinum* L. (Bennett *et al.*, 2000; Peška *et al.*, 2019). For example, the haploid genome size of bulb onion was estimated as ~16,400 Mega bases (Mb)/1C, which is the largest among all cultivated diploid crops, five times greater than the human genome (*c.*3000 Mb) and 120 times greater than *Arabidopsis* (*c.*125 Mb) (Ricroch *et al.*, 2005). Large genome size is usually associated with repetitive sequences (Kelly and Leitch, 2011), and at least 95% of the onion genome consists of repetitive sequences (Vitte *et al.*, 2013). Therefore, the complete genome sequencing of bulb onion became a significant challenge, and mapping and genomic-assisted breeding in *Allium* crops are lagging compared with other crops, such as tomato (*Solanum lycopersicum*), wheat (*Triticum aestivum*) and rice (*Oryza sativa*). This chapter describes the first *de novo* assembly of the genome of a doubled haploid *A. cepa*. In addition, genome organization in Alliums, including chloroplast, mitochondrial and nuclear genomes, are presented. Genetic resources of *Allium* species, including double haploids (DHs), alien chromosome addition lines and cytoplasmic substitution lines that have been used in different genetic studies to understand the gap between phenotype and genotype characters in *Allium* are discussed. The information in this chapter will be an important tool, facilitating onion breeding and research.

2. Unique Resources for Genetics and Breeding in *Allium*

Monosomic addition lines (MALs) have been useful genetic resources for studying important

*Address for correspondence: shigyo@yamaguchi-u.ac.jp

 Edible Alliums: Botany, Production and Uses (eds H.D. Rabinowitch and B. Thomas)
DOI: 10.1079/9781789249996.0015

agronomic traits in cultivated Alliums and for assigning linkage groups to physical chromosomes since Peffley *et al.* (1985) first published their pioneer research. For example, a complete set of *A. fistulosum*–*A. cepa* MALs (2n=2x+1=17, genomes + chromosomes FF+1A-FF+8A) was identified in a backcross population (BC_2) of amphidiploid hybrids between *A. fistulosum* (FF) and shallot (*A. cepa* Aggregatum group, AA) (Shigyo *et al.*, 1996). The MALs were further examined by genomic *in situ* hybridization (GISH) to confirm the integrity of chromosomal constitutions (Shigyo *et al.*, 1998). Subsequently, 48 shallot chromosome-specific molecular markers (Shigyo *et al.*, 1997a) were developed together with qualitative data for morphological identifications (Shigyo *et al.*, 1997b). All these morphological characters seem to be associated with the genetic effects of shallot additional chromosome(s) on the integral diploid background of *A. fistulosum*. The reddish-brown sheaths in FF+5A indicated high flavonoid and anthocyanin content in this line (Shigyo *et al.*, 1997c). Indeed, integrated metabolome and transcriptome analysis of the complete set of MALs provides further evidence that chromosome 5A in shallots harboured many candidate genes associated with flavonoid biosynthesis (Abdelrahman *et al.*, 2019). FF+5A exhibited high expression of many up- and downstream flavonoid and anthocyanin-related genes, which was consistent with the accumulation of many flavonoids and anthocyanins-related compounds compared with other MALs. Likewise, the transcriptome and phytochemical analysis of saponins in MALs indicated that FF+2A exhibited an accumulation of alliospiroside, a saponin with potent anti-fungal activity against *Fusarium* pathogens, and a valuable trait for disease resistance (Abdelrahman *et al.*, 2017). Furthermore, Yaguchi *et al.* (2013) examined antioxidant activity and polyphenol accumulation in the MALs, revealing that the FF+2A and FF+6A lines exhibited high polyphenol accumulation and the highest antioxidative activity. Hence, the additional chromosomes 2A and 6A have anonymous genes related to the upregulation of polyphenol production. Shigyo and co-workers carried out such approaches in six interspecific combinations, *A. cepa* – *A. fistulosum* (Hang *et al.*, 2004; Yaguchi *et al.*, 2009), *A. cepa* – *A. roylei* Stearn (Vu *et al.*, 2012), *A. fistulosum* – *A. roylei* (Ariyanti *et al.*, 2015), *A. fistulosum* – *A. galanthum* Kar. et Kir. (Tashiro *et al.*, 2000), *A. fistulosum* – *A. oschaninii* O. Fedtsch. (Tashiro *et al.*, 2000), and *A. fistulosum* – *A. vavilovii* M. Pop. et Vved. (Tashiro *et al.*, 2000) together with further approaches for disomic additions (Shigyo *et al.*, 2003; Yaguchi *et al.*, 2008) and multiple alien chromosome additions (Masuzaki *et al.*, 2007), some of which are applicable for elucidating a certain epistatic regulation related to the biosynthesis of secondary metabolites (Masuzaki *et al.*, 2006a, b).

2.1. Cytoplasmic male sterility (CMS)

CMS is a maternally inherited trait that suppresses the production of viable pollen grains, whereas the female gametes remain undisturbed, therefore they can serve as seed parents in hybrid seed production (Yamashita *et al.*, 2007). On the other hand, a single nuclear locus named *Ms* is known to restore male fertility of CMS in onions. In a recent study by Yu and Kim (2021), inheritance patterns in segregating populations suggested that both *Ms* and *Ms2* loci were redundantly involved in fertility restoration in onions' male-fertility restoration. *Ms2* locus was located at the end of chromosome 2 with 70 cM distance from the *Ms* locus (Yu and Kim, 2020). Whether causal genes for *Ms* and *Ms2* loci are paralogs is not confirmed yet; however, the transcription levels of *orf725*, a CMS-associated gene in onions, were significantly downregulated in male-fertile individuals of segregating populations, thus the causal gene for *Ms2* might be able to restore male fertility by suppressing transcription of *orf725* or degrading transcripts (Yu and Kim, 2020).

Allium galanthum (G) cytoplasm induces CMS in bulb onion (Havey, 1999), Japanese bunching onion (Yamashita, 1999; Yamashita and Tashiro, 1999, 2004; Yamashita *et al.*, 2005) and shallot (Yamashita and Tashiro, 1999). In shallot, the G cytoplasm was introduced through continuous backcrossing to shallot, the recurrent parent, to obtain interspecific F_1 hybrids (Yamashita and Tashiro, 1999). *A. roylei* (R)-originated CMS lines were identified in the BC_2 progenies of a single amphidiploid (possessing R cytoplasm) between *A. roylei* (female parent) and *A. cepa*

(Vu *et al.*, 2011). Breeding onion F_1 hybrid varieties takes 10–12 years to establish the A (male-sterile line), B (male-sterile maintainer) and C (restorer) lines (Shigyo and Kik, 2008), while alloplasmic lines possessing either G or R cytoplasm save breeders time and nullify the need for the B line. Moreover, since the CMS system is conditioned by the incompatibility between the wild species' cytoplasm and the *A. cepa* nucleus, any onion population acts as a male sterility maintainer. Unfortunately, no G or R CMS onion lines have been used in actual onion F_1 seed production. It should be noted, however, that compared to the S-type cytoplasmic male sterility, there are concerns about nectar production, a preferred factor by flower-visiting insects, and/or interaction of pollen production with the environment. Further research on the ecology and physiology of cytoplasm in wild species is needed.

2.2. Inbred lines

The traditional breeding of onion inbred lines is long and difficult due to its biennial life cycle, inbreeding depression and high heterozygosity. Thus, DH lines provide several advantages, including complete homozygosity and reduced DNA methylation and elimination of deleterious alleles compared with inbred lines (Bohanec, 2002), and consequently exhibit vigorous vegetative growth and morphological uniformity (Hyde *et al.*, 2012; Khan *et al.*, 2020). Single- and two-step protocols have been used to generate gynogenic haploids in onion. The former involves culturing whole flower buds, ovaries or ovules to the embryo or plantlet stage (Bohanec and Jakše, 1999; Bohanec, 2002), whereas the latter includes pre-culture of flower buds on basal media with or without plant growth regulators; then, isolation of ovary or ovule and subsequent subculturing on regeneration media with growth regulators (Martinez *et al.*, 2000; Michalik *et al.*, 2000). DH lines may be induced *in vitro* by culturing haploid cells from the male (androgenesis) or female (gynogenesis) gametophytes. However, the efficiency of androgenic generation DHs is limited in some *Allium cepa* genotypes due to defective pollen or anthers, and/or the production of albinos during culture (Gurel *et al.*, 2008). Gynogenesis has been the most commonly used method for producing haploids in *A. schoenoprasum* L. (Keller, 1990), leek (Kaska *et al.*, 2013; Alan *et al.*, 2016), *A. altaicum* (Keller, 1990), *A. giganteum* (Susek *et al.*, 2002), shallot (Sulistyaningsih *et al.*, 2006), *A. fistulosum* (Ibrahim *et al.*, 2016), *A. tuncelianum* (Yarali and Yanmaz, 2016) and others.

DH lines offer very promising tools to understand the gap between phenotype and genotype, and DH chromosomal doubling can shorten the time, offer homozygous pure lines, and provide valuable materials for genomic analysis. For example, comparative metabolome analysis of shallot and bulb onion DHs (DHA and DHC, respectively) and F_1 hybrid revealed genotype-specific metabolites for each. DHA accumulated several stress-related metabolites, which is consistent with the biotic and abiotic stress shallot tolerance compared with bulb onion (Abdelrahman *et al.*, 2015). Similarly, Khosa *et al.* (2016) developed transcriptome data from reproductive and vegetative organs of onion DH line CUDH 2107 to identify tissue-specific expressed genes involved in pollen fertility.

The Cornell onion breeding programme compared 20 DH lines with commercial hybrids and open-pollinated cultivars (OP) developed from the same source germplasm over two years, under field conditions (Hyde *et al.*, 2012). Results indicated a similar vegetative vigour of both DH and OPs, with minimal inbreeding depression. A comparison between two sets of hybrids produced using male DH lines and two different commercial females showed increased vegetative growth vigour of DH lines and their derived hybrids probably due to elimination of deleterious lethal genes during gynogenesis, an additional benefit for onion breeding strategies (Hyde *et al.*, 2012).

DH lines lack allelic exchanges within a locus, thus genetic effects are controlled by single alleles or interactions between loci, thus facilitating detection of QTL traits (Duangjit *et al.*, 2014). Hence, onion DH lines from the four Spanish cultivars, including highly pungent landrace BGHZ1354, sweet cultivar Fuentes de Ebro, and two Valenciana-type commercial varieties, Recas and Rita (Fayos *et al.*, 2015), served for the development of linkage maps. More comprehensive research and improved DH *in vitro* gynogenesis production methods are needed and would be useful to produce DH in onion and other *Allium* species.

2.3. Wild species

Wild relatives have great potential to expand the supply of usable genetic variation and useful traits, and are thus important for introgression breeding (Hao *et al.*, 2020). *Allium roylei* (RR, 2n=16) possess several desired traits, including resistance to Botrytis leaf blight and downy mildew (Abdelrahman *et al.*, 2014). Similarly, *Allium fistulosum* (AF) exhibits several important agronomical traits for onion improvement, thus great attention has been given to the introgression of *A. fistulosum* genes to *A. cepa*. However, the interspecific hybrid is largely sterile, and heteromorphic bivalents are present (Emsweller and Jones, 1935; Albini and Jones, 1990). In addition, the backcross (BC_1) of the interspecific hybrid with *A. cepa* usually yields a small progeny, due to nucleo-cytoplasmic incompatibility (Van der Valk *et al.*, 1991). Using *A. roylei* as a bridging species helps to circumvent these problems. Hence, *A. cepa* (CC) × *A. roylei* (CR), *A. cepa* × *A. fistulosum* (CF) and *A. fistulosum* × *A. roylei* (FR) (Khrustaleva and Kik, 1998) were bridge-crossed with (CC × FR) where *A. cepa* served as the female parent and FR the male parent (Khrustaleva and Kik, 1998). Thereafter, genome organization of the interspecific hybrids and trihybrid population were analysed using multi-colour GISH (Khrustaleva and Kik, 1998).

In a follow-up study, trihybrid genotypes from the CC × RF population, where only one homeolog of a chromosome pair underwent interspecific recombination, were used for GISH analysis of chromosomes 5 and 8 recombinants. Visualization of recombination points and the physical positions of recombination were integrated into AFLP linkage maps of both chromosomes, thus concluding that in *Allium*, recombinations predominantly occur in the proximal half of chromosome arms. Of the PstI/MseI markers, 57.9% are located near the centromeric region, thus suggesting genes' presence in this region (Khrustaleva *et al.*, 2005).

Investigations of trihybrid CC × RF population for root traits inheritance revealed QTLs for the rooting system, thus breeding for an improved rooting system in onion is possible (De Melo, 2003). Analyses of trihybrid CC × CF population for the genetic basis for response to arbuscular mycorrhizal fungus *Glomus intraradices* indicated that the combination of three genomes extends the genetic variation for plant growth and the response to mycorrhiza (Galván Vivero *et al.*, 2011). One QTL for the number of stem-borne roots of mycorrhizal plants was found on linkage group 9, associated with *A. fistulosum* alleles. This QTL co-segregated with QTLs for average performance and total dry weight of mycorrhizal and non-mycorrhizal plants. Another QTL for Fusarium basal rot (FBR) resistance from *A. roylei* was identified on a distal region of chromosome 2, and one QTL from *A. fistulosum* was identified on the long arm of chromosome 8 (Galván Vivero, 2009). These QTLs showed an additive effect, and together accounted for 31% and 40% of the total variation for FBR incidence and severity at harvest; and 31% and 29% after storage, respectively.

3. Genomes

Plant cells have three genomes: mitochondrial, plastid and nuclear. The number of genome copies per organelle depends on the cell type, the age of tissue and species. Hence, each *Arabidopsis thaliana* (L.) Heynh. a cell has two copies of the nuclear genome (gDNA) (Lutz *et al.*, 2011). The number of mitochondria and plastid DNAs (mtDNA and ptDNA, respectively), however, vary with cell types. *Arabidopsis* root cells have ~400 mtDNA (Kato *et al.*, 2008), whereas maize anther cells have a 20–40-fold higher number of mtDNA (Warmke and Lee, 1978). In *Arabidopsis* and sugar beet, the copy number of ptDNAs remains at ~1700 per single gDNA molecule, and the ratio of ptDNA to gDNA remains constant even as the ploidy level of the cell changes (Lutz *et al.*, 2011).

3.1. Chloroplast genome

Originated from photosynthetic bacteria, the chloroplast (cp), is the key organelle for carbon fixation and photosynthesis in green plants

(Cho *et al.*, 2015). The cps genomes and genetic information are maternally inherited in most angiosperms (Song *et al.*, 2017). The highly conserved genome structure and gene content served for the generation of genetic markers for DNA barcoding, molecular identification, phylogenetic classification and screening of genetic resources and breeding (Hu *et al.*, 2016). The *Allium* cp whole-genome DNA sequencing initiation (Von Kohn *et al.*, 2013) resulted in cp genomes sequencing of *A. obliquum* L., *A. prattii* C.H. Wright, *A. victorialis* L. Sp. Pl. 1: 295. 1753, *A. cepa*, and *A. sativum* (http://www.ncbi.nlm.nih.gov/genome/organelle/ accessed September 2022). Recently, the cp genomes of *A. tuberosum* Rottl. ex Spreng., *A. fistulosum*, *A. sativum*, and *A. cepa* have been sequenced using NGS technologies (Huo *et al.*, 2019), and comparisons with cp genomes of *A. cepa* CMS-T and CMS-S, *A. victorialis*, *A. obliquum* and *A. prattii* revealed that the length of cp DNAs ranged from 152 (*A. obliquum*) to 154 kb (*A. prattii*) and that the genomic structure and gene organization are well conserved (Huo *et al.*, 2019).

The *Allium* cp genome contained large and small single-copy regions (LSC and SSC, respectively) separated by large inverted repeats (IRs) (Fig. 15.1). The nine cp genomes presented typically quadripartite structures, with two IRs (ranged from 26,370 to 26,564 bp) separated by the LSC (81,588 to 83,392 bp) and SSC (17,853 to 18,066 bp) regions (Huo *et al.*, 2019) (Fig. 15.1). Additionally, all *Allium* cp genomes showed similar gene order and content with 140–141 genes consisting of eight rRNA genes, five to ten pseudogenes, 37–38 tRNA genes, and 88–89 protein-coding genes, located in the IR regions (Huo *et al.*, 2019) (Fig. 15.1). The GC contents of nine complete cp genomes were quite similar in each region or category, including complete cp genome, LSC, IR, SSC, coding sequences (CDSs), tRNA, rRNA, and pseudogene. The highest was observed in rRNA (>55%) and the lowest in SSC (>29%). The numbers and distributions of three repeat types in the nine *Allium* cp genomes, including 394 repeats with 131 tandem repeats, 154 dispersed repeats, and 109 palindromic, were similar and conserved. The numbers of SSRs differ among the nine *Allium* genomes and vary from 73 in *A. tuberosum* to 96 in *A. cepa* CMS-N and CMS-T (Huo *et al.*, 2019). Finally, the phylogenetic analysis of *Allium* species and *Agapanthus coddii* F.M. Leight. produced two distinct branches with extraordinarily strong support (100% and 1.00). In the genus *Allium*, nine accessions were divided into two sister clades. The first clade contained *A. prattii* and *A. victorialis*; the second clade included seven accessions, in which *A. cepa* CMS-T and CMS-N were grouped in a sister branch and then clustered step-by-step with *A. cepa* CMS-S, *A. fistulosum*, *A. obliquum*, *A. sativum* and *A. tuberosum* (Huo *et al.*, 2019).

3.2. Mitochondrial genome

Scheffler (2008) stipulates that the plant mitochondrial genome harbours important information, such as mode of gene expression, organizational diversity, evolution and nuclear-cytoplasmic interactions. Collectively, these make the mitochondrial genome one of the most interesting genomes for molecular biologists and plant genetics. Plant mitochondrial genomes typically have many long and short repeated sequences and intra- and intermolecular recombination may create various DNA molecules in this organelle, and recombination may create a novel gene that causes CMS (Tsujimura *et al.*, 2019). The complete onion mitochondrial genome sequence was reported for a CMS-S inbred line (Kim *et al.*, 2016), of which ~10% consisted of repetitive sequences with short repeats of <100 base pairs. Additionally, the gene-encoding cytochrome c biogenesis protein ($ccmF_N$) was split into two genes (Kim *et al.*, 2016), as recorded in 30 other *Allium* species. In a follow-up study, comparative analysis of mitochondrial genome sequences of two recently diverged cytoplasms, CMS-S and CMS-T-like, resulted in the identification of *orf725*, a chimeric gene consisting of *cox1* with other sequences, as the casual gene responsible for CMS in onions (Kim *et al.*, 2018). In addition, CMS-T-like cytoplasm has recently diverged from the normal cytoplasm by the introduction of *orf725* (Kim *et al.*, 2018). More recently, NGS sequencing of the mitochondrial genome of the bulb onion variety Momiji-3, possessing CMS-S-type cytoplasm, revealed that the three circles are as a result of recombination. In addition, mitochondrial genome structure was visualized using pulsed-field gel electrophoresis (Tsujimura *et al.*, 2019). Further, mapping the

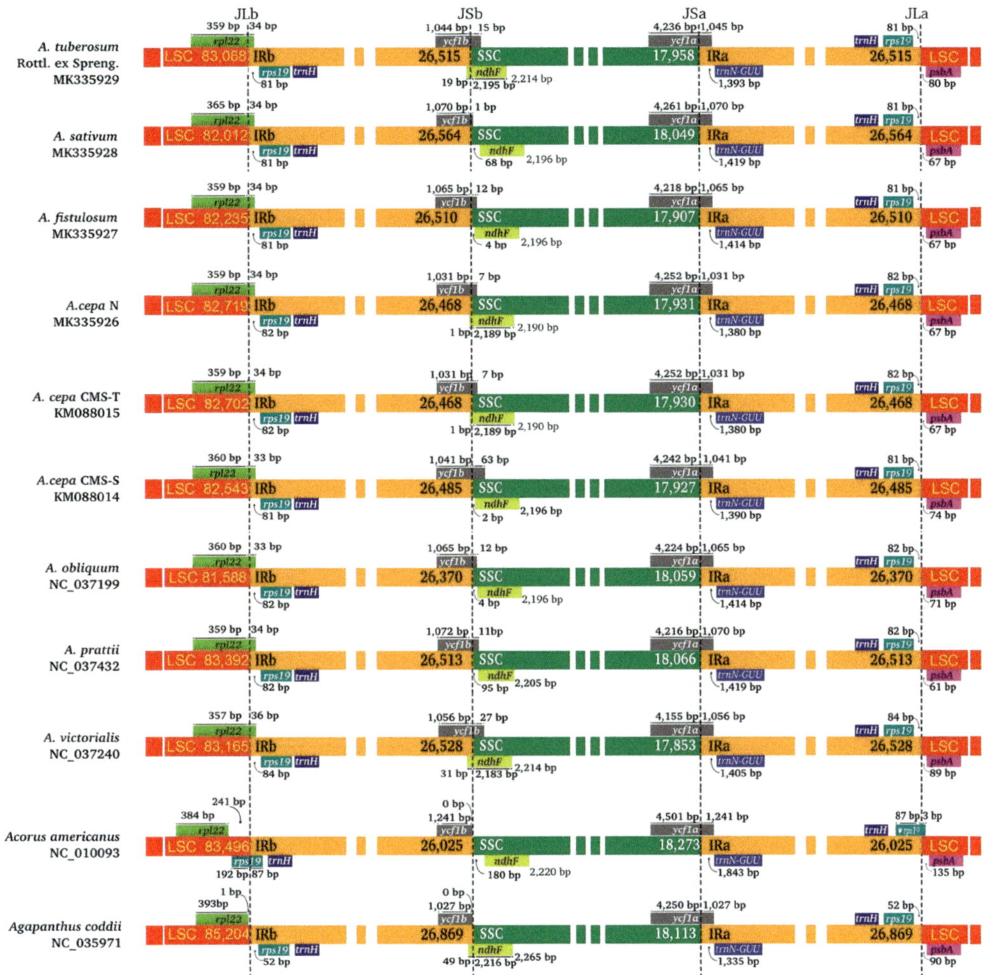

Fig. 15.1. Comparison of the large single-copy (LSC), inverted repeat (IR), and small single-copy (SSC) boundary regions among the nine chloroplast genomes of different *Allium* species. The numbers above the gene features indicate the distances from the end of the gene to the boundary sites. (Adapted from Huo *et al.*, 2019 (Creative Commons 4.0))

transcript data to mitochondrial genome indicated that a new functional gene was absent from the Momiji-3 mitochondrial genome f and confirmed *orf725* candidacy gene for CMS. Recently, Hui *et al.* (2020) published the complete sequence of the onion's yellow stripe mutant mitochondrial genome. Compared with normal onions, 556 SNPs and 279 small indel numbers were detected, out of which 14 SNPs and two small indels in the exon region affected the translation protein. It should be noted, however, that studies on purified *Allium* mitochondrial DNA have not yet been performed, thus the number of differences between onion genotypes remains unclear.

3.3. Nuclear genome

A study of *Allium* gDNA in 42 indicated that 4C nuclear DNA contents ranged between 41.19 and 142.78 pg, irrespective of the basic chromosome number, ploidy or taxonomic group (Labani and Elkington, 1987). Similarly, there was no correlation between the proportion of the C-banded region in the karyotype and the increase in DNA nuclear content. For example, *A. ursinum* L. has the largest DNA content of 71.39 pg/4C DNA/genome but showed the lowest proportions of C-banded chromosomal region in the genus, while *Allium* species with moderate DNA amounts (21.11–42.45 pg)/4C DNA/genome showed the highest proportions of C-banded chromosomal region (Vosa, 1976; Al-Sheikh Hussain, 1977).

In general, DNA amounts in *Allium* species showed high divergence. For instance, the DNA values (2C) in triploid *A. carinatum* L. vary between 32 ± 67 and 50 ± 28 pg (Nagl and Fusenig, 1979; Labani and Elkington, 1987), and in leek between 24 ± 1 and 60 ± 58 pg (Ranjekar *et al.*, 1978; Labani and Elkington, 1987). Such contradictory data highlighted a need for clarification about the reliability of the nuclear DNA values published in the *Allium* genus.

Baranyi and Greilhuber (1999) reported that the 2C DNA values of the investigated *Allium* species were normally distributed, and only 29 of the 60 2C values published deviate less than 10% from the above results, thus highlighting a need for improving a general agreement in the standardization and preparative procedures of cytophotometric genome size determination in *Allium* species.

Jakše *et al.* (2008) measured the nuclear DNAs of two onion bacterial artificial chromosomes (BACs) to investigate the nature and distribution of repetitive sequences and to estimate gene densities. The complete sequences from two BACs were AT-rich (64.8%) and revealed long tracts of degenerated retroviral elements and transposons, similar to other larger plant genomes. A total of 499,997 bp of onion nuclear DNA yielded an estimated mean density of one gene per 168 kb, among the lowest reported to date. Methyl filtration was highly effective relative to random shotgun reads in reducing frequencies of anonymous sequences from 82% to 55% and increasing non-organellar protein hits from 4% to 42% (Jakše *et al.*, 2008). Ricroch *et al.* (2005) evaluated the genome size of 30 *Allium* species including major vegetable crops and their wild allies. The 2C DNA value of gDNA was measured and compared with data derived from the Angiosperms (http://www.rbgkew.org.uk). Overall, genome size (1C DNA) values varied from 7 pg in *A. altyncolicum* to 31.5 pg in *A. ursinum*, showing 3.3-fold variation and a coefficient of variation of 27.4%. The GC content, evaluated in 24 accessions representing 23 *Allium* species, displayed values ranging from 38.5% in leek to 41.2% in *A. cernuum*, with a median value of 39.9% (Ricroch *et al.*, 2005). Using different molecular markers, the comparison of molecular phylogenies led Fritsch and Friesen (2002) to propose a new idea for the classification of *Allium*, with 67 sections and 14 subgenera. A variation in cytogenetic characters such as nuclear DNA amount and chromosome numbers, when compared to the classification of the genus *Allium*, does not reveal clear discontinuities between the taxonomic groups (Ohri *et al.*,1998); but the variation mirrors the great diversity observed in morphology, life form, ecology and breeding systems (Hanelt *et al.*, 1992).

4. Linkage, Cytogenetic and Physical Mapping

The initial efforts have been succeeded by the construction of an *Allium* species genetic map using restriction fragment length polymorphism (RFLP) markers of the intraspecific onion cross BYG15-23 × AC43 (King *et al.*, 1998). This map was subsequently extended with the single-nucleotide polymorphism (SNP), simple sequence repeat (SSR) markers and homologs of transposable elements derived from the sequencing of expressed sequence tag (EST) in onion (Kuhl *et al.*, 2004; Martin *et al.*, 2005). These markers enabled partial map constructions in other intraspecific onion crosses to develop the map-based genetic analyses of various traits, including fertility restoration and bulb colour development (Gökçe *et al.*, 2002; McCallum *et al.*, 2006; Khar *et al.*, 2008). More recently, another genetic map based on *A. fistulosum* genomic SSR markers and onion EST-derived SNP and SSR markers, provided further insight into comparative studies between the genomes of the two crops (Tsukazaki *et al.*, 2010). The key resource that enabled alignment of *Allium* genetic maps to physical chromosomes and facilitated comparison between species was the complete set of *A. fistulosum* – *A. cepa* monosomic addition lines developed by Shigyo *et al.* (1996). These were initially applied to anchor AFLP-based maps in the interspecific *A. cepa* × *A. roylei* cross (van Heusden *et al.*,

2000), and subsequently to anchor the BYG15-23 x AC43 map (Martin *et al.*, 2005), and anchor SSR-based maps in *A. fistulosum* (Tsukazaki *et al.*, 2008) to physical chromosomes, and more recently to assign many more onion EST-derived anchor markers used in *A. fistulosum* maps (Tsukazaki *et al.*, 2010).

Further, a large number of molecular and phenotypic markers associated with many candidate genes coding for valuable economic traits have also been assigned to *A. cepa* chromosomes (Martin *et al.*, 2005; Masuzaki *et al.*, 2006a; Yaguchi *et al.*, 2008), providing an important guide for functional and quantitative trait locus (QTL) studies. Recently, RobustDb, a generic online genomics database that presented information mainly for garlic maps, led to the development of marker data (Bhasi *et al.*, 2010). Likewise, the VegMarks database (http://vegmarks.nivot.affrc.go.jp/ accessed September 2022) contained detailed information regarding *A. fistulosum* markers. A total of 107 markers comprising three *A. fistulosum* genomic SSRs, 31 SNP markers, and the 73 additional onion EST-SSRs derived from onion ESTs were evaluated in the population previously used to construct an AFLP-based linkage map (McCallum *et al.*, 2012). Additionally, an interspecific *A. cepa* × *A. roylei* map was augmented with additional genetic markers to increase correspondences among *Allium* maps. Then, 11 linkage groups spanning one Morgan were formed using LOD 5 cut-off. Using the monosomic addition set, the anchor loci were assigned and mapped in the interspecific cross with the consequent additional landmarks for aligning genetic linkage maps in *A. cepa* and *A. fistulosum*. The AlliumMap database (https://www.oniongenome.wur.nl, accessed October 2022) contains 512 correspondences between markers in various *Allium* maps (McCallum *et al.*, 2012). This database represents an integrated point to access details of the genetic markers and sequence resources employed across multiple studies in cultivated *Allium* to resolve common questions of crop evolution and economic trait regulation across these major *Allium* crops. The genotyping-by-sequencing (GBS) provides a high degree of complexity reduction followed by synchronized SNP discovery and genotyping for species with complex genomes (Jo *et al.*, 2017). GBS analysis conducted on F_2 (NW-001 X NW-002) population, provided 1,851,428 SNPs. After filtering, 10,091 high-fidelity SNP markers that satisfied the criteria of segregation ratio were used to construct an onion genetic map (eight linkage groups) and spanned a genetic length of 1383 centiMorgans (cM) with an average marker interval of 8.08 cM. The newly developed linkage maps will be valuable tools for the genetic mapping of important agronomic traits and marker-assisted selection in onion breeding programmes worldwide.

Likewise, the F_2 (doubled haploid (DH) H6 (red bulb) and the inbred line SP3B (yellow bulb)) population was used to construct another SNP-based genetic linkage map to identify the QTLs associated with the total anthocyanin content of onion bulbs using GBS analysis (Choi *et al.*, 2020). Hence, an onion genetic map with eight linkage groups consisting of 319 GBS-based SNP loci and 34 high-resolution melting markers was constructed (Choi *et al.*, 2020). Based on this linkage map, two QTLs, *qAC4.1* and *qAC4.2*, for anthocyanin content and one major QTL, *qAS7.1*, for anthocyanin synthesis, were identified. This major QTL located on chromosome 7 with a phenotypic variation of 87.61% can be putatively identified as *dihydroflavonol 4-reductase* (*DFR*) gene responsible for bulb colouring in onion. Recently, an ultra-high-density linkage map of transcriptome-based unigene markers was assigned on *A. cepa* chromosomes using RNA sequence data from 96 lines of the F_2 mapping population (a doubled haploid shallot DHA × a doubled haploid onion DHC) (Fujito *et al.*, 2021). SNPs with at least four reads on 25,462 unigenes were anchored on eight *A. cepa* chromosomes. In total, multiple SNPs were identified on 22,184 unigenes and a single SNP site on 3278. The chromosome marker information was made public via the web database *Allium* TDB (http://alliumtdb.kazusa.or.jp/ accessed September 2022), which is valuable in genome sequencing and useful for validation of the original genetic map in *Allium*. In a recent study, an ultra-sensitive dual-colour Tyr-FISH technique was developed to construct a reliable and easy method for *in situ* mappings of short unique DNA sequences on onion chromosomes (Kudryavtseva *et al.*, 2021), which allows more accurate determination for physical distances between markers and can be applied for the faster integration of genetic and cytogenetic maps. For instance, Tyr-FISH was applied to physically locate molecular markers tightly linked to the nuclear male-fertility (Ms) restoration locus on onion (Khrustaleva *et al.*, 2016). Relatively short genomic amplicons and a cDNA clone were successfully visualized on chromosome 2 near the centromere, a region of low recombination, which partially explained the identified molecular markers tightly linked to the Ms locus in different segregating progenies (Khrustaleva *et al.*, 2016). Likewise, Tyr-FISH mapping on *A. cepa* chromosomes was performed using EST-clones API20 and API66, previously mapped in the linkage groups corresponding to *A. cepa* chromosomes 1 and 5, respectively (Romanov *et al.*, 2014). Results indicated that API66 possesses an insert with high homology to the sucrose transportase gene, and API20 to the light harvesting-like protein 3 encoding gene (Romanov *et al.*, 2014). In addition, Tyr-FISH analysis of EST-clone API66 revealed 25 hybridization sites on the mitotic metaphase chromosomes of *A. cepa*, and the signals were dispersed along all chromosomes (Romanov *et al.*, 2014). It worth noting that the sucrose transportase gene belongs to a large super-family of genes in plants, which explained multiple sites of API66 hybridization (Romanov *et al.*, 2014). Future research involves the creation of more molecular cytogenetic markers based on mapping chromosome-specific repeats and genes/markers can be further applied in onion breeding and genome assembly at the chromosome level. Expansion of FISH applications in plants with the development of new technologies as oligo-FISH mapping, CAS-FISH for living cells, RNA-FISH for measuring gene transcription, and a haplotype-specific FISH for crossing over visualization will be the gold-standard method for mapping physical chromosomes and detecting their organization.

5. Novel Analysis Methods for Large Genome Size

Irrespective of ploidy, genome sizes refer to the amount of DNA enclosed in a haploid genome set expressed in terms of the number of base pairs (1 kb = 1000 bp), megabases (1 Mb = 1,000,000 bp), the DNA mass in picograms (1 pg = 10^{-12} g = 1000 Mb of DNA), or as C-value ('C' stands for constant, referring to the fact that the genome size is constant from cell to cell in a given organism) equal to the mass of DNA in pg) in a haploid set of chromosomes (Maloy and Hughes, 2013), regardless of ploidy level, which means one-third or a quarter of the total mass of DNA in the cell (Maloy and Hughes, 2013).

The genome size depends on differences in the rates at which insertions and deletions (indels) occur, transposon activity, segmental duplications and the efficiency of natural selection in eliminating or promoting such changes (Sung *et al.*, 2016). Large genomes have higher mutational liability, even in non-coding areas, thus a large fraction of non-coding regions is likely to evolve neutrally concerning indels (Mueller, 2015).

Bulb onion has a large haploid genome size of 16,400 Mb, and unique telomeric sequences with the lowest GC content (32%) among angiosperms (Shigyo *et al.*, 2018; Abdelrahman *et al.*, 2020), yet it is the largest of all diploid crops (Brenchley *et al.*, 2012; Marcussen *et al.*, 2014). Enormous genome-size differences exist among closely related *Allium* species. For example, bulb onion has 28% more nuclear DNA than *A. fistulosum* and 7% more than garlic nuclear DNA.

The bulb onion genomic fragments cloned into BACs demonstrated that most sequences were similar to retro-element and transposons, indicating that the onion genome exhibited low gene densities and high frequencies of repetitive DNA as major components of large chromosomes (Suzuki *et al.*, 2001). The C_0T reannealing kinetics indicated that about 40% of the onion genome is highly repetitive (>1000x copies) and 40% has 100–1000 copies (Stack and Comings, 1979). Overall, at least 95% of the *A. cepa* genome consists of repetitive sequences, most of which are dispersed repeats (Shibata and Hizume, 2002). Due to the size of the genome and its repetitive nature, developing an onion reference genome assembly is challenging.

Although the accessibility of the structural genomic resources in *Allium* species is limited, using next-generation sequencing (NGS; e.g., Roche, Illumina, and ABI Solid platforms) technologies and large computational abilities, sequencing, assembly and annotation of large genomes has become feasible. The advantages of third-generation sequencing (TGS) technologies such as Pacific Biosciences over short-read sequencing as Illumina's HiSeq, NovaSeq, NextSeq,

MiSeq instruments and Ion Torrent sequencers (Thermo Fisher Scientific) include greater read lengths, facilitates the assembly of large and complex genomes such as onion (Abdelrahman *et al.*, 2018b), i.e. reads of up to 600 bases (Goodwin *et al.*, 2016) and >10 kb (Rothberg *et al.*, 2011), respectively. Consequently, improved *de novo* assembly, mapping certainty, transcript isoform identification, and detection of structural variants become feasible (Depledge *et al.*, 2019). Concomitantly, the ever-diminishing computational power costs and the growing availability of on-demand cloud-based computing steadily increase assembly and annotation efficiency. Hence, large-scale and cost-effective sequencing, assembly and annotation of onion DNA are becoming feasible (Abdelrahman *et al.*, 2020).

The three key approaches to large genome sequencing like the onion genome include shotgun sequencing, where DNA is fragmented into short reads, which are then sequenced. Since a large proportion of onion reads are repeated sequences, shot-gun sequencing is unsuitable for its large and complex genome. Suzuki *et al.* (2001) were the first to Sanger sequence large genomic fragments cloned into BACs. Two entire BAC clones, and the random ends of genomic fragments cloned into BACs, yielded 298 kb of random-end sequences and 202 kb from the entire BAC clones (Jakse *et al.*, 2008), with AT content of 35.7%, which is close to the estimate of 32% by Kirk *et al.* (1970). Additionally, only 5% of these random genomic sequences showed significant similarities to non-organellar proteins, 25% were highly similar to transposons or retrotransposons, and 70% showed no significant hits to the databases and were primarily degenerated retro-elements (Jakse *et al.*, 2008). Additional sequence survey of onion random genomic sequences was conducted using DNA isolated from DH line 15197. The DNA was trimmed, and random fragments were sequenced using Sanger and 454 technologies, generating 6590 reads, among which 82% had no significant hits in the databases, 4% matched nuclear-encoded proteins and 14% matched transposons (Jakse *et al.*, 2008).

The second approach efficiently samples expressed regions. Sequencing random cDNAs thus revealed the collection of genes (the transcriptome) expressed in any given tissue, at a specific developmental stage, or after treatment (Abdelrahman *et al.*, 2017b, 2020). However, the normalization of the cDNA reduces the frequencies of highly expressed genes and increases the number of sequencing reads from rarer transcripts (Abdelrahman *et al.*, 2017a).

A free access web-based tool named *Allium* TDB (http://alliumtdb.kazusa.or.jp/) contains several transcriptional data of *Allium* species. Likewise, the transcriptome sequencing on vegetative and reproductive tissues of homozygous DH CUDH2107 was used to develop a multi-organ reference transcriptome catalogue (Khosa *et al.*, 2016). In general, the 271,665 contigs of transcripts were generated using the Trinity pipeline. This dataset was analysed for gene ontology (GO), and the bulb onion transcripts were grouped into 95 functional groups, among which Binding Domains was the most abundant GO term followed by Carbohydrate-Binding (Khosa *et al.*, 2016).

The third approach involves enrichment for lower-copy regions by removing repetitive DNAs, and reduced representation sequencing involves selection against methylated DNA (Rabinowitch *et al.*, 1999; Peterson *et al.*, 2002; Springer *et al.*, 2004; Shagina *et al.*, 2010). In bulb onion, the sequencing of these fragments is an efficient approach to enrich genic regions in its enormous genome. Hence, the complete onion DH 15197 sequencing of methyl-filtered DNA fragments revealed that among the 2712 methyl-filtered fragments, 3% matched transposons, 55% were anonymous and 42% were similar to non-organellar proteins (Jakse *et al.*, 2008).

The first insight into the genome sequencing of *A. cepa* is now ongoing through the SEQUON project (www.oniongenome.wur.nl) using Illumina HiSeq 2500, intending to sequence the onion nuclear DNA and provide *de novo* genome assembly and *ab initio* annotation applied to the doubled haploid onion line DHCU066619, for a whole-genome shotgun sequencing of an onion genome (Finkers *et al.*, 2021). Initially, Illumina HiSeq 2500 sequencing of three small libraries yielded 769 Gb sequence data, whose analyses provided an estimated genome size of ~13.6 Gb. Secondly, the MaSurCa-based assembly resulted in 10.8 Gb in 6.2M contigs with a contig N50 of 2.7 kb. This assembly was further scaffolded using 18.1M PacBio RS II reads. Finally, the hybrid Illumina/PacBio assembly was further improved using Dovetail Chicago and subsequent HiRise scaffolding, thus indicating no misassemblies in the Illumina/PacBio hybrid contigs. The combination of these three technologies resulted in an assembly of 14.9 Gb in 92.9K scaffolds with a scaffold N50 size of 436 kb. With an estimated genome size of 164 Gb/1C. To further organize genome assembly, three intraspecific genetic linkage maps (Duangjit *et al.*, 2013; Choi *et al.*, 2020; Fujito *et al.*, 2021) and two interspecific genetic linkage maps (Scholten *et al.*, 2016) were used to anchor scaffolds into pseudomolecules using AllMaps (Tang *et al.*, 2015). A final assembly was of 14.9 Gb with an N50 of 461 kb, out of which only 2.2 Gb (~6.7%) was ordered into eight pseudomolecules using five different genetic linkage maps (Finkers *et al.*, 2021); the remainder of the genome is presented in 89.8 K scaffolds. Additionally, 72.4% of the genome was successfully identified as repetitive sequences consisting mainly of retrotransposons. The *ab initio* gene prediction indicated 540,925 putative gene models, which is far more than expected, possibly due to the presence of pseudogenes. Of these models, 86,073 showed similarity to UNIPROT published proteins. No gene-rich regions were found, and most genes were uniformly distributed across the genome (Finkers *et al.*, 2021). The analysis of synteny with garlic showed co-linearity but also major rearrangements between both species. This assembly is the first high-quality genome sequence available for the study of onion and will be a valuable resource for further research.

Similarly, garlic chromosome-level reference genomes and annotations by combining SMRT sequencing with PacBio Sequel Nanopore, 10x Genomics, Illumina HiSeq paired-end, and high-throughput chromosome conformation capture (Hi-C) sequencing resulted in a genome of 16.24 Gb (contig N50 length = 194 kb, scaffold N50 length = 725 kb (Sun *et al.*, 2020)). Then, 252.5 Gb of long reads obtained from Nanopore sequencing were mapped to the assembly. Thereafter, the Hi-C technology was applied to assemble a chromosome-level genome. A total of 6.34 billion reads from four Hi-C libraries were used, and approximately 87.5% of the assembled sequences were mapped to eight pseudomolecules with a super-scaffold N50 length of 1.69 Gb (Sun *et al.*, 2020). The final assembly covered 96.1% of the garlic genome (approximately 16.9 Gb), as estimated by the k-mer analysis. In addition, a total of 38 alliinase genes were expressed in various tissues, out of which four tandem duplicated genes (*Asa5G05557.1*, *Asa5G05559.1*, *Asa5G05560.*1 and *Asa5G0 5561.1*) produced highly abundant transcripts in bulb tissues (Sun *et al.*, 2020). In addition, their expression levels increased during bulb growth and expansion, suggesting that these four genes exhibited functional redundancy and may be responsible for the biosynthesis of allicin at a high concentration in garlic bulbs (Sun *et al.*, 2020).

6. Prospects

Omics technologies, such as metagenomics, metabolomics, proteomics, transcriptomics and genomics have great potential to open new avenues in research and improvement of *Allium* crops where genome sequence information is limited. Recently, NGS platforms have been routinely used for the discovery of markers such as SSRs and SNPs of various bulb onion populations. For example, a high-throughput genotyping approach, such as diversity arrays (DArTseq), was utilized for population structure analyses of a large garlic population containing 417 accessions (Egea *et al.*, 2017). Similarly, the double-digest restriction site-associated DNA sequencing (ddRAD-seq) approach was used for SNP discovery in inbred lines of Korean short-day onion, facilitating structure analyses and genetic relationship studies (Lee *et al.*, 2018). Although genomic resources in *Allium* species have been restricted to date, the recent rapid progress of NGS technology is making whole-genome sequencing remarkably easier than before. The genome sequencing of *A. cepa* is now ongoing through the SEQUON project (https://www.oniongenome.wur.nl, accessed October 2022) by using Illumina HiSeq 2500 sequencer. More recently, HiFi reads (a type of data produced using the circular consensus sequencing (CCS) mode on one of the PacBio Sequel Systems) provide base-level resolution with >99.9% single-molecule read accuracy. HiFi reads can accurately detect all types of variants, from single nucleotide to structural variants, with high precision and recall and phase haplotypes, even in hard-to-sequence regions of complex genomes such as onion.

Transcriptome sequencing is an efficient approach to generate sequence information from expressed regions of the genome and has been widely used in the Alliums. Several workers use RNA-seq for studying organ development, marker discovery, male sterility, flavonoid biosynthesis, and the abiotic stress response in different *Allium* species (Abdelrahman *et al.*, 2015, 2019). In context, *Allium* TDB (http://alliumtdb.kazusa.or.jp/; accessed September 2022) contains many useful transcript libraries of root, stem, bulb and leaves for various *Allium* species. Similarly, a draft reference transcript for onion, using long-read sequencing technology will help in whole-genome sequencing in Alliums (Sohn *et al.*, 2016). A transcriptome catalogue for homozygous double haploid line CUDH2107 for understanding the genetic and molecular basis of various traits was developed (Khosa *et al.*, 2016).

After the establishment of omics platforms, plant metabolism research has transitioned from the study of individual gene functions to metabolic system research. Mass spectrometry (MS) is a major instrument that has high sensitivity and broad metabolite detection capabilities, and thus facilitates the detection of hundreds of metabolites having diverse bioactivities in the alliaceous crops (Abdelrahman *et al.*, 2020, 2021). In the future, the individual bio-resource-specific metabolic patterns can be used for molecular breeding of *Allium* crops while the broad metabolic profiles of *Allium* bio-resources can be used for integrated omics approaches. The integration of metabolomics and transcriptomics will provide insight into the molecular mechanism of *Allium* metabolite biosynthesis.

Image techniques such as GSH and FSH using fluorescent imaging systems are an important approach for *Allium* breeding and genetic studies, thus further developments in *Allium* imaging analysis are needed. Recently, the selection of fusarium basal rot-resistant onion was carried out using digital image analysis to quantify symptom development in the basal plate of dormant bulbs (Mandal and Cramer, 2021). Analysis with confocal microscopy identified bright blue-green autofluorescence from *Fusarium oxysporum*-infected tissue, effectively differentiating diseased from healthy tissue. Visual scoring of the fusarium basal rot symptom was combined with stereo fluorescence microscopic images captured using a green fluorescence protein dual filter to quantify accurately fusarium basal rot severity in the basal plate tissue (Mandal and Cramer, 2021). The new developed method could be used for developing resistant cultivars for onion breeding programmes in the near future.

Targeted genome-editing technologies, especially clustered, regularly interspaced, short palindromic repeats (CRISPR)/(CRISPR)-associated protein 9 (Cas9), have great potential to aid in the breeding of crops that can produce high yields under biotic/abiotic stress (Abdelrahman *et al.*, 2018a). This is due to their high accuracy and low risk of off-target effects, compared with conventional methods. The use of the CRISPR/Cas9 system is commonly used for targeted mutagenesis in crop plants, including gene knockouts, modifications and the activation and repression of target genes. However, there are no reports regarding genome editing in onion, probably due to the complex genome structure with many repetitive sequences, affecting the efficacy of the CRISPR/Cas9 system. Genomic selection (GS) is being increasingly applied in plant breeding programmes to enhance the genetic gain of economically important traits. For example, the GS scheme was proposed to prevent inbreeding depression in onions by avoiding the co-selection of closely related plants and combining the shortening of generation time (Sekine and Yabe, 2020). Ten years of breeding programmes to evaluate the efficiency of different selection schemes in onion, including general phenotypic selection, self-crossing phenotypic selection, general genomic selection, and inbreeding avoiding genomic selection were compared (Sekine and Yabe, 2020). General GS with shortening of generation time yielded the highest genetic gains among the selection schemes; however, inbreeding depression increased rapidly in later years. The proposed GS combining shortening of generation time with updating of the prediction model was superior to the others in later years, as it yielded relatively high genetic gain while maintaining significantly low levels of inbreeding. These results suggested that GS can be beneficial in onion breeding, and an optimal scheme should be selected depending on the selection period.

References

Abdelrahman, M. (2020) *Allium* functional genomic development for future climatic changes. In: Kole, C. (ed.) *Genomic Designing of Climate-Smart Vegetable Crops*. Springer, Cham, Switzerland, pp. 345–358.

Abdelrahman, M., Hirata, S., Ito, S.-I., Yamauchi, N. and Shigyo, M. (2014) Compartmentation and localization of bioactive metabolites in different organs of *Allium roylei*. *Bioscience, Biotechnology, and Biochemistry* 78, 1112–1122.

Abdelrahman, M., Sawada, Y., Nakabayashi, R., Sato, S., Hirakawa, H. *et al.* (2015) Integrating transcriptome and target metabolome variability in doubled haploids of *Allium cepa* for abiotic stress protection. *Molecular Breeding* 35, 195.

Abdelrahman, M., Abdel-Motaal, F., El-Sayed, M., Jogaiah, S., Shigyo, M. *et al.* (2016) Dissection of *Trichoderma longibrachiatum*-induced defense in onion (*Allium cepa* L.) against *Fusarium oxysporum* f. sp. *cepa* by target metabolite profiling. *Plant Science* 246, 128–138.

Abdelrahman, M., El-Sayed, M., Sato, S., Hirakawa, H., Ito, S.I. *et al.* (2017a) RNA-sequencing-based transcriptome and biochemical analyses of steroidal saponin pathway in a complete set of *Allium fistulosum-A. cepa* monosomic addition lines. *PLOS ONE* 12, 0181784.

Abdelrahman, M., Mahmood, H.Y.A.H., El-Sayed, M., Tanaka, S. and Tran, L.S. (2017b) Isolation and characterization of Cepa2, a natural alliospiroside A, from shallot (*Allium cepa* L. Aggregatum group) with anticancer activity. *Plant Physiology and Biochemistry* 116, 167–173.

Abdelrahman, M., Al-Sadi, A.M., Pour-Aboughadareh, A., Burritt, D.J. and Tran, L.P. (2018a) Genome editing using CRISPR/Cas9–targeted mutagenesis: an opportunity for yield improvements of crop plants grown under environmental stresses. *Plant Physiology and Biochemistry* 131, 31–36.

Abdelrahman, M., Jogaiah, S., Burritt, D.J. and Tran, L.P. (2018b) Legume genetic resources and transcriptome dynamics under abiotic stress conditions. *Plant, Cell & Environment* 41, 1972–1983.

Abdelrahman, M., Hirata, S., Sawada, Y., Hirai, M.Y., Sato, S. *et al.* (2019) Widely targeted metabolome and transcriptome landscapes of *Allium fistulosum–A. cepa* chromosome addition lines revealed a flavonoid hot spot on chromosome 5A. *Scientific Reports* 9, 3541.

Abdelrahman, M., Ariyanti, N.A., Sawada, Y., Tsuji, F., Hirata, S. *et al.* (2020) Metabolome-based discrimination analysis of shallot landraces and bulb onion cultivars associated with differences in the amino acid and flavonoid profiles. *Molecules* 25, 5300.

Abdelrahman, M., Hirata, S., Mukae, T., Yamada, T., Sawada, Y. *et al.* (2021) Comprehensive metabolite profiling in genetic resources of garlic (*Allium sativum* L.) collected from different geographical regions. *Molecules* 26, 1415.

Alan, A.R., Toprak, F.C. and Kaska, A. (2016) Production and evaluation of gynogenic leek (*Allium ampeloprasum* L.) plants. *Plant Cell, Tissue and Organ Culture* 125, 249–259.

Albini, S.M. and Jones, G.H. (1990) Synaptonemal complex spreading in *Allium cepa* and *Allium fistulosum*. III. The F_1 hybrid. *Genome* 33, 854–866.

Al-Sheikh Hussain, L. (1977) Cytology and species relationships of *Allium* section Codonoprasum (Rchb.) Endl. PhD thesis, University of Sheffield, UK.

Ariyanti, N.A., Hoa, V.Q., Khrustaleva, L.I., Hirata, S., Abdelrahman, M. *et al.* (2015) Production and characterization of alien chromosome addition lines in *Allium fistulosum* carrying extra chromosomes of *Allium roylei* using molecular and cytogenetic analyses. *Euphytica* 206, 343–355.

Baranyi, M. and Greilhuber, J. (1999) Genome size in *Allium*: in quest of reproducible data. *Annals of Botany* 83, 687–695.

Bennett, M.D., Bhandol, P. and Leitch, I.J. (2000) Nuclear DNA amounts in angiosperms and their modern uses – 807 new estimates. *Annals of Botany* 86, 859–909.

Bhasi, A., Simon, P., Senalik, D., Kumar, B., Manikandan, V. *et al.* (2010) RoBuST: an integrated resource of genomics information for plants in the root and bulb crop families Apiaceae and Alliaceae. *BMC Plant Biology* 10, 161.

Bohanec, B. (2002) Double-haploid onion. In: Rabinowitch, H.D. and Currah, L. (eds) *Allium Crop Science: Recent Advances*. CAB International, Wallingford, UK, pp. 145–157.

Bohanec, B. and Jakše, M. (1999) Variations in gynogenic response among long-day onion (*Allium cepa* L.) accessions. *Plant Cell Reports* 18, 737–742.

Brenchley, R., Spannagl, M., Pfeifer, M., Barker, G.L., D'Amore, R. *et al.* (2012) Analysis of the bread wheat genome using whole-genome shotgun sequencing. *Nature* 491, 418, 705–10.

Cho, K.S., Yun, B.K., Yoon, Y.H., Hong, S.Y., Mekapogu, M. *et al.* (2015) Complete chloroplast genome sequence of tartary buckwheat (*Fagopyrum tataricum*) and comparative analysis with common buckwheat (*F. esculentum*). *PLOS ONE* 10, e0125332.

Choi, Y., Kim, S. and Lee, J. (2020) Construction of an onion (*Allium cepa* L.) genetic linkage map using genotyping-by-sequencing analysis with a reference gene set and identification of QTLs controlling anthocyanin synthesis and content. *Plants* 9, 616.

De Melo, P.E. (2003) The root systems of onion and *Allium fistulosum* in the context of organic farming: a breeding approach. PhD thesis, Wageningen University, The Netherlands.

Depledge, D.P., Srinivas, K.P., Sadaoka, T., Bready, D., Mori, Y. *et al.* (2019) Direct RNA sequencing on nanopore arrays redefines the transcriptional complexity of a viral pathogen. *Nature Communications* 10, 754.

Duangjit, J., Bohanec, B., Chan, A.P., Town, C.D. and Havey, M.J. (2013) Transcriptome sequencing to produce SNP-based genetic maps of onion. *Theoretical and Applied Genetics* 126, 2093–2101.

Duangjit, J., Welsh, K., Wise, M.L., Bohanec, B. and Havey, M.J. (2014) Genetic analyses of anthocyanin concentrations and intensity of red bulb color among segregating haploid progenies of onion. *Molecular Breeding* 34, 75–85.

Egea, L.A., Mérida-García, R., Kilian, A., Hernandez, P. and Dorado, G. (2017) Assessment of genetic diversity and structure of large garlic (*Allium sativum*) germplasm bank, by diversity arrays technology genotyping-by-sequencing platform (DArTseq). *Frontiers in Genetics* 8, 98.

Emsweller, S.L. and Jones, H.A. (1935) An interspecific hybrid in *Allium*. *Hilgardia* 9, 265–273.

Fayos, O., Vallés, M.P., Garcés-Claver, A., Mallor, C. and Castillo, A.M. (2015) Doubled haploid production from Spanish onion (*Allium cepa* L.) germplasm: embryogenesis induction, plant regeneration and chromosome doubling. *Frontiers in Plant Science* 6, 384.

Finkers, R., van Kaauwen, M., Ament, K., Burger-Meijer, K., Egging, R. *et al.* (2021) Insights from the first genome assembly of onion (*Allium cepa*). *bioRvix* 03.05.434149.

Fujito, S., Akyol, T.Y., Mukae, T., Wako, T., Yamashita, K.-I. *et al.* (2021) Construction of a high-density linkage map and graphical representation of the arrangement of transcriptome-based unigene markers on the chromosomes of onion, *Allium cepa L. BMC Genomics* 22, 481.

Galván Vivero, G.A. (2009) Resistance to Fusarium basal rot and response to arbuscular mycorrhizal fungi in *Allium*. PhD thesis, Wageningen University, The Netherlands.

Galván Vivero, G.A., Kuyper, T.W., Burger, K., Paul Keizer, L.C., Hoekstra, R.F. *et al.* (2011) Genetic analysis of the interaction between *Allium* species and arbuscular mycorrhizal fungi. *Theoretical and Applied Genetics* 122, 947–960.

Gokce, A., McCallum, J., Sato, Y. and Havey, M. (2002) Molecular tagging of the Ms locus in onion. *The American Society for Horticultural Science* 127, 576–582.

Goodwin, S., McPherson, J.D. and McCombie, W.R. (2016) Coming of age: ten years of next-generation sequencing technologies. *Nature Reviews Genetics* 17, 333–351.

Govaerts, R., Kington, S., Friesen, N., Fritsch, R.M., Snijman, D.A. *et al.* (2005–2020) World checklist of Amaryllidaceae. Available at: http://apps.kew.org/wcsp/ (accessed 25 July 2022).

Hanelt, P. (1992) Infrageneric grouping of Allium: the Gatersleban approach. In: Hanelt, P., Hammer, K. and Knupfler, H. (eds) *The Genus Allium: Taxonomic Problems and Genetic Resources*. Institut für Pflanzengenetik und Kulturpflanzenforschung, Gatersleben, Germany, pp. 107–123.

Hang, T.T.M., Shjgyo, M., Yamauchi, N. and Tashiro, Y. (2004) Production and characterization of alien chromosome additions in shallot (*Allium cepa* L. Aggregatum group) carrying extra chromosome(s) of Japanese bunching onion (*A. fistulosum* L.). *Genes & Genetic Systems* 79(5), 263–269.

Hao, M., Zhang, L., Ning, S., Huang, L., Yuan, Z. *et al.* (2020) The resurgence of introgression breeding, as exemplified in wheat improvement. *Frontiers in Plant Science* 11, 252.

Havey, M.J. (1999) Seed yield, floral morphology, and lack of male-fertility restoration of male-sterile onion (*Allium cepa*) populations possessing the cytoplasm of *Allium galanthum*. *Journal of the American Society of Horticultural Science* 124, 626–629.

Hu, H., Hu, Q., Al-Shehbaz, I.A., Luo, X., Zeng, T. *et al.* (2016) Species delimitation and interspecific relationships of the genus *Orychophragmus* (Brassicaceae) inferred from whole chloroplast genomes. *Frontiers in Plant Science* 7, 1826.

Hui, L., Chen, W., Pan, M., Li, W., He, L. *et al.* (2020) Characterization of the complete mitochondrial genome of yellow stripe mutant in onion (*Allium cepa* L.). *Mitochondrial DNA Part B* 5, 2986–2987.

Huo, Y.M., Gao, L.M., Liu, B.J., Yang, Y.Y., Kong, S.P. *et al.* (2019) Complete chloroplast genome sequences of four *Allium* species: comparative and phylogenetic analyses. *Scientific Reports* 9, 12250.

Hyde, P.T., Earle, E.D. and Mutschler, M.A. (2012) Doubled haploid onion (*Allium cepa* L.) lines and their impact on hybrid performance. *Horticultural Science* 47, 1690–1695.

Ibrahim, A.M., Kayat, F.B., Susanto, D., Kashiani, P. and Arifullah, M. (2016) Haploid induction in spring onion (*Allium fistulosum* L.) via gynogenesis. *Biotechnology* 15, 10–16.

Jakše, J., Meyer, J.D.F., Suzuki, G., McCallum, J., Cheung, F. *et al.* (2008) Pilot sequencing of onion genomic DNA reveals fragmented transposable elements, low gene densities, and significant gene enrichment after methyl filtration. *Molecular Genetics and Genomics* 280, 287–292.

Jo, J., Purushotham, P.M., Han, K., Lee, H.R., Nah, G. *et al.* (2017) Development of a genetic map for onion (*Allium cepa* L.) using reference-free genotyping-by-sequencing and SNP assays. *Frontiers in Plant Science* 8, 1606.

Kaska, A., Toprak, F.C. and Alan, A.R. (2013) Gynogenesis induction in leek (*Allium ampeloprasum* L.) breeding materials. *Current Opinion in Biotechnology* 24S, S42–S47.

Kato, N., Reynolds, D., Brown, M.L., Boisdore, M., Fujikawa, Y. *et al.* (2008) Multidimensional fluorescence microscopy of multiple organelles in *Arabidopsis* seedlings. *Plant Methods* 4, 9–10.

Keller, J. (1990) Culture of unpollinated ovules, ovaries, and flower buds in some species of the genus *Allium* and haploid induction via gynogenesis in onion (*Allium cepa* L.). *Euphytica* 47, 241–247.

Kelly, L.J. and Leitch, I.J. (2011) Exploring giant plant genomes with next-generation sequencing technology. *Chromosome Research* 19, 939–953.

Khan, P.S.S.V., Vijayalakshmi, G., Raja, M.M., Naik, M.L., Germanà, M.A. *et al.* (2020) Doubled haploid production in onion (*Allium cepa* L.): from gynogenesis to chromosome doubling. *Plant Cell, Tissue and Organ Culture (PCTOC)* 142, 1–22.

Khar, A., Jakse, J. and Havey, M. (2008) Segregations for onion-bulb colors reveal that red is controlled by at least three loci. *American Society for Horticultural Science* 133, 42–47.

Khosa, J.S., Lee, R., Bräuning, S., Lord, J., Pither-Joyce, M. *et al.* (2016) Doubled haploid 'CUDH2107' as a reference for bulb onion (*Allium cepa* L.) research: development of a transcriptome catalogue and identification of transcripts associated with male fertility. *PLOS ONE* 11, e0166568.

Khrustaleva, L.I. and Kik, C. (1998) Cytogenetical studies in the bridge cross *Allium cepa* × (*A. fistulosum* × *A. roylei*). *Theoretical and Applied Genetics* 96, 8–14.

Khrustaleva, L.I., de Melo, P.E., van Heusden, A.W. and Kik, C. (2005) The integration of recombination and physical maps in a large-genome monocot using haploid genome analysis in a trihybrid *Allium* population. *Genetics* 169, 1673–1685.

Khrustaleva, L.I., Jiang, J. and Havey, M.J. (2016) High-resolution tyramide-FISH mapping of markers tightly linked to the male-fertility restoration (Ms) locus of onion. *Theoretical and Applied Genetics* 129, 535–545.

Kim, B., Kim, K., Yang, T.J. and Kim, S. (2016) Completion of the mitochondrial genome sequence of onion (*Allium cepa* L.) containing the CMS-S male-sterile cytoplasm and identification of an independent event of the ccmF N gene split. *Current Genetics* 62, 873–885.

Kim, B., Yang, T.J. and Kim, S. (2018) Identification of a gene responsible for cytoplasmic male-sterility in onions (*Allium cepa* L.) using comparative analysis of mitochondrial genome sequences of two recently diverged cytoplasms. *Theoretical and Applied Genetics* 132, 313–322.

King, J., Bradeen, J., Bark, O., McCallum, J. and Havey, M. (1998) A low-density genetic map of onion reveals a role for tandem duplication in the evolution of an extremely large diploid genome. *Theoretical and Applied Genetics* 96, 52–62.

Kirk, J.T.O., Rees, H. and Evans, G. (1970) Base composition of nuclear DNA with the genus *Allium*. *Heredity* 25, 507–512.

Kudryavtseva, N., Ermolaev, A., Karlov, G., Kirov, I., Shigyo, M. *et al.* (2021) A dual-color Tyr-FISH method for visualizing genes/markers on plant chromosomes to create integrated genetic and cytogenetic maps. *International Journal of Molecular Sciences* 22, 5860.

Kuhl, J., Cheung, F., Yuan, Q., Martin, W., Zewdie, Y. *et al.* (2004) A unique set of 11,008 onion expressed sequence tags reveals expressed sequence and genomic differences between the monocot orders Asparagales and Poales. *Plant Cell* 16, 114.

Labani, R.M. and Elkington, T.T. (1987) Nuclear DNA variation in the genus *Allium* L. (Liliaceae). *Heredity* 59, 119–128.

Lee, J.H., Natarajan, S., Biswas, M.K., Shirasawa, K., Isobe, S. *et al.* (2018) SNP discovery of Korean short day onion inbred lines using double digest restriction site-associated DNA sequencing. *PLOS ONE* 13(8), e0201229.

Lutz, K.A., Wang, W., Zdepski, A. and Michael, T.P. (2011) Isolation and analysis of high quality nuclear DNA with reduced organellar DNA for plant genome sequencing and resequencing. *BMC Biotechnology* 11, 54.

Maloy, S. and Hughes, K. (2013) *Brenner's Encyclopedia of Genetics*, 2nd edn. Elsevier, New York, pp. 52–54.

Mandal, S. and Cramer, C.S. (2021) Comparing visual and image analysis techniques to quantify fusarium basal rot severity in mature onion bulbs. *Horticulturae* 7, 156.

Marcussen, T., Sandve, S.R., Heier, L., Spannagl, M., Pfeifer, M. *et al.* (2014) Ancient hybridizations among the ancestral genomes of bread wheat. *Science* 345, 1250092–1250092.

Martin, W., McCallum, J., Shigyo, M., Jakse, J., Kuhl, J. *et al.* (2005) Genetic mapping of expressed sequences in onion and in silico comparisons with rice show scant colinearity. *Molecular Genetics and Genomics* 274, 197–204.

Martínez, L.E., Agüero, C.B., López, M.E. and Galmarini, C.R. (2000) Improvement of in vitro gynogenesis induction in onion (*Allium cepa* L.) using polyamines. *Plant Science* 156, 221–226.

Masuzaki, S., Shigyo, M. and Yamauchi, N. (2006a) Complete assignment of structural genes involved in flavonoid biosynthesis influencing bulb color to individual chromosomes of the shallot (*Allium cepa* L.). *Genes & Genetic Systems* 81, 255–263.

Masuzaki, S., Shigyo, M. and Yamauchi, N. (2006b) Direct comparison between genomic constitution and flavonoid contents in Allium multiple alien addition lines reveals chromosomal locations of genes related to biosynthesis from dihydrokaempferol to quercetin glucosides in scaly leaf of shallot (*Allium cepa* L.). *Theoretical and Applied Genetics* 112, 607–617.

Masuzaki, S., Yaguchi, S., Yamauchi, N. and Shigyo, M. (2007) Morphological characterisation of multiple alien addition lines of Allium reveals the chromosomal locations of gene(s) related to bulb formation in *Allium cepa* L. *J. Horticultural Science and Biotechnology* 82, 393–396.

McCallum, J., Clarke, A., Pither-Joyce, M., Shaw, M., Butler, R. *et al.* (2006) Genetic mapping of a major gene affecting onion bulb fructan content. *Theoretical and Applied Genetics* 112, 958–967.

McCallum, J., Baldwin, S., Shigyo, M., Deng, Y., van Heusden, S. *et al.* (2012) AlliumMap – a comparative genomics resource for cultivated *Allium* vegetables. *BMC Genomics* 13, 168.

Michalik, B., Adamus, A. and Nowak, E. (2000) Gynogenesis in Polish onion cultivars. *Journal of Plant Physiology* 156, 211–216. DOI: 10.1016/S0176-1617(00)80308-9.

Mueller, R.L. (2015) Genome biology and the evolution of cell-size diversity. *Cold Spring Harbor Perspectives in Biology* 7, a019125.

Nagl, W. and Fusenig, H.P. (1979) Types of chromatin organization in plant nuclei. *Plant Systematics and Evolution, Supplement* 2, 221–233.

Ohri, D., Fritsch, R.M. and Hanelt, P. (1998) Evolution of genome size in *Allium* (Alliaceae). *Plant Systematics and Evolution* 210, 57–86.

Peffley, E.B., Corgan, J., Horak, K. and Tanksley, S. (1985) Electrophoretic analysis of *Allium* alien addition lines. *Theoretical and Applied Genetics* 71, 176–184.

Peška, V., Mandáková, T., Ihradská, V. and Fajkus, J. (2019) Comparative dissection of three giant genomes: *Allium cepa*, *Allium sativum*, and *Allium ursinum*. *International Journal of Molecular Sciences* 20, 733.

Peterson, D., Schulze, S., Sciara, E., Lee, S., Bowers, J. *et al.* (2002) Integration of Cot analysis, DNA cloning, and high-throughput sequencing facilitates genome characterization and gene discovery. *Genome Research* 12, 795–807.

Rabinowicz, P., Schutz, K., Dedhia, N., Yordan, C., Parnell, L. *et al.* (1999) Differential methylation of genes and retrotransposons facilitates shotgun sequencing of the maize genome. *Nature Genetics* 23, 305–308.

Ranjekar, P.K., Pallotta, D. and Lafontaine, J.G. (1978) Analysis of plant genomes. V. Comparative study of molecular properties of DNAs of seven *Allium* species. *Biochemical Genetics* 16, 957–970.

Ricroch, A., Yockteng, R., Brown, S.C. and Nadot, S. (2005) Evolution of genome size across some cultivated *Allium* species. *Genome* 48, 511–520.

Romanov, D.V., Kiseleva, A.V. and Khrustaleva, L.I. (2014) Physical mapping of genes on the chromosomes of onion (*Allium cepa* L.) using EST-clones and Tyramide-FISH. *Izvestiya of Timiryazev Agricultural Academy* 1, 105–114.

Rothberg, J.M., Hinz, W., Rearick, T.M., Schultz, J., Mileski, W. *et al.* (2011) An integrated semiconductor device enabling non-optical genome sequencing. *Nature* 475, 348–352.

Scheffler, I.E. (2008) *Mitochondria*, 2nd edn. Wiley-Liss, New York.

Scholten, O.E., van Kaauwen, M.P.W., Shahin, A., Hendrickx, P.M., Keizer, L.C.P. *et al.* (2016) SNP-markers in *Allium* species to facilitate introgression breeding in onion. *BMC Plant Biology* 16, 187.

Sekine, D. and Yabe, S. (2020) Simulation-based optimization of genomic selection scheme for accelerating genetic gain while preventing inbreeding depression in onion breeding. *Breeding Science* 70, 594–604.

Shagina, I., Bogdanova, E., Mamedov, I.Z., Lebedev, Y., Lukyanov, S. and Shagin, D. (2010) Normalization of genomic DNA using duplex-specific nuclease. *Biotechniques* 48, 455–459.

Shibata, F. and Hizume, M. (2002) The identification and analysis of the sequences that allow the detection of *Allium cepa* chromosomes by GISH in the allodiploid A. wakegi. *Chromosome* 111, 184–191.

Shigyo, M. and Kik, C. (2008) Onion. In: Prohens, J. and Nuez, F. (eds) *Vegetables II: Fabaceae, Liliaceae, Umbelliferae and Solanaceae* (Handbook of Plant Breeding). Springer, pp. 121–159.

Shigyo, M., Tashiro, Y., Isshiki, S. and Miyazaki, S. (1996) Establishment of a series of alien monosomic addition lines of Japanese bunching onion (*Allium fistulosum* L.) with extra chromosomes from shallot (*A. cepa* L. Aggregatum group). *Genes & Genetic Systems* 71, 363–371.

Shigyo, M., Lino, M., Isshiki, S. and Tashiro, Y. (1997a) Morphological characteristics of a series of alien monosomic addition lines of Japanese bunching onion (*Allium fistulosum* L.) with extra chromosomes from shallot (*A. cepa* L. Aggregatum group). *Genes and Genetic Systems* 72 , 181–186.

Shigyo, M., Tashiro, Y., Lino, M., Terahara, N., Ishimaru, K. and Isshiki, S. (1997b) Chromosomal locations of genes related to flavonoid and anthocyanin production in leafsheath of shallot (*Allium cepa* L. Aggregatum group). *Genes and Genetic Systems* 72 , 149–156.

Shigyo, M., Wako, T., Kojima, A., Yamauchi, N. and Tashiro, Y. (2003) Transmission of alien chromosomes from selfed progenies of a complete set of Allium monosomic additions: the development of a reliable method for the maintenance of a monosomic addition set. *Genome* 46, 1098–1103.

Shigyo, M., Khar, A. and Abdelrhaman, M. (2018) The *Allium* genomes. In: Kole, C. (ed.) *Compendium of Plant Genomes*. Springer, Cham, Switzerland.

Sohn, S.H., Ahn, Y.K., Lee, T.H., Lee, J.E., Jeong, M.H. *et al.* (2016) Construction of a draft reference transcripts of onion (*Allium cepa*) using long-read sequencing. *Plant Biotechnology Reports* 10, 383390.

Song, Y., Chen, Y., Lv, J., Xu, J., Zhu, S. *et al.* (2017) Development of chloroplast genomic resources for *Oryza* species discrimination. *Frontiers in Plant Science* 8, 1854.

Springer, N., Xu, X. and Barbazuk, W.B. (2004) Utility of different gene enrichment approaches toward identifying and sequencing the maize gene space. *Plant Physiology* 136, 3023–3033.

Stack, S.M. and Comings, D.E. (1979) The chromosomes and DNA of *Allium cepa*. *Chromosoma* 70, 161–181.

Sulistyaningsih, E., Aoyagi, Y. and Tashiro, Y. (2006) Flower bud culture of shallot (*Allium cepa* L. Aggregatum group) with cytogenetic analysis of resulting gynogenic plants and somaclones. *Plant Cell, Tissue and Organ Culture* 86, 249–255.

Sun, X., Zhu, S., Ningyang, L., Cheng, Y., Zhou, J. *et al.* (2020) A chromosome-level genome assembly of garlic (*Allium sativum*) provides insights into genome evolution and allicin biosynthesis. *Molecular Plant* 13, 1–12.

Sung, W., Ackerman, M.S., Dillon, M.M., Platt, T.G., Fuqua, C. *et al.* (2016) Evolution of the insertion-deletion mutation rate across the tree of life. *G3* (Bethesda) 8, 2583–2591.

Susek, A., Javornik, B. and Bohanec, B. (2002) Factors affecting direct organogenesis from flower explants of *Allium giganteum*. *Plant Cell, Tissue and Organ Culture (PCTOC)* 68, 27–33.

Suzuki, G., Ura, A., Saito, N., Do, G., So, B. *et al.* (2001) BAC FISH analysis in *Allium cepa*. *Genes & Genetic Systems* 76, 251–255.

Tang, H., Zhang, X., Miao, C., Zhang, J., Ming, R. *et al.* (2015) ALLMAPS: robust scaffold ordering based on multiple maps. *Genome Biology* 16, 3.

Tashiro, Y., Tsutsumi, M. and Shigyo, M. (2000) Production and gene analyses of alien monosomic addition lines of *Allium fistulosum* L. with extra chromosomes from wild species in section *Cepa* of *Allium*. *Acta Horticulturae* 521, 211–218.

Tsujimura, M., Kaneko, T., Sakamoto, T., Kimura, S., Shigyo, M. *et al.* (2019) Multichromosomal structure of the onion mitochondrial genome and a transcript analysis. *Mitochondrion* 46, 179–186.

Tsukazaki, H., Yamashita, K., Yaguchi, S., Masuzaki, S., Fukuoka, H. *et al.* (2008) Construction of SSR-based chromosome map in bunching onion (*Allium fistulosum*). *Theoretical and Applied Genetics* 117, 1213–1223.

Tsukazaki, H., Yamashita, K., Yaguchi, S., Yamashita, K., Hagihara, T. *et al.* (2010) Direct determination of the chromosomal location of bunching onion and bulb onion markers using bunching onion-shallot monosomic additions and allotriploid-bunching onion single alien deletions. *Theoretical and Applied Genetics* 122, 501–510.

Van der Valk, P., de Vries, S.E., Everink, J.T., Verstappen, F. and de Vries, J.N. (1991) Pre- and post-fertilization barriers to backcrossing the interspecific hybrid between *Allium fistulosum* L. and *A. cepa* L. with *A. cepa*. *Euphytica* 53, 201–209.

Van Heusden, A.V., Shigyo, M., Tashiro, Y., van Ginkel, R.V. and Kik, C. (2000) AFLP linkage group assignment to the chromosomes of *Allium cepa* L. via monosomic addition lines. *Theoretical and Applied Genetics* 100, 480–486.

Vitte, C., Estep, M.C., Leebens-Mack, J. and Bennetzen, J.L. (2013) Young, intact and nested retrotransposons are abundant in the onion and asparagus genomes. *Annals of Botany* 112, 881–889.

Von Kohn, C., Kiełkowska, A. and Havey, M.J. (2013) Sequencing and annotation of the chloroplast DNAs and identification of polymorphisms distinguishing normal male-fertile and male-sterile cytoplasms of onion. *Genome* 56, 737–742.

Vosa, C.G. (1976) Heterochromatic handing patterns in *Allium*, II. Ueterochromatic variation in species of the *Paniculatum* group. *Chromosoma (Berl.)* 57, 119–133.

Vu, H.Q., Iwata, M., Yamauchi, N. and Shigyo, M. (2011) Production of novel alloplasmic male sterile lines in *Allium cepa* harbouring the cytoplasm from *Allium roylei*. *Plant Breeders* 130, 469–475.

Vu, H.Q., El-Sayed, M.A., Ito, S., Yamauchi, N. and Shigyo, M. (2012) Discovery of a new source of resistance to *Fusarium oxysporum*, cause of Fusarium wilt in *Allium fistulosum*, locatedon chromosome 2 of *Allium cepa* Aggregatum group. *Genome* 55(11), 797–807.

Warmke, H.E. and Lee, S.L. (1978) Pollen abortion in T cytoplasmic male-sterile corn (*Zea mays*): a suggested mechanism. *Science* 200, 561–563.

Yaguchi, S., McCallum, J., Shaw, M., Pither-Joyce, M., Onodera, S. *et al.* (2008) Biochemical and genetic analysis of carbohydrate accumulation in *Allium cepa* L. *Plant and Cell Physiology* 49, 730–739.

Yaguchi, S., Hang, T.T.M., Tsukazaki, H., Vu, Q.H., Masuzaki, S., Wako, T. *et al.* (2009) Molecular and biochemical identification of alien chromosome additions in shallot (*Allium cepa* L. Aggregatum group) carrying extra chromosome(s) of bunching onion (*A. fistulosum* L.). *Genes and Genetic Systems* 84(1), 43–55.

Yaguchi, S., Matsumoto, M., Date, R., Harada, K., Maeda, T. *et al.* (2013) Biochemical analyses of the antioxidative activity and chemical ingredients in eight different *Allium* alien monosomic addition line. *Bioscience, Biotechnology, and Biochemistry* 77, 2486–2488.

Yamashita, K. (1999) Development of cytoplasmic male sterile lines of Japanese bunching onion (*Allium fistulosum* L.) and shallot (*A. cepa* L. Aggregatum group) with cytoplasms of wild species. PhD thesis, Kagoshima University.

Yamashita, K. and Tashiro, Y. (1999) Possibility of developing a male sterile line of shallot (*Allium cepa* L. Aggregatum group) with cytoplasm from *A. galanthum* Kar. et Kir. *Journal of the Japanese Society of Horticultural Science* 68, 256–262.

Yamashita, K., Wako, T., Ohara, T., Tsukazaki, H. and Kojima, A. (2005) Improvement of rust resistance in bunching onion (*Allium fistulosum* L.) by recurrent selection. *Journal of the Japanese Society of Horticultural Science* 74(6), 444–450.

Yamashita, K.I., Shigyo, M., Masuzaki, S.I., Yaguchi, S., Hang, T.T.M. *et al.* (2007) The application of alien-chromosome addition lines and cytoplasmic substitution lines to studies on genetics and breeding in *Allium cepa*. *Genes Genomes Genomics* 1, 27–34.

Yarali, F. and Yanmaz, R. (2016) The effects of media composition on *in vitro* gynogenic embryo induction in *Allium tuncelianum* (Kollaman) Ozhatay, Matthew, Siraneci). *International Journal of Agricultural Science* 8, 3247–3251.

Yu, N. and Kim, S. (2020) Identification of *Ms2,* a novel locus controlling male-fertility restoration of cytoplasmic male-sterility in onion (*Allium cepa* L.), and development of tightly linked molecular markers. *Euphytica* 217, 191.

16 Metabolomics of *Allium*

Kavita Sharma[1,3]*, Sarah Almulla[1], Neelima Mahato[2], Samjhana Pradhan[2], Sameena Mateen[3] and Srinath Pashikanti[1,3]

[1]Department of Chemistry, Idaho State University, Idaho, USA; [2]School of Chemical Engineering, Yeungnam University, Gyeongsan, South Korea; [3]Biomedical and Pharmaceutical Sciences, College of Pharmacy, Kasiska Division of Health Sciences, Idaho State University, Pocatello, Idaho, USA

1. Introduction

Metabolomics is the study of the chemical process that involves small molecule substrates, intermediates and products of cell metabolism. It aims to construct unbiased observations using analytical tools and data analysis (Dias *et al.*, 2012).

Allium belongs to the Amaryllidaceae family and is one of the largest monocot genera, comprised of about 1000 different species, most of which contain various secondary metabolites that have a multitude of benefits. These complicated polymorphous groups of vascular plants include some of the most economically important crops like onion (*A. cepa*), garlic (*A. sativum*), shallot (*A. cepa* aggregatum group), leek (*A. ampeloprasum*) and the Japanese bunching onion (*A. fistulosum*) (Shigyo *et al.*, 2018).

Many *Allium* spp., and particularly the species bulb onion, are rich in sulphur-containing compounds. These sulphur-containing metabolites factor into *Allium*'s plant physiology and are the reason behind some of *Allium*'s important traits like colour, flavour and pungency, thus cultivated and consumed since ancient times. Additionally, these bioactive components possess nutritional and medicinal properties, and fitness-enhancing constituents, including anti-microbial properties (Chapter 3, this volume). Several studies have been conducted to identify the metabolome of the economically and medicinally important genus.

This chapter is focused on the metabolomics of *Allium* through reviewing published literature on what makes the studying of metabolomics possible. These studies showcase the most common methods used to extract, isolate and characterize secondary metabolites and how they have been adapted and compared to one another to hopefully aid future metabolomic researchers in their decision of which one to use. This review also focuses the majority on the two most commonly produced and researched species of *Allium* – bulb onion and garlic (Chapter 6, this volume). Both species have provided a massive database of information on the secondary metabolites that are present in each one and can hopefully act as a starting position for future researchers looking to either expand on bulb onion or garlic, perhaps someone looking to explore the other *Allium* species. Future research into the metabolomics of *Allium* could lead to breakthroughs in cancer medicine, nutritional diseases and pharmaceutical uses.

Onion and garlic have both been reported to possess active components with the ability to affect human health that can reduce the risk of diabetes and cardiovascular diseases, having

*Address for correspondence: kavitasharma@isu.edu

 Edible Alliums: Botany, Production and Uses (eds H.D. Rabinowitch and B. Thomas)
DOI: 10.1079/9781789249996.0016

anti-microbial and anti-fungal properties, and protecting against infections (El-Saber Batiha *et al.*, 2020; Chapter 3, this volume). Hence, thiosulphinates are also responsible for their pungent aroma and taste, and can also act as epigenetic modulators (Chhabria *et al.*, 2015).

These sulphur metabolites are rather unstable, which necessitates the development of suitable analytical methods (Dias *et al.*, 2012). Lately, attention has been given to the study of the polar fraction of the *Allium* metabolome that includes compounds like anthocyanins, flavonoids, saponins etc., which are often glycosylated (Lanzotti, 2012). It is also rich in non-sulphur components and a study of phytochemicals has identified *Allium* as one of the richest plant sources of antioxidants and anti-inflammatory components, demonstrating its efficacy in the prevention of metabolic disorders (Liberati-Čizmek *et al.*, 2019; see also Chapter 3, this volume). Many studies have been done to develop a methodology to properly extract and analyse these unstable metabolites. Reviewing these studies will allow for current and future chemists, biochemists and biologists to gain insight on what is already known about *Allium* and inspire them to continue to build and explore the world of *Allium* metabolomics.

2. The Secondary Metabolites of *A. cepa* and *A. sativum*

2.1. *Allium cepa* L.

Multiple studies on onion attempted to discover what compounds are responsible for its interesting flavour, aroma and medicinal properties. A notable group of onion metabolites is the phenolic compounds, which often receive plenty of attention for their contribution to the biological properties of medicinal plants. These include phenolic acids, flavonoids and anthocyanins. Common phenolic acids contained by onion are ferulic acid, gallic acid and protocatechuic acid. Various forms of onions' anthocyanins, cyanidin-glucopyranoside and peonidin-glucopyranoside, are often useful for their ability to improve micro-circulation and night vision (Teshika *et al.*, 2019). About 80% of the flavonoids found in onions are quercetins, especially quercetin glucoside and quercetin di-glucoside (Dias *et al.*, 2012). These flavonoids play a major role in the overall antioxidant activity of dietary plants. Quercetin has been found to have anti-HIV properties and studies have shown that it can reduce the risk of cardiovascular diseases by protecting LDL cholesterol from oxidation (Chapter 3, this volume).

Bulb onion has been known to be an excellent source for a multitude of biologically active organosulphuric compounds similar to alliin, which fascinate chemists, biochemists and biologists. Upon crushing the cells, the alliin, which acts as a substrate, mixes with the enzyme alliinase to generate a multitude of sulphur volatiles like allicin and methiin. Hence, onion has become a great resource for naturally occurring sulphur. A great deal of studies have been done on their benefits to daily human life (Table 16.1), and Marrelli *et al.* (2019) reviewed the effect of bulb onion metabolites on the treatment of obesity.

2.2. Garlic (*Allium sativum* L.)

Garlic contains hundreds of secondary metabolites including the sulphur-containing compounds ajoenes, thiosulphinates, vinyldithiins, sulfides etc. (El-Saber Batiha *et al.*, 2020; Chapter 6, this volume). The three major cysteine sulfoxides present in garlic are *S*-propyl-cysteine-sulfoxide (PCSO), allicin and *S*-methyl cysteine-sulfoxide (MCSO). These three are the main odoriferous molecules of freshly minced garlic homogenates, and depending on the water content, temperature and the alliinase enzyme, these molecules can produce more than 50 metabolites (El-Saber Batiha *et al.*, 2020).

Allicin is highly unstable, thus it easily decomposes at room temperature to form the more stable ajoenes and vinyldithiins (El-Saber Batiha *et al.*, 2020).

Sulphuric bioactive compounds were found to have multiple pharmacological activities (El-Saber Bariha *et al.*, 2020; Chapter 3, this volume), mainly anti-bacterial, anti-fungal, antiviral, anti-protozoal, antioxidant, anti-inflammatory, anti-cancer, anti-obesity and anti-hypertensive. The main metabolites in garlic that attributed to these were allicin, diallyl disulfide (DADS), diallyl trisulfide (DATS) and ajoenes (El-Saber Bariha *et al.*, 2020).

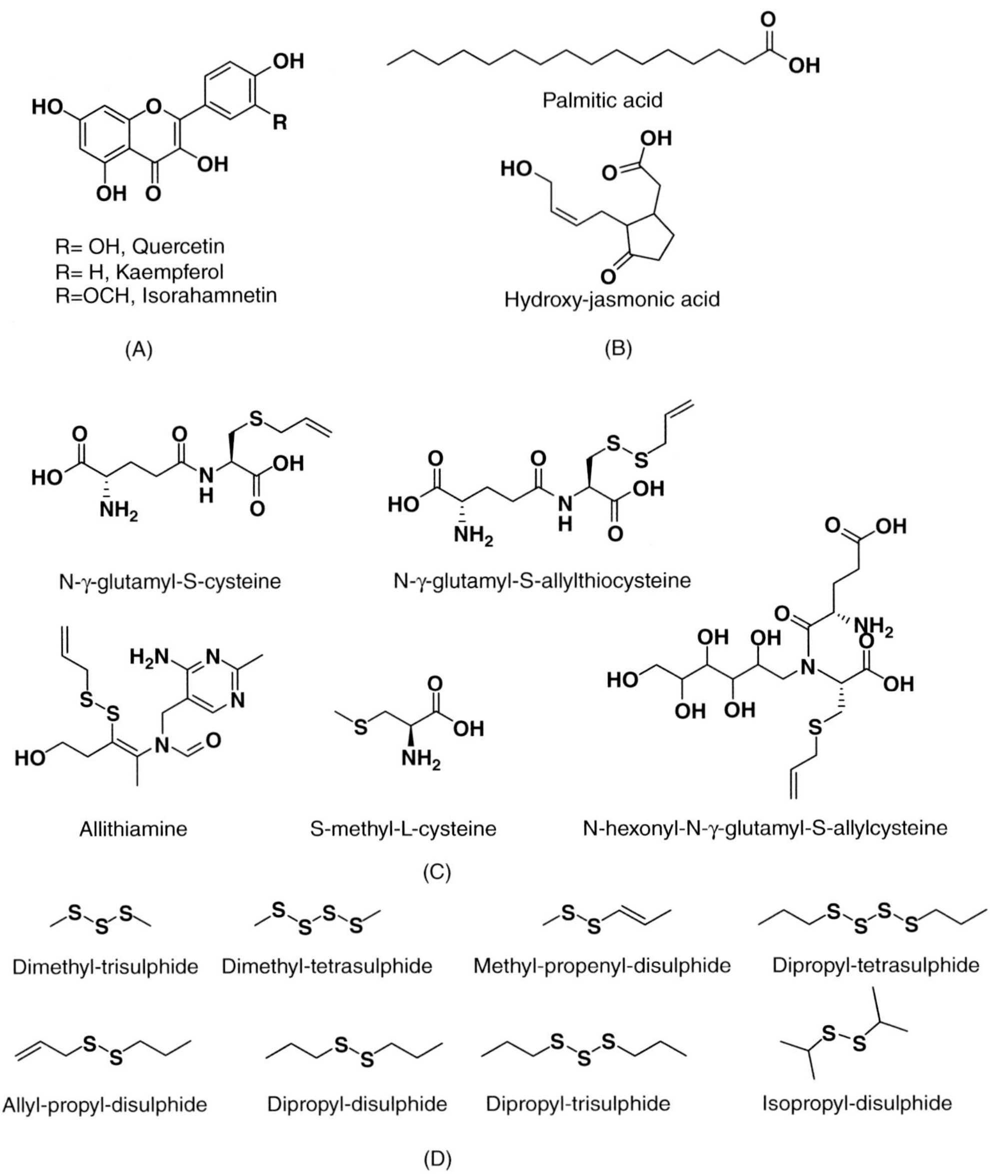

Fig. 16.1. The important bioactive compounds present in Allium: (A) flavonoids; (B) fatty acids; (C, D) sulphur-containing compounds.

3. Metabolic Techniques for Sample Extraction, Isolation and Characterization

Metabolomics is an approach using the different separation and spectrometric methods to analyse and characterize the different metabolites of any organism. Lately, this has developed into a powerful means of understanding the different aspects such as food value, disease signatures, and a general understanding of the metabolism. Such techniques have been used to analyse the constituents of different *Allium* species.

Table 16.1. The beneficial effects of allium extract.

Identified species extract	Application/function	Reference
Allium ascalonicum L.	Protective against immune-suppressive cyclosporine A nephrotoxicity in rats	Wongmekiat *et al.* (2008)
Allium sativum	Drug-resistant pathogenic bacteria (e.g. *E. coli, P. aeruginosa, B. subtilis*)	Gull *et al.* (2012)
Allium sativum L.	Improve endothelial dysfunction in an insulin-dependent model of uncontrolled diabetes	Baluchnejad-Mojarad *et al.* (2003)
Allium sativum, Allium cepa	Confer significant protection against atherogenic diet-induced atherosclerosis	Lata *et al.* (1991)
Allium sativum L.	Significant increases in antioxidant enzymes and suppressed glutathione depletion and lipid peroxidation in hepatic tissue	Kim *et al.* (2013)
Allium sativum	Prevention of thrombosis	Ali (1995)
Allium sativum	Effective inhibitor of tumour growth	Ejaz *et al.* (2003)
Allium cepa L.	Anti-platelet effects in rat platelets	Ro *et al.* (2015)
Allium cepa L.	Modulating or preventing weight gain or related diseases	Marrelli *et al.* (2019)
Allium cepa L.	Decreasing post-prandial hyperglycemia is to retard absorption of glucose by the inhibition of carbohydrate hydrolysing enzymes, α-amylase, and α-glucosidases, in the digestive organs	Kim *et al.* (2011)
Allium cepa L.	Cardioprotective against myocardial injury	Kharadi *et al.* (2016)
Allium cepa L.	Inactivation of PKC-ε induced by phosphorylating ERK1/2 is responsible for the neuroprotective effect	Lee and Jung (2016)
Allium sativum and *Allium cepa* L.	Potent nephroprotective property and can improve renal structural and serum biomarkers	Sherkatolabba-sieh *et al.* (2017)
Allium sativum	Decrease the glucose levels, activities of liver enzymes, oxidative stress markers, iNOS gene expression, and NO production increased significantly in diabetic rats	Ziamajidi *et al.* (2018)
Allium tuberosum	Preventive and anti-inflammatory effect against vascular disease and has potential for supporting prevention against the early process of atherosclerosis	Hur *et al.* (2017)

3.1. Extraction methods

In metabolomic case studies for Alliums, various sampling, collection, extraction, isolation and then characterization techniques have been used across different experiments. Each one has its advantages and disadvantages. Choosing the correct methods for each step is vital to having economical, reproducible and accurate metabolomic results. Common extraction methods for metabolomics include super fluid extraction (SFE) and subcritical water extraction (SWE) but for this review we will be taking an in-depth look at studies that used microwave-assisted extraction (MAE), solid phase microextraction (SPME) and headspace-solid phase microextraction (HS-SPME), and discuss the effectiveness of each method in *Allium* metabolomic research.

Originally used as a digestion method for various biological or chemical samples (Destandau *et al.*, 2013), MAE has been adapted for the extraction of organic compounds through microwave irradiation. As time went on, MAE became a widely accepted technique for extracting natural products for analysis.

While solvents with high dielectric constants are often used, there are cases where only the sample matrix will be heated. This is done so that the pre-heated solutes are released in a cold solvent, as it prevents the degradation of thermolabile compounds (Destandau *et al.*, 2013). A particularly useful asset of this technique is its ability to enhance the recovery of secondary metabolites and aroma compounds.

Fair amounts of plant secondary metabolites occur in the cytoplasm or the cell walls. Then, when the water in the core of the sample begins to heat, cell walls may rupture. This cell disruption can then shorten the diffusion path and allow for the mass transfer of the solvent into the plant sample. This will also allow for the secondary metabolites to mix with the solvent, which ultimately ends in a successful extraction (Destandau *et al.*, 2013).

Solid-phase microextraction (SPME) is a technique used to prepare samples for a multitude of operations such as collection, extraction, analyte enrichment and isolation from sample matrices. SPME is considered a powerful tool for solvents' free volatile extraction and with no heat application (Kataoka, 2017). The non-exhaustive technique and the process consists of two basic steps. First, the analytes are portioned between the extraction phase and the sample matrix. Second, subsequent desorption of the concentrated extracts into a chosen analytical instrument. Each phase of the SPME process can be performed in a single step, hence the SPME methods allow for quantitative and semi-quantitative data results (Kataoka, 2017).

The SPME is a common method of extraction in the different studies of *Allium* metabolomics. Farag *et al.* (2017) used SPME as a means to extract 20 headspace volatile components from fresh and pickled onion samples belonging to both sulphur and non-sulphur compounds. They concluded that SPME was a superior extracting method as they were able to maintain the onions' original good aroma without forming artifacts commonly generated when steam distillation was applied (Farag *et al.*, 2017).

Headspace solid-phase microextraction (HS-SPME) successfully extracted various volatiles for analyses of phytochemical profiles and antimicrobial activities of onion and garlic bulbs. GC/MS analyses revealed that while it was useful to unravel a phthalate ester, which could improve the overall quality of commercial garlic, it showed inconsistencies, and thus that HPLC-MS was the superior method for exploring the *Allium* metabolites (Farag *et al.*, 2017).

3.2. Characterization methods

HPLC is suitable for analysing high to medium polar plant-specific metabolites like flavonoids, terpenoids and alkaloids (Shigyo *et al.*, 2018). HPLC is often coupled with high-resolution analytical equipment such as mass spectrometry (MS), time-of-flight (TOF) and Fourier transform (FT), yet liquid chromatography-mass spectrometry (LC-MS) is the most common and comprehensive technique when it comes to measuring a wide range of metabolites (Salem *et al.*, 2020).

Gas chromatography (GC) isolates, detects and determines chemical components, mainly gases or organic molecules, in a sample mixture

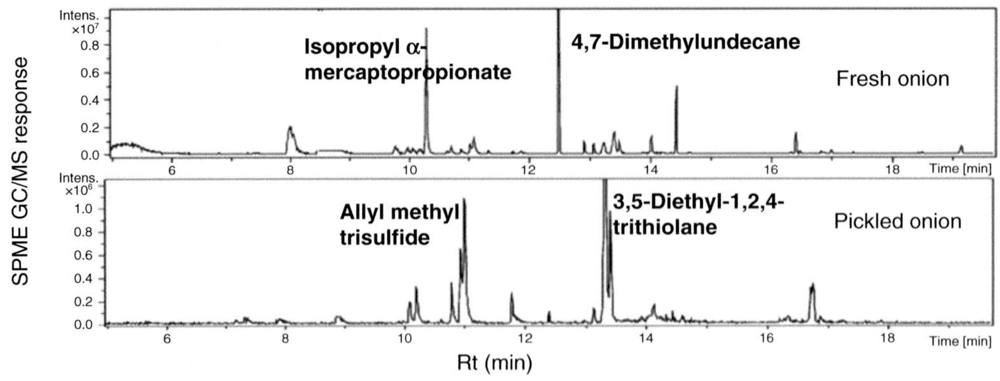

Fig. 16.2. Solid-phase microextraction (SPME)/GC-MS chromatogram of headspace volatiles collected from fresh and pickled onion bulb.

(Harris, 2016). Similarly to HPLC, GC is often coupled with a high-resolution analysis method, specifically mass spectrometry.

Mass spectrometry (MS) is a highly sensitive characterization technique that measures the masses and number of ions in the gas phase. It offers both qualitative and quantitative information as well as distinguishes different substances with the same retention time (Harris, 2016). MS paired with chromatography is one of the most used techniques in the characterization for metabolomics. When compared to other methods, LC-MS provided better coverage of the sulphur compounds and other metabolite classes (flavonoids and phenolic acids), making it the superior method for studying *Allium* metabolomics (Farag *et al.*, 2017). Some of the major bioactive compound's fragmentation by ultra-high performance liquid chromatography-quadrupole time-of-flight mass spectrometry (UPLC-QTOF-MS) are represented in Fig. 16.3 (Farag *et al.*, 2017).

The GC-MS is one the most popular methods for metabolomics analysis. GC-MS allows for polar metabolites to be derivatized and then rendered volatile and ultimately separated by GC. GC-MS was originally thought to be the most suitable technique for accurate determination of only primary metabolites, but it is currently noted as a very useful technique for secondary metabolites, too (Salem *et al.*, 2020).

Nuclear magnetic resonance spectroscopy (NMR) is commonly used by many fields of chemistry. It is an absorption process that requires the spinning of the nuclei in an external magnetic field. Since it is non-destructive, non-biased and easily quantifiable, NMR requires little to no chromatographic separation or chemical derivatization (Emwas *et al.*, 2019) and thus has become one of the most commonly used analytical methods in metabolomics. Not restricted to biofluid or tissue extract analysis, NMR spectroscopy is suitable for the study of intact tissues, organs and a multitude of other solid (or semi-solid) samples. NMR spectroscopy can produce spectra of different nuclei, thus allowing for a study of different metabolite classes (i.e. nitrogen-containing and phosphorus-containing metabolites). A significant disadvantage of NMR, however, is its lesser sensitivity (between 10 and 100 times less) compared to other popular methods like LC-MS or GC-MS.

Soininen *et al.* (2014) reported the quantitative metabolite profiling of edible onion species by NMR and showed the importance of ^{1}H NMR in the quantification and identification of the most abundant metabolites.

An investigation into the different metabolites contributing to the classifications of Italian garlic using the high-resolution magic angle spinning (HRMAS) NMR method revealed various bioactive compounds among the samples of garlic. Carbohydrates, organic acids, amino acids, fatty acids and organosulphur compounds were all found within the garlic samples (Ritota *et al.*, 2012).

4. Characterizing the Metabolome of Different Members of the *Allium* Genus

MAE extracts were used for comparative analyses of the different phytochemical compounds in *Allium* species, to evaluate their nutritional and therapeutic value and contribution towards the preservation of natural resources (Sadeghi *et al.*, 2017). Herein, the authors report the metabolic profiling of *Allium* spp. aimed at characterizing eight hydroalcoholic extracts from oven-dried bulbs and other commercial materials belonging to the *Allium* genus, inclusive of white and red garlic cultivars *A. triquetrum*, *A. roseum* and great-headed garlic (GHG, *A. ampeloprasum*). Earlier, a comparative study of the nutritional properties of GHG v. garlic revealed a strong potential of the former as a functional food (Najda *et al.*, 2016). Recently, profiling of black garlic (slowly long-cooked garlic) by LC coupled to tandem mass spectrometry revealed that thermal treatment changes significantly the composition in garlic of the amino acids family, followed by organosulphur compounds (Molina-Calle *et al.*, 2017a, b). The powerful and useful method for analysing the polar fraction of a variety of complex matrices such as foodstuff and essential oils, fourier-transform ion cyclotron resonance mass spectrometry (FT-ICR MS) coupled with electrospray ionization (ESI) was employed to maximize the identification of phytochemicals and ensure an efficient survey of the quality and freshness of these edible complex mixtures. Also, the characterization of *Allium* extracts has been complemented by targeted spectrophotometric and chromatographic methods, focusing on a CIEL*a*b* colorimetric analysis and on the separation, identification and quantification of polyphenols by using an HPLC-PDA method.

The HPLC-PDA analyses revealed gallic acid, catechin and carvacrol as the most abundant and largely distributed phenols, whose antioxidant and antiseptic properties suggest a beneficial use for the investigated *Allium* extracts. These results provide new insight into *Allium* species metabolome profiling and show the great potential of the present methodologies for comparative foodstuff characterization.

Farag *et al.* (2017) studied the composition and anti-microbial properties of red onions and garlic using a microwave, freeze-drying or air-drying, and further analysed the extracts to understand the composition of volatile and

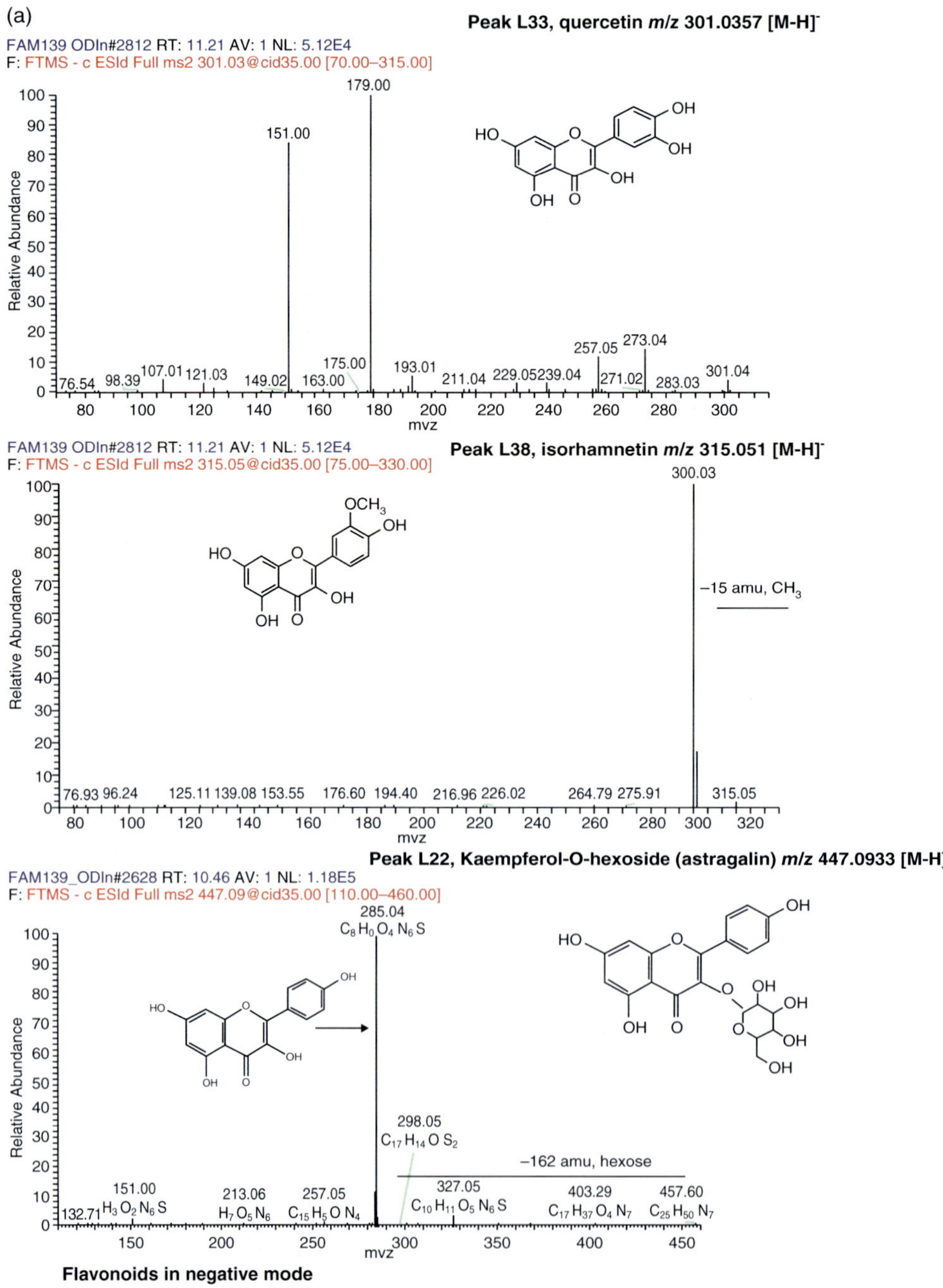

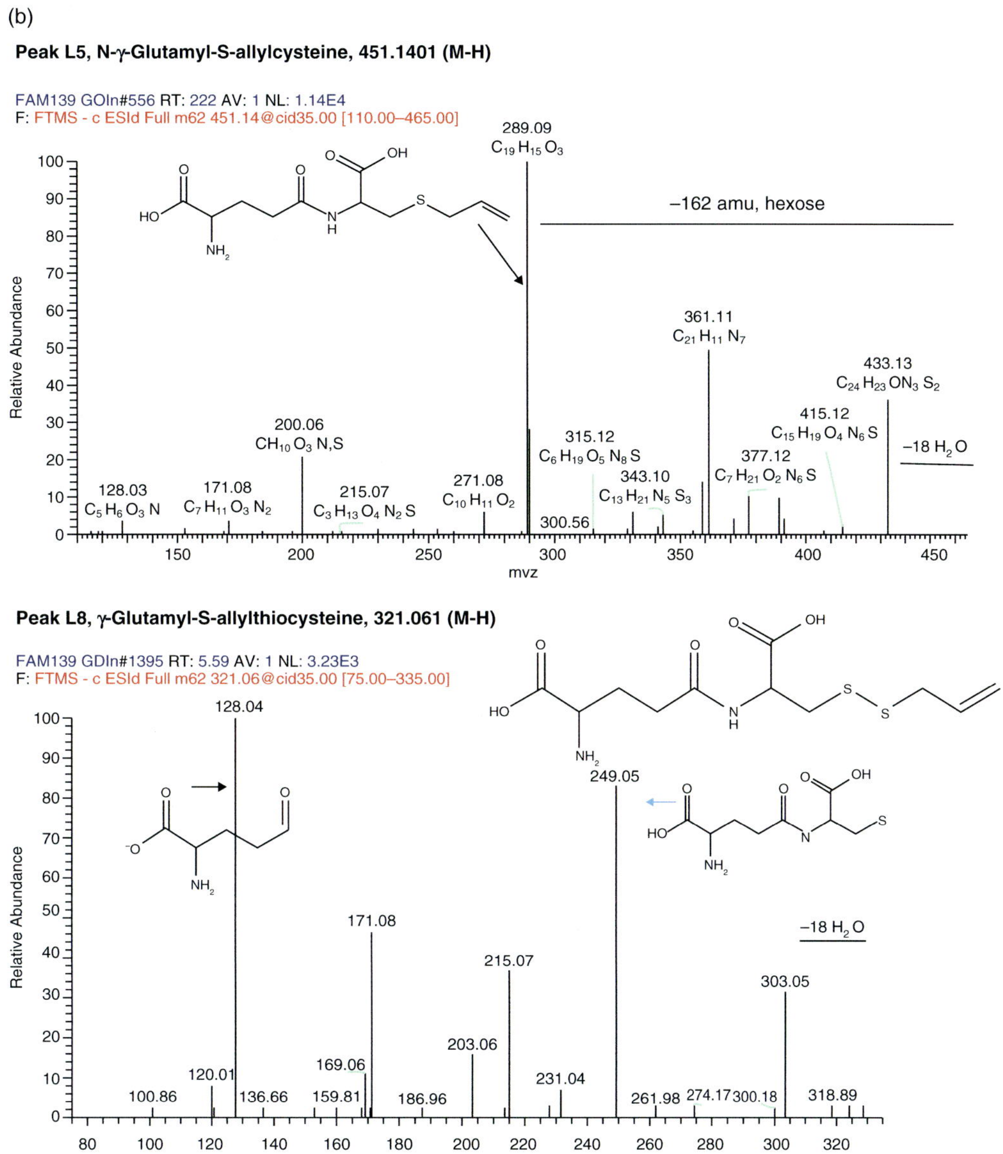

Fig. 16.3. Mass spectra for (a) flavonoids in negative mode; (b) peptides in negative mode.

non-volatile metabolites. Volatiles were collected using SPME coupled to GC/MS. Forty-two volatiles were identified including 30 sulphur compounds, four nitriles, three esters and four aromatics. Profiling of the polar non-volatile metabolites using ultra-performance liquid chromatography coupled to high-resolution MS (UPLC/MS) detected 51 metabolites including dipeptides, fatty acids, flavonoids and phenolic compounds. Major peaks in GC/MS or UPLC/MS shaping the differences between garlic and red onions were assigned to sulphur compounds and flavonoids. Whereas sulphur conjugates amounted to the major forms in garlic, flavonoids predominated in the chemical composition of red onion. As for the impact of drying on *Allium* metabolites, notable and clear separations among specimens were revealed using principal component analysis (PCA). The PCA scores plot of the UPLC/MS dataset revealed closer metabolite composition of microwave-dried specimens to freeze-dried ones, and distance from air-dried bulbs, observed in both plants. Compared to GC/MS, the UPLC/MS derived PCA model was more consistent and better in assessing the impact of drying on *Allium* metabolism. A phthalate derivative was identified exclusively in a commercial garlic preparation via GC/MS of yet unknown origin. The freeze-dried samples of both *Allium* species displayed stronger anti-microbial properties compared to air-dried specimens. Garlic was, in general, more active than the red onion.

The colour intensity of the bulb is related to the pungent smell of the garlic. In this regard, in some Latin American countries, there is high preference of dark purple garlic as compared to the lighter counterparts. Reportedly, the GC-MS-based untargeted metabolome study of the metabolome of dark and light purple garlic showed distinct metabolite signatures with a high content of organosulphur compounds and better medicinal properties in the dark purple varieties. Moreover, organic acids and fatty acids such as malonic acid and docosanoic acid were also the key distinct components between the dark and light purple varieties (Leite *et al.*, 2021). Similarly, Diretto *et al.* (2017) conducted LC-HRMS, together with bioinformatics including PCA, HCL and correlation network analyses on three fractions of garlic cloves from purple and white ecotypes, including the tunic (SS), internal (IS) and external (ES). The study showed that the SS fraction in purple garlic and the ES fraction of white varieties had high concentration of sulphur metabolites. Likewise, the ES fraction of the purple ecotype had high saponin content (40 times higher concentration) with five saponins including desgalactotigonin-rhamnose, proto-desgalactotigonin, proto-desgalactotigonin-rhamnose, voghieroside D1, sativoside B1-rhamnose and sativoside R1 being exclusive to the purple variety.

Other than mass spectrometry, NMR is a powerful tool to analyse metabolic profiles. Saviano *et al.* (2019) employed this technique to identify the active components of onion and described for the first time a comparative analysis of the metabolome of white cv. San Pietro onion (SP); red cv. Tropea (T) and yellow cv. Montoro (M) onions. The study highlighted clear differences between onion varieties for the fresh and nine months-stored batches. Higher levels of fructo-oligosaccharides (FOS) were recorded in the fresh WP; T showed a high content of citrate, glucose and amino acids, while M had many sulphur components. In the stored samples (M and R), carbohydrates and sulphur components decreased, while in SP, the levels of free monosaccharides increased. Linoleic acid was present in high amounts in the apolar extracts of fresh M and SP cultivars.

5. Metabolomics to Chemical Characterization and Physiological Relevance

Metabolomics is an appropriate approach for the identification of medicinally important metabolites like the organosulphur compounds. However, once identified, downstream analyses are necessary to confirm the physiological roles.

Block *et al.* (2017) introduced a novelty to the subject by discussing the synthesis, synthesis-reactivity and physiological activity of fluorinated analogues of garlic organosulphur compounds. They also report about antithrombotic and anti-angiogenesis activity of difluoroallicin (S-(2-fluoroallyl) 2-fluoroprop-2-ene-1-sulfinothioate) and S-2-fluoro-2-propenyl-l-cysteine, and also on the synthesis of 1,2-bis(2-fluoroallyl)disulfane, 5-fluoro-3-(1-fluorovinyl)-3,4-dihydro-1,2-dithiin, trifluoroajoene ((E,Z)-1-(2-fluoro-3-((2-fluoroallyl)sulfinyl)prop-1-en-1-yl)-2-(2-fluoroallyl) disulfane), and a bis(2-fluoroallyl)polysulfane mixture. All tested organosulphur compounds showed effective inhibition of either fibroblast growth factor (FGF) or vascular endothelial growth factor (VEGF)-mediated angiogenesis (anti-angiogenesis activity) in the chick chorioallantoic membrane (CAM) or the mouse Matrigel® models. No embryo mortality was observed. Difluoroallicin demonstrated greater inhibition ($p < 0.01$) versus organosulphur compounds tested. Difluoroallicin demonstrated dose-dependent inhibition of angiogenesis in the mouse Matrigel® model, with maximal inhibition at 0.01 mg/implant. Allicin and difluoroallicin showed an effective anti-platelet effect in suppressing platelet aggregation compared to other organosulphur compounds tested. In platelet/fibrin clotting (anti-coagulant activity), difluoroallicin showed concentration-dependent inhibition of clot strength compared to allicin and the other organosulphur compounds tested.

Siyo *et al.* (2017) reported on the activation of the unfolded protein response (UPR) as the mechanism of cytotoxicity of the small sulphur-rich compound, ajoene's analogue bisPMB. Ajoene interferes with biological processes and is cytotoxic to cancer cells in the low micromolar range. BisPMB is a synthetic ajoene analogue that has better cytotoxic properties than ajoene. The authors studied the pathways and processes affected by bisPMB by performing a DNA microarray analysis of bisPMB-treated WHCO1 oesophageal cancer cells. The most significantly enriched biological pathways as determined by gene ontology, KEGG and ingenuity pathway analysis (web-based bioinformatics application) were those involving protein processing in the endoplasmic reticulum (ER) and the unfolded protein response. In support of these pathways, bisPMB inhibited global protein synthesis and increased levels of ubiquitinated proteins. BisPMB also induced alternate splicing of the transcription factor XBP-1, increased the expression of the ER stress sensor GRP78 and induced expression of the ER stress marker C/EBP homologous protein (CHOP)/growth arrest and DNA damage-inducible gene 153 (GADD153). CHOP expression was found to be the key to bisPMB-mediated cytotoxicity, as its silencing with siRNA rendered the cells resistant to bisPMB. The MAPK proteins JNK and ERK1/2 were activated following bisPMB treatment. However, JNK activation was not critical in the cytotoxicity of bisPMB, and ERK1/2 activation was found to play a pro-survival role. Overall, the ajoene analogue bisPMB appears to induce cytotoxicity in WHCO1 cells by activating the unfolded protein response through CHOP/GADD153.

Aged black garlic (ABG) is a garlic preparation with a sweet and sour taste and no strong odour. Lately, it has been introduced to Asian markets as a functional food, and there is a rising interest in its potential health benefits. It no longer contains allicin but is enriched with several downstream metabolites. Ryu and Kang (2017) reviewed the physiological activities and constituents of ABG. Thorough *in vitro* and *in vivo* studies have demonstrated that it has a variety of biological functions such as antioxidant, anti-cancer, anti-inflammatory, anti-diabetic, anti-obesity, cardioprotective, anti-allergic and hepatoprotective effects. When compared to raw garlic, however, ABG shows lower anti-coagulation, anti-inflammatory, immunomodulatory and anti-allergic effects.

S-allyl cysteine (SAC) is a constituent of aged garlic extract that has demonstrated efficacy as an antioxidant and anti-inflammatory agent. ABG is enriched in SAC, a beneficial compound in treating pulmonary fibrosis in rats (Tsukoa *et al.*, 2017). As there are currently no other effective treatments, the development of updated strategies is essential for improving disease outcomes. SAC was administered to rats as a single intratracheal instillation of bleomycin (2.5 mg/kg) at 0.15% SAC-containing diet from seven days before instillation up until the conclusion of the experiment (14 days post-instillation). Consequently, a significant reduction in collagen mRNA expression and protein deposition (33.3 ± 2.7 μg/mg and 28.2 ± 2.1 μg/mg tissue in vehicle- and SAC-treated rats, respectively) was recorded, and fibrotic area decreased, as assessed histologically. In the rats' lungs, SAC also attenuated the increased expression of transforming growth factor-β1 (TGF-β1), a central regulator of myofibroblast recruitment, activation and differentiation. While bleomycin instillation increased the number of myofibroblasts within the lung mesenchymal area, this change was significantly reduced by SAC treatment. SAC may exert efficacy as an anti-fibrotic by attenuating myofibroblast differentiation through TGF-β1-mediated fibroproliferative processes. Thus, the results indicate SAC may be useful for the prevention or treatment of pulmonary fibrosis.

Kodera *et al.* (2017) reported on the biological properties of the closely related S-1-propenyl-L-cysteine (S1PC). S-1-propenyl-l-cysteine (S1PC) is a stereoisomer of S-allyl-l-cysteine (SAC), an important sulphur-containing amino acid that plays a role in the beneficial pharmacological effects of aged garlic extract (AGE).

The existence of S1PC in garlic preparations has been known since the 1960s. However, only recently, the biological, chemical, pharmacological and pharmacokinetic properties of S1PC were analysed. S1PC was found in trace amounts in raw garlic, but the concentration increased almost up to the level similar to SAC through an ageing process of AGE. S1PC showed immunomodulatory effects *in vitro* and *in vivo*, and reduced blood pressure in a hypertensive animal model. A pharmacokinetic study revealed that S1PC was readily absorbed after oral administration in rats and dogs with the bioavailability of 88–100%. Additionally, S1PC had little inhibitory influence on human cytochrome P450 activities, even at a concentration of 1 mM. Based on these findings, S1PC was suggested to be another important, pharmacologically active and safe component of AGE, similar to SAC.

Pérez-Torres *et al.* (2016) discuss the way antioxidant properties of AGE affect cardiovascular function (CF) in rats exhibiting metabolic syndrome (MS). Control rats receiving saline solution (C + SS), MS rats (30% sucrose in drinking water from weaning) receiving saline solution (MS + SS), control rats receiving AGE (C + AGE 125 mg/kg/12 h), and MS rats treated with AGE (MS + AGE) were studied. MS + SS had shown increased levels of triglycerides, insulin, systolic blood pressure, leptin, HOMA index, and advanced glycation end products. AGE returned their levels to control values ($p < 0.01$). Cholesterol was reduced by AGE ($p = 0.05$). Glutathione and GPx showed a decrease in MS + SS rats and were increased with AGE ($p = 0.05$). Lipid peroxidation was increased in MS + SS and AGE reduced it ($p = 0.001$). Vascular functioning was considerably hampered by MS (increased vasocontraction and reduced vasodilation) and AGE treatment improved it ($p = 0.001$). Coronary vascular resistance was increased in MS rats and AGE decreased it ($p = 0.001$). The cardiac performance was not modified by MS but AGE treatment increased it. NO measured in the perfusate liquid from the heart and serum citrulline, nitrites/nitrate were decreased in MS, and AGE treatment increased them ($p < 0.01$). In conclusion, AGE reduced MS-induced cardiovascular risk, through its antioxidant properties.

Albrecht *et al.* (2017) reported a facile synthesis and purification procedure to achieve a good yield of highly pure allicin. Allicin is a reactive sulphur species (RSS) and defence substance from garlic. The compound acts as a broad-spectrum antibiotic that is effective against multiple drug-resistant (MDR) strains. The authors discuss a detailed method for allicin synthesis based on diallyl-disulfide (DADS) oxidation by H_2O_2 using acetic acid as a catalyst. Albrecht *et al.* (2017) reported the progress of diallyl-disulfide synthesis and the reaction mechanism was analysed by HPLC and the identity and purity of the products were verified with LC-MS and ^{1}H-NMR.

6. Conclusions and Future Perspectives

Several noteworthy studies have been performed for deciphering the compounds present in the different species of *Allium*. These functional studies, both chemical and biological, have elucidated the medicinal properties of several bioactive components like allicin, ajoenes, SAC etc. However, many studies have not performed method validation, which appears to be a drawback. Those need to be addressed. Several advancements in the present technologies have made it possible to analyse in environmentally friendly, fast and inexpensive ways. The reduction of the packing particle size improved; consequently, the efficiency and resolution of the analysis. Moreover, new and advanced chromatographic techniques, like ultra-performance liquid chromatography (UPLC) and ultra-fast liquid chromatography have improved resolution and contributed to such advances.

Mass spectrometry (MS) is, arguably, the most important analytical spectroscopic tool of modern times, especially when coupled to GC and LC. While it has been widely used for determinations, when coupled to GC, this methodology generates technical artifacts. Nevertheless, these techniques allow for obtaining large amounts of information related to the identification of new compounds, varietal and species characterization, and metabolomic studies. Alliaceous crops are the fourth most abundant group of commercially produced non-leguminous vegetables after potatoes, cassava and tomatoes (FAO, 2020). Like garlic, the beneficial health effects attributed to onions are essential for both breeders and consumers. Onion consumption has been associated with decreased cardiovascular events because of their hypocholesterolemic, hypolipidemic, anti-hypertensive, anti-diabetic, anti-thrombotic and anti-hyperhomocysteinemia effects, as well as with many other biological activities including anti-microbial, antioxidant, anti-carcinogenic, anti-mutagenic, anti-asthmatic, immunomodulatory and prebiotic activity (Corzo-Martínez *et al.*, 2007; Chapter 3, this volume).

Garlic and onion have phytochemicals categorized as functional foods, including fructans, flavonoids and organosulphur compounds. *Allium* crops, specifically garlic and onion, display an increase in importance since 2000, due to an increase in *per capita* consumption. One of the reasons may be that consumers are aware of the potential of these vegetables to enhance health and prevent chronic diseases. Comparative genomic approaches have been widely used to understand the genetics background of the differences between species of a genus, and such studies are of growing interest as improved sequencing technologies, enabling ever broader and more detailed surveys of germplasm. Online databases that integrate the genetic map, sequence, marker and germplasm are now key tools for exploiting such data. Due to *Allium*'s economic significance, there is a clear and pressing need for such resources. Such studies would enable further improvement of nutritional and medicinal properties. Metabolomic studies of these vegetables would lead to the identification of several lead compounds that would be useful for the development of drugs. Further omics studies for validating the effect of such dietary components under physiological components would be useful in revealing the medicinal potential of such components.

References

Albrecht, F., Leontiev, R., Jacob, C. and Slusarenko, A.J. (2017) An optimized facile procedure to synthesize and purify allicin. *Molecules* 22, 770. DOI: 10.3390/molecules22050770.

Ali, M. (1995) Mechanism by which garlic (*Allium sativum*) inhibits cyclooxygenase activity: effect of raw versus boiled garlic extract on the synthesis of prostanoids. *Prostaglandins, Leukotrienes & Essential Fatty Acids* 53, 397–400. DOI: 10.1016/0952-3278(95)90102-7.

Baluchnejadmojarad, T., Roghani, M., Homayounfar, H. and Hosseini, M. (2003) Beneficial effect of aqueous garlic extracts on the vascular reactivity of streptozotocin-diabetic rats. *Journal of Ethnopharmacology* 85, 139–144. DOI: 10.1016/s0378-8741(02)00372-0.

Block, E., Bechand, B., Gundala, S., Vattekkatte, A., Wang, K. *et al.* (2017) Fluorinated analogs of organosulfur compounds from garlic (*Allium sativum*): synthesis, chemistry and anti-angiogenesis and antithrombotic studies. *Molecules* 22, 2081. DOI: 10.3390/molecules22122081.

Chhabria, S.V., Akbarsha, M.A., Li, A.P., Kharkar, P.S. and Desai, K.B. (2015) In situ allicin generation using targeted alliinase delivery for inhibition of MIA PaCa-2 cells via epigenetic changes, oxidative stress and cyclin-dependent kinase inhibitor (CDKI) expression. *Apoptosis* 20(10), 1388–1409. DOI: 10.1007/s10495-015-1159-4.

Corzo-Martínez, M., Corzo, N. and Villamiel, M. (2007) Biological properties of onions and garlic. *Trends in Food Science and Technology* 18(12), 609–625. DOI: 10.1016/j.tifs.2007.07.

Destandau, E., Michel, T. and Elfakir, C. (2013) Microwave-assisted extraction. In: Rostagno, M.A. and Prado, J.M. (eds) *Natural Product Extraction: Principles and Applications*. Royal Society of Chemistry, UK, 113–156.

Dias, D.A., Urban, S. and Roessner, U. (2012) A historical overview of natural products in drug discovery. *Metabolites* 2, 303–336. DOI: 10.3390/metabo2020303.

Diretto, G., Rubio-Moraga, A., Argandoña, J., Castillo, P., Gómez-Gómez, L. *et al.* (2017) Tissue-specific accumulation of sulfur compounds and saponins in different parts of garlic cloves from purple and white ecotypes. *Molecules* 22(8), 1359. DOI: 10.3390/molecules22081359.

Ejaz, S., Woong, L.C. and Ejaz, A. (2003) Extract of garlic (*Allium sativum*) in cancer chemoprevention. *Experimental Oncology* 25, 93–97.

El-Saber Batiha, G., Magdy Beshbishy, A., Wasef, L.G., Elewa, Y.H.A., Al-Sagan, M.A. *et al.* (2020) Chemical constituents and pharmacological activities of garlic (*Allium sativum* L.): a review. *Nutrients* 12, 872. DOI: 10.3390/nu12030872.

Emwas, A.H., Roy, R., McKay, R.T., Tenori, L., Saccenti, E. *et al.* (2019) NMR spectroscopy for metabolomics research. *Metabolites* 9, 123. DOI: 10.3390/metabo9070123.

FAO (2020) FAOSTAT. Available at: http://www.fao.org/faostat/en/#data (accessed 26 July 2022).

Farag, M.A., Ali, S.E., Hodaya, R.H., El-Seedi, H.R., Sultani, H.N. *et al.* (2017) Phytochemical profiles and antimicrobial activities of *Allium cepa* red cv. and *A. sativum* subjected to different drying methods: a comparative MS-based metabolomics. *Molecules* 22, 761. DOI: 10.3390/molecules22050761.

Gull, I., Saeed, M., Shaukat, H., Aslam, S.M., Samra, Z.Q. *et al.* (2012) Inhibitory effect of *Allium sativum* and *Zingiber officinale* extracts on clinically important drug resistant pathogenic bacteria. *Annals of Clinical Microbiology and Antimicrobials* 11, 1–6. DOI: 10.1186/1476-0711-11-8.

Harris, D.C. (2016) Principles of chromatography and mass spectrometry. In: Harris, D.C. (ed.) *Exploring Chemical Analysis*, 5th edn. W.H. Freeman, New York, pp. 465–469.

Hur, H.J. and Lee, A.S. (2017) Protective effect of Allium tuberosum extract on vascular inflammation in tumor necrosis factor-α-induced human vascular endothelial cells. *European Journal of Cancer Prevention* 22, 228. DOI: 10.15430/JCP.2017.22.4.228.

Kataoka, H. (2017) Sample preparation for liquid chromatography. In: Fanali, S., Haddad, P.R., Poole, C.F. and Marja-Liisa Riekkola, M.L. (eds) *Liquid Chromatography: Applications*. Elsevier, pp. 1–37.

Kharadi, G.B., Patel, K.J., Purohit, B.M., Baxi, S.N. and Tripathi, C.B. (2016) Evaluation of cardioprotective effect of aqueous extract of *Allium cepa* Linn. bulb on isoprenaline-induced myocardial injury in Wistar albino rats. *Research in Pharmaceutical Sciences* 11, 419. DOI: 10.4103/1735-5362.192494.

Kim, I., Kim, H.R., Kim, J.H. and Om, A.S. (2013) Beneficial effects of *Allium sativum* L. stem extract on lipid metabolism and antioxidant status in obese mice fed a high-fat diet. *Journal of the Science of Food and Agriculture* 93, 2749–2757. DOI: 10.1002/jsfa.6094.

Kim, S.H., Jo, S.H., Kwon, Y.I. and Hwang, J.K. (2011) Effects of onion (*Allium cepa* L.) extract administration on intestinal α-glucosidases activities and spikes in postprandial blood glucose levels in SD rats model. *International Journal of Molecular Sciences* 12, 3757–3769. DOI: 10.3390/ijms12063757.

Kodera, Y., Ushijima, M., Amano, H., Suzuki, J.I. and Matsutomo, T. (2017) Chemical and biological properties of S-1-Propenyl-L-cysteine in aged garlic extract. *Molecules* 22, 570. DOI: 10.3390/molecules22040570.

Lanzotti, V. (2012) Bioactive polar natural compounds from garlic and onions. *Phytochemistry Reviews* 11, 179–196. DOI: 10.1007/s11101-012-9247-3.

Lata, S., Saxena, K.K., Bhasin, V., Saxena, R.S., Kumar, A. *et al.* (1991) Beneficial effects of *Allium sativum*, *Allium cepa* and *Commiphora mukul* on experimental hyperlipidemia and atherosclerosis: a comparative evaluation. *Journal of Postgraduate Medicine* 37, 132.

Lee, B.K. and Jung, Y.S. (2016) *Allium cepa* extract and quercetin protect neuronal cells from oxidative stress via PKC-ε inactivation/ERK1/2 activation. *Oxidative Medicine and Cellular Longevity* 2016, 2495624. DOI: 10.1155/2016/2495624.

Leite, V., Reis, M. and Pinto, F. (2021) Untargeted metabolomics reveals metabolic changes linked to bulb purpling in garlic (*Allium sativum* L.). *ACS Food Science & Technology* 1(2), 242–248. DOI: 10.1021/acsfoodscitech.0c00106.

Liberati-Čizmek, A.M., Biluš, M., Brkić, A.L., Barić, I.C., Bakula, M. *et al.* (2019) Analysis of fatty acid esters of hydroxyl fatty acid in selected plant food. *Plant Foods for Human Nutrition* 74, 235–240. DOI: 10.1007/s11130-019-00728-8.

Maccelli, A., Cesa, S., Cairone, F., Secci, D., Menghini, L. *et al.* (2020) Metabolic profiling of different wild and cultivated *Allium* species based on high-resolution mass spectrometry, high-performance liquid chromatography-photodiode array detector, and color analysis. *Journal of Mass Spectrometry* 55, 4525. DOI: 10.1002/jms.4525.

Marrelli, M., Amodeo, V., Statti, G. and Conforti, F. (2019) Biological properties and bioactive components of *Allium cepa* L.: focus on potential benefits in the treatment of obesity and related comorbidities. *Molecules* 24, 119. DOI: 10.3390/molecules24010119.

Molina-Calle, M., de Medina, V.S., Priego-Capote, F. and de Castro, M.D.L. (2017a) Establishing compositional differences between fresh and black garlic by a metabolomics approach based on LC–QTOF MS/MS analysis. *Journal of Food Composition and Analysis* 62, 155–163. DOI: 10.1016/j.jfca.2017.05.004.

Molina-Calle, M., Sánchez de Medina, V., Calderón-Santiago, M., Priego-Capote, F. and Luque de Castro, M.D. (2017b) Untargeted analysis to monitor metabolic changes of garlic along heat treatment by LC–QTOF MS/MS. *Electrophoresis* 38, 2349–2360. DOI: 10.1002/elps.201700062.

Najda, A., Błaszczyk, L., Winiarczyk, K., Dyduch, J. and Tchórzewska, D. (2016) Comparative studies of nutritional and health-enhancing properties in the 'garlic-like' plant *Allium ampeloprasum* var. *ampeloprasum* (GHG-L) and *A. sativum*. *Scientia Horticulturae* 201, 247–255. DOI: 10.1016/j.scienta.2016.01.044.

Pérez-Torres, I., Torres-Narváez, J.C., Pedraza-Chaverri, J., Rubio-Ruiz, M.E., Díaz-Díaz, E. *et al.* (2016) Effect of the aged garlic extract on cardiovascular function in metabolic syndrome rats. *Molecules* 21, 1425. DOI: 10.3390/molecules21111425.

Ritota, M., Casciani, L., Han, B.-Z., Cozzolino, S., Leita, L. *et al.* (2012) Traceability of Italian garlic (*Allium sativum* L.) by means of HRMAS-NMR spectroscopy and multivariate data analysis. *Food Chemistry* 135, 684–693. DOI: 10.1016/j.foodchem.2012.05.032.

Ro, J.Y., Ryu, J.H., Park, H.J. and Cho, H.J. (2015) Onion (*Allium cepa* L.) peel extract has anti-platelet effects in rat platelets. *Springerplus* 4, 17. DOI: 10.1186/s40064-015-0786-0.

Ryu, J.H. and Kang, D. (2017) Physicochemical properties, biological activity, health benefits, and general limitations of aged black garlic: a review. *Molecules* 22, 919. DOI: 10.3390/molecules22060919.

Sadeghi, A., Hakimzadeh, V. and Karimifar, B. (2017) Microwave assisted extraction of bioactive compounds from food: a review. *International Journal of Food Sciences and Nutrition* 7, 19–27. DOI: 10.5923/j.food.20170701.03.

Salem, M.A., Perez de Souza, L., Serag, A., Fernie, A.R., Farag, M.A. *et al.* (2020) Metabolomics in the context of plant natural products research: from sample preparation to metabolite analysis. *Metabolites* 10, 37. DOI: 10.3390/metabo10010037.

Saviano, G., Paris, D., Melck, D., Fantasma, F., Motta, A. *et al.* (2019) Metabolite variation in three edible Italian *Allium cepa* L. by NMR-based metabolomics: a comparative study in fresh and stored bulbs. *Metabolomics* 15, 105. DOI: 10.1007/s11306-019-1566-6.

Sherkatolabbasieh, H., Hagh-Nazari, L., Shafiezadeh, S., Goodarzi, N., Zangeneh, M.M. *et al.* (2017) Ameliorative effects of the ethanolic extract of *Allium saralicum* RM Fritsch on CCl4-induced nephrotoxicity in mice: a stereological examination. *Archives of Biological Sciences* 69, 535–543. DOI: 10.2298/ABS160914129S.

Shigyo, M., Khar, A. and Abdelrahman, M. (eds) (2018) *The Allium Genomes*. Springer, Cham, Switzerland, pp. 53–66.

Siyo, V., Schäfer, G., Hunter, R., Grafov, A., Grafova, I. *et al.* (2017) The cytotoxicity of the ajoene analogue BisPMB in WHCO1 oesophageal cancer cells is mediated by CHOP/GADD153. *Molecules* 22, 892. DOI: 10.3390/molecules22060892.

Soininen, T.H., Jukarainen, N., Auriola, S.O., Julkunen-Tiitto, R., Karjalainen, R. and Vepsäläinen, J.J. (2014) Quantitative metabolite profiling of edible onion species by NMR and HPLC–MS. *Food Chemistry* 165, 499–505. DOI: 10.1016/j.foodchem.2014.05.132.

Teshika, J.D., Zakariyyah, A.M., Zaynab, T., Zengin, G., Rengasamy, K.R. *et al.* (2019) Traditional and modern uses of onion bulb (*Allium cepa* L.): a systematic review. *Critical Reviews in Food Science and Nutrition* 59(supp 1), S39–S70. DOI: 10.1080/10408398.2018.1499074.

Tsukioka, T., Takemura, S., Minamiyama, Y., Mizuguchi, S., Toda, M. *et al.* (2017) Attenuation of bleomycin-induced pulmonary fibrosis in rats with S-allyl cysteine. *Molecules* 22, 543. DOI: 10.3390/molecules22040543.

Wongmekiat, O., Leelarugrayub, N. and Thamprasert, K. (2008) Beneficial effect of shallot (*Allium ascalonicum* L.) extract on cyclosporine nephrotoxicity in rats. *Food and Chemical Toxicology* 46, 1844–1850. DOI: 10.1016/j.fct.2008.01.029.

Zhang, Z., Yang, M.J. and Pawliszyn, J. (1994) Solid-phase microextraction: a solvent-free alternative for sample preparation. *Analytical Chemistry* 66, 844A–853A. DOI: 10.1021/ac00089a001.

Ziamajidi, N., Behrouj, H., Abbasalipourkabir, R. and Lotfi, F. (2018) Ameliorative effects of *Allium sativum* extract on iNOS gene expression and NO production in liver of streptozotocin+nicotinamide-induced diabetic rats. *Indian Journal of Clinical Biochemistry* 33, 147–153. DOI: 10.1007/s12291-017-0656-3.

17 Molecular Breeding in *Allium*

Anil Khar*
Division of Vegetable Science, ICAR-Indian Agricultural Research Institute, New Delhi, India

1. Introduction

Alliums, belonging to the Amaryllidaceae family, comprise about 1000 species (Govaerts *et al.*, 2021). Among them, onion, garlic, shallot and bunching onion are the major ones. These Alliums are grown throughout the world for their culinary and medicinal properties. Thought to be under cultivation for the last 5000 years, onion and garlic have been the subject of research in both the agricultural and medicinal fields. Earlier, their improvement was limited to increased productivity and then hybrid development. The earliest report of use of molecular markers in onion was the identification of *roylei* linked markers (De Vries *et al.*, 1992b). Later research consisted of diversity assessments using non-PCR- and PCR-based markers.

The biennial onion is a highly cross-pollinated crop and suffers from high inbreeding depression. Improvement of garlic is mainly based on clonal selection. It is unlike many other crops where accelerated improvement has been possible due to genomics-assisted breeding along with conventional tools (Poland, 2015). The above limitations supplemented with a huge genome size of approximately 16 Gb made Alliums non-ideal candidates for the genomics revolution.

Availability of homozygous inbred populations, development of molecular markers and QTL mapping of genes of interest are a prerequisite to enable genomic-assisted breeding for various traits in bulb onion. Hence, the characterization of a large number of important trait-related genes laid a sound foundation for genomics-assisted breeding. High-throughput marker assays have been recently made available to the breeders, and subsequently, large-scale genotyping became economically and feasible time-wise. All of these together speeded up the development and implementation of genomics-based breeding. Recent reports of low-coverage sequencing of onion genome (Fu *et al.*, 2019), high-quality onion genome (Finkers *et al.*, 2021), chromosome-level genome assembly of garlic (Sun *et al.*, 2020) and fertility restoration in garlic (Chapter 6, this volume) have opened up new vistas of genome revolution in Alliums. In the foreseeable future, genomics-assisted breeding would be a possible strategy to breed onion and garlic cultivars with high-yield potential, nutritionally enriched, and with biotic and abiotic stress-resistance/-tolerance. This chapter summarizes the research work done in onion and garlic that may be of help to breeders to draw up an effective genomics-assisted breeding scheme.

*Address for correspondence: anil.khar@gmail.com

 Edible Alliums: Botany, Production and Uses (eds H.D. Rabinowitch and B. Thomas)
DOI: 10.1079/9781789249996.0017

2. Molecular Markers

Molecular markers are neutral, reproducible across laboratories, and experience no environmental effect on their expression. Hence, they are used extensively for the assessment of genetic diversity, population structure, linkage maps, QTL mapping and introgression of genes. Despite the economic significance of onion, garlic and leek as major vegetable crops, worldwide, genomic resources are rudimentary. The enormous genome sizes of onions, i.e. 16.3 GB per 1C nucleus (Arumuganathan and Earle, 1991), biennial generation time, severe inbreeding depression and a small community of researchers have slowed down the pace of the genomic revolution compared to contemporaries, i.e. potato and tomato. Similarly, the large genome size of garlic coupled with asexual propagation and, in leek, polyploidy, and less acreage, have played a negative role in generating interest among researchers. Use of restriction fragment length polymorphisms (RFLPs) (McCallum *et al.*, 2001), amplified length polymorphisms (AFLPs) (Van Heusden *et al.*, 2000), target region amplified polymorphic (TRAP) markers (Kisha and Cramer, 2011), simple sequence repeats (SSRs), and single nucleotide polymorphisms (SNPs) (Jakse *et al.*, 2005; McCallum *et al.*, 2008) have been reported. RAPD markers were utilized for diversity assessment (Sangeeta Kutty *et al.*, 2006; dos Santos *et al.*, 2012) at earlier stages but lately, SSR markers have been in vogue. Recently, with the high-throughput genomics and economical SNP genotyping, SNP marker resources in onion have been enriched by Scholten *et al.* (2016) through transcriptome sequencing and using a novel genotyping-by-sequencing approach (GBS) (Labate *et al.*, 2020).

2.1. Onion

Molecular markers such as SSRs and SNPs are useful in onion as they are co-dominant and are easy for high-throughput analysis. The development of genomic SSR makers from onion was first reported by Fischer and Bachmann (2000) but complex amplification requirements made them less popular for genotyping. Further, SSRs and SNPs were developed from expressed sequence tags (EST) (Kuhl *et al.*, 2004; Martin *et al.*, 2005). These SSRs have been reproducible for mapping (McCallum *et al.*, 2006) and cultivar discrimination (Jakse *et al.*, 2005). McCallum *et al.* (2008) converted SNPs from EST-SSR marker sequencing and suggested that re-sequencing of EST markers will be helpful for SNP identification.

Molecular studies have revealed that Indian populations are genetically a distinct group and a potential gene reservoir for broadening the genetic base (McCallum *et al.*, 2008). Khar *et al.* (2011) supported the hypothesis regarding the novelty of Indian genetic resources. A new set of genomic SSRs with their applicability in determining diversity within and between onion populations was published by Baldwin (2012a). This marker resource was important since these were genomic SSRs and, before this, SSR markers based on expressed sequences (ESTs) were only available, and have been utilized for diversity estimations (Hanci and Gökçe, 2016; Rivera *et al.*, 2016). Havey and Ghavami (2018) employed SNPs to estimate genetic relationships among onion populations from Europe, America, eastern Asia and India. Onion populations from Europe, America and eastern Asia clustered together, supporting the view of Goldman *et al.* (2001) that onion was introduced from Europe to America and Asia. The gene pool of onions from India is divergent from the European pool and may have directly been introduced from Central Asia. Transcriptome sequencing of two inbred lines (Duangjit *et al.*, 2013) led to SNP identification and these are being used for diversity estimation worldwide (Havey and Ghavami, 2018; Taylor *et al.*, 2019; Villano *et al.*, 2019).

2.2. Garlic

Despite being cultivated for millennia and known for its culinary and medicinal benefits, genomic resources in garlic have not increased exponentially. Compared to onion, the history of molecular marker development in garlic is very rudimentary. Assessment of garlic diversity using isozymes (Pooler and Simon, 1993), RAPD (Ipek *et al.*, 2003; Khar *et al.*, 2008), ISSR (Jabbes *et al.*, 2011), RAPD and ISSR (Shaaf *et al.*, 2014), SRAP (Chen *et al.*, 2013), AFLPs (Ovesna *et al.*, 2011; García-Lampasona *et al.*, 2012) and locus-specific markers (Ipek *et al.*, 2008) has been reported. Most of these markers are dominant, cumbersome and unstable. Micro-satellite markers are the markers of choice as they have high polymorphism, co-dominant, genomic abundance and can be validated in different laboratories. Ma *et al.* (2009) were the first to report eight SSR markers in garlic. Zhao *et al.* (2011) employed them for estimating diversity, population structure and core collection. Jo *et al.* (2012) estimated diversity in 120 accessions from five different countries employing seven out of these eight SSRs. Finally, Cunha *et al.* (2012) employed 16 new SSRs using (CT)8- and (GT)8-enriched library, ten of which showed polymorphism. Chen *et al.* (2014) used the same set of markers and reported eight polymorphic. Khar (2012) used 99 SSRs from onion (Fischer and Bachmann, 2000; McCallum *et al.*, 2008), eight garlic SSRs (Ma *et al.*, 2009) and 31 primers mined from garlic EST database (Kim *et al.*, 2009a). He reported 18 polymorphic SSRs for estimation of genetic diversity in garlic. Recently, da Cunha *et al.* (2014) were able to assess the genetic diversity and structure of Brazilian accessions using 17 SSR markers developed by Ma *et al.* (2009) and Cunha *et al.* (2012). Sensing the paucity of SSR markers in garlic, a deep transcriptome sequencing of garlic (Liu *et al.*, 2015) led to the identification of 194 amplifiable SSRs in garlic. For the first time, the Diversity Arrays Technology 'genotyping-by-sequencing' platform (DArTseq) was employed by Egea *et al.* (2017) for high-throughput genotyping of 417 garlic accessions. Clustering was observed based on variety and geographical origin and around 31% of redundant accessions were identified. Barboza *et al.* (2018) characterized and compared SSRs from garlic, related Alliums and non-Allium transcriptomes. A total of 14 selected SSRs clustered in the garlic accession based on flowering behaviour, botanical variety and ecophysiological characteristics. Now, the report of chromosome-level genome assembly of garlic (Sun *et al.*, 2020) will usher this crop into the genomic revolution of the garlic improvement programmes.

3. Male Sterility

Cytoplasmic male sterility leads to the inability of plants to produce viable pollen grains and has been reported in more than 140 species. Among vegetables, the CMS system is commercially utilized in pepper, onion, carrot, broccoli, cabbage and radish. Hybrid onions became a reality with the discovery of 'S' cytoplasm in Italian Red (Jones and Emsweller, 1936) and the single restorer of fertility locus (*Ms*) (Jones and Clarke, 1943). Another CMS system, 'T', was discovered (Berninger, 1965) and three independent loci were known to control fertility restoration (Schweisguth, 1973). CMS-S has been more popular due to its stability in diverse environments (Havey, 2000) and the ease in maintaining the sterility system.

In the CMS-S system, the male-sterile line (*Smsms*) is propagated through crossing with a maintainer line that contains normal (N cytoplasm) and homozygous recessive *Ms* locus (*msms*). Identification of the maintainer and restorer lines is rather difficult as both carry N cytoplasm and are phenotypically fertile. Hence, a test cross of male-fertile line with male-sterile line is done and progeny testing in the second year confirms whether the male-sterile line is maintainer or restorer. Onion is a highly cross-pollinated crop with a biennial generation time. Moreover, the high frequency of dominant alleles at the *Ms* locus (Havey and Randle, 1996) makes it laborious to go through the conventional breeding procedure to identify maintainer lines. Additionally, the interaction of cytoplasm and nuclear genes makes it possible that the fertile lines may have either N or S cytoplasm. To answer these questions and to accelerate the breeding process, marker-assisted selection (MAS) plays an important role.

Cytoplasm identification is the key step in onion hybrid development. Molecular markers capable of identifying cytoplasm types enable onion breeders to save time and effort. Sato (1998) developed molecular markers to distinguish normal and CMS-S mitotypes based on the variation of relative copy numbers of mt DNA whereas Havey (1995) used polymorphic sequences of the chloroplast genome for marker development. Engelke *et al.* (2003) reported molecular markers for the identification of CMS-T mitotype. This marker was designed from the chive mtDNA sequence containing *orfA501* (Engelke and Tatlioglu, 2000). A simple PCR marker, MKFR, was designed by Kim *et al.* (2009a) based on the novel chimeric gene *orf725*. This marker is useful for the detection of

CMS-S, -T and -N cytoplasm in onion. Von Kohn *et al.* (2013) sequenced chloroplast DNA of N and S-cytoplasmic onions and identified an indel in *acc*D and two polymorphic restriction enzyme sites for the identification of N and S cytoplasm. Besides cytoplasm S-, N- and T-, Kim *et al.* (2015) hypothesized the presence of Cytotype X and Y based on the comparative analysis of chloroplast genome of male-fertile and male-sterile (S and T) cytoplasm.

Development of PCR markers exhibiting linkage disequilibrium with restorer of fertility (*Ms*) locus is of paramount importance. Gökçe *et al.* (2002) were the first to identify AOB272 (0.9 cM), a restriction fragment length polymorphism (RFLP) marker for *Ms* locus identification. Martin *et al.* (2005) reported another marker AGF136 positioned on the opposite side of the AOB272 marker. Thereafter, Bang *et al.* (2011) derived two markers, OPT and Psao, from gene sequences encoding putative oligopeptide transporter (OPT) and an EST probe encoding photosystem I subunit O (PsaO). Although OPT marker was tightly linked to the *Ms* locus at a distance of 1.5 cM, it showed little linkage disequilibrium with the *Ms* locus. Thereafter, Bang *et al.* (2013) reported a new CAPS marker, ACms.1100, based on the random amplification of polymorphic DNA (RAPD) and bulk segregant analysis. Other CAPS markers, jnurf05 and jnurf17, were reported by Park *et al.* (2013). These markers were linked closely to the *Ms* locus at a distance of 0.05 cM. Although these markers were useful for the selection of *Ms* locus in segregating families, not many reports of their utilization in open-pollinated populations are available. The *Ms* locus has been mapped to chromosome 2 (Gökçe *et al.*, 2002) and markers linked to *Ms* locus have been identified.

Most of the studies to detect linkage disequilibrium of *Ms* locus with markers have been carried out in F_2 or BC_1 families. The efficiency of markers to detect *Ms* locus in open-pollinated populations is unknown. Havey (2013) identified three SNPs and concluded that these SNPs must be tightly associated with *Ms* locus to remain in linkage disequilibrium with genotypes at *Ms* among randomly selected plants from open-pollinated populations, However, he cautioned that due to recombination between these SNPs and *Ms* locus, the success of these SNPs to identify correct genotypes will vary among onion populations. Two SCAR markers, based on AFLP and hi TAIL-PCR techniques, were developed by Yang *et al.* (2013) to identify *Ms* locus. Kim (2014) employed 14 PCR markers, reported to be in linkage disequilibrium with *Ms* locus, and observed 11 recombinants from 4273 segregating plants. Based on their results, he identified a new PCR marker, jnurf12, that matched perfectly with male-fertility phenotypes in all recombinants. But this marker was unsuitable due to multiple banding patterns in diverse breeding lines. Hence, a new PCR marker, jnurf13, was reported that was observed to be in linkage disequilibrium with *Ms* locus in different breeding lines. Two new PCR-based markers, AcSKP1 (Huo *et al.*, 2015) and AcPMS1 (Kim *et al.*, 2015), have been developed to identify and tag *Ms* locus in onion. AcSKP1 marker perfectly predicted the genotypes at the *Ms* locus in the four OP populations according to the phenotypes of testcross progenies. Kim *et al.* (2015) performed a combination of BSA and RNA-seq to identify candidates for the *Ms* locus in onion. *AcPMS1*, a gene involved in DNA mismatch repair was identified as the best candidate.

A preliminary search for the application of these markers for the identification of cytoplasm and *Ms* locus revealed that research on these lines has gained momentum since 2015. Until 2015, primers reported by Sato (1998) and Engelke *et al.* (2003) were mostly used for cytoplasm determination (Ragassi *et al.*, 2012; Saini *et al.*, 2015; Ferreira and Santos, 2018). Recently, new primers for cytoplasm (Kim *et al.*, 2009b; von Kohn *et al.*, 2013) and *Ms* locus (Huo *et al.*, 2015; Kim *et al.*, 2015) are being utilized for the identification of cytoplasm and *Ms* locus in onions from Brazil (Ferreira and Santos, 2018), India (Khar and Saini, 2016; Manjunathagowda and Anjanappa, 2020) and South Africa (Gazendam *et al.*, 2018). However, some caution in the use of these markers needs to be applied. Khar and Saini (2016) studied the feasibility of reported PCR markers for the identification of normal and sterile cytoplasm and *Ms* locus in Indian onion populations. They observed that PCR markers for cytoplasm are accurate in predicting N and S cytoplasm, but PCR markers for *Ms* locus are showing recombination. Khrustaleva *et al.* (2016) applied Tyramide FISH for the physical mapping of *Ms* locus markers in onion. They reported that the markers were assigned to the locations close to the centromere of chromosome 2, which is a region of lower recombinations. Hence, the identified molecular markers linked to the *Ms* locus are unable to identify the precise genotype due to the reduced recombination near this gene.

4. Onion Quality

Breeder-friendly molecular markers are needed for key bulb characteristics, such as single centres, desirable skin and flesh colours, pungency, maturity, as well as significant health-enhancing attributes provided by the thiosulfinates, flavonoids and fructans. Genetic analyses have revealed several major QTL for these important traits. Three loci on chromosomes 3, 5 and 8 govern the amounts of fructans, fructose, glucose and sucrose (Galamarini *et al.*, 2001; Havey *et al.*, 2004; McCallum et *al.*, 2006; Raines *et al.*, 2009). Two markers (ACM033 and ACABE58) are linked to Frc — a major locus on chromosome 8 responsible for differences in fructan content. However, for marker-assisted selection, tighter links are required. Sucrose transporter and acid invertase are important candidate genes involved in carbohydrate metabolism and map to chromosome 3 (high trait linkage) and 5, respectively (Galmarini *et al.*, 2001; Havey *et al.*, 2004; McCallum *et al.*, 2006; Raines *et al.*, 2009). Two enzymes (1-sucrose-sucrose fructosyltransferase and fructan 6-fructosyltransferase) involved in fructan biosynthesis have been mapped to chromosome 6, but no trait linkage was detected (Fujishima *et al.*, 2005). Genes encoding plastidic ferredoxin-sulphite reductase (SiR) and plastidic ATP sulfurylase (ATPS) are closely linked (1–2 cM) on chromosome 3. Lachrymatory factor synthase, the enzyme responsible for the tearing property and associated health benefits of onion, was mapped on chromosome 5 close to a dry matter QTL, suggesting that linkage drag might exist for these two traits (Masamura *et al.*, 2012).

5. Genomics and Bioinformatics Resources

Next-generation sequencing provides rapid and cost-effective ways to develop the genomic resources needed for molecular analyses and breeding in onion and other Alliums. Transcriptome and genome sequencing identifies candidate genes and molecular markers and provides an opportunity for genomic-assisted selection to be more extensively used in onion breeding (Baldwin *et al.*, 2012b; Duangjit *et al.*, 2013; Lee *et al.*, 2013; Kim *et al.*, 2014). These new tools along with conventional methods have the potential to significantly enhance the scope and pace of new onion and other Alliums cultivar development.

5.1. Linkage maps

Difficulties in developing, maintaining, exchange of genetic stocks, high heterozygosity, large genome size, and dearth of genomic data have limited the mapping of important traits in onion. First, a genetic map of an *Allium* species, using RFLP markers, was developed by King *et al.* (1998) in an intraspecific onion cross 'BYG15-23xAC43'. This map was further supplemented with SNPs and EST-derived SSR markers (Kuhl *et al.*, 2004; Martin *et al.*, 2005). Further, these markers led to the construction of genetic maps for the identification of linked markers to fertility restoration (Gökçe *et al.*, 2002), colour (Khar *et al.*, 2008) and other bulb traits (McCallum *et al.*, 2008). Breeding systems of *A. fistulosum* facilitated the development of several larger mapping pedigrees and detailed genetic maps based on SSR and AFLP markers (Tsukazaki *et al.*, 2008). These maps were used to conduct QTL analysis for seedling vigour (Ohara *et al.*, 2009). Tsukazaki *et al.* (2010) reported an *A. fistulosum* map based on *A. fistulosum* genomic SSR markers and onion EST-derived SNP and SSR markers. The only *Allium* relative that has produced fertile hybrids with onion, *A. roylei* (Van der Meer and de Vries, 1990), has been used to develop an interspecific map (Van Heusden *et al.*, 2000) and backcross progenies with valuable disease resistance (Scholten *et al.*, 2007). The cross of *A. roylei* with *A. fistulosum* enabled the development of bridge crosses between *A. cepa*, *A. roylei* and *A. fistulosum* (Khrustaleva and Kik, 1998) and the potential of introgressing *A. fistulosum* genes in onion. The development of *A. cepa* alien monosomic addition lines (AMALs) enabled the alignment of *Allium* genetic maps to physical chromosomes and facilitated comparative analysis (Shigyo *et al.*, 1996). AMALs have initially been applied to anchor AFLP-based maps in the interspecific *A. cepa* x *A. roylei* cross (Van Heusden *et al.*, 2000) and subsequently to anchor the 'BYG15-23 x AC43' map (Martin *et al.*, 2005). Subsequently, they were used to anchor SSR-based maps in *A. fistulosum* (Tsukazaki *et al.*, 2008) to physical chromosomes, and to assign many more onion EST-derived anchor markers used in *A. fistulosum* maps (Tsukazaki *et al.*, 2010).

In garlic, the first genetic map (Zewdie *et al.*, 2005) was developed using 37 markers forming nine linkage groups. A male fertility locus was also placed on the map. Based on AFLP markers, Ipek *et al.* (2005) developed the first low-density genetic map. Due to the lack of flowering, research on molecular mapping was limited in garlic. Studies on the molecular characterization of the nucleotide-binding site (NBS) encoding resistance genes, and induction analysis of a putative candidate gene linked to Fusarium basal rot resistance in *Allium sativum* (Rout *et al.*, 2014), led to the identification of 28 AsRGA (*A. sativum* resistance gene analogs) sequences from a resistant garlic genotype CBT-As153, that can form the basis of Fusarium basal rot resistance. The identified AsRGAs can act as a valuable resource towards the development of RGA-based molecular markers for genetic mapping in garlic and will pave the way towards cloning of novel R-genes against *F. oxysporum* f. sp. *cepae*.

5.2. Transcriptome sequencing

RNA sequencing is valuable for gene identification, expression profiling and variant discovery to identify candidate genes and markers (SSR and SNP) for marker-assisted breeding (Martin *et al.*, 2013; Mutz *et al.*, 2013). Transcriptome sequencing in bulb onion has been achieved by McCallum *et al.* (2001) followed by Kuhl *et al.* (2004). Recently, deep transcriptome sequencing was used to increase the wealth of sequence data and provide candidate sequences that may affect several useful traits in bulb onion. Baldwin *et al.* (2012b) used GS-FLX sequencing on normalized cDNA libraries prepared from shoots of long-day double haploid (CUDH 2150) and short-day inbred (Nasik Red). Later, Duangjit *et al.* (2013) followed Roche 454 FLX RNA sequencing from pooled samples of different tissues in two long-day populations (OH1 and 5225) and identified single nucleotide polymorphisms (SNPs). Recently, Illumina sequencing of cDNA libraries prepared from leaf tissues of short-day double haploid (H6) and long-day inbred (SP3B) lines was used to produce a high-quality *de novo* assembled transcriptome (Kim *et al.*, 2014). Targeted metabolic profiling coupled with transcriptome studies led Abdelrahman *et al.* (2015) to conclude that metabolic traits and their relevant genes for abiotic stress are strongly linked in shallot and the F_1 hybrid with onion. Hence, more emphasis for development towards abiotic stress must be diverted towards shallot.

Sun *et al.* (2012) analysed the transcriptome of garlic and discovered the genes involved in sulphur metabolism by using Illumina sequencing. A total of 127,933 unigenes were generated by *de novo* assembly and were compared with the sequences in public databases. These unigenes data will provide the foundation for research on gene expression, genomics and functional genomics in garlic. Liu *et al.* (2015) performed transcriptome analysis using Illumina pair-end sequencing. Out of 200 randomly selected EST-SSRs, 194 markers were successfully amplified in garlic. The EST-SSRs developed in this study represent the first large-scale development of SSR markers for garlic. Kamenetsky *et al.* (2015) integrated transcriptome catalogue and organ-specific profiling of gene expression in fertile garlic from multiplexed gene libraries, using RNA collected from various plant organs, including inflorescences and flowers. The *de novo* transcriptome of fertile garlic represents a new resource for research and breeding of this important crop, as well as for the development of effective molecular markers for useful traits, including fertility and seed production, resistance to pests, and nutraceutical characteristics. Similarly, Shemesh-Mayer *et al.* (2015) tried to reveal the mechanism underlying flower and pollen development in garlic by transcriptome and proteome analysis. Morphophysiological studies, combined with transcriptome and proteome analyses and quantitative PCR validation, enabled the identification of genes and specific processes involved in gametogenesis in male-sterile and fertile garlic genotypes.

5.3. Genome sequencing

Bacterial artificial chromosome (BAC) libraries have been a valuable tool for marker development and sequencing specific genomic regions (Farrar and Donnison, 2007). An onion BAC library with 48,000 clones covering 0.39 of the genome has been developed (Suzuki *et al.*, 2002). A genetic map, using reference-free genotyping-by-sequencing and SNP assays, has been developed (Jo *et al.*, 2017). Segregating F_2 'NW-001' × 'NW-002' population, and multiple parental lines were used for GBS analysis. Linkage maps and developed SNP markers will be valuable for genetic mapping of important agronomic traits and marker-assisted selection. More recently, the first genome assembly of onion using four TrueSeq genomic DNA libraries prepared from a doubled haploid line DHCU066619 were sequenced using Illumina HiSeq 2500 (Finkers *et al.*, 2021).

Ipek *et al.* (2005) demonstrated linkage and developed the first low-density genetic map by using AFLP and gene-specific markers. Likewise, Zewdie *et al.* (2005) demonstrated genetic linkage among expressed regions in the garlic genome. Thirty-seven markers formed nine linkage groups covering 415 centimorgans (cM) with an average distance of 15 cM between loci; the other 16 loci remained unlinked. A male fertility locus was placed on the map. This first genetic map of garlic is a seminal step toward the genetic improvement of garlic and eventual marker-assisted breeding.

5.4. Bioinformatics-driven resources for onion breeding

Development of genomic resources and utilization of these resources for marker-assisted breeding and gene-related queries is possible only when a comprehensive database is available. In Alliums, no such database exists. Mostly, queries are addressed to the NCBI database. VegMarks database contains information about *Allium fistulosum* markers (VegMarks, 2021). McCallum *et al.* (2012) published another database, AlliumMap, that provided comparative genomic resources among cultivated *Allium* vegetables. A new database, Onion Genomic Resource (OGR), with three-tier architecture, catalogues the genomic developments specific to the onion. The database contains gene annotation, miRNAs information, and other genes related to the onion. This database is open resource and freely available at http://webtom.cabgrid.res.in/ogr/ (accessed September 2022) (Shukla *et al.*, 2016). An online EST database (GarlicESTdb) for gene discovery and future gene annotation is available (Kim *et al.*, 2009a). Putative annotation, single nucleotide polymorphisms (SNP) and tandem repeat datasets can be searched and downloaded. The GarlicESTdb web application is freely available at: https://oak.kribb.re.kr/handle/201005/8987 (accessed 31 July 2022). Another online database, RobustDb, was presented by Bhasi *et al.* (2010) and contains garlic map and marker data. But most of these databases are not being maintained and are now defunct. A robust database for onion and garlic containing information about genome, genes, markers and SSRs linked to the gene of interest is required.

Alliums are one of the most important groups among vegetables for their culinary value and nutraceutical properties. Due to enormous genome size and complex breeding systems, advances in genomics and related areas have not been encouraging. With the publication of the onion and garlic genomes in the last year, renewed interest to gain insights into the gene function, markers linked to biotic and abiotic stress, processing quality and other important horticultural traits has been rekindled. A comprehensive strategy and a research group on an international level are required to address all the issues in the wake of climate change and increasing population.

References

Abdelrahman, M., Sawada, Y., Nakabayashi, R., Sato, S., Hirakawa, H. *et al.* (2015) Integrating transcriptome and target metabolome variability in doubled haploids of *Allium cepa* for abiotic stress protection. *Molecular Breeding* 35, 1–11.

Abubakar, L., Ado, S.G., Suberu, H.A. and Magaji, M.D. (2006) Screening of onion (*Allium cepa* L.) cultivars for resistance to purple blotch (*Alternaria porri* L.) disease. *Biological and Environmental Sciences Journal for the Tropics (BEST)* 3, 30–36.

Arumuganathan, K. and Earle, E.D. (1991) Nuclear DNA content of some important plant species. *Plant Molecular Biology Reporter* 9, 208–218.

Baldwin, S., Pither-Joyce, M., Wright, K., Chen, L. and McCallum, J. (2012a) Development of robust genomic simple sequence repeat markers for estimation of genetic diversity within and among bulb onion (*Allium cepa* L.) populations. *Molecular Breeding* 30, 1401–1411.

Baldwin, S., Revanna, R., Thomson, S., Pither-Joyce, M., Wright, K. *et al.* (2012b) A toolkit for bulk PCR-based marker design from next-generation sequence data: application for development of a framework linkage map in bulb onion (*Allium cepa* L.). *BMC Genomics* 13, 637.

Bang, H., Cho, D.Y., Yoo, K.S., Yoon, M.K., Patil, B.S. *et al.* (2011) Development of simple PCR-based markers linked to the *Ms* locus, a restorer-of-fertility gene in onion (*Allium cepa* L.). *Euphytica* 179, 439–449.

Bang, H., Kim, S., Park, S.O., Yoo, K.S. and Patil, B.S. (2013) Development of a codominant CAPS marker linked to the *Ms* locus controlling fertility restoration in onion (*Allium cepa* L.). *Scientia Horticulturae* 153, 42–49.

Barboza, K., Beretta, V., Kozub, P.C., Salinas, C., Morgenfeld, M.M. *et al.* (2018) Microsatellite analysis and marker development in garlic: distribution in EST sequence, genetic diversity analysis, and marker transferability across *Alliaceae*. *Molecular Genetics and Genomics* 293, 1091–1106.

Berninger, E. (1965) Contribution a l'étude de la strérilité male de l'oignon (*Allium cepa* L.). *Ann Amélior Plant* 15, 183–199.

Bhasi, A., Simon, P., Senalik, D., Kumar, B., Manikandan, V. *et al.* (2010) RoBuST: an integrated resource of genomics information for plants in the root and bulb crop families *Apiaceae* and *Alliaceae*. *BMC Plant Biology* 10, 161.

Chen, S., Zhou, J., Chen, Q., Chang, Y., Du, J. *et al.* (2013) Analysis of the genetic diversity of garlic (*Allium sativum* L.) germplasm by SRAP. *Biochemical Systematics and Ecology* 50, 139–146.

Chen, S., Chen, W., Shen, X., Yang, Y., Qi, F., Liu, Y. and Meng, H. (2014) Analysis of the genetic diversity of garlic (*Allium sativum* L.) by simple sequence repeat and inter simple sequence repeat analysis and agro-morphological traits. *Biochemical Systematics and Ecology* 55, 260–267.

Cunha, C.P., Hoogerheide, E.S., Zucchi, M.I., Monteiro, M. and Pinheiro, J.B. (2012) New microsatellite markers for garlic, *Allium sativum* (*Alliaceae*). *American Journal of Botany* 99, e17–e19.

da Cunha, C.P., Resende, F.V., Zucchi, M.I. and Pinheiro, J.B. (2014) SSR-based genetic diversity and structure of garlic accessions from Brazil. *Genetica* 142, 419–431.

De Vries, J.N., Wietsma, W.A. and De Vries, T. (1992a) Introgression of leaf blight resistance from *Allium roylei* Stearn into onion (*A. cepa* L.). *Euphytica* 62, 127–133.

De Vries, J.N., Wietsma, W.A. and Jongerius, M.C. (1992b) Linkage of downy mildew resistance genes *Pd1* and *Pd2* from *Allium roylei* Stearn in progeny of its interspecific hybrid with onion (*A. cepa* L.). *Euphytica* 64, 131–137.

dos Santos, M.D.D.M., Ragassi, C.F., Fonseca, M.E.D.N., Buzar, A.G.R., Oliveira, V.R. *et al.* (2012) Genetic diversity of tropical-adapted onion germplasm assessed by RAPD markers. *Horticultura Brasileira* 30, 112–118.

Duangjit, J., Bohanec, B., Chan, A.P., Town, C.D. and Havey, M.J. (2013) Transcriptome sequencing to produce SNP-based genetic maps of onion. *Theoretical and Applied Genetics* 126, 2093–2101.

Egea, L.A., Mérida-García, R., Kilian, A., Hernandez, P. and Dorado, G. (2017) Assessment of genetic diversity and structure of large garlic (*Allium sativum*) germplasm bank, by diversity arrays technology 'genotyping-by-sequencing' platform (DArTseq). *Frontiers in Genetics* 8, 98.

Engelke, T. and Tatlioglu, T. (2000) Mitochondrial genome diversity in connection with male sterility in *Allium schoenoprasum* L. *Theoretical and Applied Genetics* 100, 942–948.

Engelke, T., Terefe, D. and Tatlioglu, T. (2003) A PCR-based marker system monitoring CMS-(S), CMS-(T) and (N)-cytoplasm in the onion (*Allium cepa* L.). *Theoretical and Applied Genetics* 107, 162–167.

Farrar, K. and Donnison, I.S. (2007) Construction and screening of BAC libraries made from Brachypodium genomic DNA. *Nature Protocols* 2, 1661.

Ferreira, R.R. and Santos, C.A.F. (2018) Partial success of marker-assisted selection of 'A' and 'B'onion lines in Brazilian germplasm. *Scientia Horticulturae* 242, 110–115.

Finkers, R., van Kaauwen, M.P., Ament, K., Burger-Meijer, K., Egging, R.J. *et al.* (2021) Insights from the first genome assembly of onion (*Allium cepa*). *Genetics* 11(9).

Fischer, D. and Bachmann, K. (2000) Onion microsatellites for germplasm analysis and their use in assessing intra- and interspecific relatedness within the subgenus *Rhizirideum*. *Theoretical and Applied Genetics* 101, 153–164.
Fu, J., Zhang, H., Guo, F., Ma, L., Wu, J. *et al.* (2019) Identification and characterization of abundant repetitive sequences in *Allium cepa*. *Scientific Reports*, 9(1), 1–7.
Fujishima, M., Sakai, H., Ueno, K., Takahashi, N., Onodera, S. *et al.* (2005) Purification and characterization of a fructosyltransferase from onion bulbs and its key role in the synthesis of fructo-oligosaccharides *in vivo*. *New Phytologist* 165, 513–524.
Galmarini, C.R., Goldman, I.L. and Havey, M.J. (2001) Genetic analyses of correlated solids, flavor, and health-enhancing traits in onion (*Allium cepa* L.). *Molecular Genetics and Genomics* 265, 543–551.
García-Lampasona, S., Asprelli, P. and Burba, J.L. (2012) Genetic analysis of a garlic (*Allium sativum* L.) germplasm collection from Argentina. *Scientia Horticulturae* 138, 183–189.
Gazendam, I., Greyling, M.M. and Laurie, S.M. (2018) The application of molecular markers to accelerate the recovery of cytoplasmic and nuclear male sterility in South African onion (*Allium cepa* L.) hybrid parental lines. *The Journal of Agricultural Science* 10, 95–109.
Gökçe, A.F. and Havey, M.J. (2002) Linkage equilibrium among tightly linked RFLPs and the *Ms* locus in open-pollinated onion populations. *Journal of the American Society for Horticultural Science* 127, 944–946.
Goldman, I.L., Schroeck, G. and Havey, M.J. (2001) History of public onion breeding programs in the United States. *Plant Breeding Reviews* 20, 67–104.
Govaerts, R., Kington, S., Friesen, N., Fritsch, R., Snijman, D.A. *et al.* (2005–2019) World checklist of Amaryllidaceae. Available at: https://apps.kew.org/wcsp/ (accessed 31 July 2022).
Hanci, F. and Gökçe, A.F. (2016) Molecular characterization of Turkish onion germplasm using SSR markers. *Czech Journal of Genetics and Plant Breeding* 52, 71–76.
Havey, M.J. (1995) Identification of cytoplasms using the polymerase chain reaction to aid in the extraction of maintainer lines from open-pollinated populations of onion. *Theoretical and Applied Genetics* 90, 263–268.
Havey, M.J. (2000) Diversity among male-sterility-inducing and male-fertile cytoplasms of onion. *Theoretical and Applied Genetics* 101, 778–782.
Havey, M.J. (2013) Single nucleotide polymorphisms in linkage disequilibrium with the male-fertility restoration (*Ms*) locus in open-pollinated and inbred populations of onion. *Journal of the American Society for Horticultural Science* 138, 306–309.
Havey, M.J. and Ghavami, F. (2018) Informativeness of single nucleotide polymorphisms and relationships among onion populations from important world production regions. *Journal of the American Society for Horticultural Science* 143, 34–44.
Havey, M.J. and Randle, W.M. (1996) Combining abilities for yield and bulb quality among long- and intermediate-day open-pollinated onion populations. *Journal of the American Society for Horticultural Science* 121, 604–608.
Havey, M.J., Galmarini, C.R., Gökçe, A.F. and Henson, C. (2004) QTL affecting soluble carbohydrate concentrations in stored onion bulbs and their association with flavor and health-enhancing attributes. *Genome* 47, 463–468.
Huo, Y.M., Liu, B.J., Yang, Y.Y., Miao, J., Gao, L.M. *et al.* (2015) AcSKP1, a multiplex PCR-based co-dominant marker in complete linkage disequilibrium with the male-fertility restoration (*Ms*) locus, and its application in open-pollinated populations of onion. *Euphytica* 204, 711–722.
Ipek, M., Ipek, A. and Simon, P.W. (2003) Comparison of AFLPs, RAPD markers, and isozymes for diversity assessment of garlic and detection of putative duplicates in germplasm collections. *Journal of the American Society for Horticultural Science* 128, 246–252.
Ipek, M., Ipek, A., Almquist, S.G. and Simon, P.W. (2005) Demonstration of linkage and development of the first low-density genetic map of garlic, based on AFLP markers. *Theoretical and Applied Genetics* 110, 228–236.
Ipek, M., Ipek, A. and Simon, P.W. (2008) Rapid characterization of garlic clones with locus-specific DNA markers. *Turkish Journal of Agriculture and Forestry* 32, 357–362.
Jabbes, N., Geoffriau, E., Le Clerc, V., Dridi, B. and Hannechi, C. (2011) Inter simple sequence repeat fingerprints for assessing genetic diversity of Tunisian garlic populations. *The Journal of Agricultural Science* 3, 77–85.
Jakse, J., Martin, W., McCallum, J. and Havey, M.J. (2005) Single nucleotide polymorphisms, indels, and simple sequence repeats for onion cultivar identification. *Journal of the American Society for Horticultural Science* 130, 912–917.

Jo, J., Purushotham, P.M., Han, K., Lee, H.R., Nah, G. *et al.* (2017) Development of a genetic map for onion (*Allium cepa* L.) using reference-free genotyping-by-sequencing and SNP assays. *Frontiers in Plant Science* 8, 1606.

Jo, M.H., Ham, I.K., Moe, K.T., Kwon, S.W., Lu, F.H. *et al.* (2012) Classification of genetic variation in garlic (*Allium sativum* L.) using SSR markers. *Australian Journal of Crop Science* 6, 625–631.

Jones, H.A. and Clarke, A. (1943) Inheritance of male sterility in the onion and the production of hybrid seed. *Proceedings of the American Society for Horticultural Science* 43, 189–194.

Jones, H.A. and Emsweller, S.L. (1936) A male-sterile onion. *Proceedings of the American Society for Horticultural Science* 34, 582–585.

Kamenetsky, R., Faigenboim, A., Mayer, E.S., Michael, T.B., Gershberg, C. *et al.* (2015) Integrated transcriptome catalogue and organ-specific profiling of gene expression in fertile garlic (*Allium sativum* L.). *BMC Genomics* 16, 12. DOI: 10.1186/s12864-015-1212-2.

Khar, A. (2012) Cross amplification of onion derived microsatellites and mining of garlic EST database for assessment of genetic diversity in garlic. *Acta Horticulturae* 969, 289–295. DOI: 10.17660/ActaHortic.2012.969.37.

Khar, A. and Saini, N. (2016) Limitations of PCR-based molecular markers to identify male-sterile and maintainer plants from Indian onion (*Allium cepa* L.) populations. *Plant Breeding* 135, 519–524.

Khar, A., Asha Devi, A. and Lawande, K.E. (2008) Analysis of genetic diversity among Indian garlic (*Allium sativum* L.) cultivars and breeding lines using RAPD markers. *Indian Journal of Genetics and Plant Breeding* 68, 52–57.

Khar, A., Lawande, K.E. and Negi, K.S. (2011) Microsatellite marker based analysis of genetic diversity in short day tropical Indian onion and cross amplification in related *Allium* spp. *Genetic Resources and Crop Evolution* 58, 741–752.

Khrustaleva, L.I. and Kik, C. (1998) Cytogenetical studies in the bridge cross *Allium cepa* × (*A. fistulosum* × *A. roylei*). *Theoretical and Applied Genetics* 96, 8–14.

Khrustaleva, L.I., Jiang, J. and Havey, M.J. (2016) High-resolution tyramide-FISH mapping of markers tightly linked to the male-fertility restoration (*Ms*) locus of onion. *Theoretical and Applied Genetics* 129, 535–545.

Kim, D.W., Jung, T.S., Nam, S.H., Kwon, H.R., Kim, A. *et al.* (2009) GarlicESTdb: an online database and mining tool for garlic EST sequences. *BMC Plant Biology* 9, 61. DOI: 10.1186/1471-2229-9-61.

Kim, S. (2014) A codominant molecular marker in linkage disequilibrium with a restorer-of-fertility gene (*Ms*) and its application in reevaluation of inheritance of fertility restoration in onions. *Molecular Breeding* 34, 769–778.

Kim, S., Lee, E.T., Cho, D.Y., Han, T., Bang, H. *et al.* (2009) Identification of a novel chimeric gene, *orf725*, and its use in development of a molecular marker for distinguishing among three cytoplasm types in onion (*Allium cepa* L.). *Theoretical and Applied Genetics* 118, 433–441.

Kim, S., Kim, M.S., Kim, Y.M., Yeom, S.I., Cheong, K. *et al.* (2014) Integrative structural annotation of de novo RNA-Seq provides an accurate reference gene set of the enormous genome of the onion (*Allium cepa* L.). *DNA Research* 22, 19–27.

Kim, S., Kim, C.W., Park, M. and Choi, D. (2015) Identification of candidate genes associated with fertility restoration of cytoplasmic male-sterility in onion (*Allium cepa* L.) using a combination of bulked segregant analysis and RNA-seq. *Theoretical and Applied Genetics* 128, 2289–2299.

King, J.J., Bradeen, J.M., Bark, O., McCallum, J.A. and Havey, M.J. (1998) A low-density genetic map of onion reveals a role for tandem duplication in the evolution of an extremely large diploid genome. *Theoretical and Applied Genetics* 96, 52–62.

Kisha, T.J. and Cramer, C.S. (2011) Determining redundancy of short-day onion accessions in a germplasm collection using microsatellite and targeted region amplified polymorphic markers. *Journal of the American Society for Horticultural Science* 136, 129–134.

Kuhl, J.C., Cheung, F., Yuan, Q., Martin, W., Zewdie, Y. *et al.* (2004) A unique set of 11,008 onion expressed sequence tags reveals expressed sequence and genomic differences between the monocot orders Asparagales and Poales. *Plant Cell* 16, 114–125.

Labate, J.A., Glaubitz, J.C. and Havey, M.J. (2020) Genotyping by sequencing for SNP marker development in onion. *Genome* 63, 607–613.

Lee, R., Baldwin, S., Kenel, F., McCallum, J. and Macknight, R. (2013) FLOWERING LOCUS T genes control onion bulb formation and flowering. *Nature Communications* 4, 2884. DOI: 10.1038/ncomms3884.

Liu, T., Zeng, L., Zhu, S., Chen, X., Tang, Q. *et al.* (2015) Large-scale development of expressed sequence tag-derived simple sequence repeat markers by deep transcriptome sequencing in garlic (*Allium sativum* L.). *Molecular Breeding* 35, 204.

Ma, K.H., Kwag, J.G., Zhao, W., Dixit, A., Lee, G.A. *et al.* (2009) Isolation and characteristics of eight novel polymorphic microsatellite loci from the genome of garlic (*Allium sativum* L.). *Scientia Horticulturae* 122, 355–361.

Manjunathagowda, D.C. and Anjanappa, M. (2020) Identification and development of male sterile and their maintainer lines in short-day onion (*Allium cepa* L.) genotypes. *Genetic Resources and Crop Evolution* 67, 357–365.

Martin, L., Fei, Z., Giovannoni, J. and Rose, J.K.C. (2013) Catalyzing plant science research with RNA-seq. *Frontiers in Plant Science* 4, 66.

Martin, W.J., McCallum, J., Shigyo, M., Jakse, J., Kuhl, J.C. *et al.* (2005) Genetic mapping of expressed sequences in onion and in silico comparisons with rice show scant colinearity. *Molecular Genetics and Genomics* 274, 197–204.

Masamura, N., McCallum, J., Khrustaleva, L., Kenel, F., Pither-Joyce, M. *et al.* (2012) Chromosomal organization and sequence diversity of genes encoding lachrymatory factor synthase in *Allium cepa* L. *G3* 2, 643–651.

McCallum, J., Leite, D., Pither-Joyce, M. and Havey, M.J. (2001) Expressed sequence markers for genetic analysis of bulb onion (*Allium cepa* L.). *Theoretical and Applied Genetics* 103, 979–991.

McCallum, J., Clarke, A., Pither-Joyce, M., Shaw, M., Butler, R. *et al.* (2006) Genetic mapping of a major gene affecting onion bulb fructan content. *Theoretical and Applied Genetics* 112, 958–967.

McCallum, J., Thomson, S., Pither-Joyce, M., Kenel, F., Clarke, A. *et al.* (2008) Genetic diversity analysis and single-nucleotide polymorphism marker development in cultivated bulb onion based on expressed sequence tag–simple sequence repeat markers. *Journal of the American Society for Horticultural Science* 133, 810–818.

McCallum, J., Baldwin, S., Shigyo, M., Deng, Y., van Heusden, S. *et al.* (2012) AlliumMap-A comparative genomics resource for cultivated Allium vegetables. *BMC Genomics* 13, 1–8.

Mutz, K.O., Heilkenbrinker, A., Lönne, M., Walter, J.G. and Stahl, F. (2013) Transcriptome analysis using next-generation sequencing. *Current Opinion in Biotechnology* 24, 22–30.

Ohara, T., Tsukazaki, H., Song, Y.S., Wako, T., Yamashita, K.I. *et al.* (2009) Mapping of quantitative trait loci controlling seedling growth in bunching onion (*Allium fistulosum* L.). *Journal of the Japanese Society for Horticultural Science* 78, 436–442.

Ovesná, J., Kučera, L., Horníčková, J., Svobodová, L., Stavělíková, H. *et al.* (2011) Diversity of S-alk (en) yl cysteine sulphoxide content within a collection of garlic (*Allium sativum* L.) and its association with the morphological and genetic background assessed by AFLP. *Scientia Horticulturae* 129, 541–547.

Park, J., Bang, H., Cho, D.Y., Yoon, M.K., Patil, B.S. *et al.* (2013) Construction of high-resolution linkage map of the *Ms* locus, a restorer-of-fertility gene in onion (*Allium cepa* L.). *Euphytica* 192, 267–278.

Poland, J. (2015) Breeding-assisted genomics. *Current Opinion in Plant Biology* 24, 119–124.

Pooler, M.R. and Simon, P.W. (1993) Characterization and classification of isozyme and morphological variation in a diverse collection of garlic clones. *Euphytica* 68, 121–130.

Ragassi, C.F., dos Santos, M.D.D.M., Fonseca, M.E.D.N., Oliveira, V.R., Buzar, A.G.R. *et al.* (2012) Genotyping of polymorphisms associated with male-sterility systems in onion accessions adapted for cultivation in Brazil. *Horticultura Brasileira* 30, 409–414.

Raines, S., Henson, C. and Havey, M.J. (2009) Genetic analyses of soluble carbohydrate concentrations in onion bulbs. *Journal of the American Society for Horticultural Science* 134, 618–623.

Rivera, A., Mallor, C., Garcés-Claver, A., García-Ulloa, A., Pomar, F. *et al.* (2016) Assessing the genetic diversity in onion (*Allium cepa* L.) landraces from northwest Spain and comparison with the European variability. *New Zealand Journal of Crop and Horticultural Science* 44, 103–120.

Rout, E., Nanda, S., Nayak, S. and Joshi, R.K. (2014) Molecular characterization of NBS encoding resistance genes and induction analysis of a putative candidate gene linked to Fusarium basal rot resistance in *Allium sativum*. *Physiological and Molecular Plant Pathology* 85, 15–24.

Saini, N., Hedau, N.K., Khar, A., Yadav, S., Bhatt, J.C. *et al.* (2015) Successful deployment of marker assisted selection (MAS) for inbred and hybrid development in long-day onion (*Allium cepa* L.). *Indian Journal of Genetics and Plant Breeding* 75, 93–98.

Sangeeta Kutty, M., Veere Gowda, R. and Anand, L. (2006) Analysis of genetic diversity among Indian short-day onion (*Allium cepa* L.) cultivars using RAPD markers. *The Journal of Horticultural Science and Biotechnology* 81, 774–778.

Sato, Y. (1998) PCR amplification of CMS-specific mitochondrial nucleotide sequences to identify cytoplasmic genotypes of onion (*Allium cepa* L.). *Theoretical and Applied Genetics* 96, 367–370.

Scholten, O.E., Van Heusden, A.W., Khrustaleva, L.I., Burger-Meijer, K., Mank, R.A. *et al.* (2007) The long and winding road leading to the successful introgression of downy mildew resistance into onion. *Euphytica* 156, 345–353.
Scholten, O.E., van Kaauwen, M.P., Shahin, A., Hendricks, P.M., Keizer, L.P. *et al.* (2016) SNP-markers in *Allium* species to facilitate introgression breeding in onion. *BMC Plant Biology* 16, 187.
Schweisguth, B. (1973). Étude d'un nouveau type de stérilité male chez l'oignon, *Allium cepa* L. *Annal Amélior Plant* 23, 221–233.
Shaaf, S., Sharma, R., Kilian, B., Walther, A., Özkan, H. *et al.* (2014) Genetic structure and eco-geographical adaptation of garlic landraces (*Allium sativum* L.) in Iran. *Genetic Resources and Crop Evolution* 61, 1565–1580.
Shemesh-Mayer, E., Ben-Michael, T., Rotem, N., Rabinowitch, H.D., Doron-Faigenboim, A. *et al.* (2015) Garlic (*Allium sativum* L.) fertility: transcriptome and proteome analyses provide insight into flower and pollen development. *Frontiers in Plant Science* 6, 271.
Shigyo, M., Tashiro, Y., Isshiki, S. and Miyazaki, S. (1996) Establishment of a series of alien monosomic addition lines of Japanese bunching onion (*Allium fistulosum* L.) with extra chromosomes from shallot (*A. cepa* L. *Aggregatum* group). *Genes & Genetic Systems* 71, 363–371.
Shukla, S., Iquebal, M.A., Jaiswal, S., Angadi, U.B., Fatma, S. *et al.* (2016) The onion genomic resource: a genomics and bioinformatics driven resource for onion breeding. *Plant Gene* 8, 9–15.
Sun, X., Zhou, S., Meng, F. and Liu, S. (2012). *De novo* assembly and characterization of the garlic (*Allium sativum*) bud transcriptome by Illumina sequencing. *Plant Cell Reports* 31, 1823–1828.
Sun, X., Zhu, S., Li, N., Cheng, Y., Zhao, J. *et al.* (2020) A chromosome-level genome assembly of garlic (*Allium sativum*) provides insights into genome evolution and allicin biosynthesis. *Molecular Plant* 13(9), 1328–1339.
Suzuki, G., Do, G.S. and Mukai, Y. (2002) Efficient storage and screening system for onion BAC clones. *Breeding Science* 52, 157–159.
Taylor, A., Teakle, G.R., Walley, P.G., Finch-Savage, W.E., Jackson, A.C. *et al.* (2019) Assembly and characterizasion of a unique onion diversity set indentifies resistance to Fusarium basal rot and improved seedling vigour. *Theoretical and Applied Genetics* 132, 3245–3264.
Tsukazaki, H., Yamashita, K.I., Yaguchi, S., Masuzaki, S., Fukuoka, H. *et al.* (2008) Construction of SSR-based chromosome map in bunching onion (*Allium fistulosum*). *Theoretical and Applied Genetics* 117, 1213–1223.
Tsukazaki, H., Honjo, M., Yamashita, K.I., Ohara, T., Kojima, A. *et al.* (2010) Classification and identification of bunching onion (*Allium fistulosum*) varieties based on SSR markers. *Breeding Science* 60, 139–152.
Van der Meer, Q.P. and De Vries, J.N. (1990) An interspecific cross between *Allium roylei* Stearn and *Allium cepa* L., and its backcross to *A. cepa*. *Euphytica* 47, 29–31.
Van Heusden, A.W., Shigyo, M., Tashiro, Y., Vrielink-van Ginkel, R. and Kik, C. (2000a) AFLP linkage group assignment to the chromosomes of *Allium cepa* L. via monosomic addition lines. *Theoretical and Applied Genetics* 100, 480–486.
Van Heusden, A.W., Van Ooijen, J.W., Vrielink-van Ginkel, R., Verbeek, W.H.J., Wietsma, W.A. *et al.* (2000b) A genetic map of an interspecific cross in *Allium* based on amplified fragment length polymorphism (AFLP™) markers. *Theoretical and Applied Genetics* 100, 118–126.
VegMarks (2021) VegMarks, a DNA database for vegetables. Available at: https://vegmarks.nivot.affrc.go.jp/VegMarks/app/page/home (accessed 17 May 2021).
Villano, C., Esposito, S., Carucci, F., Iorizzo, M., Frusciante, L. *et al.* (2019) High-throughput genotyping in onion reveals structure of genetic diversity and informative SNPs useful for molecular breeding. *Molecular Breeding* 39, 1–11.
Von Kohn, C., Kiełkowska, A. and Havey, M.J. (2013) Sequencing and annotation of the chloroplast DNAs and identification of polymorphisms distinguishing normal male-fertile and male-sterile cytoplasms of onion. *Genome* 56, 737–742.
Yang, Y.Y., Huo, Y.M., Miao, J., Liu, B.J., Kong, S.P. *et al.* (2013) Identification of two SCAR markers co-segregated with the dominant *Ms* and recessive *ms* alleles in onion (*Allium cepa* L.). *Euphytica* 190, 267–277.
Zewdie, Y., Havey, M.J., Prince, J.P. and Jenderek, M.M. (2005) The first genetic linkages among expressed regions of the garlic genome. *Journal of the American Society for Horticultural Science* 130, 569–574.
Zhao, W.G., Chung, J.W., Lee, G.A., Ma, K.H., Kim, H.H. *et al.* (2011) Molecular genetic diversity and population structure of a selected core set in garlic and its relatives using novel SSR markers. *Plant Breeding* 130, 46–54.

18 Future Prospects for *Allium* Crops

Brian Thomas[1]* and Haim D. Rabinowitch[2]*
[1]*School of Life Sciences, University of Warwick, Coventry, UK;* [2]*The RH Smith Institute of Plant Science and Genetics in Agriculture, The Hebrew University of Jerusalem, Rehovot, Israel*

1. Strength in Diversity

Allium crops form one of the most abundant and varied groups of edible vegetables in the world, e.g. bulb onions rank second to tomatoes in global production. Alliaceous crops make a major contribution to healthy diets in numerous popular flavoured dishes, worldwide. With a few exceptions (i.e. bulb onion and garlic), comparatively speaking, alliums are also one of the least studied of the major crop groups. The reasons for this include their diversity, their breeding systems (or lack of them in sterile species), often outbreeding, and being subject to inbreeding depression. This contrasts with the major arable crops such as cereals and oilseeds, where a limited number of species are grown at a scale that attracts funding for underpinning research.

This research deficit is particularly apparent for the ~30 species sometimes called the more minor *Allium* crops, e.g. shallots, leeks, bunching onion, Chinese chives, chives and more. Shortage of public funds and lack of interest by seed companies greatly reduce expectations for an imminent leap.

International bodies such as FAO, the World Bank, USDA, the EU and others could address this issue by joining global forces and local strengths, e.g. by establishing 'universal multi-disciplinary crop-specific task teams' capable of advancing the complex crop's performance by utilizing genetic resources, modern genomics, physiology and state-of-the-art agronomic technologies through innovation (see Chapter 8, this volume).

Diversity is an important source of innovation, and has the potential for collecting, characterizing and usage of the natural biodiversity of *Allium* for human nourishment (see Chapter 2, this volume). It is still available in the main centres of *Allium* evolution, in local and neighbouring production fields, as well as in national and private collections.

The tasks could be through new publicly funded expeditions and screening of existing gene pools to identify adaptations to climate change as well as better tolerance to diseases and pests, richness in valuable nutraceutical properties, fertility, productivity, pungency, storability, and more.

Of special importance is keeping ability. In the tropics and subtropics, only short-to-intermediate photoperiod types of onion form bulbs. Most of these genotypes deteriorate fast as dormancy is very short. Hence, heavy post-harvest losses are common. The introduction of long-keeping genes from long-photoperiod genotypes and the development of long-keeping short-photoperiod hybrids are possible as in Israeli-bred, short-photoperiod hybrids. Additionally,

*Addresses for correspondence: Brian.thomas@warwick.ac.uk; Haim.rabinowitch@mail.huji.ac.il

 Edible Alliums: Botany, Production and Uses (eds H.D. Rabinowitch and B. Thomas)
DOI: 10.1079/9781789249996.0018

more efficient and less costly conservation technologies would need to be developed and optimized to make this task viable, and economically feasible. The establishment of a global diversity-evaluation network (similar to the European ECP/GR for germplasm database) would contribute greatly to the advancement of *Allium* species science and production worldwide.

Such an approach is particularly important for garlic, where wild species and locally domesticated plants from the wild in Central Asian countries and Xinjiang of China have the highest potential level of intraspecific variation for future improvement programmes. In addition to tolerance to biotic and abiotic stress, great variability in nutraceutical substances, vigour, size, shape, photoperiod response and dormancy are of paramount importance for crop improvement. Of utmost scientific and economic importance is the capacity of many genotypes from these gene pools to bloom and set viable seeds (see Chapter 6, this volume).

For countries challenged by the means required for protecting their natural heritage, an international convention should be adopted.

2. Sustainable Production

The main twin goals of allium improvement are increasing food production to meet an increasing population while maintaining an environment free from unwanted chemicals. In this context, it is worth noting that onions have a relatively low carbon footprint compared to other food crops.

Precision agriculture, using robots with artificial vision and AI can ensure that fertilizers, growth regulators (such as gibberellic acid for flowering enhancement) and water are targeted most efficiently to individual plants. Holistic approaches such as studying and manipulating the rhizosphere microbiome, including mycorrhiza and nitrogen-fixing organisms, and its effects on different *Allium* species, should be developed.

Reducing chemical inputs for pests and diseases protection is also of high priority. Diversity on the one hand and modern molecular biology tools (e.g. gene editing, engineering and more) on the other are currently the major tools for improvement. Deploying genetic resistances sustainably, so that they do not break down under disease pressure, will require multiple sources of resistance, deployment strategies and surveillance.

3. Nanotechnology

Nanotechnology will become a powerful tool for a more in-depth and accurate scientific understanding of plant physiology and molecular processes, while it is expected to reduce both the degradation losses of the applied compounds and their impact on the environment in the production fields. This is accompanied by better protection, reducing costs and benefitting crop production. Specific nanoscale carriers should be developed for the efficient delivery and controlled release of encapsulated plant growth regulators, fertilizers and pesticides to the heavily waxed vertical allium's foliage.

4. Genetics

Much of the recent progress in crop research has been driven by the availability of genome sequences. Sequencing has become progressively cheaper and faster, but one of the key difficulties with alliums is the size of the genome and the amount of repetitive DNA, making assembly and annotation very challenging. Consequently, alliums have lagged behind other crops in this regard. However, that is likely to change quite quickly (see Chapter 15, this volume) now that the onion and garlic (Sun *et al.*, 2020) genomes have been sequenced. Increased sequence information can facilitate approaches such as genome-wide association study (GWAS) and geographically weighted regression (GWR).

More sequence information would allow genotyping techniques to be refined to properly identify variability at the molecular level and facilitate the further development of molecular markers.

Additionally, interspecific hybridization, either by classical means or somatic breeding, should be enhanced for the exchange of genes between the ~1000 *Allium* species to improve resistance to pests and diseases, flavour, nutraceutics level and other quality traits, production season, productivity, response to vernalization, and more.

4.1. Molecular markers

Molecular marker systems, with subsequent genetic maps, and QTL mapping used to link traits and genetics are common for the major crops and are used both in basic studies and for applied work. The main differences between traditional and molecular breeding work are cost and time for the development of new varieties, as genotypes assessment is performed at the seedling stage rather than phenotypic expression of the mature plant in classical breeding. However, very few markers are available for bulb onion and fewer for garlic chives, bunching onion and leek. At the time of writing, little genetic work has been carried out on the niche crop, with nothing publicly available on other minor alliaceous crops.

In bulb onion, molecular markers have been used mainly for diversity analysis, varietal identification, selections for colour and quality improvement, and male sterility. Further development of markers for important traits such as flavour (pungency, sweetness), level of nutraceutics, bolting resistance, long shelf life, and tolerance to both abiotic and biotic stresses will greatly benefit onion culture. A similar approach should be applied to other major and minor alliaceous crops.

4.2. Genome editing

Knowledge of the sequence detail and diversity will also lead to the more efficient application of new generations of methods for directed breeding through targeted modification of gene sequences, e.g. CRISPR, TALEN, PPR (pentatricopeptide repeat which can precisely edit both RNA and DNA), engineered minichromosomes, and more. The combination of gene sequences and transformation methods would be particularly powerful in enabling the transfer of elite traits between varieties adapted to different environments or geographical locations. The transformation would also be important in *Allium sativum*, where fertility and seed production have been largely eliminated during cultivation, in Chinese chives that often develops seeds apomictically, and more.

5. Microbiome and Endophytes

Modern science perceives plants not as autonomous entities, but as holobionts made of the plant host plus the interactions with its associated microbiota, thus markedly increasing the functional adaptability of plants to their biotic and abiotic environment. For instance, arbuscular mycorrhizal fungi (AMF) play a significant role in the development of environmentally friendly agriculture based on soil characteristics and environmental factors, including biotic and abiotic stresses. Alliums are particularly dependent on soil mycorrhiza, due to their simple, very superficial, meagre and inefficient root system, which makes them vulnerable to drought, compared to most other species. Since root branching is rare, alliums have low root densities and are thus very inefficient in absorbing nutrients from the growing medium, and are thus prone to nutrients deficiencies. AMF play an important role in allium nutrition and especially concerning the absorption of phosphorus. Similarly, nitrogen-fixing organisms may play a significant role in lowering chemical fertilization, costs and damage to the environment. Employing modern microbiomics technologies to enhance our understanding of the complementary relationships between these organisms together with selection for efficiency, in both plants and fungi, should greatly improve productivity and support the minimization of synthetic fertilization.

Endophytes, often fungi and bacteria, thrive between living plant cells where they maintain symbiotic to bordering on pathogenic relationships with the plant. Effects such as host growth enhancement, nutrient acquisition and improving the plant's ability to tolerate abiotic stresses were reported. In-depth science into the synergistic relationships between the host and these micro-organisms should contribute greatly to crop economics and the environment. Biological control through beneficial micro-organisms is gaining momentum. Several biological control agents (BCAs) (e.g., *Bacillus, Pantoea, Streptomyces, Trichoderma, Clonostachys, Pseudomonas, Burkholderia*, and certain yeasts) are already available for prevention/slowing down leaf, root and fruit-storage diseases. These should also be adopted to allium culture.

6. Seeds

The availability of high-quality seeds that result in uniform seedlings with high vigour and consistent adult characteristics provides the backbone of efficient production. In many crops, and specifically in alliums, the development of hybrid seeds is central to this.

F_1 hybrids have transformed production in species such as onion, shallot, bunching onion and leek, but are less established for other alliaceous crops, e.g. chives. There is a relatively high number of well-regarded commercial bulb onion hybrid cultivars available, all displaying good yields. As a highly heterozygous biennial crop, developing new cultivars through traditional breeding techniques is a time-consuming process with the average hybrid cv. taking 14 or more years to produce. Implementation of molecular markers and state-of-the-art genetic technologies should accelerate the process (e.g. by selecting the homozygous *msms* at the seedlings stage, solely) and reduce the time to market, thus benefitting growers and consumers alike from the novel excelling genotypes.

Some crops such as garlic and shallot have traditionally been propagated vegetatively, although in recent years shallots have been grown from seeds, and hybrid shallots are now available in Israel and Holland. The restoration of fertility is essential for the classical breeding of various *Allium* species, like the most relevant commercial varieties of garlic. The challenge for garlic is greater as the availability of fertile garlic (see Chapter 6, this volume) is a precondition to hybrid seed production.

Seed vigour is important in crop establishment and has been demonstrated recently to be a trait that shows genetic variation and hence can be a target for breeding through the development of linked markers.

Climate change is now unequivocally occurring. With increased atmospheric CO_2 concentration, the higher temperatures lead to a widespread risk of drought and plant growth effects such as florogenesis induction. In most alliums, florogenesis requires cold induction, hence global warming may adversely affect seed production and more so hybrid seed production where nicking between two-parent lines is a must. Cold storage of mother bulbs provides an adequate solution where the bulb-to-seed production system is common (see Chapter 13, this volume). However, where seed-to-seed production is practised, reduced requirement for vernalization may be accompanied by heavy bolting in the production fields. Hence, cultural adaptation is required.

For sterile minor alliums, but also for fertile species, induced genetic variation by molecular tools combined with the development of artificial seeds, can benefit and accelerate the crops' improvement and availability to consumers.

7. Adaptation to Climate Change

There is a complicated relationship between agriculture and climate change as agriculture is one of the most vulnerable sectors to rising/decreasing temperatures and is directly affected by climate change. On the other hand, agriculture is the second most important source of greenhouse gases and is the largest contributor of particulate emissions in the USA and other developed agricultural countries. Climate change is likely to bring less stable and more extreme weather patterns. Crops will need to be resilient to a range of temperatures and water availability. Largely in alliaceous crops, salinity may become an increasing problem, especially in low-lying and coastal regions. On the other hand, bolting may become less of an issue, and consequently, earlier sowing/planting becomes feasible with a marked impact on field establishment, crop-weed competition, yield and quality.

Ideally, the ability to transfer genes that confer resistance to particular stresses between cultivars would increase the overall resilience of the crop. This will depend on progress in molecular breeding and genetic research.

8. Enhanced Nutraceutics

Alliums are a source of compounds that can have beneficial effects as part of the diet (see Chapter 3, this volume). A deeper understanding of biosynthetic pathways, as well as regulatory-related genes and networks across the range of *Allium* crops, undoubtedly will benefit the well-being of consumers. To achieve improved nutraceutical levels, interdisciplinary approaches are required including genomics, metabolomics, transcriptomics, proteomics, nutrigenomics and metagenomics, in combination with studies on genetic diversity. Increasing the nutritional benefits of alliums will also require collaborations between plant, nutritional and medical scientists.

9. Expanding Markets

The geographical range of onion and garlic, the two major *Allium* crops, has expanded in the past to cover most of the inhabited world. There is a close historical connection between local cuisines and the *Allium* species grown in any specific region. As food becomes ever more international, there is increasing opportunity for more minor crops, e.g. chives or bunching onions, to expand their range. Collections of genetic diversity can be a source of material that can be screened for suitability to new geographical areas of production.

For countries challenged by the means required for protecting their genetic diversity and natural heritage, an international convention should be adopted and better, more efficient and less costly conservation technologies need to be developed and optimized. For a few *Allium* species where seed storage is not applicable, improving techniques for cryopreservation to make it workable for more species is essential for their long-term preservation.

10. Conclusions

Digital and cyber technologies, omics research and gene editing are the three major areas that are going to transform the way we think about how food is produced. The research opportunities to improve the production, quality and sustainability of *Allium* crops are numerous. Progress is likely to be incremental and crop-specific but will involve the application of new approaches and technologies with general applicability. These could include widespread and cost-efficient genome sequencing, the use of CRISPR-Cas and more advanced methods that do not have the restrictions of current GMOs, the wider application of omics technologies, precision agriculture using artificial intelligence, robotics and machine vision. These will run hand-in-hand with long-standing generic objectives for improving crop performance, such as integrated crop management, enhancing the plant–soil interface, improving seed quality, maintaining post-harvest quality and reducing waste. The potential for *Allium* crops and the associated research is immense.

References

Sun, X., Zhu, S., Li, N., Cheng, Y., Zhao. J. *et al.* (2020). A Chromosome-level genome assembly of garlic (*Allium sativum*) provides insights into genome evolution and allicin Biosynthesis. *Molecular Plant* 13, 1328–1339.) https://doi.org/10.1016/j.molp.2020.07.019

Index

Note: The page numbers in italics and bold represents figures and tables respectively.